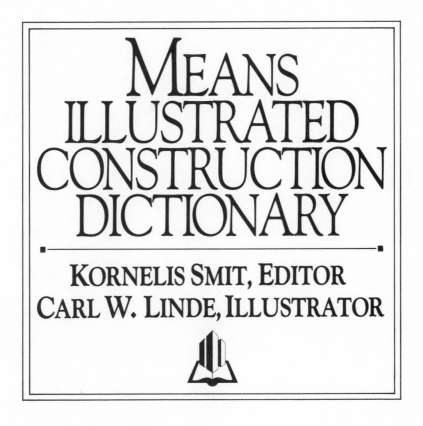

MEANS
ILLUSTRATED
CONSTRUCTION
DICTIONARY

KORNELIS SMIT, EDITOR
CARL W. LINDE, ILLUSTRATOR

PUBLISHER
E. Norman Peterson, Jr.

EDITOR-IN-CHIEF
William D. Mahoney

CONTRIBUTING EDITORS

Joyce A. Baron
William T. Cashin
Michael P. Gardner
Arlene Giblin
F. William Horsley
Johnathan C. Horsley

William D. Mahoney
Melville J. Mossman
John J. Moylan
Richard J. Ossolinski
James P. Treichler
Rory M. Woolsey

TECHNICAL ASSISTANTS

Paul W. Burr
Catherine R. Carey
Cheryl A. Carlson
John E. Corbett
Marcia A. Crosby
Patricia M. Dorr
Marie Doten

Sheila Droege
Paul G. Gibbons
Gayla M. Godby
Tania E. Howes
Ann L. Linde
Carol A. Logue
Helen A. Marcella
Elisabeth S. McGrath

Kevin A. McKearin
Laurel A. Parsons
Ann E. Rehill
Marion E. Schofield
Diane E. Silva
Denis J. Sullivan
James K. Vinci

1st Edition
©COPYRIGHT
1985

R. S. MEANS COMPANY, INC.
CONSTRUCTION CONSULTANTS & PUBLISHERS
100 CONSTRUCTION PLAZA
P.O. Box 800
KINGSTON, MA 02364-0800
(617) 747-1270

4th Printing

Library of Congress Catalog Card Number Pending
ISBN 0-911950-82-6

FOREWORD

This entirely new reference is written in the simple everyday language as it is used by the construction trades and professions in the United States today. Many slang, regional, and colloquial terms are included since they are the vocabulary of the construction industry today. Every word or phrase is explained in non-technical terms without sacrificing the accuracy and clarity of the definition. Illustrations or examples are used where necessary to clarify a definition.

Archaic and historic architectural terms are purposely left out: this is a modern construction dictionary with up-to-date technology and terminology.

In compiling the list of terms defined in this dictionary, the editors found that many words are unique to the construction industry and can also have an entirely different meaning depending upon the trade being referred to. Meticulous attention was given to details and research. You will find many new terms, particularly new products and systems, that have never been compiled in any previously published dictionary. To identify and define these terms, the editors contacted many groups, associations, societies, and manufacturers.

In order to help keep definitions consistent throughout the industry, many of these groups, associations, societies, and manufacturers have assisted in the production of this book by granting permission to reproduce specific text and graphics from their publications. The following is a list of these contributors and publications.

American Association of Cost Engineers (AACI)
American Ceramic Society (ACS)
American Concrete Institute (ACI)
The American Institute of Architects (AIA)
American Society of Plumbing Engineers (ASPE)
American Institute of Steel Construction, Inc. (AISC)
The Asphalt Institute
Brick Institute of America
Builders Hardware Manufacturers Association, Inc.
Deep Foundations Institute
The Gypsum Association
International Institute of Lath and Plaster
International Masonry Institute
The Lincoln Electric Company
North Castle Books, Publishers of *Moving The Earth*
Portland Cement Association
Random Lengths
United States League of Savings Institutions (USL)

We particularly thank The American Institute of
Architects who have granted us permission to reproduce
portions of the publication, *AIA Glossary of Construction
and Industry Terms* copyrighted 1982 by the American
Institute of Architects under permit number 85015.
Also, special thanks to the American Concrete Institute
for permission to reproduce portions of the publication
Cement and Concrete Terminology third printing June
1984; and Random Lengths for permission to reproduce
portions of their publication *Terms of the Trade* second
edition 1984.

Kornelis Smit, Editor

 ABBREVIATIONS

The abbreviations listed below are those most commonly used in the construction industry. Alternative forms (usually nonstandard) are shown in parentheses.

a acre
A area
A&E architect-engineer
AAMA Architectural Aluminum Manufacturers Association
ABC aggregate base course, Associated Builders and Contractors.
ABS acrylonitrile butadiene styrene
ABT air blast transformer, (about)
ac, a-c, a.c. alternating current
a.c. asphaltic (a.c. paving)
AC air conditioning, alternating current (on drawings), armored cable (on drawings), asbestos cement
ACB asbestos-cement board, air circuit breaker
ACC accumulator
Access. accessory
ACD automatic closing device
ACI American Concrete Institute
ACM asbestos-covered metal
ACS American Ceramic Society
ACSR aluminum cable steel reinforced, aluminum conductor steel reinforced
Acst acoustic
Actl actual
a.d. air dried
AD access door, air dried, area drain, as drawn
ADD addendum (on drawings), addition (on drawings)
Addit. additional
ADF after deducting freight (used in lumber industry)
ADH adhesive
adj adjust, adjustable, adjoining, adjacent
ADS automatic door seal
AG above grade
AGA American Gas Association
AGC Associated General Contractors.
Aggr aggregate

AGL above ground level
AH ampere hour (also, amp hr)
AIA American Institute of Architects
AIC ampere interrupting capacity
AIEE American Institute of Electrical Engineers
AIMA Acoustical and Insulating Materials Association
AISC American Institute of Steel Construction
AISI American Iron and Steel Institute
AITC American Institute of Timber Construction
AL aluminum (also, alum)
ALLOW allowance (also, Allow)
ALM alarm
ALS American Lumber Standards
ALT alternate
ALTN alteration
ALY alloy
AMB asbestos millboard
AMD air-moving device
amp ampere
ANL anneal
ANSI American National Standards Institute
AP access panel
APC acoustical plaster ceiling
APF acid-proof floor
Appd approved
Approx approximate
Apt apartment
APW Architectural Projected Window
AR as required, as rolled
ARC W, ARC/W arc weld
ARS asbestos roof shingles
ART. artificial
AS automatic sprinkler
ASA American Standards Association
asb asbestos
ASBC American Standard Building Code

ASC asphalt surface course

ASCE American Society of Civil Engineers

ASEC American Standard Elevator Codes

ASHRAE American Society of Heating, Refrigeration and Air Conditioning Engineers

ASI American National Standards Institute

ASME American Society of Mechanical Engineers

asph asphalt

ASR automatic sprinkler riser

ASSE American Society of Sanitary Engineering

ASTM American Society for Testing and Materials

AT asphalt tile, airtight

ATB asphalt-tile base

ATC acoustical tile ceiling, architectural terra cotta

ATF asphalt-tile floor

atm atmosphere, atmospheric

aux auxiliary

av, ave, avg average

A/W all-weather

AW actual weight

AWG American wire gauge

A.W.W.I. American Wood Window Institute

A DEFINITIONS

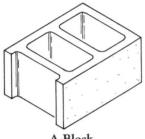

A-Block

abaciscus, abaculus (1) A tessara used in mosaic tile. (2) A small abacus.

abacus The top component of a column capital.

abamurus A masonry buttress for the support of a wall.

abandon catch basin Destroy or fill in existing catch basin.

abate (1) To cut away in stone or to beat down in metal in order to create figures or a pattern in relief. (2) To decrease concentrations such as pollutants into an outfall.

abatement In lumber industry, the amount of wood lost as waste during the process of sawing or planing.

abat-jour A sloped opening, in a roof or wall, designed to direct daylight downward.

abatsons Acoustical reflectors for directing sound downward.

abatvent Wall louvers that restrict wind from entering a building, but that admit light and air.

abatvoix An acoustical reflector for a single voice such as behind and over a church pulpit.

ABC extinguisher A fire extinguisher for type A, B, and C fires.

A-block A hollow masonry unit, with one closed end, that is commonly used at wall openings.

Abney level A hand-held level used for measuring vertical angles.

above-grade subfloors A floor above ground level, but one having no headroom below.

abrade To scrape or wear a surface by friction or striking.

Abrams' law The rule stating that with given materials, curing, and testing conditions, concrete strength is inversely related to the ratio of water to cement. Low water to cement ratios produce high strengths.

abrasion The process of wearing away a surface by friction.

abrasion resistance index A comparison of the abrasion resistance of a given material to that of rubber. The index is used principally in aggregate handling equipment.

abrasive A hard material used for wearing or polishing a surface by friction. Also, the material that is adhered to or embedded in a surface such as sandpaper or a whetstone.

aggregate The aggregate used in making a surface abrasive, such as concrete slabs.

floor A floor with an abrasive adhered to or embedded in the surface to prevent slipping.

floor tile Floor tile with an abrasive adhered to the surface.

nosing A strip of anti-skid abrasive adhered to or attached to the nosing of a stair tread.

stair tread A stair tread which has an abrasive surface.

terrazzo A terrazzo floor that has an abrasive surface rather than a high polish.

abreuvoir The joint filled with mortar and located between masonry units.

absolute humidity See **humidity**.

absolute pressure See **pressure**.

absorbed moisture Moisture that has been absorbed by a solid such as masonry.

absorbent A material that has an affinity for certain substances and attracts these substances from a liquid or gas with which it is in contact, thus changing the

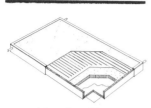

Absorber Plate

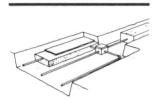

Absorption Bed or Field

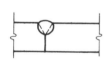

Abut

physical or chemical properties (or both) of the material.

absorber (1) A device, containing liquid, for absorbing refrigerant vapor or other vapors. (2) In a refrigeration system, the component on the low-pressure side used for absorbing refrigerant vapors.

absorber plate That part of some solar-energy system which collects the solar energy.

absorbing well See **dry well**.

absorption The process by which a liquid is drawn into the pores of a permeable material. (2) The process by which solar energy is collected on a surface. (3) The increase in weight of a porous object resulting from immersion in water for a given time, expressed as a percent of the dry weight.

absorption air conditioning An air cooling and dehumidifying system powered by solar or other energy collected on absorbing plates.

absorption bed or field A network of trenches, which may contain coarse aggregate and distribution pipe, and which is used to distribute septic tank effluent into the surrounding soil.

absorption coefficient See *sound absorption coefficient*.

absorption field See *absorption bed*.

absorption loss (1) Water losses that occur until soil particles are sufficiently saturated, such as in filling a reservoir for the first time. (2) Water losses that occur until the aggregate in a concrete mix is saturated.

absorption rate, initial rate of absorption The weight of water absorbed when a brick or concrete masonry unit is partly immersed in water for one minute, expressed in grams or ounces per minute.

absorption refrigeration system (1) A cooling system in which the refrigerant vapors that enter an absorber are released into a generator by application of heat. (2) A cooling system operated by heat from solar absorbers.

absorption trench See *absorption bed*.

absorption-type liquid chiller A system using an absorber, condenser, and associated accessories to cool a secondary liquid.

ABS plastic pipe Acrylonitrile-butadiene-styrene plastic pipe which is resistant to heat, impact, and chemicals.

abstract of title A deed for a parcel of land showing encumbrances and a history of ownership.

abut To join at one edge or end without overlapping.

abutment (1) The surface at which one member abuts another. (2) The structure which supports the end of a bridge or anchors the cables of a suspension bridge.

abutment piece In structural framing, the horizontal member that distributes the load of vertical members, and is thus the sole plate of a partition.

abuttals The properties bounding a parcel of land or body of water.

abutting joint A joint between two pieces of wood in which the grain of one is at an angle to the grain of the other (usually 90 degrees).

abutting tenon In carpentry, one of two wood members which meet in a common mortise.

Abyssinian well See **well point**.

acceleration An increase in velocity or rate of change. In construction, particularly, the speeding up of the setting or hardening process of concrete. The process of acceleration allows

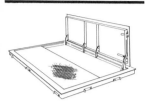

Access Door

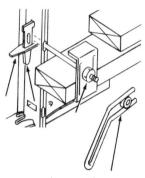

Accessories

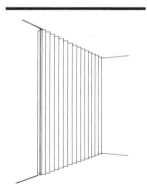

Accordion Partition

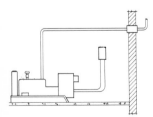

AC Generator

forms to be stripped sooner or floors to finished earlier.

accelerator An additive which, when added to a concrete, mortar or, grout mix, speeds the rate of hydration and thereby causes it to set or harden sooner.

accepted bid The proposal or bid prepared by a contractor and accepted by an owner or the owner's representative as the basis for entering into a construction contract.

access The means of approach to a building, area, or room. Also a port or opening through which equipment may be inspected or repaired.

access connection A ramp or roadway for entering or exiting an arterial highway.

access door or panel A means of access for the inspection, repair, or service of concealed systems, such as air conditioning equipment.

access eye A cleanout provided at drain pipe bends for rodding.

access flooring A raised flooring system with removable panels that allows access to the area below. This type of flooring is frequently used in computer rooms and provides easy access to cables.

accessible That which is able to be easily removed, repaired, or serviced without damaging the finish of a building.

accessories In the placing of concrete, the items used to assemble scaffolding, shoring, and forms, other than the forms themselves and the wales and frames.

accessory building A secondary building on the same lot adjacent to the main building.

accident A sudden unexpected event, identifiable as to time and place, resulting in personal injury or property damage.

accolade Ornamental treatment over an arch, doorway, or window formed by two ogee curves meeting in the middle.

accordion door A folding door, usually fabric faced, hung from an overhead track and folding like the bellows of an accordion.

accordion partition A retractable partition having the same features as an accordion door.

accouple To join, tie, or couple together.

accouplement The pairing of pilasters or columns, as in a colonnade or buttress.

accrued depreciation (1) The total reduction of the value of property as stated on a balance sheet for accounting or tax purposes. (2) The reduction of the value of property by wear and tear.

accumulator (1) A pressure vessel whose volume is used to maintain a constant pressure. (2) In refrigeration, a storage chamber for low side refrigerant. An accumulator is also called a "surge drum" or a "surge header."

acetone A highly flammable organic solvent used with lacquers, paint thinners, paint removers, and resins.

acetylene A gas which, when compressed and mixed with oxygen, is used for welding and cutting metal.

acetylene torch The torch used when welding and cutting with compressed acetylene and oxygen.

AC generator A generator that produces alternating current.

acid and alkali-resistant grout or mortar A grout or mortar that is highly resistant to prolonged exposures to alkaline compounds, acid liquids, or gasses.

acid etch or aciding A method of cleaning the latence from concrete by washing it with an acid solution and rinsing with water.

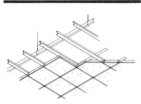

Acoustical Ceiling

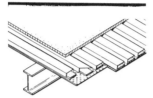

Acoustical Metal Deck

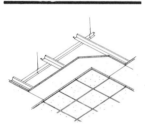

Acoustical Tile

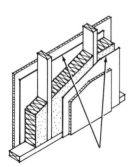

Acoustical Wallboard

acid polishing The polishing of a surface, particularly glass, by acid treatment.

acid-proof floor A floor that resists deterioration caused by exposure to acid.

acid resistance A measurement of the ability of a surface to resist the attack of acids.

acid-resistant brick Brick that resists deterioration caused by exposure to acid. This type of brick should be laid with acid resistant mortar.

acid soil Soil with a pH value of less than 6.6.

acid steel Steel made with a silica flux or in a silica lined furnace.

acoustical A term used to define systems incorporating sound control.

barrier A building system that restricts sound transmission.

block A masonry block, with sound absorbing qualities, that is usually defined in terms of its N.R.C. (noise reduction coefficient) rating.

board A construction material in board form that restricts or controls the absorption and transmission of sound.

booth See *acoustical enclosure*.

ceiling A ceiling system constructed of sound-control materials. The system may include lighting fixtures and air diffusers.

correction Special planning, shaping, and equipping of a space to produce the optimum reception of sound for an audience.

door A door constructed of sound absorbing materials and installed with gaskets around the edges.

enclosure An enclosure constructed of acoustical materials for speaking, listening, and recording, as in a record store or a phone booth.

glass A pane of glass composed of two lites with dead air space between them for sound absorption. This type of glass is frequently used for borrow lights and other interior applications.

metal deck A metal decking including a sound-absorbing material at a small additional cost per square foot.

panel Modular units composed of a variety of sound-absorbing materials for ceiling or wall mounting.

partition A term applied particularly to movable, demountable, and operable partitions with sound-absorbing characteristics.

phone booth See *acoustical enclosure*.

plaster A wall and ceiling plaster having sound-absorbing characteristics.

reduction factor A value expressed in decibels that defines the reduction in sound intensity that occurs when sound passes through a material.

room See *acoustical enclosures*.

sealant A caulking or joint sealant with acoustical characteristics.

sprayed-on material A fibrous material with acoustical properties applied to a surface by spraying through a nozzle.

tile A term applied to modular ceiling panels in board form with sound-absorbing properties. This type of tile is sometimes adapted for use on walls.

transmission factor The reciprocal of the sound reduction factor.

wall tile See *acoustical tile*.

wallboard Wallboard with sound absorbing properties.

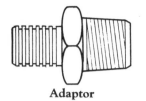

Adaptor

window wall Double-glazed window walls with acoustical framing. This type of wall system is used particularly at airports.

acoustic block See **acoustical block**.

acoustics The science of sound transmission, absorbency, generation, and reflectance. In construction, the effects of these properties on the acoustical characteristics of an enclosure.

acquiescence A term frequently used when owners of adjacent property agree on a boundary between their properties, if the original boundary is difficult or impossible to establish.

acre A common unit of land-area measurement equal to 160 square rods or 43,560 square feet.

acre foot A unit of volume measurement equal to an area of one acre times one foot thick. The acre foot is used to measure the volume of water or of ore deposits.

acropodium An elevated pedestal or plinth bearing a statue.

acropolis The citadel of an ancient Greek city, usually surrounding the temple to the patron deity.

acrylic See **acrylic resin**.

acrylic fiber Fibers produced from polymerized acrylonitrile, a liquid derivative of natural gas. A tough economical fiber commonly used in commercial and residential carpets and draperies.

acrylic plastic glaze A clear plastic sheet that is bonded to glass and that increases the ability of the glass to resist breaking and shattering.

acrylic resin (acrylate resin) A clear, tough thermoplastic resin used in construction in sheet and corrugated form, as an adhesive, and as the main ingredient in some caulking and sealing compounds.

actinic glass Glass that filters out radiation from sunlight.

activated alumina A form of aluminum oxide that absorbs moisture readily and is therefore used as a drying agent.

activated charcoal (activated carbon) A material obtained principally as a by-product of the paper industry and used in filters for absorbing smoke, odors, and vapors.

activated rosin flux A flux with a rosin base which enhances the flow of solder.

activated sludge The sludge settled out of oxygenated sewage.

activated sludge process A sewage-treatment method involving the introduction of oxygen and bacteria into sewage. The by-products are water and activated sludge.

active earth pressure The horizontal component of pressure exerted by earth on a wall.

active leaf In a double-leaf door, the leaf to which the latching or locking mechanism is attached.

activity In CPM (Critical Path Method) scheduling, a task or item of work required to complete a project.

activity arrow In CPM (Critical Path Method) scheduling, a graphic representation of an activity.

activity duration In CPM (Critical Path Method) scheduling, the estimated time required to complete an activity.

acute angle An angle of less than 90 degrees.

adamant plaster A quick-setting gypsum plaster usually applied over a base coat of plaster.

adaptor Any device designed to match the size or characteristics of one item to that of another, particularly in the plumbing, air conditioning, and electrical trades.

Addition

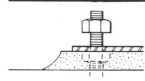

Adjustable Wrench

Adjusting Nut

addendum, addenda (pl.) Alteration or clarification of the plans or specifications provided to the bidder by the owner or by the owner's representative prior to bid time. An addendum becomes part of the contract documents when the contract is exectuted.

addition Any construction added to an existing structure that increases the height or floor area.

additional services Professional services, provided by the architect or engineer of a project, which were not included in the original owner-architect agreement.

additive A substance that is added to a material to enhance or modify its characteristics, such as the curing time, plasticity, or color of concrete, or the volatility of diesel fuel.

additive alternate A specific alternate option for construction specifications or plans that results in a net increase in the base bid.

additive constant In stadia surveying, a correctional constant to be added to each calculated distance.

address system An electronic audio system with microphone and speakers either permanently installed or mobile. Wiring for a permanent system should be done prior to any finish work.

adherend An appliance or part which is held to another by an adhesive.

adhesion The binding together of two surfaces by an adhesive.

adhesion-type ceramic veneer Ceramic tile or other veneer attached to a backing by mortar, grout, or adhesive only, with no anchors.

adhesive A general term that refers to any substance which binds two surfaces together. In construction, the term is used principally in the wallboard and roofing trades.

adiabatic A condition in which there is no heat gain or heat loss.

adiabatic curing The curing of concrete or mortar, particularly a test cylinder, in a controlled environment with no heat gain or heat loss.

adiabatic process A thermodynamic process occurring in the absence of heat gain or heat loss.

adit The entrance or approach to a building or the entrance to a mine.

adjustable base anchor An attachment to the base of a door frame above a finished floor.

adjustable clamp A temporary clamping device which can be adjusted for position or size.

adjustable doorframe A door frame with a jamb that can be adjusted to accommodate different wall thicknesses.

adjustable multiple-point suspension scaffold See **scaffold**.

adjustable square (double square) A carpenter's tool used for marking and scribing lumber. Usually, an adjustable square incorporates a level bubble.

adjustable wrench A wrench with a jaw that can be adjusted to fit different sized nuts or bolt heads.

adjusted base cost The total estimated cost or estimated unit cost for a project after adding or deducting addenda or alternatives.

adjustment factor A constant used in any calculation, usually a multiplier.

adjusting nut A threaded nut, used for alignment of an object, that is often coupled with a locking nut to secure it in position.

adjusting screw A screw, used for alignment of an object, that is often coupled with a locking nut to secure it in position.

admixture An ingredient other than cement, aggregate, or water that is added to a concrete or mortar mix to to affect the physical or chemical characteristics of the concrete. The most common admixtures affect plasticity, air entrainment, and curing time.

adobe An aluminous clay used to make unfired brick.

adobe blasting Same as **mud-capping**.

adobe brick A large, roughly formed, unfired brick made from adobe and straw.

adsorbed water Water that is held on the surface of materials by electrochemical forces. This water, such as that on the surfaces of aggregate in a concrete mix, has a higher density, and thus different physical properties from those of the free water in the mix.

adsorbent A material that has the ability to extract certain substances from gasses, liquids, cr solids by causing them to adhere to its surface without changing the physical properties of the adsorbent. Activated carbon, silica gel, and activated alumina are economical materials frequently used for this application.

adsorption The process of extracting specific substances from the atmosphere or from gasses, liquids, or solids by causing them to adhere to the surfaces of an adsorbent without changing the physical properties of the adsorbent.

advanced slope grouting A method of grouting in which the grout is forced horizontally through preplaced aggregate.

advance slope method A method of placing concrete where the sloped face of the fresh concrete moves forward as the concrete is placed.

adverse possession Occupation of property by one not the true owner openly, notoriously, and continuously.

advertisement for bids The published, public notice soliciting bids for construction. Usually, it is required by law that the advertisement be published in newspapers of general circulation in the area, when public funds are to be used for construction.

adz A long-handled tool with a curved blade set perpendicular to the handle and used for dressing lumber.

adz-eye hammer A claw hammer with a long eye for receiving the handle.

aeolian Wind blown or transported.

aerate To introduce air, as into soil or water, by natural or mechanical means.

aerated concrete See **cellular concrete**.

aeration The process of introducing air into a substance or area by natural or mechanical means.

aeration plant A sewage-treatment plant in which air is introduced into the sewage to accelerate the decomposition.

aerator A mechanical device for the introduction of air into a material such as soil, water, or sewage.

aerator fitting A pipe fitting by which air can be introduced into a flow of water.

aerial Pertaining to, caused by, or present in the air.

ladder An extension ladder capable of reaching higher places and often mounted on a vehicle such as a fire truck.

lift A term commonly applied to mobile working platforms that are elevated hydraulically or mechanically.

photography Photography performed from a vehicle in flight.

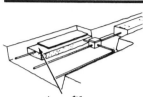

Aerofilter

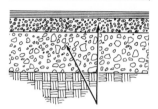

Aggregate

sewer A sewer pipe or line supported above grade on bents or pedestals.

survey A survey of the earth's surface based upon aerial photographs and ground-control points.

tramway A cable and tower system for transporting materials and personnel. A tramway is used economically in mining and dam building operations for transporting great volumes of material over rough terrain.

aerodynamic instability A harmonic motion occurring in a structure during high winds and endangering structural integrity. The term was used to define the failure of the Tacoma Narrows bridge.

aerofilter A bed of coarse aggregate used for filtering sewage.

aerosol paints Paints packaged in a pressurized can with a spray nozzle attached.

affidavit of non-collusion A sworn statement by the bidders on the same project that the prices on their proposals were arrived at independently without consultation between or among them.

affinity A tendency for two substances to unite chemically or physically.

A-frame (1) A structural system or hoisting system with three members erected in the shape of an upright capital "A". (2) A building with a steep gable roof that extends to the ground.

after-cooler A system that removes the heat from compressed gas after compression in a refrigeration system.

afterfilter, final filter In air conditioning, a filter located at the outlet end of the system.

after-flush The water that drains from a water closet after flushing and before the valve is closed by the incoming water pressure.

aftertack, residual tack A defect in paint that causes the painted surface to become tacky under certain environmental conditions.

age hardening A term used to describe a hardening process of metals at room temperature.

agent A person authorized by another person to act on the first person's behalf. An architect is frequently the owner's agent.

agglomeration The gathering together of small particles into larger particles, particularly to accelerate settlement in suspensions.

aggregate Granular material such as sand, gravel, crushed gravel, crushed stone, slag, and cinders. Aggregate is used in construction for the manufacturing of concrete, mortar, grout, asphaltic concrete, and roofing shingles. It is also used for leaching fields, drainage systems, roof ballast, landscaping, and as a base course for pavement and grade slabs. Aggregate is classified by size and gradation.

abrasive An anti-skid aggregate worked into the surface of a concrete floor.

coarse Aggregate that is larger than 1/8 inch and is retained on the No. 8 sieve.

coarse-graded Aggregate with a continuous grading from coarse through fine, with a predominance of coarse particles.

concrete The fine and course aggregate used in manufacturing concrete. Both are usually washed and graded.

exposed A concrete surface with the aggregate exposed. This can be economically attained by applying a retarder to the surface before the concrete has set, and subsequently removing the cement paste to the desired depth.

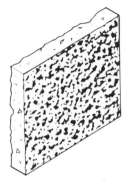

Aggregate Panel

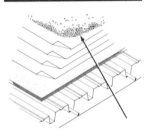

Aggregate, roof

fine Aggregate smaller than 1/8 inch. Fine aggregate passes through the No. 8 sieve.

heavyweight The aggregate produced from materials with high specific gravity, such as linomite, iron ore tailings, and magnetite.

interlock A term applying to a situation where the aggregate from one side of a concrete joint projects between the aggregate of the other side of the joint, thus resisting shear.

lightweight One of several materials used to decrease the unit weight of concrete, thereby reducing the structural load and cost of a building. The materials most commonly used are perlite and vermiculite. The use of lightweight aggregate is costly, but sometimes necessary in construction.

macadam A crushed aggregate of uniform size, placed to a specified depth, over which hot asphalt is distributed.

marble A material produced from crushed marble of varying colors. It is used extensively for decorative landscaping, highway shoulders, and built-up roofing.

masonry Washed sand used in a mortar mix.

open-graded An aggregate in which a skip between the sieve gradations has been deliberately performed so that the the voids are not filled with intermediate sized particles.

panel A precast concrete panel with exposed aggregate.

plaster Natural or manufactured washed sand.

roof (1) The aggregate used for a tar-and-gravel application. (2) The ballast used for membrane-type roofing.

spreader A particular piece of equipment used for placing aggregate to a desired depth on a roadway or parking lot.

stone An aggregate consisting of crushed stone.

testing Any of a number of tests performed to determine the physical and chemical characteristics of an aggregate. Common tests are for abrasion, absorption, specific gravity, and soundness.

well-graded An aggregate that incorporates sizes from the maximum to the minimum specified so as to fill most of the voids. This type of aggregate is used for asphaltic concrete mixes and for base courses.

aggregate bin A structure designed for storing and dispensing aggregate, which is loaded from the top and emptied from the bottom.

aging The chemical and physical changes in a material incurred by the passage of time. Concrete generally increases in strength with age, whereas rubber deteriorates with prolonged oxidation.

agitating speed The rate at which a concrete or mortar mixer rotates the drum or blades to agitate mixed materials to prevent segregation or setting.

agitation The moving of blades through, or the rotation of, a drum containing concrete or mortar to prevent segregation or setting of mixed mortar or concrete.

agitator A mechanical device used to maintain plasticity and to prevent segregation, particularly in concrete and mortar.

aglite A lightweight aggregate produced from expanded clay and used in concrete mixes.

A-grade wood A plywood surface that is smooth and paintable. Neatly executed patches are permitted.

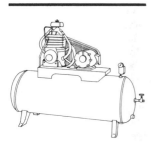

Air Compressor

Air Diffuser

agreement (1) A meeting of the minds. (2) A promise to perform, between signatories of a document. (3) In construction, the specific documents setting forth the terms of the contracts between architect, owner, engineer, construction manager, contractor, and others.

agreement form A standard printed form used by the signatories to an agreement with blank spaces to fill in the specific information pertinent to a particular contract.

agricultural lime A granular hydrated lime used for soil conditioning.

agricultural pipe drain A drainage system of porous, perforated, or open-jointed pipe laid in a trench with porous fill.

aiguille (1) A drill used for boring holes in masonry or cut stone. (2) A drill for boring blast holes in rock.

air (1) In construction, shortened term for "air conditioning." (2) The oxygen used in an oxygen/acetylene torch system.

air alternator A system that switches air from one side of an ice chest to the other.

air balancing The process of adjusting a heating or air conditioning duct system for equal distribution to all areas.

air base In aerial surveying, the distance between control stations in overlapping photographs.

air blown mortar Another term for **shotcrete**.

airborne transmission A term that refers to sound traveling through air in a structure.

air brick A brick manufactured to ventilate a masonry structure.

air brush A device with a nozzle for applying paint with compressed air.

air chamber In water piping, a vertical pipe containing entrapped air to absorb the pressure shock when a valve is closed suddenly.

air change The volume of air in an enclosure that is being replaced by new air. The number of air changes per hour is a measure of ventilation.

air cleaner A device, often hung from a ceiling, for removing impurities from the environment. The device may have a mechanical or electrostatic filter.

air compressor A machine that extracts air from the atmosphere and compresses it into a holding chamber. The compressed air is used for operating pneumatic tools. Air compressors are classified by CFM (cubic feet per minute) of compressed air they can produce.

air conditioner A system that may control temperature, humidity, and/or the cleanliness of air within an enclosure.

air conditioning The process of controlling the temperature, humidity, and cleanliness of air and of distributing it within an enclosure.

air content The volume of air present in a concrete or mortar mix, expressed as a percentage of the total volume. A controlled air content prevents cracking of concrete in the freeze/thaw cycle.

air curtain A directed narrow stream of air flowing across an opening and deterring the transfer of air, contaminants, and insects from one side to the other. Air curtains are commonly used on grocery store refrigerator cases and mall store fronts.

air diffuser An outlet in an air-supply duct for distributing and blending air in an enclosure. Usually, a round, square, or rectangular unit mounted in a suspended ceiling.

air distributing ceiling A suspended ceiling system with small perforations in the tiles for controlled distribution of the air from a pressurized plenum above.

air drain An empty space left between a foundation wall and a parallel wall to prevent the fill from laying directly against the foundation wall.

air entraining agent An admixture for concrete or mortar mixes which causes minute air bubbles to form within the mix. Air entrainment is desirable for workability of the mix and prevention of cracking in the freeze/thaw cycle.

air escape In plumbing, a valve for automatically discharging excess air from a water line.

air filter A device for removing undesirable gaseous or solid particles from ambient air. All HVAC systems include some type of air filter.

air gap In plumbing, the distance between the outlet of a faucet and the overflow level of the fixture.

air grating A fixed metal grating, particularly in masonry foundation walls, for ventilation.

air hammer A portable, pneumatic percussion tool used for breaking and hammering.

air-handling troffer A ceiling lighting unit that incorporates an air diffuser.

air leakage The air that escapes from a system or enclosure through cracks, joints, and couplings.

airlock (1) An airtight chamber such as that used in tunnel and caisson excavation. (2) An entrance room between areas of different pressures, such as the entrance to an air-supported structure. (3) In plumbing, air trapped in a system and preventing flow.

air make-up unit A system for introducing fresh conditioned air into an enclosure from which air is being exhausted.

air meter A device for measuring the air content of a concrete or mortar mix.

air-mixing plenum In an air conditioning system, a chamber in which fresh air is mixed with recirculated air.

air permeability test A procedure for determining the fineness of powdered material such as cement.

air pocket A void filled with air, such as in a concrete form when placing concrete or in a water piping system.

air purge valve A device for eliminating trapped air from a piping system.

air regulator An instrument for regulating the flow or pressure of air in a system.

air rights The legal property rights to space above a given elevation, such as a building over a roadway.

air shaft A roofless enclosed area within a building permitting light and ventilation.

air slacking The absorption of moisture and carbon dioxide from the air by lime or cement.

air-supported storage tank cover A non-rigid tank cover, usually for domestic water, supported by a slightly higher atmospheric pressure inside than outside.

air-supported structure A non-rigid structure supported by a slightly higher atmospheric pressure inside than outside. The difference in pressure is created by fans.

air terminal The top of a lightning protection system on a building.

air tool Any of a number of percussion tools using pneumatic pressure for operation.

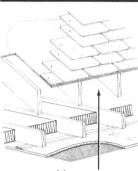

Airway

air tube system A tubular conveying system using air pressure to move capsules containing paperwork from one station to another.

air void The space occupied by entrained air in concrete or mortar.

air wall See **air curtain**.

air washer A water spraying mechanism for cleaning and humidifying air in a ventilation system.

airway The air space between thermal insulation and sheathing on a roof.

air well See **air shaft**.

aisle A longitudinal open passageway between sections of seats.

aisleway Any open passageway permitting access and traffic to flow between sections within a building.

Alabama marble A fine-grained, warm, white or cream marble.

alabaster A fine-grained translucent gypsum in white or pale hues.

Alaska yellow cedar A wood native to the Pacific Northwest with particularly good resistance to decay.

albarium A white lime used for stucco.

alclad A product with an aluminum or aluminum alloy core having an aluminum or aluminum alloy coating metallurgically bonded to the surface. The coating is anodic to the core, thus protecting it physically and electrolytically against corrosion.

alcove A recess or partly enclosed extension opening into a larger room.

alder A light-colored, lightweight hardwood often stained to simulate cherry or walnut, or used as plywood core.

algicide A chemical for resisting or preventing the growth of algae.

alidade A sighting apparatus often with a plane table, used for determining and plotting horizontal and/or vertical angles.

alienation The act of transferring title of real property from one owner to another.

aliform Any object shaped like a wing.

aligning punch A tool used for aligning holes in structural steel.

alignment (1) The adjustment of elements in a plane such as structural steel. (2) The plan or horizontal orientation of a structure or roadway.

alite The primary constituent of portland cement clinker. Alite is composed of tricalcium silicate and small amounts of magnesium oxide, aluminum oxide, ferric oxide, and other materials.

alkali Water-soluable salts of alkali metals, such as sodium and potassium, which occur in concrete and mortar mixes. The presence of alkaline substances may cause expansion and subsequent cracking.

alkali resistance The ability, particularly of paint, to resist attack by alkaline materials.

alkali soil A soil having a pH value of 8.5 or higher, and thus harmful to some plant life.

alkyd paint A paint, using an alkyd resin base, and producing a quick-drying, hard surface.

alkyd plastics Thermoset plastics with good heat and electrical insulation properties and commonly used in paints, lacquers, and molded electrical parts where temperatures will not exceed 400 degrees F.

alkyd resin A synthetic resin used as a binder in lacquers, adhesives, paints, and varnishes.

Alligator Wrench

Allen head A screw or bolt with a hexagonal recess to receive an Allen head wrench.

Allen wrench A section of hexagonal stock bent 90 degrees which is used in an Allen head screw or bolt.

alligatoring Rough cracking on a painted surface, usually caused by applying another coat before the first is dry, or by exposing a painted surface to extreme heat.

alligator shears, lever shears A shop tool used for shearing sheet metal.

alligator wrench Another name for a **pipe wrench**.

all-in aggregate See **bank run gravel**.

allover A term used to describe a repeating pattern on a surface.

allowable bearing value, allowable soil pressure The bearing capacity of a soil, in pounds per square foot (psf), determined by its characteristics such as shear, compressibility, water content, and cohesion. The higher the allowable bearing value of a soil, the smaller the footing required to support a structural member.

allowable load The ultimate load divided by a safety factor.

allowable pile bearing load The allowable load used to design a pile cluster to support a structure.

allowable stress The maximum stress allowed by code for members of a structure, depending upon the material and the anticipated use of the structure.

allowance (1) In bidding, an amount budgeted for an item for which no exact dollar amount is available. (2) A contingency for unforeseen costs. (3) The classification of connected parts or members according to their tightness or looseness.

alloy A homogeneous mixture of two or more metals to reduce cost or develop certain desirable properties.

alloy steel Steel with one or more added materials other than carbon and the allowed impurities, which gives it desired physical or chemical characteristics.

alluvium Any material such as gravel, sand, silt, or stones that has been deposited by running water.

alpha gypsum A specially processed calcined gypsum with extremely high compressive strength.

alteration Construction within a structure or to its exterior closure which does not change the overall dimensions. Alteration includes remodeling and retrofitting.

alternate bid An amount stated in a bid which can be added or deducted by an owner if the defined changes are made to the plans or specifications of the base bid.

alternating current An electric current that reverses direction at regular intervals. In the U.S. most current for domestic use reverses direction at 60 cycles per second.

alternator A machine that develops alternating current by mechanical rotation of its rotor.

altitude The angular distance of a celestial body above the horizon.

alum A double sulfate added to plaster as a hardener and accelerator.

alumina Aluminum oxide, found in the clay used in brick and clay tile.

aluminous cement See **cement**.

aluminum A silver-colored, nonmagnetic, lightweight metal used extensively in the construction industry. It is used in sheets, extrusions, foils, and castings. Aluminum is usually used in alloy form for greater strength. Sheets are often anodized for greater corrosion resistance and surface hardness. Because of

Aluminum Door

Aluminum Window

**American Standard
I Beam**

**American Standard
Channel**

its light weight and good electrical conductivity, aluminum is used extensively for electrical cables.

aluminum brass Brass with an aluminum alloy to increase corrosion resistance.

aluminium bronze A copper-aluminum alloy that is used in castings and can also be cold worked.

aluminum door A glazed door with aluminum stiles and rails.

aluminum foil A very thin aluminum sheet used extensively for thermal reflection and moisture protection.

aluminum oxide See **alumina**.

aluminum paint A paint containing aluminum paste with good heat-, light-, and corrosion-resistance properties.

aluminum window A glazed window with aluminum sash and muntins.

ambient noise The total noise level from all sources in a given area, either within a building or in an outside environment.

ambient temperature The temperature of the environment surrounding an object.

American basement The floor of a building partly above and partly below grade. Also called a "walk-out basement."

American black walnut A close-grained, dark-colored hardwood often used for carving and furniture making.

American bond See **common bond**.

American standard beam A hot-rolled steel I beam designated by the prefix S before the size and weight.

American standard channel A hot-rolled steel channel designated by the prefix C before the size and weight.

American standard pipe threads The thread size and pitch

commonly used in the U.S.A. for connecting pipe and fittings.

American table of distances A code for storage of explosives giving distances from cap storage, dwellings, and other buildings and materials.

American wire gauge, American standard wire gauge, Brown and Sharpe gauge The standard in the U.S.A. for specifying and manufacturing wire and sheet metal sizes, particularly electrical wire and metal flashing.

amino plastic A group of thermoset plastics derived from ammonia and often used in making electrical insulators.

ammeter An instrument for measuring the rate of ampere flow through an electric circuit.

ammonal An inexpensive explosive used for heavy blasts in dry holes.

ammonia A highly efficient, inexpensive gas used in manufacturing fertilizers and as a refrigerant in large refrigeration systems such as ice rinks.

ammonium chloride A white crystalline compound used as a soldering flux and in the manufacture of iron cement.

amorphous In rock, having no crystalline structure.

amount of mixing The mixing action employed to combine the ingredients of concrete or mortar, measured in time or revolutions.

ampacity A designation of the current carrying capacity of an electrical wire, expressed in amperes.

ampere The electromotive force required to move one volt of electricity across one ohm of resistance. A measure of electrical current.

amplitude The maximum plus or minus oscillation or vibration from the mean position.

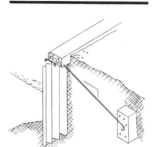

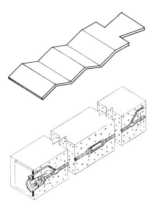

Anchor, Anchorage

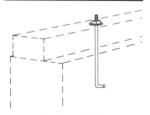

Anchor Bolt

Anchor Rod

amyl acetate, banana oil A solvent, for paints and lacquers, with a banana odor.

amylin A starch-like, water-soluable compound used as a wallpaper adhesive.

anchor, anchorage (1) A device to prevent movement when in tension, such as a tie-back for sheet piling. (2) In masonry composite wall construction, the tension connection between components. (3) In prestressed or posttensioned concrete, the end connection for the tendons. (4) A timber connector. (5) The metal devices that secure metal door and window frames to masonry.

anchorage bond stress The forces on a deformed reinforcing steel bar divided by the product of the perimeter and the embedded length.

anchorage deformation, anchorage slip In prestressing concrete members, the deformation of an anchor or slippage of tendons when the prestressing device is released.

anchorage loss See **anchorage deformation.**

anchorage zone (1) In pretensioning, the area of the member in which the stresses in the tendon anchor are developed. (2) In posttensioning, the area adjacent to the anchorage which develops secondary stresses.

anchor block A block of wood in a masonry wall to provide a means of attaching wood members.

anchor bolt, foundation bolt, hold-down bolt A threaded bolt, usually embedded in a foundation, for securing a sill, framework, or machinery.

anchored-type ceramic veneer A ceramic veneer that is attached to a backing by nonferrous anchors and grout. The minimum thickness of veneer and grout is 1 inch.

anchor log A log buried in the ground to act as a deadman.

anchor plate A plate attached to an object to which accessories or structural members may be attached by welding, screwing, nailing, or bolting.

anchor rod A threaded metal rod attached to hangers and used to support pipe and ductwork.

ancillary One of a group of buildings having a secondary or dependent use, such as an annex.

anechoic room An enclosure whose interior surfaces do not reflect sound.

anemometer An instrument for measuring the velocity of air flow.

angle The figure or measurement of a figure formed when two planes diverge from a common line. In construction, a common name for an L-shaped metal member. The term ''angle iron'' is sometimes referred to simply as an ''angle.''

angle bar Another name for *angle iron.*

angle bead A metal or wood strip set at the corner of a wall board or plaster wall to serve as a guide and for protection. Angle beads are most commonly made of nonferrous or galvanized perforated sheet metal.

angle beam Another word for *angle iron.* Usually, the term refers to one of heavier stock.

angle block, glue block A small block of wood used to fasten or stiffen the joint of two adjacent wood members, usually at right angles.

angle board A board cut at an angle and used as a jig for erecting masonry or cutting other members at a constant angle.

angle bond A metal tie, used to bond masonry, which projects into each wall at a corner.

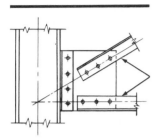

Angle Framing

Angle Iron

angle brace A piece of material temporarily or permanently secured across an angle to make it rigid, such as a strip of wood nailed across the corners of a window frame to keep it square during installation.

angle brick A brick cast with an oblique angle on one of its corners

angle buttress One of two buttresses forming a right angle at the corner of a structure.

angle cleat, angle clip A short section of angle iron used to attach structural members, such as precast panels, to structural steel.

angle closer A special brick or a portion of a brick used to close the bond on the outside corner of a brick wall.

angle collar A cast-iron pipe angle fitting with a bell-type connection at each end.

angle divider A square with an adjustable hinged blade used for setting or bisecting angles.

angle dozer A bulldozer whose blade can be set at an angle so as to push the excess material to one side only.

angle fillet A strip of triangular cross section used to cover the joint of two surfaces meeting at an angle of less than 180 degrees.

angle float A trowel with two surfaces meeting at right angles. An angle float is used for finishing plaster or concrete in a corner.

angle framing Light-gauge framing with angle iron.

angle gauge A template used to set or maintain an angle during construction.

angle hip tile See *arris hip tile*.

angle iron, angle bar An L-shaped steel structural member classified by the thickness of the stock and the length of the legs.

angle lacing A system of connecting two structural components with angle irons.

angle of repose The maximum angle above horizontal at which a material of given density and moisture content will remain without sliding. This measurement is used specifically for earth cuts and fills and stockpiled aggregates.

angle paddle A hand tool used to finish plaster.

angle post Another word for "corner post."

angle rafter See **hip rafter**.

angle ridge See **hip rafter**.

angle section See *angle iron*.

angle staff, staff angle See *angle bead*.

angle stile See *angle fillet*.

angle strut An angle iron erected to carry a compression load.

angle tie See *angle brace*.

angle trowel See *angle float*.

angle valve A valve for controlling flow in a pipe with the inlet at right angles to the outlet.

angular aggregate An aggregate made of crushed material with sharp edges, as opposed to screened gravel with round edges.

anhydrite An additive used in the manufacture of portland cement to control the set.

anhydrous calcium sulfate, dead-burnt gypsum Gypsum from which all the water of crystallization has been removed.

anhydrous gypsum plaster A high-grade finish plaster with most of the water of crystallization removed.

anhydrous lime See **lime**.

annealing The process of subjecting a material, particularly glass or metal, to heat and then slow cooling to relieve internal stress, which reduces brittleness and increases toughness.

Apartments

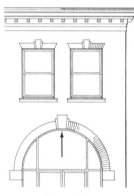

Apex Stone

annex A secondary structure either near or adjoining a primary structure.

annular ring nail A nail with a series of thread-like rings on its shank to give it good holding power. This type of nail is used for attaching gypsum board to wood studs.

anode The conductor rod used in an electrical system to protect underground tanks and pipes from electrochemical action.

anodic coating The surface finish resulting from anodizing aluminum. The finish may be clear or colored.

anodize The process of creating electrolytically a hard, noncorrosive film of aluminum oxide on the surface of a metal. This film can be either clear or colored.

antechamber An entrance, vestibule, or foyer.

anticorrosive paint A paint containing corrosive-resistant pigments such as zinc chromate, lead chromate, or red lead. This type of paint is used as a primer on iron and steel products.

antimonial lead, hard lead, regulus metal A lead alloy containing antimony that is used in sheet form for roofing, cladding, and tank lining. This alloy is harder than pure lead.

antimony oxide A white pigment added to plastics and paints to produce flame-retardant properties.

antiquing A method of applying paint exposing a second color from underneath used on trim and furniture.

antisiphon trap In a drainage system, a plumbing trap that provides a water seal to prevent siphonage.

antislip paint A paint with coarse particles mixed in to roughen the surface. This type of paint is used on steps, ramps, walkways, and porches.

antismudge ring The frame around a ceiling-mounted air diffuser that reduces the formation of rings of dirt around the diffuser.

antistatic agent An additive that reduces the development of static electricity on the surface of plastics or on carpetings.

apartment A room or group of rooms within a building designed as a separate dwelling.

aperture In construction, any opening left in a wall for a door, window, or for ventilation.

apex The highest point, peak, or top of any structure.

apex stone, key stone, saddle stone The highest stone or block in an arch, gable, dome, or vault. Such stones are often decorative.

appentice, pent, pentice A subordinate structure, like a shed, built against another structure.

appliance panel An electrical service panel with circuit breakers or fuses specifically designed for service to appliances.

application A means of erecting wall board, ceiling tile, or other building materials by using an adhesive.

bond The measurement of the strength of adhesion between two adhered surfaces.

butyl An adhesive with a butyl base used in caulking and sealants.

cement A common term for mastics used in flooring and roofing applications.

failure The separation of two adhered surfaces by physical or chemical means.

mortar A mixture used for ceramic wall or ceiling tile with an adhesive additive.

Applied Trim

Apron

neoprene A liquid neoprene compound applied to concrete foundation walls for waterproofing.

spreader A trowel with notched edges for applying adhesive.

wall clips Special clips for installing wall board with adhesive in lieu of nails or screws.

application for payment A formal written request for payment by a contractor for work completed on a contract and, if allowed for in the contract, materials stored on the jobsite or in a warehouse.

applied trim Strips or moldings applied to, as opposed to manufactured with, door and window frames and wood paneling.

applique A nonstructural decorative object attached to a member or structure.

appraisal An estimate of the value of a property generally made by a professional. The estimate is developed from market value, replacement cost, income produced, or a combination of these.

apprentice A person who works with a skilled craftsman for a number of years to learn the trade. An apprentice is generally rated by the number of years served.

approach ramp (1) An access for vehicles to a highway. (2) A sloped access for the handicapped to a building, in lieu of stairs.

approach-zone district An area defined by a zoning ordinance or bylaw to be within the flight path of an airport. Building heights and the type of industry allowed are often restricted within such an area.

approved In construction, materials, equipment, and workmanship in a system, or a measurable portion thereof, which have been accepted by an authority having jurisdiction. Usually, the term refers to approval for payment, approval for continuation of work, or approval for occupancy.

approved equal Material, equipment, or method of construction which has been approved by the owner or the owner's representative as an equivalent to that specified in the contract documents.

appurtenance Something added on to a main structure or system. Also, a condition added to the deed to a property, such as a right-of-way.

apron (1) A piece of finished trim placed under a window stool. (2) A slab of concrete extending beyond the entrance to a building, particularly at an entrance for vehicular traffic. (3) The piece of flat wood under the base of a cabinet. (4) Weather protection paneling on the exterior of a building. (5) A splashboard at the back of a sink. (6) At an airport, the pavement adjacent to hangars and appurtenant buildings.

apron flashing The flashing that diverts water from a vertical surface on a building to a sloped roof, such as that around a chimney. Also, flashing that leads water from a roof into a gutter.

apron lining A board or veneer that covers the apron piece of a staircase.

apron molding The piece of flat wood under the base of a cabinet.

apron piece, pitching piece A piece of lumber protruding from a wall to support the rough stringers of a wooden

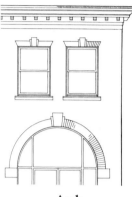

Arch

staircase at the top or at a landing.

apron rail A horizontal rail in a door at the height of the lock.

apron wall A distinct exterior wall panel extending from a window sill to the window below.

aquastat An electrical control activated by the temperature of water.

aqueduct A water-conveying system relying on gravity for flow. An aqueduct may be underground or elevated, and covered or open if elevated.

aquifer A geological waterbearing formation which can provide water for wells.

arbitration A method of settling disputes between parties of a contract by presenting information to recognized authorities. Parties agree in advance to binding arbitration of disputes, either as a clause in the contract or at the occurrence of a dispute. This method of avoiding litigation can save both time and money.

arbor (1) An enclosure of closely planted trees, vines, or shrubs which are either self-supporting or supported on a framework. (2) The rotating shaft of a circular saw or shaper.

arc The electrical discharge between two electrodes. When the electrodes are surrounded by gas in a lamp, they become a bright, economical light source. (2) Any portion of a circle or the angle which it makes.

arcade A row of arches on columns or piers alongside a covered walkway, usually lined with shops.

arc cutting A method of cutting metal with an electric welding machine. The metal is melted by the heat produced by the arc

between the electrode and the metal.

arc gouging A shaping or grooving of metal by arc cutting.

arch A curved or flat structure spanning an opening. The shape and size of arches are limited by the materials used and the support provided.

arch band The visible ribs in a vaulted ceiling.

arch bar A metal strap or beam that supports an arch.

arch beam A beam whose upper surface is arched.

arch brace A curved brace in a wooden roof truss that gives the roof an arched appearance from the inside.

arch brick, compass brick, featheredge brick, radial brick, radiating brick, radius brick, voussoir brick One of a number of bricks manufactured to construct curved surfaces, such as arches and round manholes.

arch buttress See *flying buttress*.

arch center The falsework used to support masonry during construction of an arch.

arch corner bead A job cut corner bead used to finish an arch.

arching The bridging of shear stresses in a soil mass across an area of low shear strength to adjacent areas of higher shear strength.

architect Designation reserved, usually be law, for a person or organization professionally qualified and duly licensed to perform architectural services, including but not necessarily limited to analysis of project requirements, creation and development of the project design, preparation of drawings, specifications, and bidding requirements, and general administration of the construction contract.

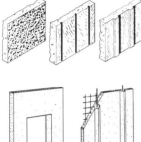

Architectural Concrete

Architectural Millwork

architect's approval Permission granted by the architect, acting as the owner's representative, for actions and decisions involving materials, equipment, installation, change orders, substitution of materials, or payment for completed work.

architect-engineer An individual or firm offering professional services as both architect and engineer. The term is generally used in federal government contracts.

architectural Pertaining to a class of construction, particularly in home building, of higher quality than average. The term often pertains to the ornamental features of a structure.

architectural area of buildings The sum of the adjusted areas of the several floors of a building, computed in accordance with AIA Document D101 and Handbook Chapter B-5.

architectural bronze An alloy of 57% copper, 40% zinc, 2.75% lead, and 0.25% tin used for ornamental forgings and extrusions.

architectural concrete Structural or non-structural concrete that will be permanently exposed to view and therefore requires special attention to uniformity of materials, forming, placing, and finishing. This type of concrete is frequently cast in a mold and has a pattern on the surface.

architectural door A grade classification of door which designates higher than standard specifications for material and appearance.

architectural fee The cost of architectural services to an owner, usually a percentage of the total contract. The fee varies according to the services provided and the complexity of the project.

architectural glass Glass with a configurated surface to obscure vision or to diffuse light.

architectural millwork, custom millwork Millwork manufactured to meet specifications of a particular job, as distinguished from stock millwork.

architectural terra-cotta Hard-burned, glazed or unglazed clay building units, plane or ornamental, machine extruded or hand molded, and generally larger in size than brick of facing tile.

architectural volume The sum of the products of the architectural areas of a building (using the area of a single story for multistory portions having the same area on each floor) and the height from the underside of the lowest floor construction system to the average height of the surface of the finished roof above the various parts of the building. (Ref: AIA Document D101 and Handbook Chapter B-5.)

architecture The art and science of designing and building structures.

arch stone The wedge-shaped masonry units used in building an arch.

arch truss A roof truss having a curved upper chord and a straight lower chord.

archway A passageway through an arch, particularly a long one.

arc light A light produced by an arc across two electrodes.

arc welding The joining of metal parts by fusion. Heat is produced by electricity passing between an electrode and the metal and is usually accompanied by a filler metal and/or pressure.

are An area equal to 100 square meters.

area (1) A measurement of a given planar region or of the surface of a solid. (2) A particular part of a

Areaway

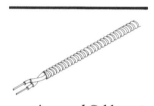

Armored Cable

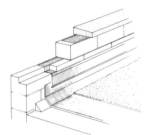

Arriss Fillet

building that has been set aside for a specific purpose.

area drain A catch basin or other device designed to collect surface water.

area light A light source used to illuminate a significant area, either inside or out.

area method A construction cost-estimating system employing unit square foot costs multiplied by the adjusted gross floor area of a building.

area wall A masonry wall surrounding or partly surrounding an open area, particularly one below grade, such as an areaway at the entrance to a basement.

areaway An open area located below grade and adjacent to a building to admit light, air, or access to a basement or crawl space.

areaway grating A steel or cast iron grating placed over an areaway, usually at grade level.

arenaceous In classifying soil, a term referring to material that is composed mostly of sand.

argillaceous In classifying soil, a term referring to material that is composed mostly of clay or shale.

argillite A shale composed mostly of clay and silt.

Arizona marble A stone mined in a variety of colors, the best known of which are Apache Gold, Geronimo, and Navajo block.

Arkansas marble A stone mined in a variety of colors, the best known of which are Ozark Famosa, Ozark Fleuri, and Ozark Rouge.

arkose A sandstone containing a high percentage of feldspar and used as a building stone.

armature The rotating part of a motor or generator consisting of copper wire wound around an iron core.

arm conveyor A belt with protruding arms or angles to carry materials into a building.

armored cable, metal-clad cable An electrical conduit of flexible steel cable wrapped around insulated wires.

armored concrete Concrete with a surface treatment containing steel or iron and used in areas with heavy, steel-wheeled traffic.

armored faceplate A metal faceplate mortised into the edge of a door to protect the lock mechanism.

armored front A tamperproof metal plate that covers the set screws of a mortise lock.

armored plywood Plywood that is faced on one or both sides with metal cladding.

armor plate A metal plate that is installed on the lower part of a door to protect it from kicks and scratches. The plate is usually not less than 36 inches high. Also called a "kick plate."

aromatic red cedar A highly aromatic, fine-textured wood with a distinctive red and white color used for fence posts and rails and for moth-proofing closets.

arrester A wire screen at the top of a chimney or incinerator to prevent burning material from flying out. Also called a "spark arrester."

arriss fillet, doubling piece, tilting fillet A triangular piece of wood used under roofing to tilt it away from a vertical surface, such as a parapet or chimney. Also see a **cant strip**.

arris gutter An archaic term used to describe a V-shaped wooden gutter fixed to the eaves of a building.

arris hip tile, angle hip tile An L-shaped roof tile manufactured to fit over the hip of a roof.

arris rail A triangular wooden rail with the broadest section used as a base.

arris tile See *arris hip tile*.

arrissing tool A special float used to round the edge of freshly placed concrete.

arrow diagram A CPM (Critical Path Method) diagram where arrows represent activities in a project.

article A subdivision of a document such as a contract document.

artificial turf A synthetic material, designed to simulate a natural grass surface, used to form playing surfaces for indoor or outdoor sports arenas, as on football fields.

asbestine A fine powdered talc used in mixing paint.

asbestos, asbestos fiber A flexible, noncombustible, inorganic fiber used primarily in construction as a fireproofing and insulating material. Because the airborne fibers associated with asbestos have been demonstrated to constitute a health hazard, many new controls have been placed on its use. Great care should be exercised in the use of asbestos, and local and state building codes should be consulted before installing or removing this material.

asbestos blanket Asbestos fibers woven or bonded into a flexible blanket to be used as a fire or heat barrier.

asbestos-cement board, asbestos-cement wallboard, asbestos sheeting A dense, rigid board made from asbestos fibers bonded together with portland cement and used in building construction where fire or heat protection is required. The material is manufactured in flat and corrugated sections.

asbestos-cement pipe A strong, light, noncorrosive pipe made from asbestos fibers and portland cement and used in construction for water pipe drainage and air ducts.

asbestos curtain, fire curtain, safety curtain A fireproof curtain that closes the stage of a theatre from the auditorium in case of fire on the stage.

asbestos duct Asbestos-cement pipe used as an air duct.

asbestos encapsulation An airtight enclosure of asbestos fibers with sealants or film that prevent fibers from becoming airborne and creating a health hazard.

asbestos felt A building product made by saturating asbestos felt material with asphalt or some other binder. This material is used extensively in roofing and sheathing systems.

asbestos formboard Rigid asbestos cement shaped into formboard for cementitious decks.

asbestos joint runner, pouring rope Asbestos fibers woven into a rope and used to wrap around a bell joint as a retainer for molten lead caulking.

asbestos plaster A fireproof plaster with high insulating properties made from asbestos fibers and bentonite.

asbestos removal A special trade that has developed since the health hazards of airborne asbestos have been revealed. Applies principally to ceiling tile and pipe insulation.

asbestos roofing Rigid asbestos-cement sheets, either flat or corrugated, with good insulating and fireproofing qualities.

asbestos roof shingle Asbestos roofing material in shingle form.

asbestos runner See *asbestos joint runner*.

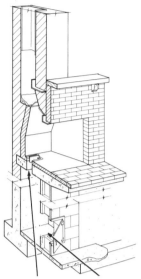

Ash Dump and Pit

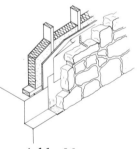

Ashlar Masonry

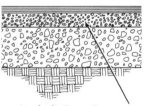

Asphalt Base Course

asbestos waterproofing A foundation waterproofing in sheet form made of PVC with an asbestos sheet backing.

ash A strong, long-grained hardwood with excellent bending qualities. This wood is used in veneers, trim, and flooring.

ash dump An opening in the bottom of a firebox or fireplace into which ashes are swept, falling into an ashpit below.

ashpit A cleanout under a fire place, usually at the base of a chimney, where ashes are removed.

ashlar (1) Any squared building stone. The term usually refers to thin stone used as facing. If the horizontal courses are level, it is called "coursed ashlar," and, if they are broken, it is called "random ashlar." (2) Short vertical studs between the ceiling joists and the rafters.

ashlar brick, rock-faced brick A brick with a broken face resembling stone.

ashlar line A horizontal line on the exterior face of a masonry wall.

ashlar masonry A stone masonry wall or veneer composed of rectangular units bonded with mortar. See **ashlar**.

ashlar veneer A nonstructural wall facing composed of ashlar masonry.

aspect The orientation of a building with respect to the points of a compass.

aspect ratio In any configuration, the ratio of the long dimension to the short dimension.

aspen A smooth-grained, white hardwood used for trim and veneer.

asphalt A dark-brown to black bitumen pitch that melts readily. It appears in nature in asphalt beds and is also produced as a by-product of the petroleum industry.

asphalt, blown Asphalt that has had air blown through it at high temperatures to give it workability for roofing, pipe coating, foundation waterproofing, and other purposes.

asphalt base course A bottom paving course consisting of coarse aggregate and asphalt.

asphalt block A manufactured paving block made from asphaltic concrete and aggregate. The block is typically manufactured in squares, rectangles, and hexagons and comes in dark gray or black colors.

asphalt cement Asphalt that has been refined to meet the specifications for use in paving and other special uses. It is classified by penetration. See **penetration**.

asphalt coating A term that refers to the asphaltic coating of corrugated metal pipe. Coatings can be inside, outside, or just on the invert.

asphalt color coat An asphalt surface treatment which has been impregnated with aggregate of a specified color.

asphalt concrete See **asphaltic concrete**.

asphalt curb An extruded or hand-formed berm made from asphaltic concrete.

asphalt cut-back An asphalt that has been liquified by an additive for a specific use.

asphalt cutter Any of a variety of machines designed to cut asphalt pavement.

asphalt emulsion Liquid asphalt in which water has been suspended. When the water evaporates, the asphalt hardens. Asphalt emulsion is used in paving as a tack coat to bind one course to another.

**Asphalt Prepared
Roofing**

asphalt expansion joint
Premolded felt or fiberboard
impregnated with asphalt and
used extensively as an expansion
joint for cast-in-place concrete.

asphalt felt Felt impregnated
with asphalt and used in roofing
and sheathing systems.

asphalt filler A liquid asphalt
used for filling joints and cracks
in pavement and floors.

asphalt flashing cement A semisolid asphaltic material used to
apply flashing.

asphalt floor tile Any of a
number of economical, resilient
floor tiles with an asphalt base.
These tiles are particularly
suited for application to
concrete floors on grade, as they
are not affected by moisture.
For wood subfloors, they are
usually laid on a felt
underlayment.

asphalt joint filler See *asphalt
filler*.

asphalt lamination A lamination
of sheet material, such as paper
or felt, using asphalt as the
adhesive.

asphalt leveling course A course
of asphaltic concrete pavement
of varying thickness spread on
an existing pavement to
compensate for irregularities
prior to placing the next course.

asphalt lined pipe See *asphalt
coated pipe*.

asphalt, liquid An asphaltic
material having a fluid
consistency at normal
temperatures. The common
types specified for pavements
are cutback, rapid curing (RC),
medium curing (MC), and slow
curing (SC), which are blended
with petroleum solvents, and
emulsion, which is blended with
water.

asphalt overlay One or more
courses of asphaltic concrete
placed over existing pavement.

The process of overlaying
usually includes cleaning, a tack
coat, and a leveling course.

asphalt paint An economical,
liquid-asphaltic product used
principally for weatherproofing.

asphalt paper A paper that has
been coated or saturated with
asphalt and is used as a moisture
barrier.

asphalt pavement Any pavement
made from one or more layers
of asphaltic concrete.

asphalt pavement sealer A
material applied to asphalt
pavement after compaction to
protect it from deterioration
from weather or petroleum
products.

asphalt pavement structure All
the successive courses of asphalt
products placed above the
subgrade in a pavement.

asphalt paving See *asphaltic
concrete*.

asphalt penetration A measure
of the hardness or consistency
of asphalt, expressed as the
distance a needle of standard
diameter will penetrate a sample
with given time, load, and
temperature conditions.

asphalt plank A reinforced,
premolded, structural plank
made of asphalt impregnated
fibers and mineral filler. The
material comes in 3 to 8 foot
lengths, 6 to 12 inches wide.

*Asphalt prepared roofing,
asphaltic felt, cold-process
roofing, prepared roofing,
rolled roofing, rolled strip
roofing, roofing felt, sanded
bituminous felt, saturated felt,
self-finished roofing felt* A
roof covering that comes in rolls
and is manufactured from
asphalt impregnated felt with a
harder surface of asphalt applied
to the surface. All or part of the
weather side may be covered
with aggregate of various sizes
and colors.

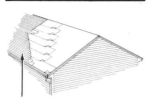

Asphalt Shingles

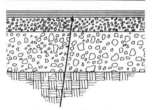

Asphalt Surface Course

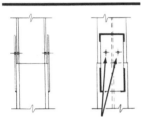

Assembling Bolt

asphalt prime coat A tack coat, usually an emulsion, to help one course adhere to another in pavement construction.

asphalt primer A liquid asphalt of low viscosity which is applied to a nonbitiminous surface such as concrete to prepare it for an asphalt course.

asphalt seal coat A thin asphalt surface treatment used to waterproof and improve the texture of the wearing surface of a pavement, particularly that of an asphaltic macadam. Depending upon the purpose, a seal coat may or may not include aggregate.

asphalt shingles, composition shingles, strip slates Roofing felt, saturated with asphalt and coated on the weather side with a harder asphalt and aggregate particles, which has been cut into shingles for application to a sloped roof.

asphalt soil stabilization Treatment of existing soil with asphalt penetration, mixing, and compaction often used as a base for asphaltic concrete pavement.

asphalt surface course The top or wearing course of asphaltic concrete pavement.

asphalt surface treatment The application of liquid asphalt to any asphaltic pavement with or without adding aggregate.

asphalt tack coat A light coat of asphalt, usually an emulsion, added to an existing pavement to create a bond between it and another course.

asphaltic A term used for materials containing asphalt. Also, a term used interchangeably with asphalt in the construction industry. It is usually more correct to use "asphaltic" than to use "asphalt," as in asphaltic macadam.

asphaltic concrete A mixture of liquid asphalt and graded aggregate used as a paving

material for roadways and parking lots. It is usually spread and compacted in layers over a prepared base while hot.

asphaltic macadam A term generally referring to a penetration method of paving whereby the aggregate is placed first and then liquid asphalt is sprayed into the voids, followed by the addition of a finer-graded aggregate. Penetration macadam usually needs a seal coat to prevent damage by water infiltration.

asphaltic mastic, mastic asphalt A viscous asphaltic material used as an adhesive, a waterproofing material, and a joint sealant.

aspiration In an air conditioning system, the introduction of room air into the flow of air from a diffuser.

aspirator A device that draws a stream of gas or liquid into it by means of the suction created by liquid or gas passing through an orifice. An aspirator is used for mixing air with a stream of water or for mixing a controlled amount of a chemical with water.

assembled occupancy For design purposes, the maximum number of people who will occupy a room or hall at one time.

assembling bolt A threaded bolt used for temporarily holding structural steel members together during welding.

assessed valuation The value of a property assigned by a municipality for real estate tax purposes. The valuation may be higher or lower than the market value.

assessment (1) A tax on property. (2) A charge for specific services, such as sewer or water, by a government agency.

assignment With respect to a contract, a document stating that payment for work completed or

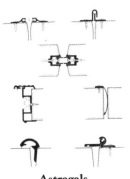

Astragals

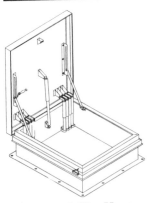

Automatic Fire Vent

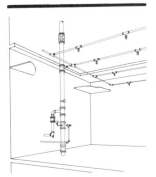

Automatic Sprinkler System

materials delivered must be made to someone other than the company or person specified in the contract.

astragal A molding attached to one of a pair of doors or casement windows to cover up the joint between two stiles.

atmospheric pressure The pressure (14.7 psi) exerted by the earth's atmosphere at sea level under standard conditions.

attenuation The sound reduction process in which sound energy is converted to heat or motion energy. Any sound absorbing system.

Atterberg limits Terms defining the properties of soils at different water contents.

Atterberg test Laboratory tests to determine the Atterberg limits.

attic tank A domestic water-storage tank installed above the highest plumbing fixture in a building to provide water pressure by gravity.

attic ventilator An electric fan, frequently thermostatically controlled, to move hot air out of an attic.

auger (1) A carpenter's hand tool for boring holes in wood. (2) A hand-held or rotary power tool with a helical cutting edge for drilling holes in soil. Augers are used for taking soil samples, drilling for caissons, or drilling for cast-in-place piles.

autoclave A chamber in which steam at high pressure is used to cure precast concrete members.

automatic Descriptive of any device which is activated by the environment or a predetermined condition, such as a ventilator that opens at a given temperature.

automatic air vent A vent that opens and closes at a predetermined temperature.

automatic batcher A batcher that blends all the ingredients of a mixture at predetermined weights upon receiving a single command.

automatic circuit breaker See *circuit breaker*.

automatic door A power-operated door that opens and closes at the approach of a person or vehicle.

automatic door bottom An attachment to the bottom of a door which is actuated when the door is closed to seal the bottom of the door at the threshold.

automatic fire pump A pump in a standpipe or sprinkler system that turns on when the pressure drops below a predetermined minimum.

automatic fire vent A device in the roof of a building that operates automatically to control fire or smoke.

automatic operator A remote operating device. The term usually refers to the opening and closing of doors by electronically actuated switches.

automatic smoke vent See *automatic fire vent*.

automatic sprinkler system A system that is designed to provide instant and continuous spraying of water over large areas in the case of fire.

automatic transfer switch In an electrical system, a switch that automatically transfers the load to another circuit when the voltage drops below a predetermined level.

auxiliary rafter, cushion rafter A rafter used to strengthen the main rafter, usually at area of greatest load.

auxiliary reinforcement In a prestressed concrete member, the reinforcing steel other than the prestressing steel.

auxiliary rim lock A surface-mounted lock in addition to the main lock on a door.

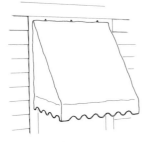

Awning

average bond stress The force exerted on a steel reinforcing bar divided by the product of the perimeter times the embedded length.

avodire, white mahogany A light-colored wood used for interior trim, plywood, and paneling.

award The formal communication accepting a bid or proposal for services, construction, materials, or equipment.

awl A hand tool used for piercing holes, particularly in leather, and often fitted with a needle for sewing heavy materials.

awning A projection over a door or window, often retractable, for protection against rain and sun.

awning blind A blind that is hinged at the top.

awning window A window that is hinged at the top.

ax, axe A sharp-edged hand tool for splitting wood and hewing timber.

axed brick, rough axed brick Brick shaped by an ax so as to create rough surfaces.

axhammer A hand tool for dressing stone.

axial flow fan See **centrifugal fan**.

axial force diagram In statics, a graphic representation of the axial loads acting at each section of a structural member.

axial load, axial force The longitudinal force acting on a structural member.

axis A straight line representing the center of symmetry of a plane or solid object.

azimuth The horizontal angle measured clockwise from north to an object.

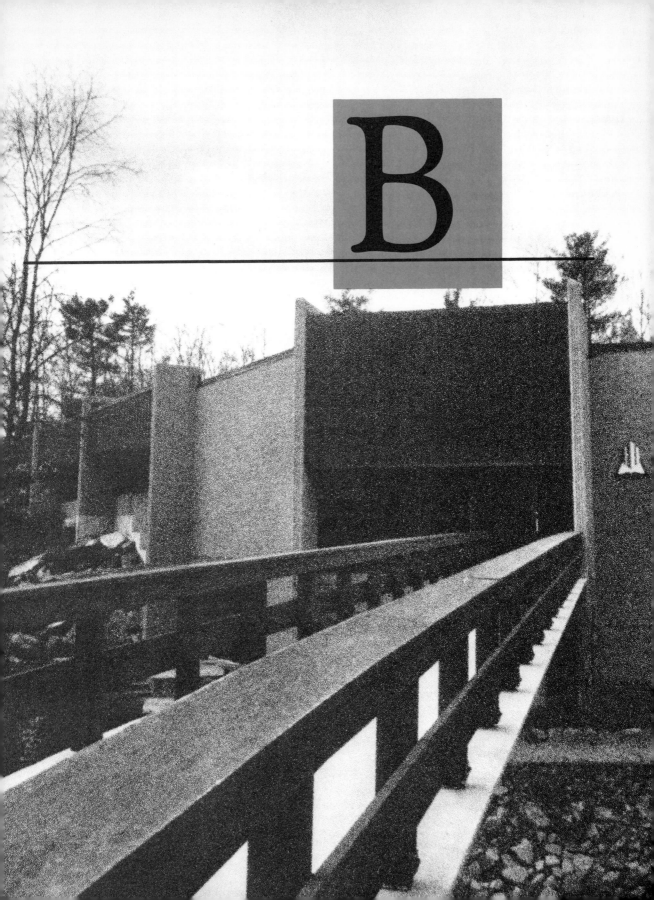

# ABBREVIATIONS

The abbreviations listed below are those most commonly used in the construction industry. Alternative forms (usually nonstandard) are shown in parentheses.

B1S banded one side, bead one side

B2E banded two ends

B2S banded two sides, bead two sides, bright two sides

B2S1E banded two sides and one end

B3E beveled on three edges

B4E beveled on four edges

B&B in the lumber industry, grade B and better

b&cb beaded on the edge and center

B&O back-our punch

B&S beams and stringers, bell and spigot, Brown and Sharpe gauge

B beam, boron, brightness

BA bright annealed

bat. batten

bbl, brl barrel

BC building code

BCM broken cubic meter

BCY broken cubic yard

bd. in the lumber industry, board

BF, bd. ft. in the lumber industry, board foot

bdl bundle

BET. between

Beth. B Bethlehem beam

bev in the lumber industry, beveled

bev sid beveled siding

BFP backflow preventer

bg bag

BG below ground

Bh Brinell hardness

Bhn Brinell hardness number

BHP brake horsepower

BL building line

B/L bill of lading

bldg building

blk block, black

BLKG blocking

BLO blower

BLR boiler

blt built, borrowed light

B/M, BOM bill of materials

b.m. in the lumber industry, board measure

BM bench mark, beam

BMEP brake mean effective measure

b of b back of board

BP blueprint, baseplate, bearing pile, building paper

bpd barrels per day

BPG beveled plate glass

BR bedroom

brc brace

brcg bracing

BRG bearing

BRK brick

BRKT, bkt bracket

BRS brass

Br Std, BS British Standard

BRZ bronze

BRZG brazing

BSMT basement

BSR building space requirements

BTB bituminous treated base

Btr., btr better

Btu British Thermal Unit

BTUH Btu per hour

but. buttress

BW butt weld

BX interlocked armored cable

B DEFINITIONS

Backfilling

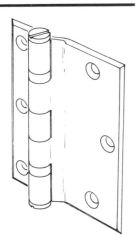

Backflap Hinge, Flap Hinge

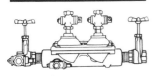

Backflow Preventer

babbitt An antifriction alloy composed of tin and lesser amounts of copper and antimony. Babbit is used in bushings and bearings.

back (1) That part, area, or surface that is farthest from the front. (2) The portion behind or opposite that is intended for use or view. (3) The reverse (scale). (4) That portion which offers strength or support from the rear. (5) The extrados of an arch or vault sometimes concealed in the surrounding masonry. (6) In slate, tile, etc., the side opposite the bed. (7) The surface of wallboard which receives the plaster. (8) The side of a piece of lumber or plywood opposite the face. The back is the side with the lower overall quality or appearance.

back arch A concealed arch which carries the inner part (or backing) of a wall where a lintel carries the exterior facing.

backband A rabbeted molding used to surround the outside edge of a casing.

back boxing Thin boards used in construction of doublehung windows to enclose a box for the balance weights next to the wall, and to keep the box free of mortar. See **back lining**.

back-brush To paint over a freshly painted surface with a finishing return stroke.

back check The mechanism in a hydraulic door closer or door check which lessens the speed with which the door may be opened.

back clip A special clip used on the back of gypsum board in certain applications, as in some demountable partitions, to hold the board in place by insertion into slots or other receptacles in the framing.

back coating Asphalt coating applied to the back side of shingles or roll roofing.

back-draft damper A damper whose blades are gravity-controlled and allow the passage of air in only one direction.

back edging A process by which glazed ceramic pipe is cut by first chipping away the glaze and then chipping the underlying pipe until it is cut through.

backfill Earth, soil, or other material used to replace previously excavated material, as around a newly constructed foundation wall.

back fillet The return of a fillet molding.

backfilling (1) The process of placing backfill. (2) Rough masonry laid behind a facing or between two faces. (3) Brickwork laid in spaces between structural timbers. See **nogging**.

back flap, back fold, back shutter That leaf of a folding window shutter which falls behind the exposed leaf, sometimes in a recess in the casing.

backflap hinge, flap hinge A hinge which has a flat plate or strap that is fastened to the face of a door or shutter. This type of hinge is used especially when the stile is too thin to accept a butt hinge.

backflow (1) The unintentional reversal of the normal and intended direction of flow. Backflow is sometimes caused by back-siphonage. (2) The flow of water or other liquids, mixtures, or substances into the distributing pipes of a potable water supply system from a source other than the intended source.

backflow preventer A device or

33

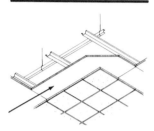

Backhoe

Backing Board

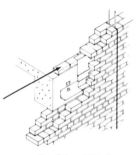

Backing Brick

means to prevent backflow into the potable water system.

backflow valve See **backwater valve.**

back gutter A shallow gutter installed on the upslope side of a chimney on a sloping roof. A back gutter is used to divert water around the chimney.

backhoe A powered excavating machine used to cut trenches with a boom-mounted bucket drawn through the ground toward the machine. The bucket is raised and swung to either side to deposit the excavated material.

backing (1) The bevel applied to the upper edge of a hip rafter. (2) Positioning furring onto joists to create a level surface on which to lay floorboards. (3) Furring applied to the inside angles of walls or partitions to provide solid corners for securing wallboard. (4) The first coat of plaster on lath. (5) The unseen or unfinished inner face of a wall. (6) Coursed masonry applied over an extrados of an arch. (7) Interior wall bricks concealed by the facing bricks. (8) The wainscoting between a floor and a window. (9) The material under the pile or facing of a carpet. (10) The stone used for random rubble walls.

backing board (1) In a suspended acoustical ceiling, gypsum board to which acoustical tiles are secured. (2) Gypsum wallboard or other material secured to wall studs prior to paneling to provide rigidity, sound insulation, and fire resistance.

backing brick A lower quality of brick used where it will be concealed by face brick or other masonry.

backjoint A rabbet in masonry such as that over a fireplace to receive a wood nailer.

backlighting Illumination from behind an object.

back lining, back jamb (1) In a weighted sash window, the thin wood strip that closes the jamb of a cased frame to provide a smooth surface for the operation of the weighted sash and, where applicable, prevents abrasion of brickwork by the sash weights. (2) The framing piece which constitutes the back recess for box shutters.

back lintel A lintel used to support the backing of a masonry wall, hence not visible on the face.

back-mop To apply hot bituminous material, either by mop or mechanical applicator, to the underside of roofing felt during the construction of a built-up roof.

back-nailing Nailing the layers, or plies, of a built-up roof to the substrate so as to help prevent slippage; performed in addition to hot mopping.

back nut (1) A threaded nut which helps to create a watertight joint, as on the long thread of a pipe connector, and whose one dished side accepts a grommet. (2) A locknut.

back-paint To apply paint to the reverse or unseen side of an object, not for appearance, but for protection against weather.

backplastering Plaster applied to one face of a lath system following application and subsequent hardening of plaster applied to the opposite face.

backplate A wood or metal plate which functions as backing for a structural member.

back pressure Hydraulic or pneumatic pressure in a direction opposite the normal and intended direction of flow through a pipe, conduit, or duct; usually caused by a restriction to the flow.

back-pressure valve See **check valve**.

back primed Back-painted woodwork; used primarily for

Backsplash

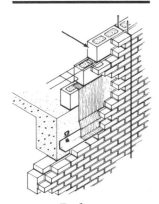

Backup

exterior shingles or siding, or exterior trim.

back putty, bed glazing The glazing compound applied between the glass and its sash or frame.

backsaw A hand saw used in finish carpentry work. The back (non-cutting) edge of the saw is stiffened with a steel or brass strip. A backsaw is used for cutting mitered joints and other joinery work.

backset The offset or horizontal distance between the face or front of a door back to the center of the key hole or central axis of the knob.

back sight In surveying, a sight on an established survey point or line.

back siphonage The flowing back of used, contaminated or polluted water from a plumbing fixture or vessel into the potable water supply often due to negative pressure in a pipe. See **backflow**.

backsplash A protective panel, apron, or sheet of waterproof material positioned on a wall behind a sink, counter, or lavatory.

backup (1) That part of a masonry wall behind the exterior facing. (2) Any material or substance placed into a joint to be sealed to reduce its depth and/or to inhibit sagging of the sealant. (3) An overflow due to a blockage in a piping system.

backup strip, lathing board A narrow strip of wood secured to a corner of a wall or partition to provide a base on which to nail the ends of lathing.

back veneer The ply on veneer plywood opposite the face veneer and usually of lesser quality.

back vent In plumbing, a venting device installed on the downstream side of a trap to protect it from siphonage.

backwater valve, backflow

valve A check valve in a drainage pipe which prevents reversal of flow.

badger (1) A tool used to remove excess mortar or other deposits from the inside of pipes or culverts after they have been laid. (2) A badger plane.

badger plane A large, wooden, hand plane for rabbeting, whose mouth is fashioned obliquely from side to side to allow its use close to corners.

baffle (1) A tray or partition employed on conveying equipment to direct or change the direction of flow. (2) An opaque or translucent plate-like protective shield used against direct observation of a light source; i.e. a light baffle. (3) A plate-like device for reducing sound transmission. (4) Any construction whose function is to change the direction of flow of a liquid.

bag, sack A quantity of portland cement; 94 pounds in the United States, 87.5 pounds in Canada, 112 pounds in the United Kingdom, and 50 kilograms in most other countries. For other than portland cement, different weights per bag are commonly used.

bag plug An inflatable drain stopper usually used at the lowest point in a piping system; used during testing of the system's integrity.

bag trap A plumbing trap, shaped like an S, whose inlet and outlet are in alignment.

baked finish A film of paint, or varnish baked at temperatures above 150 degrees F, to produce a tough, durable finish.

bakelite A plastic developed for use in electrical fittings, door handles, pulls, etc. Bakelite has high chemical and electrical resistance.

balance arm A supporting arm at the side of a projected window

Balanced Sash

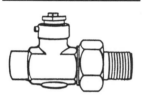

Balancing Plug Cock

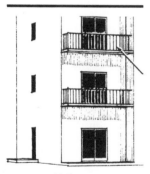

Balcony

which allows the sash to be opened without an appreciable change in its center of gravity.

balance beam, balance bar A counterbalance consisting of a long beam attached to that moveable structure, such as a drawbridge or gate, whose weight it offsets during opening and closing.

balanced circuit A three-wire electric power circuit whose main conductors all carry substantially equal currents, either alternating or direct, and in which there exist substantially equal voltages between each main conductor and neutral.

balanced earthwork Cutting and filling in which the amount of one is equal to the amount of the other after swelling and compaction factors are applied.

balanced load (1) In an electric circuit, such as a three-wire system, a load connected in such a manner that the currents on each side are equal. (2) In reinforced concrete beam design, a load that would cause crushing of concrete and yielding of tensile steel simultaneously.

balanced moment Moment capacity at simultaneous crushing of concrete and yielding of tension steel. See also **balanced load**.

balanced reinforcement An amount and distribution of reinforcement in a flexural member such that the allowable tensile stress in the steel and allowable compressive stress in the concrete are attained simultaneously.

balanced sash A sash in a double hung window requiring very little effort to raise or lower because its weight is counterbalanced with weights or pretensioned springs.

balanced step, dancing step, dancing winder One of a series of winders which do not radiate from a common center but rather

are balanced, i.e., their narrow ends are nearly equal in width to that of the straight portion of the adjacent stair flight to provide the line of traffic with a relatively even tread width.

balancing (1) Adjusting the mass distribution of a rotor to diminish journal vibrations and control the forces on the bearings from eccentric loading. (2) In an HVAC system, adjusting the system to produce the desired level of heating and cooling in each area of a building.

balancing plug cock See **balancing valve**.

balancing valve, balancing plug cock A pipe valve used to control the flow rather than to shut it off.

balata (1) A rubberlike, nonelastic latex gum used to make industrial belting and gaskets. (2) A tropical American tree that yields a latex-like sap.

balaustre, canary wood A heavy, hard, and glossy South American wood whose color can be yellowish brown, purplish brown, or orange.

balcony (1) A platform which protrudes from a building and can be cantilevered or supported from below and usually protected by a railing or balustrade. (2) A gallery protruding over the main floor of an auditorium; usually provides additional seating. (3) In a theater, an elevated platform used as part of a permanent stage setting.

balk, baulk In building construction, a squared timber whose dimensions are at least 4 inches by 4-1/2 inches.

ballast (1) A layer of coarse stone, gravel, slag, etc., over which concrete is placed. (2) The crushed rock or gravel of a railroad bed upon which ties are set. (3) The transformer-like device that limits the current flowing through the gas within a fluorescent lamp or a high intensity discharge lamp, and that provides the lamp with the proper starting voltage.

ballast factor The ratio of the luminous output of a lamp when functioning on a ballast to its luminous output when functioning under standardized rating conditions.

ballast noise rating The degree of noise created by a fluorescent lamp ballast, represented by the letters A (the quietest) to F (the loudest).

ballast-tamper A portable machine used to compact the ballast under railroad ties; usually powered by electricity or compressed air.

ball-bearing hinge A butt hinge having ball bearings positioned between the knuckles to reduce friction.

ball breaker Same as **wrecking ball**.

ball catch A door fastener in which a spring-tensioned metal ball engages the striking plate to keep the door closed until force is applied.

ball-check valve A device used to stop the flow of liquid in one direction while allowing flow in an opposite direction. The pressure against a spring-loaded ball opens the valve in one direction of flow. Pressure from the other direction forces the ball against a seat, closing the valve and preventing flow.

ball cock A float valve incorporating a spherical float; used to control the height of water, as in a water closet. See **float valve**.

ball float The somewhat spherical floating device by which a ball valve is controlled.

balling up In welding, the globules of flux or of molten brazing filler metal which result from

inadequate wetting of the metal being welded.

ball joint A flexible mechanical joint that allows the axis of one part to be set at an angle to the other by virtue of the design of the two components. One possesses a fixed spherical shell to accommodate the ball-shaped end of the other.

balloon framing, balloon frame A style of wood framing in which the vertical structural members, i.e., the posts and studs, are single, continuous pieces from sill to roof plate. The intermediate floor joists are supported by girts spiked to or let into the studs. The elimination of cross grains in the studding reduces differential shrinkage.

balloon-payment loan An installment loan demanding a final payment significantly larger than any of the previous annuity payments; often employed temporarily, until long term financing arrangements are finalized.

ball peen hammer A hammer having a hemispherical peen on one end of its head.

ball test (1) A test to determine the consistency of freshly mixed concrete by measuring the depth of penetration of a cylindrical metal weight or plunger which is dropped into it. (2) A test to determine whether a drain is circular and unobstructed, performed by rolling a specifically sized ball through the drain. See **Kelley ball**.

ball valve A spherically shaped gate valve which provides a very tight shut-off for fluids in a high-pressure piping system.

balsa, corkwood A soft, porous wood which, at a density of 7-10 pounds per cubic foot, makes it lighter in weight than any other wood. Balsa is used as the core

**Baluster, Banister
Balustrade**

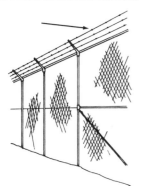

Band Saw

Barbed Wire, Barbwire

material in lightweight sandwich panels and for models.

baluster, banister (1) One of a series of short, vertical supporting elements for a handrail or a coping. (2) Any vase-shaped supporting member or column. (3) The roll on the side of an Ionic capital.

balustrade A complete railing system, including a top rail, balusters, and sometimes a bottom rail.

banana oil See **amyl acetate**.

band (1) A group of small bars or wire encircling the main reinforcement in a member to form a peripheral tie. A band is a group of bars distributed in a slab, wall, or footing. (2) A horizontal ornamental feature of a wall, such as a flat frieze or fascia, usually having some kind of projecting molding at its upper and lower edges.

band clamp A metal clamp consisting of two pieces which are bolted at their ends to hold riser pipes.

banding Wood strips or veneers attached to the exposed edges of plywood or particleboard in the construction of furniture or shelves.

band iron A thin metal strip used as a form tie or form hanger.

band saw A power saw with a continuous piece of flexible steel running around two pulleys and with teeth on one or both sides. A band saw is used to cut logs into cants, to rip lumber, and to cut curved shapes.

banister (1) A handrail. (2) A balustrade, especially one on the side of a staircase.

bank A mass of soil rising above a digging or trucking level. Excavation and loading is done at the face of a bank.

bank cubic yard A unit designating the volume of bank material measured or calculated

before removal from the bank.

banker A table or bench on which stonemasons or bricklayers shape their materials before setting them.

bank material Soil or rock in its natural position before excavation or blasting.

bank measure A determination of the volume of a mass of soil or rock in its natural position before excavation or blasting.

bank-run gravel, bank gravel, run-of-bank gravel Granular material excavated without screening, scalping, or crushing. This type of gravel is a naturally occurring aggregate comprised of cobbles, gravel, sand, and fines.

bank sand Sand which is unlike lake sand in that it has sharp edges providing a better bond and more strength when used in plastering.

bar (1) A deformed steel member used to reinforce concrete. (2) A solid piece of metal whose length is substantially greater than its width.

barb bolt, rag bolt A bolt whose jagged edges prevent it from being easily removed from any object into which it is driven.

barbed wire, barbwire Two or more wires twisted together with intermittent barbs incorporated during manufacture; used for security and livestock fencing.

bar bending The process of bending reinforcing steel into shapes required for reinforced concrete construction.

bar chair A device used to support reinforcing bars in position during the placement of concrete.

bar clamp A carpenter's clamping device consisting of a long bar with adjustable clamping jaws. A bar clamp is used to hold joinery components during gluing.

bare Descriptive of a piece of building material whose actual

Bar Joist

dimensions are smaller than those specified; scant.

barefaced tenon, bareface tenon A tenon having a "shoulder" on only one side, and used in the construction of many wood doors.

bargain and sale deed A deed in which the grantor admits that he has some interest, though not necessarily a clear and unemcumbered title to the property being conveyed. This kind of deed often contains a warranty that the grantor did not encumber the property or convey away any part of the title during his period of ownership.

barge spike, boat spike A long, square spike with a chisel point, used mostly in heavy timber construction.

barite A mineral of barium sulfate often used as an aggregate for concrete in the construction of high-density radiation shielding.

barium plaster A gypsum plaster to which barium salts have been added during manufacture used on walls of x-ray rooms.

barium sulfate See **barite**.

bar joist A light steel joist of open web construction with a single zig-zagged bar welded to upper and lower chords at the points of contact and used as floor and roof supports.

bar mat An assembly of steel reinforcement composed of two or more layers of bars placed at right angles to each other and tied together by welding or wire ties.

bar molding, bar-rail molding A rabbeted molding serving as a nosing when fixed to the edge of a countertop or bar.

barn-door hanger A type of hanger used for heavy exterior sliding or rolling doors, consisting of two pulleys secured to the top of the door in tandem, connected by a heavy metal strap. The door is moved by rolling along a

horizontal track which either hangs from the lintel or is anchored to a parallel member projecting slightly from it.

barn-door stay One of at least a pair of small pulleys which facilitates and directs the movement of a sliding barn (or other) door by rolling along a horizontal track or rail.

barometer (1) A device which measures atmospheric pressure.

Baroque A style of architecture and decoration developed in the late 17th century and characterized by the elaborate use of scrolls, curves, and other symmetrical ornamentation.

barracks A usually simple, unadorned building or group of buildings that provide temporary housing, often for military personnel.

barrel (1) A unit of weight measure for portland cement, equivalent to four bags or 376 pounds. (2) A standard cylindrical vessel with a liquid capacity of 31-1/2 gallons. (3) That part of a pipe where the bore and wall thickness remain uniform.

barrel bolt, tower bolt A cylindrical bolt mounted on a plate having a case projecting from its surface to contain and guide the bolt.

barreling, tumbling An operation by which small articles are place in a cylindrical case and painted by being tumbled in a barrel of paint.

barricade As defined by OSHA, an obstruction to deter the passage of persons or vehicles. Any of several devices to detour or restrict passage.

barrow (1) A wheelbarrow. (2) A large mound of earth or rocks intentionally placed on top of an ancient burial site for protection.

barrow run A temporary smooth ramp of planks or plywood for wheeled transport of materials at a construction site.

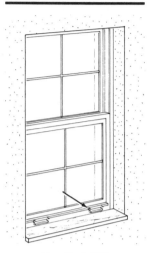

Bar Sash Lift

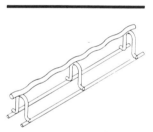

Bar Support, Bar Chair

Base Angle

bar sash lift A handle on the bottom rail of a sash for raising and lowering.

bar screen A coarse screening device, made from equally spaced parallel bars or rods, used to separate large stones from gravel or crushed stone.

bar size section A hot-rolled bar, angle, channel tee, or zee whose largest cross sectional dimension does not exceed 3 inches.

bar strainer (1) A screening device, fabricated from parallel bars or rods, used over a drain to prevent the entrance of foreign objects. (2) A bar screen.

bar support, bar chair A rigid device of formed wire, plastic, or concrete, used to support or hold reinforcing bars in proper position during concreting operations.

bar-type grating An open grate with parallel bearing bars evenly spaced and attached to a frame. The grating may be cast or welded and may have cross bars.

barway A gate with one or more sliding bars to act as large latch bolts.

barytes Same as **barite**.

basalt A dark, fine-grained igneous rock of high density widely used for paving aggregate.

base (1) The lowest part of anything upon which the whole rests. (2) A subfloor slab or "working mat" either previously placed and hardened or freshly placed, on which floor topping is placed. The base is the underlying stratum on which a concrete slab, such as pavement, is placed. (3) A board or molding used against the bottom of walls to cover their joint with the floor and to protect them from kicks and scuffs. The base is the protection covering the unfinished edge of plaster or gypsum board. (4) The lowest visible part of a building.

base anchor A fixed or adjustable metal device attached to the base of a door frame to secure it to the floor.

base angle Angle iron stock attached to the perimeter of a foundation for supporting and aligning tilt-up wall panels.

base bid The amount of money stated in the bid as the sum for which the bidder offers to perform the work described in the bidding documents, prior to adjustments for alternate bids which have been submitted.

base bid specifications The specifications listing or describing only those materials, equipment, and methods of construction upon which the base bid must be predicated, exclusive of any alternate bids.

base block A usually unadorned squared block that terminates a molded baseboard at an opening or serves as a base when attached to the foot of a door or bottom of window trim.

baseboard, mopboard, scrubboard, skirting board, washboard See **base, def. 3.**

baseboard heater A heating system using the heating elements housed in special panels placed horizontally along the baseboard of a wall.

baseboard radiator unit A heating device designed and positioned in a room so as to replace a baseboard. The heat source is commonly from hot water, steam, or electricity.

base coat (1) All plaster applied beneath the finish coat. (2) The initial coat of paint or stain applied to a surface.

base course (1) A layer of specified selected material of planned thickness constructed on the subgrade or sub-base of a pavement to serve one or more functions such as distributing loads, providing drainage, or minimizing frost action. (2) The lowest course of masonry in a wall, pier, foundation, or footing course.

base elbow A cast iron pipe elbow with a flange or baseplate for supporting it on a floor or block.

base flashing (1) In roofing, the flashing supplied by the upturned edges of a watertight membrane. (2) The metal or composition flashing used with any roofing material at the joint between the roofing surface and a vertical surface, such as a parapet or wall.

base line (1) The meticulously established reference line used in surveying or timber cruising. (2) In construction, the center or reference line of location of a highway, railway, building, or bridge.

base map A map employed in urban planning to indicate the principal outstanding physical characteristics of an area and which is thereafter used as a reference for subsequent mapping.

basement The bottom full story of a building below the first floor. A basement may be partially or completely below grade.

basement soil See **subgrade**.

base metal In joining two metal pieces, the parent metal which is actually welded, brazed, or soldered, and remains unmelted after the joining process, as opposed to the filler metal deposited during the joining operation.

base molding Trim molding applied to the upper edge of interior baseboard.

baseplate A plate used to distribute vertical loads from building columns or machinery.

base screed A galvanized metal screed with perforated or expanded flanges to provide ground for plaster and to separate areas of dissimilar materials.

base sheet The saturated and/or coated felt sheeting laid as the first ply in a built-up roof system.

base shoe, base shoe molding, floor molding shoe molding, carpet strip The molding or carpet strip covering the joint between a floor and a baseboard, often a quarter-round bead.

base shoe corner A block or a piece of molding installed in the corner of a room so as to eliminate the need to miter the base shoe.

base tee A pipe tee having an attached supporting baseplate.

base tile The bottom course of tile in a tiled wall.

base trim Any decorative molding at the base of a wall, column, or pedestal.

basic module A unit of dimension used when coordinating the sizes of building elements and other components.

basin (1) A somewhat circular, natural, or excavated hollow or depression having sloping sides and usually used for holding water. (2) A similarly shaped plumbing fixture, such as a sink.

basket crib A construction of interlocking timbers which can be arranged so as to function as a shaft liner, a protective device around a concrete pier in water, or a floating temporary foundation.

basket weave A pattern of bricks, placed flat or on edge, and arranged in a checkerboard layout.

bas-relief, basso-revievo, basso rilievo Sculpture, carving, or embossing which protrudes slightly from its background.

bastard Any non-standard item deviating from normal size, slope, fabrication, or quality.

bastard file A flat file whose grain is somewhat less than coarse and which is primarily used to smooth metal surfaces.

bastard granite A gneissic rock whose formation resembles that of

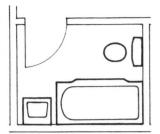

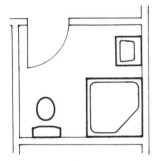

Bathroom

Batten

true granite. Bastard granite is used mostly in wall construction but is not a true granite.

bastard joint Same as **blind joint**.

bastard masonry (1) Thin blocks of facing stones used to face brick or rubble walls. Bastard masonry is dressed and built to resemble ashlar. (2) Rough, quarry-dressed ashlar stones.

bastard pointing In masonry, a type of pointing which emphasizes the joint by forming a small ridge projecting along its center.

bastard-sawn (1) Lumber which has been sawn so that the annual rings make angles of 30-60 degrees to the surface of the piece. (2) Any lumber which has been flat-sawn, plain-sawn, or slash-sawn.

bastard spruce Same as **Douglas fir**.

baswood, American linden A fine-textured soft wood used for carving, cabinet work, and paneling. Baswood is also the primary source of excelsior.

bat (1) A burned brick or shape which, because of being broken, has only one good end. (2) A piece of brick. (3) A single unit of batt insulation. (4) A piece of wood when serving as a brace.

Bataan mahogany Same as **tanguile**.

bat bolt A bolt whose butt or tang has been provided with barbs or similar protrusions to increase its grip.

batch The quantity produced as the result of one mixing operation by adding the proper weights and volumes, as in a batch of concrete.

batch box A container of known volume used to measure the constituents of concrete or mortar in proper proportions.

batch mixer A machine which mixes concrete, grout, or mortar in batches, unlike a continuous mixer.

batch plant An operating installation of equipment including batchers and mixers as required for batching and mixing concrete materials. Also called a "mixing plant" when mixing equipment is included.

bathroom A room usually equipped with a water closet, a lavatory, and a bathtub and/or shower.

batted work, broad tooled Stone whose surface has been scored downward from the top by a batting tool with which 8 to 10 narrow parallel strides per inch have been administered.

batten (1) a narrow strip of wood used to cover the joints of parallel boards or plywood when used as siding. The resultant pattern is referred to a board and batten. (2) A strip of wood placed perpendicular to several parallel pieces of wood to hold them together. (3) A furring strip fastened to a wall so as to provide a base for lathing or plastering. (4) In roofing, a strip of wood placed over boards or roof structural members to provide a base for the application of wood or slate shingles or clay tiles. (5) The steel strip which fastens the metal flooring on a fire escape.

battenboard See **coreboard**

batten door, ledged door, unframed door A door in which stiles are absent and which consists of vertical boards or sheathing secured on the back side by horizontal battens.

battened column A column composed of two longitudinal shafts fastened securely to each other by batten plates.

battened wall, strapped wall A wall to which battens have been secured.

batten plate, stay plate A steel plate connecting two parallel components, such as flanges or angles, of a built-up structural

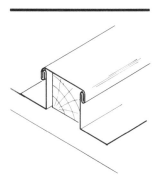

Batten Seam

Battered Wall

Bay Window

column, girder, or strut, and transmitting the shear between them.

batten roll, conical roll In metal roofing, a roll joint fabricated over a triangular wood piece.

batten seam In metal roofing, a seam fabricated around a wood strip.

batter To incline from the vertical. A slope, as of the outer side of a wall, that is wider at the bottom than at the top.

batter boards Pairs of horizontal boards nailed to wood stakes adjoining an excavation, used with strings as a guide to elevations and to outline a proposed building. The strings strung between boards can be left in place during the excavation.

batter brace, batter post (1) A bracing member positioned diagonally so as to reinforce an end of a truss. (2) An inclined timber forming a side support to a tunnel roof.

battered wall A wall which slopes backward, as by recessing or sloping masonry in successive courses.

batter level An instrument which measures the inclination of a slope.

batter pile, brace pile, spur pile (1) A pile driven somewhat diagonally so as to offer resistance to horizontal forces. (2) Any pile installed at an angle to the vertical. Also called a "raking pile".

batter rule An instrument for adjusting the inclination of a battered wall during its construction. A batter rule incorporates a rule or frame and a plumb line and bob.

batter stick A tapered board hung vertically by its wide end or used in conjunction with a

level to check the batter of a wall surface.

battery A device in which electrochemical action is employed to store and provide direct electric current.

batting tool A mason's dressing chisel used to apply a striated surface to stone.

batt insulation Thermal or sound insulating material, such as fiberglass or expanded shale, which has been fashioned into a flexible, blanket-like form and often has a vapor barrier on one side. Batt insulation is manufactured in dimensions which facilitate its installation between the studs or joists of a frame construction.

Bauhaus A school of art, design, and architecture established by Walter Gropius in Weimar, Germany after World War I, in 1919.

bay (1) In construction, the space between two main trusses or beams. (2) The space between two adjacent piers or mullions or between two adjacent lines of columns. (3) A small, well-defined area of concrete laid at one time in the course of placing large areas, such as floors, pavements, or runways. (4) The projecting structure of a bay window.

bay window A usually large window or group of windows projecting from a wall of a building and forming a recess within the building.

beacon (1) A light which indicates a location by directing its powerful beam slightly above the horizontal and rotating it so that, to a stationary observer, it appears to be flashing. A beacon is used at airports, lighthouses, etc. (2) A radio transmitter which broadcasts warning or guiding signals.

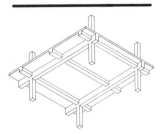

**Beam-and-Girder
Construction**

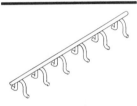

Beam Bolster

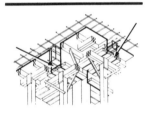

**Beam Bottom
Beam Form**

bead (1) Any molding, stop, or caulking used around a glass or panel to hold it in position. (2) A stop or strip of wood against which a door or window sash closes. (3) Strip of sheet metal which has been fabricated so as to have a projecting nosing and two perforated or expanded flanges. A bead is used as a stop at the perimeter of a plastered surface or as reinforcement at the corners. (4) A narrow, half-round molding, either attached to or milled on a larger piece. (5) A square or rectangular trim of less than one inch in width and thickness. (6) A choker ferrule, the knob on the end of a choker.

bead-and-real Half-round molding into which are incorporated alternating patterns of discs and elongated beads.

bead butt, bead and butt Thick, framed panelwork, as on a door, having one face flush with the frame and decorated with moldings (beads) on the adjoining edges, running with the grain and butting against the rail. The other side is recessed and without moldings.

bead butt and square Door paneling comprising one flush face and one bead butt.

bead molding Small, half-round, convex molding either continuous or divided to resemble a succession of beads.

bead plane A special plane having a curved cutting edge and employed in the cutting of beads in wood.

beam (1) A structural member transversely supporting a load, as a girder, rafter, or purlin. (2) The graduated horizontal bar of a weighing scale. (3) A ray of light.

beam anchor, joist anchor, wall anchor A metal tie for securing a beam, joist, or floor firmly to a wall.

beam-and-girder construction A type of floor construction in which slabs are used to distribute the load to spaced beams and girders.

beam-and-slab-floor A type of floor construction in which reinforced concrete beams are used to support a monolithic concrete floor slab.

beam bearing plate A metal plate used under the end of a beam to distribute the reaction load over a larger area. When used under a column, it is termed a "loose plate".

beam blocking Covering or enclosing a beam, joist, or girder so as to make it appear larger than it really is.

beam bolster A rod or heavy wire device for supporting steel reinforcement in the formwork for a reinforced concrete beam.

beam bottom The soffit of a beam.

beam box Same as **wall box**.

beam brick See **brick**.

beam ceiling A type of construction in which the structural and/or ornamental overhead beams are left exposed to view from the room below.

beam column A structural member which transmits an axial load as well as a transverse load.

beam compass An instrument comprising a small horizontal bar with two vertical components that can slide along it. One carries a sharp pointed tip and is held by the user in a stationary position. The other carries a pencil tip and is moved around the first to draw circles or arcs of circles for full-sized working drawings.

beam fill, beam filling Masonry, brickwork, or concrete placed between floor or ceiling joists of their supports so as to stiffen the joists and provide fire resistance.

beam form A retainer or mold constructed to give the necessary

Bearing

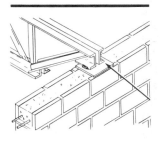

Bearing Plate

shape, support, and finish to a concrete beam.

beam hanger, beam saddle (1) A wire, strap, or other hardware device used to hang beam forms from another structural member. (2) In timber construction, a strap, wire, or stirrup used to support a beam.

beam haunch A poured concrete section which continues beyond a beam to support the sill.

beam pocket (1) A space left open in a vertical structural member to receive a beam. (2) An opening in the column or girder form where forms for an intersecting beam will be framed.

beam saddle Same as **beam hanger**.

beam side The side panels of a form for a concrete beam.

beam spread A measure of the dispersion of a beam of light. Expressed as the angle between two directions in the plane in which the candlepower is equal to a specified percent (usually 10%) of the maximum candlepower in the beam.

beam test A method of measuring the flexural strength (modulus of rupture) of concrete by testing a standard unreinforced beam.

bearer (1) As defined by OSHA, a horizontal member of a scaffold upon which the platform rests and which may be supported by ledgers. (2) Any load-supporting horizontal structural member. (3) Any device which provides support for a landing or window in a stair. (4) In balloon framing, the ribbon board upon which the joists for the second-floor rest.

bearing (1) That section of a structural member, such as a beam or truss, which rests on the supports. (2) A device used to support or steady a shaft, axle, or trunnion. (3) In surveying, the horizontal angle between a reference direction, such as true

north, and a given line. (4) Descriptive of any wall which provides support to the floor and/or roof of a building.

bar (1) A wrought-iron bar used on masonry to offer a level support for floor joists. (2) A supporting bar for a grating.

capacity The maximum unit pressure which a soil or other material will withstand without failure or without excessive settlement.

pile A pile which supports a vertical load.

plate A steel plate positioned under a beam, column, girder, or truss to distribute a load to a support member.

pressure The existing load on a bearing surface divided by the area of the bearing surface.

stratum The soil or rock stratum on which a footing or mat bears or carries the load transferred to it by a pile, caisson, or similar deep foundation unit.

wall Any wall which supports a vertical load as well as its own weight.

Beaufort scale A numerical scale used to indicate successive ranges of wind velocity (0 for calm to 12 for hurricane force).

bed (1) The mortar into which masonry units are set. (2) Sand or other aggregate on which pipe or conduit is laid in a trench. (3) To set in place with putty or similar compound, as might be performed in glazing. (4) A supporting base for engines or machinery. (5) To level or smooth a path into which a tree is to be felled.

bedding (1) A prepared base for masonry or concrete. (2) The lath or other support(s) upon which pipe is laid. (3) See **bed, def. 2**. (4) Any material used to **bed,**

Bell, Hub

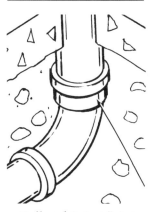

Bell-and-Spigot Joint

def. 3.

*bedding coat, bed
coat* Ordinarily, the initial coat
of joint compound on gypsum
board over tape, bead, and
other fastener heads.

bedding course The initial layer
of mortar at the bottom of a
masonry construction.

bedding dot A small area of
plaster when built out of the
face of a finished wall or ceiling
and acting as a screed for
leveling and plumbing in a
plastering operation.

bedding stone A flat slab of
marble by which bricklayers and
masons check the flatness of
rubbed bricks.

bed dowel A dowel placed in the
mortar bed for a stone to prevent
settlement before the mortar has
set.

bed joint (1) The horizontal layer
of mortar upon which a masonry
unit is laid. (2) In an arch, a
horizontal joint or one which
radiates between adjacent
voussoirs. (3) A horizontal fault
in a rock formation.

bed molding (1) A molding placed
at the angle between a vertical
surface and an overhanging
horizontal surface, such as
between a sidewall and the eaves
of a building. (2) In a band of
moldings, the lowest one. (3) In
classical architecture, a molding of
a cornice of an entablature,
located between the corona and
the frieze.

bedplate A baseplate, frame, or
platform which supports a
structural element, furnace, or
heavy machine.

bedrock Solid rock which
underlies the earth's surface soil
and which can provide, by its very
existence, the foundation upon
which a heavy structure may be
erected.

beech, beechwood Any of a
number of hardwoods of the
genus *Fagus*, having characteristics
of straight, close grain and good
workability. Beech is often used in
furniture, turnings, tools, and
similar products.

beetle, maul A heavy mallet
having a large wooden head and
used for driving pegs or wedges,
or in other applications where
material might sustain damage if
struck with a conventional sledge
hammer.

Belgian block A paving stone
shaped like a truncated pyramid
and laid with the largest face
down.

bell, hub That portion of a pipe
for a short distance is sufficiently
enlarged to receive the end of
another pipe of the same diameter
for the purpose of making a joint.

**bell-and-spigot joint, spigot-and-
socket joint** The commonly used
joint in cast iron pipe. Each piece
is made with an enlarged diameter
or bell at one end into which the
plain or spigot end of another
piece is inserted. The joint is
sealed by cement, oakum, lead, or
rubber caulked into the bell
around the spigot.

belled excavation The bell-shaped
lower portion of a shaft or footing
excavation.

belling In pier, caisson, or pile
construction, the process of
enlarging the base of a foundation
element at the bearing stratum so
as to provide more bearing area.

bell joint See **bell-and-spigot
joint**.

bellows expansion joint In a run
of piping, a joint accomplished
with flexible metal bellows which
can expand and/or contract
linearly to allow for thermally
induced linear fluctuations of the
run itself.

bell transformer A very small
transformer which supplies

Belt Sander

Benched Foundation

power, at low voltage, when demanded by a doorbell or similar device.

bell trap A bell-shaped trap sometimes used in floor drains. Its installation is prohibited for residential use by the National Plumbing Code.

bell wire Small-diameter wire of low current-carrying capacity whose insulation is rated at 30 volts or less.

belt (1) A flexible endless loop which conveys power (or materials) between the pulleys or rollers around which it passes. (2) A course of brick or stone which protrudes from a wall of similar material and is usually positioned in line with the window sills.

belt conveyor A power-driven apparatus comprising a single endless belt and the idler wheels around which it passes to transport materials or products placed on its upper surface.

belt loader An excavating machine comprising an auger or other cutting edge which digs away the earth and a conveyer belt which elevates the excavated material for loading onto a truck or elsewhere.

belt sander An electrically powered portable sanding tool in which a continuous abrasive belt is driven in only one direction. A belt sander is used to smooth surfaces.

belt tightener A pulley device incorporated into the construction of a conveyer belt for adjusting the belt tension by counterweight and pulley positioning.

bench (1) A long seat made in any of several materials, such as wood, metal, or fiberglass, with or without a back, and usually accommodating more than one person. (2) One step, level, or shelf in a cut or excavation which is performed in several layers. Also called "a berm". (3) A pretensioning bed.

bench brake A large, heavy, bench-mounted device for bending sheet metal.

bench dog A peg or pin partially inserted into a hole at an edge or end of a workbench to help secure a piece of work or prevent it from sliding off the bench.

benched foundation A foundation cut as a series of horizontal steps in an inclined bearing stratum to prevent sliding when loaded.

bench hook, side hook In carpentry, any device used to secure the work and protect the top surface of a workbench from being scarred or damaged by any slipping or sliding of the work. It usually keeps the work positioned toward the front of the bench.

benching (1) A half-round channel cast in the concrete in the bottom of a manhole to direct discharge when the flow is low. (2) Concrete laid on steeply sloping fill to prevent sliding. (3) Concrete laid on the side slopes of drainage channels where the slopes are interrupted by manholes, etc. (4) Concrete laid in a trench for a pipeline to provide firmer support.

bench mark A marked reference point on a permanent object, such as a metal disc set in concrete, whose elevation as referenced to a datum is known. A bench mark is used in surveying to determine the elevation of other points.

bench plane A plane, such as a block plane or a jack plane, used primarily in bench work on flat surfaces.

bench sander An electrically powered, stationary sanding machine, including a rotating abrasive disc or grinder or an endless abrasive belt, and mounted on a bench. A bench sander is used to smooth the surface of objects held against it.

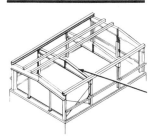

Bent

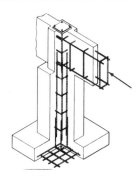

Bent Bar

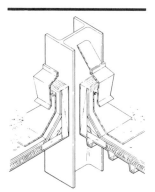

Bevel

bench stop A usually notched, adjustable metal apparatus fastened close to an end of a workbench to securely hold a piece of wood during planing.

bench table A course of masonry, wide enough to form a seat, protruding from the foot of an interior wall or column.

bench vise An ordinary vise attached to a work bench to hold a material or component while it is being worked on.

benchwork All work performed on a bench rather than on machines or in the field.

bend (1) An elbow fitting, or other short length of bent conduit, used to join two lengths of straight adjacent conduit. (2) A pipe fitting used to achieve a change in direction.

bending beam See **tie beam**.

bending moment The bending effect at any section of a beam. The bending moment is equal to the algebraic sum of moments taken about the center of gravity of that section.

bending moment diagram A graphic representation of the variation of bending moment along the length of the member for a given stationary system of loads.

bending schedule A list of reinforcement prepared by the designer or detailer of a reinforced concrete structure, showing the shapes and dimensions of every bar and the number of bars required.

bend test Subjecting a flat bar to a 180-degree cold bend so as to test its weld or steel and to check its ductility, which is verified if no cracking occurs during the test.

benefits (mandatory and customary) The employee benefits required by law (such as social security, workmen's compensation, and disability insurance) and by custom (such as

vacation, sick leave as well as those which are optional with the individual firm (such as pension plans, life insurance).

bent (1) A structural framework, transverse to the length of a structure, designed to carry lateral as well as vertical loads. (2) Any of several grasses of the genus *Agrostis* which are used where a resilient velvety texture is required.

bent approach An arrangement of a driveway such that it turns sharply through two out-of-line gateways. The bent approach is used for privacy and security.

bent bar A reinforcing bar bent to a prescribed shape, such as a truss, straight bar with hook, stirrup, or column tie. A bent bar is bent to pass from one face to the other of a member.

bentonite A clay composed principally of minerals in the montmorillonite group and characterized by high absorption and very large volume change with wetting or drying.

Berea sandstone A find-grained sandstone quarried at Amherst, Ohio.

berm (1) An artificially placed continuous ridge or bank of earth, usually along a roadside. Also called "a shoulder". (2) A ridge or bank of earth placed against a masonry wall. (3) A ledge or strip of earth placed so as to support pipes or beams. (4) Earthen dikes or embankments constructed to retain water on land which will be flood-irrigated. (5) Earthen or paved dike-like embankments for diverting runoff water.

berth A docking facility where a vessel can tie up and load or discharge cargo, fuel, etc.

Bethell process Pressure-impregnating wood with creosote preservative.

bevel (1) Any angle, except a right angle, or inclination of any line or surface that joins another. (2) An adjustable instrument used for determining, measuring, or reproducing angles. (3) In welding, an edge preparation on the edges to be welded.

bevel angle In welding, the angle created by the prepared edge of a member and a plane perpendicular to the surface of the member.

bevel board A board which has been cut to a predetermined, desired, or required angle and is employed in any angular wood construction, including roof and stair framing.

bevel chisel A wood-cutting chisel whose cutting edge is angled between its sides.

bevel collar See **angle collar.**

bevel cut Any cut made at an angle other than a right angle.

beveled edge (1) A vertical front edge on a door, cut so as to have a slope of 1/8 inch in 2 inches from a plane perpendicular to the face of the door. (2) The factory applied angle on the edge of gypsum board which creates a "vee" grooved joint when two pieces are installed together. (3) A chamfer strip incorporated into concrete forms for columns or beams so as to eliminate sharp corners on the finished product.

beveled washer A washer having a bevel on one side and used mostly in structural work to provide a flat surface for the nut wherever a threaded rod or bolt passes through a beam at an angle.

bevel joint A carpentry joint in which two wood pieces meet at any angle except a right angle.

bevel protractor A graduated semicircular drafting device incorporating an adjustable pivoted arm and used for measuring and plotting angles.

bevel siding See **clapboard.**

bevel square A tool similar to a try square except that it has an adjustable blade which can be set at any angle.

B-grade wood The classification of a somewhat inferior grade of solid surface veneer which contains visible repair plugs and tight knots.

bias The fixed voltage applied to an electrode.

bibcock, bib, bibb, bib tap (1) Any faucet or stopcock having its nozzle directed downward. (2) Any tap supplied by a horizontal pipe.

bib valve Any standard bibcock or faucet equipped with a handle which is turned to control the flow and screwed down to shut off the flow by closing a washer disk onto a seating within the valve.

bid A complete and properly signed proposal to do the work or designated portion thereof for the amount or amounts stipulated therein. A bid is submitted in accordance with the bidding documents.

bid abstract (summary) A compilation of bidders and their respective bids, usually separated into individual items, of a given project.

bid bond A form of bid security executed by the bidder or principal and by a surety to guarantee that the bidder will enter into a contract within a specified time and furnish any required performance bond, and labor and material payment bond.

bid call A published announcement that bids for a specific construction project will be accepted at a designated time and place.

Bidet

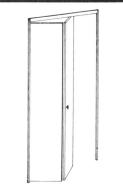

Bifolding Door

**Binder Course,
Binding Course**

bid date The date established by the owner or the architect for the receipt of bids.

bid form A form, furnished to the bidder, on which to submit his bid.

bid guarantee See **bid security**.

bid letting See **bid opening**.

bid opening The opening and tabulation of bids submitted within the prescribed bid time and in conformity with the prescribed procedures. Bid opening is preferable to bid letting.

bid price The amount, stated in the bid, for which the bidder offers to perform the work.

bid security A bid bond or deposit submitted with a bid to guarantee to the owner that the bidder, if awarded the contract, will undertake execution of the contract.

bid time The exact date and time, established by the owner or architect, for submission of bids.

bidder A person or entity who submits a bid, generally one who submits a bid for a prime contract with the owner, as distinct from a sub-bidder who submits a bid to a prime bidder. Technically, a bidder is not a contractor on a specific project until a contract exists between the bidder and the owner.

bidding documents The advertisement or invitation to bid, instructions to bidders, the bid form, other sample bidding and contract forms, and the proposed contract documents, including any addenda issued prior to receipt of bids.

bidding period The calendar period beginning at the time of issuance of bidding documents and ending at the prescribed bid time.

bidet A bathroom fixture used for hygienic washing of the genitals and posterior parts.

bifolding door A door having two leaves, each leaf consisting of two panels hinged together, so that they fold on each other when the door is opened. One free edge of each leaf, or pair of panels, is hinged at a door jamb; the other edge is supported from and guided by an overhead project.

bill of materials See **quantity survey**.

bin A storage container, usually for loose materials, such as sand, stone, or crushed rock.

binder (1) Almost any cementing material, either hydrated cement or a product of cement or lime and reactive siliceous materials. The kinds of cement and the curing conditions determine the general type of binder formed. (2) Any material such as asphalt, resin, or other materials forming the matrix of concretes, mortars, and sanded grouts. (3) That ingredient of an adhesive composition that is principally responsible for the adhesive properties which actually hold the two bodies together. (4) In paint, that non-volatile ingredient, such as oil, varnish, protein, or size, which serves to hold the pigment particles together in a coherent film. (5) A stirrup or other similar contrivance usually of small-diameter rod, which functions to hold together the main steel in a reinforced concrete beam or column.

binder course, binding course (1) A succession of masonry units between an inner and outer wall to bind them. (2) An intermediate course used in asphaltic concrete paving between the base and the surfacing material and comprising bituminously bound aggregate of intermediate size.

binding agent See **binder, def. 4**.

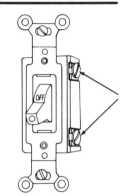

Binding Post

Biparting Door

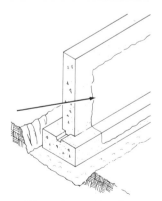

Bituminous Coating

binding beam (1) A timber or steel beam which carries the common joists in a double or framed elbow. (2) Any timber which ties together various components of a frame.

binding joist, binder See **binding beam, def. 1**.

binding piece A piece of lumber which is nailed between two parallel joists or beams to control lateral deflection.

binding post (1) A set screw which holds a conductor against the terminal of a device or equipment. (2) A post attached to an electric wire, cable, or apparatus to facilitate a connection to it.

binding rafter A purlin.

binding yarn, crimp warp, binder warp In carpeting, the natural or synthetic yarn incorporated lengthwise of the woven fabric to "bind" the tufts of pile securely.

biparting door A sliding door with two leaves that slide in the same plane and meet at the door opening.

birch A strong, fine-grained hardwood commonly used in veneer, furniture, flooring, and turned wood products.

bird screen Wire mesh used to cover chimneys, ventilators, and louvers to prevent birds from entering buildings through these accesses.

bird's-eye maple Lumber from a sugar maple tree, which has been cut to exhibit a wavy grain with many small circular or elliptical markings.

bit (1) The tool that fits into a brace or drill and is rotated to bore a hole. (2) On a soldering iron, the (usually copper) tip which heats the joint and melts the solder. (3) That part of a key which is inserted into a lock and engages the tumblers or bolt.

bit brace See *handbrace*.

bit extension A length of rod held at one end by the chuck of a brace and equipped at the other end for holding a bit. A bit extension permits the drilling of holes whose required depth is greater than the length of an ordinary bit.

bit gauge, bit stop A piece of metal projecting outward from a bit to limit its depth of penetration.

bit key Any key which has a **bit, def. 3.** Also called "a wing key".

bite In glazing, the overlap between the innermost edge of the stop, or frame, and the outer edge of the light or panel.

bitumen Any of several mixtures of naturally occurring or synthetically rendered hydrocarbons and other substances obtained from coal or petroleum by distillation. Bitumen is incorporated in asphalt and tar as used in road surfacing and waterproofing operations.

bituminous Composed of, similar to, derived from, relating to, or containing bitumen. The term "bituminous" is descriptive of asphalt and tar products.

bituminous base course, black base Bituminously bound aggregate serving as a foundation for binder courses and surface course in asphalt paving operations.

bituminous binder course See **binder course, def. 2.**

bituminous cement A class of dark substances composed of intermediate hydrocarbons. Bituminous cement is available in solid, semisolid, or liquid states at normal temperatures.

bituminous coating Any waterproof or protective coating whose base is a compound of asphalt or tar.

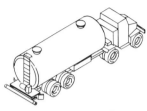

Bituminous Distributor

bituminous concrete See *asphaltic concrete*.

bituminous distributor A tank truck equipped with a perforated spray bar through which it pumps heated bituminous material, such as tar or road oil, onto the surface of a road.

bituminous emulsion (1) A suspension of tiny globules of a bituminous substance in water or an aqueous solution. (2) An invert emulsion of the above, i.e., a suspension of tiny globules of water or an aqueous solution in a liquid bituminous substance. This type of bituminous emulsion is applied to surfaces to provide a weatherproof coating. See **emulsified asphalt.**

bituminous felt See **asphalt prepared roofing**.

bituminous fiber pipe Drainage pipe made from a combination of coal tar and cellulose fiber.

bituminous grout A mixture of bituminous material and fine sand or other aggregate which, when heated, becomes liquid enough to flow into place without mechanical assistance. Bituminous grout will air cure after being poured into cracks or joints as a filler and/or sealer.

bituminous leveling course In paving operations, a course consisting of a mixture of asphalt and sand and used to level or crown a base course or existing deteriorating pavement prior to the application of a surface.

bituminous macadam A paving material comprising bituminously coated course aggregate.

bituminous paint A thick black waterproofing paint containing substantial amounts of coal tar or asphalt.

B-labeled door A door carrying a certification from Underwriters' Laboratories that it is of a construction that will pass the standard fire door test for the length of time required for a Class B opening, and that it has been prepared (cuts and reinforcement) to receive the hardware required for a Class B opening.

black japan A black, high-quality bituminous paint or varnish.

black light Electromagnetic energy of the ultraviolet spectrum. Black light is used to excite fluorescent paints and dyes so that they become visible.

black-light, fluorescent lamp A fluorescent lamp with a phosphor which emits black light.

black plate Twelve- to 32-inch wide sheets of uncoated cold-rolled steel.

blacktop See **asphaltic concrete**.

Blaine apparatus Air-permeability apparatus for measuring the surface area of a finely ground cement.

Blaine test A method for determining the fineness of cement or other fine material on the basis of the permeability to air of a sample prepared under specified conditions.

blank door (1) A recess in a wall fitted with a fixed door, used for architectural effect. (2) A door which has been fixed in position to seal an opening.

blanket (1) Soil or pieces of rock remaining or intentionally placed over a blast area to contain or direct the throw of fragments. (2) Insulation sandwiched between sheets of paper facing and used for protecting fresh concrete during curing.

encumbrance A lien or mortgage proportionately levied on every lot in a given subdivision.

insulation Faced or unfaced thermal or sound

**Blank Flange,
Blind Flange**

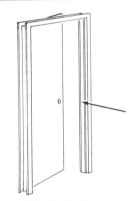

Blank Jamb

insulation, usually of fiberglass, available in densities and thicknesses which allow it to conform to the various shapes it encounters in its many applications. Blanket insulation is the same as batt insulation, except that it is supplied in continuous rolls instead of sheets. See **insulation blanket**.

blank flange, blind flange A solid plate flange used to seal off flow in a pipe.

blank jamb A vertical component of a doorframe as installed, without any preparation for hardware accommodation.

blank wall, blind wall, dead wall A wall with no openings on its entire surface.

blank window, blind window, false window (1) A recess in a wall, having the appearance of a window. (2) A window which has been sealed off but is still visible.

blast To loosen, crack or move rock or soil by utilizing explosives.

blast area The area in which explosives loading and blasting operations are being conducted.

blast cleaning Any cleaning method which is accomplished by air, liquid, abrasive, or some combination of these applied under pressure.

blast freezer A freezer room in which air, at subfreezing temperature, is circulated by blower for the purpose of freezing foods quickly.

blast-furnace slag The nonmetalic waste, developed simultaneously with iron in a blast furnace, that consists essentially of silicates and aluminosilicates of calcium and other bases.

blast heater A unit heater containing a set of heat-transfer coils through which air is forced or drawn at high velocities.

blast hole A vertical hole having a diameter of at least 4 inches and drilled to accept a charge of blasting explosives.

blasthole drill A device which bores blast holes.

blasting The process of loosening rock or hard packed material with explosives.

blasting agent According to OSHA, a material or mixture consisting of a fuel and oxidizer used for blasting, but not classified as an explosive and in which none of the ingredients is classified as an explosive, provided the finished product cannot be detonated with a No. 8 blasting cap when confined. An example is a mixture of ammonium nitrate and fuel oil or coal.

blasting cap A metallic tube closed at one end, containing a charge of one or more detonating compounds, and designed for and capable of detonation from the sparks or flame from a safety fuse inserted and crimped into the open end.

blasting mat A blanket of interwoven steel cable or interlocking steel rings and placed over a blast to contain the resultant fragments.

blast-resistant door A steel door which has been designed and fabricated to resist dynamic pressures, of short duration, up to 3,000 psi (21 MPA).

bleaching Lightening, whitening, or removing color from, either by chemical means, as with chlorine or oxalic acid, or by exposure to certain kinds of light.

bleeder A small valve used to drain fluid from a pipe, radiator, small tank, etc.

bleeder pipe, bleeder tile A usually clay pipe placed to allow water from outside a basement retaining wall to pass through the foundation into drains within the building.

Blind

Blind Door

Blind Nailing

bleeding (1) The autogenous flow of mixing water within, or its emergence from, freshly placed concrete or mortar. Bleeding is caused by the settlement of the solid materials within the mass. Also called "water grain". (2) In painting, the showing of resin or a bottom coat through the topcoat. (3) In lumber, the exuding of sap or resin.

bleeding rate The ratio of the volume of water which is released by bleeding, def. 1 to the original volume of the paste.

bleed-through, strike-through Discoloration in the face plies of wood veneer caused by seepage of cement through the veneer.

blended cement A hydraulic cement consisting essentially of an intimate and uniform blend of granulated blast-furnace slag and hydrated lime or of portland cement and granulated blast-furnace slag, portland cement and pozzolan, or portland-blast-furnace slag, cement, and pozzolan. Blended cement is produced by intergrinding portland cement clinker with the other materials or by a combination of intergrinding and blending.

blending valve A three-way valve which permits the mixing required to obtain a desired liquid temperature.

blind (1) Any panel, shade, screen, or similar contrivance used to block light or inhibit viewing. (2) An assembly of wood stiles, rails, and wood slats or louvers used in conjunction with doors and windows.

blind alley, cul-de-sac Any alley, road, or passageway having only one end open for access or exit, and closed at the other end so as to prohibit through-traffic.

blind attic An attic which is rough-floored but unfinished.

blind casing The rough window frame or sub-casing to which the trim is added.

blind door (1) A door with a louver instead of glass. (2) See **blank door**.

blind dovetail See **secret dovetail**.

blind flange See **blank flange**.

blind header (1) A brick or other masonry unit which, when laid, exhibits the appearance of a header, while in reality it is somewhat less than a whole unit. A blind header can also be a half-brick laid where only one end is visible. (2) A header brick concealed within a wall and functioning to bond adjacent tiers of bricks.

blinding (1) Applying a layer of weak concrete or other suitable material to reduce surface voids or to provide a clean, dry working surface. (2) The clogging of the openings in a screen or sieve by the material being separated. (3) Applying small chips of stone to a freshly tarred surface.

blind joint (1) A masonry joint, as that of a blind header, def. 2, which is entirely concealed. (2) In a double Flemish masonry bond, a thin joint between the adjacent ends of two stretchers. This joint bisects a header in the course directly below it.

blind mortise, stopped mortise A mortise in, but not through, a member.

blind-mortise-and-tenon joint A joint between a blind mortise and a stub tenon.

blind nailing, concealed nailing, secret nailing Nailing performed so that the nailhead cannot be seen on the face of the work.

blind pocket A pocket in the ceiling at a window head. A blind pocket is used to conceal an object when not in use, such as a pocket and a veiling to receive a

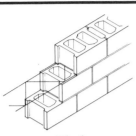

Block

Blocking

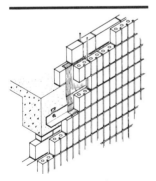

Blockwork

Venetian blind in the raised position.

blind stop A rectangular molding nailed between the outside trim and the outside sash of a window frame, serving as a stop for storm sashes or screens.

blind tenon A tenon which fits a blind mortise.

blind wall See **blank wall**.

blind window See **blank window**.

blister Usually an undesirable moisture and/or air-induced bubble or bulge which often indicates that some kind of delamination has taken place. Blisters can occur between finish plaster and the basecoat, between paint or varnish and the surface to which it has been applied, between roofing membranes or between membrane and the substrate, between reinforcing tape and the gypsum board to which it has been adhered, etc. (2) A tree disease characterized by the seeping of pitch onto the bark surface.

bloated clay See **expanded clay**.

block (1) A usually hollow concrete masonry unit or other building unit, such as glass. (2) A solid, often squared, piece of wood or other material. (3) A piece of wood nailed between joists to stiffen a floor. (4) Any small piece of wood secured to the interior angle joint to strengthen and stiffen it. (5) A pulley and its enclosure. (6) Usually one of several small rectangular divisions within a city, bounded on each side by successive streets. (7) A solid piece of wood or other material used to fill spaces between formwork members.

block and tackle A mechanical device of pulley blocks and ropes or cables and used to facilitate hoisting or moving heavy objects or loads.

block beam In structural design, a flexural member composed of individual blocks which are joined together by prestressing.

block bridging, solid bridging, solid strutting Short boards fixed between floor joists to stiffen the joists and distribute the load.

blockholing The breaking of boulders by detonating a charge of explosives that has been loaded in a drill hole.

block-in-course A kind of masonry employed in heavy construction and consisting of squared, hammer-dressed stones laid to have very close joints and forming a course not higher than 12 inches.

blocking (1) Small pieces of wood used to secure, join, or reinforce members, or to fill spaces between members. (2) Small wood blocks used for shimming. (3) A method of bonding two parallel or intersecting walls built at different times by means of offsets whose vertical dimensions are not less than 8 inches (20 centimeters). (4) The sticking together of two painted surfaces when pressed together. (5) An undesired adhesion between touching layers of material, such as occurs during storage.

blocking chisel Any rugged, broad-edged stone mason's chisel.

blocking course In masonry, a finishing course, usually of stones, placed on top of a cornice.

block plane A plane designed to be held in one hand and with a low angle of cutting blade (about 20 degrees). A block plane is used especially to clean end grain and miters.

blockwork Masonry of concrete block and mortar.

bloom (1) A thin hazy film on old paint, usually caused by weathering. A bloom can also be a similar film on glass, resulting from general atmospheric deposition of impurities, or caused more by smoke, vapor, etc. (2) Efflorescence on brickwork. (3) A hazy or other discoloration which sometimes occurs on the surface of rubber products. (4) The term given to steel which has been reduced from an ingot by being rolled in a blooming mill to a dimension of at least 6 inches square; if further reduced, it becomes a billet.

blowback The difference between the pressure at which a safety valve opens and at which it closes automatically after the release of excess pressure has occurred.

blow count (1) The number of times that an object must be struck to be driven into the soil to a desired or specified depth. (2) In soil borings, the number of times a sample spoon must be struck to be driven 6 or 12 inches. (3) In pile driving operations, the number of times a pile must be struck to be driven 12 inches. The blow count is sometimes expressed as the number of blows per unit distance of advance.

blower A fan, especially one for heavy-duty use, such as one that is used to force air through ducts to an underground excavation.

blowhole In concrete, a bug hole or small regular or irregular cavity, not exceeding 15 mm in diameter, resulting from entrapment of air bubbles in the surface of formed concrete during placement and compaction. See **gas pocket**.

blowing See **popping**.

blown asphalt Asphalt that has been processed by blowing air through it at elevated temperatures so as to impart properties for specific applications.

blown joint, blow joint A plumbing joint sealed with the use of a blowtorch.

blow-off (1) A discharge outlet on a boiler to allow for expulsion of undesirable accumulations of sediment and/or for draining of the unit. (2) In a sewer system, an outlet pipe for expelling sediment or water or for draining a low sewer.

blowtorch A small, portable gas-fired burner that generates a flame hot enough to melt soft metals. A blowtorch is used to melt lead in plumbing operations, heat soldering irons, and burn off paint.

blow up Localized buckling or breaking up of rigid pavement as a result of excessive longitudinal pressure.

blueprint Negative image reproduction having white lines on a blue background and made either from an original or from a positive intermediate print. Today the term almost always refers to Diazo Prints, which are architectural or working drawings having blue or black lines on a white background.

bluestone A hard, fine-grained sandstone or siltstone of dark greenish to bluish gray color that splits readily to form thin slabs. A variety of flagstone, bluestone is commonly used for paving walkways.

board (1) Lumber 4 in. (10 cm) to 12 in. (30 cm) wide and less than 2 in. (5 cm) thick. (2) A box office seating chart or ticket board.

board and batten A method of siding in which the joints between vertically placed boards or plywood are covered by narrow strips of wood.

board and brace A type of carpentry work consisting of boards grooved on both edges with thinner boards between them fitted to the grooves.

board butt joint A joint in shotcrete construction, achieved by sloping the gunned surface to meet a 1-inch board laid flat.

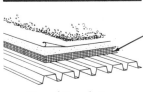

Board Insulation

Boatswain's Chair

Boiler

board foot The basic unit of measurement for lumber. One board foot is equal to a 1-inch board 12 inches in width and 1 foot in length. Thus, a 10-foot long, 12-inch wide, and 1-inch thick piece would contain 10 board feet. When calculating board feet, nominal sizes are assumed.

board insulation, insulating board, insulation board Lightweight thermal insulation, such as polystyrene, manufactured in rigid or semirigid form, whose thickness is very small relative to its other dimensions. Board insulation offers little structural strength and is usually applied under a finish material, although some types are manufactured with a finished surface on one side.

board knife A knife with a replaceable blade that has a straight cutting edge and is used to secure or trim gypsum products.

board measure A system of measuring quantities of lumber, using board foot as the basic unit.

board rule A measuring device used to find directly the number of board feet in a given board.

board sheathing A sheathing made of boards, usually tightly spaced, except that open spacing may be used in some roofs.

boardwalk A walkway made of boards or planks, often used as a promenade along a beach or shore.

boasted work A dressed stone surface with roughly parallel narrow chisel grooves of varying widths and not extended across the face of the stone.

boaster A flat, steel mason's chisel used in the dressing of stone.

boatswain's chair A seat supported by slings attached to a suspended rope, designed to accommodate one workman in a sitting position.

bob Same as **plumb bob**.

bodied linseed oil, pale-bodied oil Linseed oil whose viscosity has been thickened either by processing with heat or by the incorporation of lead, manganese, in cobalt salts or other chemicals to accelerate hardening of the oil when applied in thin coats.

bodily injury Physical injury, sickness, or disease sustained by a person.

body (1) The principal volume of a building such as the nave of a church. (2) The load-carrying part of a truck.

body coat In painting, an intermediate coat of paint applied over the priming coat but under the finishing coat.

bogie (1) A two-axial driving unit in a truck, also called a tandem. (2) A stage hanger for an overhead track from which scenery or flats are suspended.

boil (1) A wet run of material at the bottom of an excavation. (2) A swelling in the bottom of an excavation due to seepage.

boiler A closed vessel in which a liquid is heated or vaporized either by application of heat to the outside of the vessel, by circulation of heat through tubes within the vessel, or by circulating heat around liquid-filled tubes in the vessel.

boiler horsepower The gross output of a boiler in BTU/hr divided by 33,475.

boiler jacket The thermal insulation surrounding a boiler and usually covered by a more aesthetically acceptable metal or other enclosure.

boiler plate Medium-hard boiler steel which has been rolled into plates whose thickness may vary from .25 inches to 1.5 inches and from which boilers are fabricated.

boiler rating The heating capacity of a boiler, expressed in BTUs per hour. The boiler rating is not to

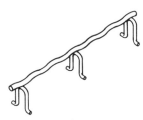

Bolster

Bolt

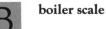

Bolt Sleeve

be confused with the horsepower rating, which is the largest rating obtained by dividing the square feet of the boiler's surface by 10.

boiler scale Disintegrated bits of the steel plate of the boiler's interior lining, which flake and fall to the boiler floor and should be periodically removed.

boiler steel A medium-hard steel which is rolled to become boiler plate.

boiserie Floor-to-ceiling interior wood paneling, usually enhanced by carving, gilding, or painting, but seldom by inlaying.

bollard (1) One of a series of short stone posts set to prevent vehicular access or to protect property from damage by vehicular encroachment. A bollard is sometimes used to direct traffic. (2) A post on a ship or wharf for securing mooring or docking lines.

bolster (1) In concrete, a continuous wire bar support used to support bars in the bottom of slabs. The top wire is corrugated at one inch centers to hold the bars in position. (2) A short wood or steel member positioned horizontally on top of a column to support beams or girders. (3) A mason's blocking chisel. (4) A piece of wood, generally a nominal four inches in cross section, placed between stickered packages of lumber or other wood products to provide space for the entry and exit of the forks of a lift truck.

bolt (1) An externally threaded cylindrical fastening device fabricated from a rod, pin, or wire, and having, at one end, a round, square, hexagonal, etc., head which projects beyond the circumference of the shank to facilitate gripping, turning, and striking. The unit is inserted through holes in fabricated or assembled parts and is secured by a corresponding, internally

threaded nut and tightened by the application of torque. (2) That protruding part of a lock which prevents a door from opening. (3) Raw material used in the manufacture of shingles and shakes. A wedge-shaped split from a short length log and taken to a mill for manufacturing. (4) Short logs to be sawn for lumber or peeled for veneer. (5) Wood sections from which barrel staves are made. (6) A large roll of cloth material of a given length, just as it comes from the loom. (7) A single package containing two or more rolls of wallpaper.

bolt sleeve A tube surrounding a bolt in a concrete wall to prevent concrete from sticking to the bolt and acting as a spreader for the formwork.

bona fide bid A bid submitted in good faith, complete and in accordance with the bidding documents, and properly signed by a person legally authorized to sign such a bid.

bond (1) The adhesion and grip of concrete or mortar to reinforcement or to other surfaces against which it is placed, including friction due to shrinkage and longitudinal shear in the concrete engaged by the bar deformations. (2) The adhesion of cement paste to aggregate. (3) Adherence between plaster coat or between plaster and a substrata produced by adhesive or cohesive properties of plaster or supplemental materials. (4) The arrangement of, or pattern formed by, the exposed faces of laid masonry units. (5) The layer of glue in a plywood joint. (6) See **(a) bid bond (b) completion bond (c) fidelity bond (d) labor and material payment bond (e) performance bond (f) statutory bond (g) surety bond.**

bond area The area of interface between two elements across

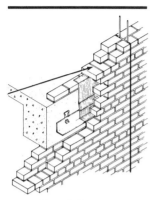

Bond Course

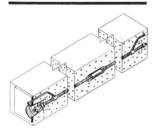

Bonded Posttensioning

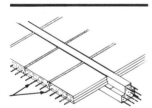

Bonded Tendon

which adhesion develops or may develop, as between concrete and reinforcing steel.

bond beam A horizontally reinforced concrete or concrete masonry beam built to strengthen and tie a masonry wall together. A bond beam is often placed at the top of a masonry wall with continuous reinforcing around the entire perimeter.

bond blister A blister formed between the base material and the coating, particularly on metal-clad products.

bond breaker A material used to prevent adhesion between freshly placed concrete and the substrate.

bond coat (1) That coat of plaster which is applied over masonry and is bonded to it. (2) A primer coat of paint or sealer.

bond course In masonry, the course consisting of units which overlap more than one wythe of masonry.

bonded member A prestressed concrete member in which the tendons are bonded to the concrete either directly (pretensioned) or by grouting (posttensioned).

bonded posttensioning A process in posttensioned construction whereby the annular spaces around the tendons are grouted after stressing in such a way that the tendons become bonded to the concrete section.

bonded roof A type of roofing guarantee offered by the manufacturer which may or may not be purchased by the owner and which covers materials and/or workmanship for a specifically stated length of time.

bonded tendon A prestressing tendon which is bonded to the concrete directly or through grouting.

bonder, header A masonry unit

which ties two or more wythes (leaves) of a wall together by overlapping.

bond face That face of a joint to which a field-molded sealant is bonded.

bond header, throughstone A bondstone or masonry unit that extends through the entire thickness of a wall.

bonding agent A substance applied to a suitable substrate to create a bond between it and a succeeding layer, as between a subsurface and a terrazzo topping or a subsequent application of plaster.

bonding capacity The maximum total contract value a bonding company will extend to a contractor in performance bonds. The total contract value is the sum of all contracts being bonded.

bonding company A firm providing a surety bond for work to be performed by a contractor payable to the owner in case of default of the contractor. The bond can be for work performance or for payment for materials and labor.

bonding conductor See **bonding jumper**.

bonding jumper A reliable conductor which is connected between metal parts to ensure good electrical conductivity between them.

bonding layer A bonding layer is often used when a new concrete slab is placed over an existing hardened concrete surface. A thin layer of mortar is spread before placing concrete for the new slab.

bond length The length of embedment of reinforcing steel in concrete required to develop the design strength. Also called "development length".

bond plaster A thin coat of specially formulated gypsum plaster applied as a bonding coat to a concrete surface before

Boom Crane

applying succeeding coats of plaster.

bond prevention (1) Procedures whereby specific tendons in pretensioned construction are prevented from becoming bonded to the concrete for a predetermined distance from the ends of flexural members. (2) Measures taken to prevent adhesion of concrete or mortar to surfaces against which it is placed.

bondstone, bonder In stone facing, a stone which extends into the backing to tie the facing wall with the backing wall.

bond strength (1) Resistance to separation of mortar and concrete from reinforcing steel and other materials with which it is in contact. (2) A collective expression for all forces such as adhesion, friction due to shrinkage, and longitudinal shear in the concrete engaged by the bar deformations that resist separation.

bond stress (1) The force of adhesion per unit area of contact between two bonded surfaces such as concrete and reinforcing steel or any other material such as foundation rock. (2) Shear stress at the surface of a reinforcing bar, preventing relative movement between the bar and the surrounding concrete.

bond timber Horizontal timbers placed in a masonry wall to bond it together and to which to nail lath, battens, and other wood members.

bonnet A wire mesh cover over the top of a chimney or vent. See also **chimney cap**.

bonus and penalty clause A provision in the contract for payment of a bonus to the contractor for completing the work before a stipulated date, and a charge against the contractor for failure to complete the work by this date.

book matching, herringbone matching Consecutive flitches of veneer from the same log, laid up side by side so that the pattern formed is almost symmetrical from the common center line. Book matching is used in decorative paneling and cabinetry.

boom A long straight member, hinged at one end, used for lifting heavy objects by means of cables and/or hydraulics. Booms can be of lattice construction or heavy tubular material.

boom crane A crane with a long slender boom, usually of lattice construction.

boom hoist A lifting device with a vertical mast and an inclined boom commonly used to hoist materials in multistory building construction.

boom jack A short member on which sheaves are mounted to guide cables to a working boom on a crane or derrick.

booster Any device which increases the power, force, or pressure produced by a system.

compressor A compressor which increases the productivity of the primary compressor by discharging into its suction line.

fan An auxiliary fan which increases the air pressure in a system, such as an HVAC system, during certain peak requirement times.

heater In plumbing, an auxiliary heater installed in a hot water piping system to increase the heat in one section.

pump An auxiliary pump installed in a system to increase or maintain the pressure or flow.

transformer A transformer installed to increase the voltage in an electric circuit.

boot A term used to describe sleeves or coverings in many construction trades, such as a

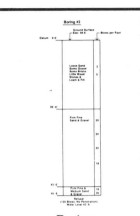

Boring

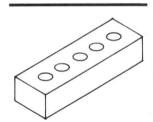

Bottle Brick

Bottom Rail

boot for pile driving, a boot for passing a pipe through a roof, and a boot for cold air return to a furnace casing.

bore (1) The bore of a pipe, valve, or fitting is its inside diameter. (2) A bore is also the circular hole left by boring.

bored latch A door latch manufactured to fit into a circular hole in a door.

bored lock A lock manufactured to fit into a circular hole in a door.

bored pile Same as cast-in-place pile.

bored well A water well constructed by drilling a hole with an auger and then inserting a casing.

boring, borehole A hole drilled in the ground to obtain samples for subsoil investigation. The borings are important to determine the load-bearing capacity of the soil and the depth of the water table.

borrow In earth moving, fill acquired from excavation from a source outside the required cut area.

borrowed light A glazed unit in an interior partition to allow light from an adjoining room or passageway.

borrow fill material See **borrow**.

borrow pit An excavation site, other than a designated cut area, from which material is taken for use nearby.

boss (1) A projecting block, usually ornamental, placed at the exposed intersection or termination of ribs or beams of a structure. (2) The enlarged portion of a shaft. (3) In masonry, a stone which is left protruding from the surface for carving in place. (4) A projection left on a cast pipe fitting for alignment or for gripping with tools.

Boston hip, Boston ridge, shingle ridge finish (1) A method of finishing a ridge or hip on a flexible shingled roof in which a

final row of shingles is bent over the ridge or hip with a lateral overlap. (2) A method of finishing a ridge or hip on a rigid shingle, slate, or tile roof in which the last rows on either side overlap at the hip or ridge. Alternate courses overlap in opposite directions.

bottle brick A hollow brick shaped so that it may be mechanically connected to adjoining bricks. There is also a void available for placing reinforcing steel.

bottom arm A long strap secured to the bottom rail of a door to attach it to a floor closer or pivot hinge.

bottom bolt A vertically mounted bolt on the bottom of a door which slides into a socket in the floor.

bottom car clearance The vertical distance between the lowest point of any mechanism attached to an elevator car and the floor of the elevator pit.

bottom dump A type of trailer, with gates on the bottom, used to haul dry materials such as sand, gravel, and crushed stone. A bottom dump is best used when material is to be distributed in layers on an embankment.

bottom lateral bracing In truss framing, the members interconnecting the bottom chords.

bottom rail The lowest bottom member of a door which connects the stiles.

boundary survey A mathematically closed diagram of the complete peripheral boundary of a site, reflecting dimensions, compass bearings and angles. It should bear a licensed land surveyor's signed certification, and may include a metes and bounds or other written description.

bow The longitudinal deflection of a piece of lumber, pipe, rod, or

Bow Window

Box Sill

the like, usually measured at its center.

bow saw A hand tool with a thin saw blade held in place between the two ends of a bow-shaped piece of metal.

bowled floor A floor which slopes downward toward a central point, as in an amphitheater.

bowstring beam See **bowstring truss**.

bowstring girder See **bowstring truss**.

bowstring roof A roof supported by bowstring trusses.

bowstring truss A roof structural member having a bow-shaped element at the top and a straight or cambered element connecting the two ends.

bow window A window that projects from a wall in the shape of an arc of a circle.

box bolt A short piece of rectangular stock that slides inside a casing, mounted on the edge of a door. The stock slides into a recess in the door casing to secure the door.

box casing In finished carpentry, the inner lining, as in a cabinet.

box chisel A special chisel used to pry open wooden boxes.

box culvert A rectangular-shaped reinforced concrete drainage structure either cast in place or precast in sections.

box drain A small rectangular-shaped drainage structure usually constructed of brick or concrete. It can either be covered or have an open grate on top.

boxed mullion A hollow mullion in a window frame built up from boards so as to appear solid and housing sash counter weights.

box frame A window frame containing hollow sections on either side in which the sash weights travel.

box gutter A rectangular-shaped

wooden roof gutter recessed in the eaves to conceal them and to protect them from falling foliage.

box-head window A window framed with a pocket in the head so that the sash can slide into it to provide maximum ventilation.

boxing (1) Enclosing or casing, as in a window frame. (2) In welding, continuing a principle fillet weld around a corner of the member.

box lewis An attachment device, used in hoisting heavy masonry units, that consists of metal components. Some or all of the components taper upward, inserted into a downward flaring hole (lewis hole) cut in the top of the masonry.

box nail A nail with a head similar to a common nail with the same penny designation of length but having a shank smaller in diameter.

box out To make a form that will create a void in a concrete wall or slab when the concrete is placed.

box scarf A joint in a rectangular-shaped roof gutter formed by beveling the ends of the two pieces to be joined.

box sill A common method of frame construction using a header nailed across the ends of floor joists where they rest on the sill.

box stair An interior staircase where the edges of the treads and risers are not revealed but closed in with a close string. Usually a box stair has a partition on both sides.

box stoop A high elevated platform at the entrance to a building with stairs running parallel to the building. The underside of the platform and stairs are enclosed.

box strike A fastening on a door frame (a strike) with an enclosed recess to receive a lock bolt.

boxwood A hard, dense, light-colored wood used for turned work and inlay.

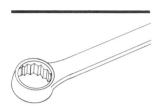

Box Wrench

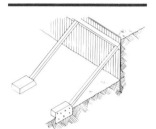

Brace (3)

Branch

box wrench A mechanic's hand tool with a closed socket, usually at each end, to fit over the head of a bolt or a nut.

brace (1) A diagonal tie that interconnects scaffold members. (2) A temporary support for aligning vertical concrete formwork. (3) A horizontal or inclined member used to hold sheeting in place. (4) A hand tool with a handle, crank, and chuck used for turning a bit or auger.

braced excavation Excavation that has the perimeter supported by sheeting.

braced frame, brace frame, brace framing A wood structural framing system where all vertical members extend for one floor only, except for corner posts. The corner posts are braced to the sill and plates.

brace rod A round steel member used as tension brace, especially to transfer wind or siesmic loads.

brace table, brace scale The markings on a carpenter's square indicating the length of the hypotenuse for right triangles with various length legs. A brace table is used by carpenters to determine the length to cut braces.

bracing See **brace**.

bracket An attachment projecting from a wall or column used to support a weight or a structural member. Brackets are frequently used under a cornice, a bay window, or a canopy.

bracket pile A steel H pile driven next to an existing foundation. A steel bracket is welded to the pile so that it extends under and supports the foundation.

bracket saw A hand saw with a narrow blade attached to the open ends of a three-sided rectangular holder used for cutting curved shapes.

bracket scaffold Scaffolding supported by brackets temporarily attached to the side of a building or column. Bolts or inserts are usually left in the previous construction to attach the brackets. This method saves the expense of shoring from the ground up.

bracket staging See **bracket scaffold**.

brad A slender, smooth wire nail with a small deep head used for finish carpentry work.

brad awl A carpenter's tool that looks like an ice pick, used to make holes for brads and wood screws.

brad pusher A special tool used to grip and insert a brad in an inaccessible location.

brake (1) A device for slowing, stopping, and holding an object. (2) A machine used to bend sheet metal.

brake drum A rotating cylinder with a machined surface, either internal or external, upon which a brake band or shoe presses in order to slow, stop, or hold an object.

brake horsepower The horsepower output of an engine or other mechanical device, as measured at the flywheel or belt by applying a mechanical break and measuring the work per unit time.

branch A term used in plumbing to define an inlet or outlet from the main pipeline, usually at a right angle to the main pipeline. The pipe may be a water supply, drain, vent stack, or any other pipe used in any mechanical piping system.

circuit A portion of the electric wiring system which extends beyond the fuse or other circuit protection device protecting that circuit.

drain A drain pipe from the plumbing fixtures or soil line of a building that runs into a main line.

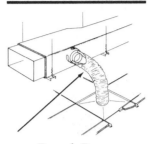

Branch Duct

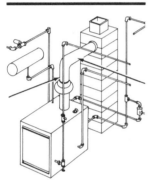

Breeching

Breezeway

duct In HVAC, a smaller duct branching from the main duct. At each branch duct, the cross sectional area of the main duct is reduced.

interval A length of soil stack or waste stack, usually one story high, within which all branches from one floor are connected.

sewer A sewer that receives sewage from a relatively small area and is connected to a main sewer or manhole.

vent (1) A vent connecting one or more individual vents to a vent stack or stack vent. (2) A vent pipe to which are connected two or more pipes that vent plumbing fixtures.

brashy A condition of wood characterized by coarse, conspicuous annual rings. Such wood has a low resistance to shock and has a tendency to fail abruptly across the grain without splintering.

brass A copper alloy with zinc as the principal alloying element.

braze To join two pieces of metal by soldering them with a nonferrous metal, such as brass.

brazed joint In plumbing, a fitting made water-tight and gas-tight by brazing.

Brazilian rosewood A dark-colored hard wood used for turned articles, paneling, and veneers.

breakdown voltage The voltage that causes the failure of existing insulation, permitting the flow of electric current.

breaker See **crusher**.

breaker ball See **headache ball**.

break-in In brick masonry, a void or socket to accept a timber.

breaking load The least load, determined by test, that would cause failure of a structural system.

break joints An arrangement of modular structural units such as masonry or plywood sheathing, so that the vertical joints of adjacent units do not line up. Also called "staggered joints".

break lines Lines used in drafting to omit part of an object so that the representation will fit on a drawing.

breast (1) The wall under a window stool to the floor level. (2) A projecting part of a wall, as at a chimney.

breast board A system of movable sheeting used to retain the face of an excavation, especially in tunnel work.

breast drill A hand-operated rotary drill having an attachment for applying pressure with the chest. A breast drill is used by carpenters for drilling horizontal holes in wood.

breast lining Interior wooden paneling applied between the window stool and the baseboard.

breast timber See **wale**.

breast work (1) The masonry work of a chimney breast. (2) The parapet of a building.

breeching The exhaust duct or pipe from a furnace or boiler to the stack or chimney.

breech fitting A Y-shaped fitting in ductwork or pipework.

breezeway A covered passage open at both ends which connects two buildings or two portions of a building.

brick A solid masonry unit of clay or shale, formed into a rectangular prism while plastic, and then burned or fired in a kiln.

adobe Large, roughly-molded, sun-dried clay brick of varying size.

arch (1) Wedge-shaped brick for special use in an arch. (2)

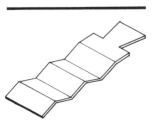

Brick Anchor

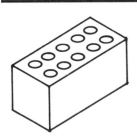

Brick, Economy

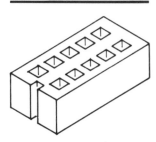

Brick, Jumbo

Extremely hard-burned brick from an arch of a scove kiln.

beam A lintel made of brick courses often held together by metal scraps.

bottle A brick shaped so that it makes a mechanical connection with the adjoining bricks. There is also a void left for inserting reinforcing steel when required.

breeze A brick made from pan breeze and portland cement, often laid into a brick wall for its good nail-holding properties.

buff standard A buff colored brick which can be either 2-1/4 x 3-3/4 x 7-3/4 inches or 2 x 4 x 8 inches, nominal.

building This is the new designation for common brick. It is a brick manufactured for building purposes, not specially treated for color or texture. Building brick in different parts of the country have different characteristics.

economy Brick with nominal dimensions of 4 x 4 x 8 inches.

engineered A brick whose nominal dimensions are 3.2 x 4 x 8 inches.

facing Brick made especially for facing purposes, often treated to produce surface texture. These brick are made of selected clay, or treated to produce a desired color.

fire Brick made of refractory ceramic material which will resist high temperatures. Fire brick is used to line fireplaces.

floor Smooth dense brick highly resistant to abrasion, used as finished floor surfaces.

gaged (1) Brick which have been ground or otherwise produced to accurate dimensions. (2) Tapered arch brick.

jumbo A generic term used to define a brick larger than the standard. Many producers use this term to designate a specific brick size manufactured by them.

Norman A brick whose nominal dimensions are 2-2/3 x 4 x 12 inches.

paving Vitrified brick especially suitable for use in pavements where resistance to abrasion is important.

Roman Brick with nominal dimensions of 2 x 4 x 12 inches.

salmon Relatively soft, under-burned brick, so named because of color. Sometimes called "chuff" or "place" brick.

SCR A patented brick with nominal dimensions of 2-2/3 x 6 x 12 inches.

sewer Low absorption abrasive-resistant brick intended for use in drainage structures.

soft-mud Brick produced by molding relatively wet clay (20 to 30 percent moisture), often by hand. When the inside of the mold is sanded to prevent sticking of the clay, the product is sand-struck brick. When the molds are wetted to prevent sticking, the product is water-struck brick.

stiff mud Brick produced by extruding a stiff but plastic clay (12 to 15 percent moisture) through a die.

brick anchor Corrugated fasteners designed to secure a brick veneer to a structural concrete wall. See also **reglet** and **dovetail anchor**.

brick and brick A method of laying brick so that units touch each other with only enough mortar to fill surface irregularities.

brick-and-half wall A brick wall having the thickness of one header plus one stretcher.

brick-and-stud work See **brick nogging**.

brickbat See **bat, def. 1.**

brick facing See **brick veneer**.

brick gauge A standard height for brick courses when laid, such as

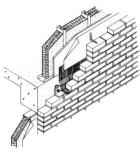

Brick Hammer

Brick Veneer

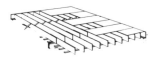

Bridged Floor

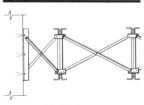

Bridging

four courses in a height of 12 inches.

brick hammer A hand tool used by masons for breaking and dressing brick. One end of the head is square and flat and the other is shaped like a chisel.

bricklayer's square scaffold A scaffold composed of framed wood squares which support a platform. A bricklayer's square scaffold is limited to light and medium duty.

brick molding The wood molding covering the gap between brick masonry and a door or window frame.

brick nogging See **nogging**.

brick trimmer A brick arch supporting a fireplace hearth or shielding a wood trimming joist from flames in front of a fireplace.

brick trowel A hand trowel with a pointed blade offset from the handle used to pick up mortar from a mortar tub and place and spread it in the joints.

brick veneer A brick facing wall laid against a structural wall but not structurally bonded to the wall and bearing no load other than its own weight.

brick whistle (1) A weep hole in a brick wall. (2) A small hole in a mortar joint at the base of a wall for draining any moisture that penetrates the wall.

bridge (1) A structure built to span an obstruction or depression and capable of carrying pedestrians and/or vehicles. (2) In an electric blasting cap, the wire which heats with current and ignites the charge. (3) The temporary structure built out over a sidewalk or roadway adjacent to a building to protect pedestrians and vehicles from falling objects.

bridgeboard A notched or cut stringer that supports the treads and risers of wooden stairs.

bridge crane A crane often used in manufacturing or assembling

heavy objects. A bridge crane requires a bridge spanning two overhead rails. A hoisting device can move laterally along the bridge and the bridge can move longitudinally along the rails.

bridge deck The load-carrying floor of a bridge which transmits the load to the beams.

bridged floor A floor supported by common joists.

bridge joint See **bridle joint**.

bridge thrust The horizontal thrust from vertical loads on an arched bridge.

bridging A system of lateral braces placed between joists to distribute the load and keep them in position.

bridle joint A type of mortise and tenon joint used when two timbers are joined at an angle of less than 90 degrees.

Briggs standard See **American standard pipe threads**.

brine In a refrigeration system, a liquid used as a heat transfer agent that remains a liquid and has a flashpoint above 150 degrees or no flashpoint. The liquid is usually a salt solution.

Brinell hardness A measurement of the hardness of metal determined in the laboratory by measuring the indentation made by a steel ball in the surface of the metal.

British thermal unit A standard measurement of heat energy which will raise the temperature of one pound of water one degree Fahrenheit.

brittle A characteristic of a material which makes it fracture easily without bending or deforming.

broach (1) To free stone block from a quarry ledge by cutting out the webbing between holes drilled close together in a row. (2) To cut wide parallel grooves in a diagonal pattern across a stone surface with the point of a chisel, finishing it

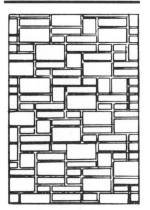

Broken Joints

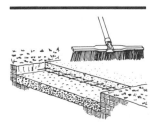

Broom Finish Concrete

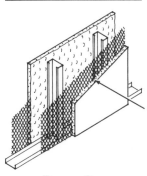

Brown Coat

for architectural use. (3) Any pointed structure, such as a steeple or spire, that is built for ornamental purposes. (4) A spire that rises directly from a tower, often without an intervening parapet. (5) A half pyramid constructed above the corners of a square tower, serving as an architectural transition from the slat of the tower to an octagonal spire.

broached spire An octagonal spire set above a square tower, with broaches braced against it at the four corners of the tower to effect a visually appealing transition.

broaching (1) A method of quarrying stone. Close holes are drilled along the breakline. A chisel, called a broach, is used to break the remaining material. The stone is removed with wedges. (2) A method of making shaped holes in metal by removing successive small pieces with a reaming tool.

broad ax A broad-bladed ax used to rough dress timber.

broad knife, stripping knife A square-edged knife with a blade shaped like a wedge, used chiefly for scraping off old paint or wallpaper.

broadloom A seamless carpet woven on a wide loom, usually 6 to 18 ft. (1.8 to 5.5 m) wide.

broad stone Same as **ashlar**.

broad tool A wide steel chisel used in the finish dressing of stone.

broad tooled See **batted work**.

brob A spike, shaped like a wedge, that is used to brace and secure the end of a timber butting against the side of another.

broken ashlar See **random work**.

broken arch A segmented arch, used largely as door or window entablature, on which the center of the arch is cut out and replaced with an ornamental figure or design.

broken-color work See **antiquing**.

broken-flight stair Same as **dogleg stair**.

broken joints Vertical joints of masonry arranged in a staggered structure with no unit placed directly on top of another, providing a more solid bond and increased structural strength.

broken joint tile A roofing tile that overlaps only the tile directly below.

broken rangework A form of stone masonry with horizontal courses of different heights, some of which are broken at intervals into multiple courses.

bronze (1) An alloy composed of copper and tin. (2) A bronze-colored alloy that uses a sizeable measure of copper to alter the properties of its chief elements, as aluminum bronze or magnesium bronze.

bronzing (1) Application of metal bronze provider to an object or substance. (2) The powdery decomposition of a paint film, caused by exposure to the elements and natural wear.

broom (1) To press a layer of roofing material against freshly applied bitumen in order to create a tight, thorough bond. (2) To flatten or spread the head of a timber pile by pounding forcefully on it. (3) To brush fresh plaster or concrete with a broom.

broom finish concrete Concrete that has been brushed with a broom when fresh in order to improve its traction or to create a distinctive texture.

brothers A two- or four-leg chain or rope sling.

Brown and Sharpe gauge See **American wire gauge**.

brown coat, floating coat (1) In two-coat wet-wall construction, the first rough coat of plaster applied as a base coat over lath or

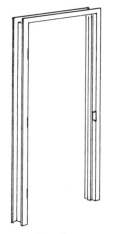

Buck

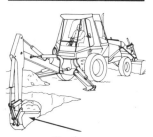

Bucket

masonry. (2) In three-coat work, the second coat of plaster applied over a scratch coat and serving as a base for the finish coat.

brown-glazed brick See **salt-glazed brick**.

brownmillerite An oxide of calcium, aluminum, and iron commonly formed in portland cement and high alumina cement mixtures.

brown-out (1) To perform an application of base-coat plaster. (2) The setting process of base-coat plaster, which darkens to a brown hue as it dries. (3) A partial loss of electrical power that dims lights. A brown-out is less severe than a blackout.

brown rot A fungus that attacks the cellulose of wood, so called for the residual brown powder left behind on the destroyed matter.

brown stain A deep brown discoloration in the sapwood of some pines, caused by fungus.

brownstone (1) Arko sandstone, dark brown or reddish brown in hue, which was used widely throughout the eastern United States in the nineteenth century for building homes and offices. (2) A house faced with brownstone.

browpiece A beam installed above a door.

brow post See **crossbeam**.

brush A hand held device made of natural or artificial bristles secured to a hue or backing, used for cleaning or painting a surface or object.

brush finish, brushed finish A finish created by applying a rotating wire brush to a surface.

brush graining A process in which a dark liquid stain is drawn across a light, dry base coat to produce an imitation effect of a wood grain.

brush rake A large apparatus, outfitted with heavy tires, that is attached to the front of a tractor

to clear land of brush and debris.

Brussels carpet (1) A carpet of worsted yarn in multiple colors, attached to a backing of tough linen thread in a pattern of uncut loops. (2) A less costly substitute for actual Brussels carpet, made in a single color of yarn.

bubble (1) Either the air bubble in a leveling tube or the tube itself. (2) A large void in gypsum board caused by air entrampment during manufacturing.

bubble tube A tube containing an air bubble which is used to level a tool or an instrument.

bubbling Trapped bubbles of air or vapors that erupt on a painted surface during application or drying.

buck (1) The wood or metal subframe of a door, installed in a wall to accommodate the finished frame. Also called a "door buck". (2) One of a pair of four-legged supporting devices used to hold wood as it is being sawed. Also called a "sawbuck" or a "sawhorse".

bucket A scoop-shaped attachment for an excavating machine that digs and transports loose earth materials, often outfitted with opening and closing mechanisms to facilitate unloading.

bucket trap A mechanical steam trap that operates on buoyancy and is designed with an inverted or upright cup that prevents the passage of steam through the system it protects.

bucket-wheel excavator A machine that dips earth and loads it onto a conveyor belt as it moves across a site, functioning by means of a rotating wheel and a number of tooth-edged buckets.

buck frame, core frame A wood frame that is built into the wall studs of a partition to accommodate a door lining. Also called a "subframe".

Buggy, Concrete Cart

Building Envelope

buckle (1) The distortion of a structural member such as a beam or girder under load, a condition brought on by lack of uniform texture as by irregular distribution of weight, moisture, or temperature. (2) A flaw or distortion on the surface of a sheet of material particularly asphalt roofing. (3) A thin tree branch bent in the shape of a U to fasten thatch onto roofs.

buckling load The load at which, in a compression member or compression portion of a member, bending progresses without increase in load.

buck opening A rough door opening.

bucksaw A wood-cutting saw consisting of a blade set in an H-shaped frame.

buck scraper A machine with a scoop which scrapes earth from a surface and is raised when full.

buff (1) To clean and polish a surface to a high luster. (2) To grind and polish a floor of terrazzo or other exposed-aggregate concrete.

buffer (1) Blasted rock left at a face to improve fragmentation and reduce scatter during a subsequent blast. (2) A loose metal used to control scattering of blast rock. (3) See **spring buffer** and **oil buffer**.

buggy, concrete cart A two- or four-wheeled cart, with or without a motor, used to transport small amounts of concrete from hoppers or mixers to forms.

bug holes Small cavities, usually not exceeding 5/8 in. (15 mm) in diameter, at the surface of formed concrete. Bug holes are caused by air bubbles trapped during placing and compacting of wet concrete.

builder's jack A temporary bracket, attached to a window sill, which projects outward and supports scaffolding.

builder's level (1) A spirit level set in a wood or metal straight edge. (2) A tilting or dump level not made to the accuracy of one used in surveying.

builder's risk insurance A special form of property insurance to cover work under construction.

builder's staging A heavy scaffold made from square timbers and usually used where heavy materials are handled.

building area The sum of the horizontal projected area of all buildings on a site. Terraces and uncovered porches are excluded, unless the stipulations of a mortgage lender or governmental program require their inclusion.

building block Any rectangular masonry unit, except a brick, used in building construction. Typical materials are burnt clay, concrete, glass, gypsum, etc.

building board Any board of building material, usually with a facing, which can be used as a finished surface on walls or ceilings.

building code The legal minimum requirements established or adopted by a governmental unit pertaining to the design and construction of buildings.

building drain That part of the lowest piping of a drainage system which receives the discharge from soil, waste, and other drainage pipes inside the walls of the building and conveys it to the building sewer.

building envelope The elements of a building which enclose conditioned areas and through which thermal energy may be transferred to or from the outside environment.

building height Generally the greatest vertical distance measured from curb or grade level to the highest level of a flat or mansard roof or to the average height of a pitched, gabled, hip, or gambrel

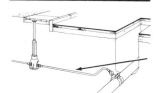

Building Main

Building Paper

Built-up

roof. Penthouses and the like are usually excluded if they do not exceed a specified height or their projected area is less than a specified percentage of the main roof.

building inspector An official of a building department, usually of a municipality, who reviews plans and inspects construction to determine if they conform to the requirements of applicable codes and ordinances, and who inspects occupied buildings for violations of the same codes and ordinances.

building insulation See **thermal insulation**.

building line The line, established by law, beyond which a building shall not extend, except as specifically provided by law. Exceptions may be terraces or uncovered porches.

building main The water supply pipe, including fittings, from the water main or other source of supply to the first distribution branch of a building.

building material Any material used in construction, such as sand, brick, lumber, or steel.

building official The officer or other designated authority charged with the administration of a building code, or a duly authorized representative.

building paper A heavy, asphalt-impregnated paper used as a lining and/or vapor barrier between sheathing and an outside wall covering or as a lining between rough and finish flooring.

building permit A written authorization for a specific project allowing construction to proceed in accordance with approved plans and specifications.

building restriction Any restriction, statutory or contractual, imposed on construction of a building or use of land.

building restriction line A line,

defined by local ordinances, beyond which a structure may not project. The line is usually parallel to the street line or other property line.

building section Any portion of a building, either a single room, a group of rooms, a floor, or a group of floors, that is within the limits of a fire division.

building services The utilities, including electricity, gas, steam, telephone, and water, supplied to and used within a building.

building storm drain A drain in a building which conveys rainwater, surface water, condensate, and similar discharge to a building storm sewer which extends to a specified point outside the building.

building trades The skilled and semi-skilled crafts used in building construction, such as carpentry, masonry, or plumbing.

building trap, main trap A fitting installed on the outlet side of a building to prevent odors and gases from passing from a sewer to the plumbing system of the building.

build up To create a mass by applying successive layers.

built-up (1) Fabricated of two or more pieces or sheets which are laminated. (2) Assembled by fastening a number of pieces or parts to each other.

built-up air casing A field-fabricated enclosure for an air-handling system with base, curbs, and drains.

built-up beam (1) A metal beam made of beam shapes, plates, and/or angles which are welded or bolted together. (2) A concrete beam made of precast units connected through shear connectors. (3) A timber beam made of smaller pieces which are fastened together.

built-up girder Same as *built-up beam*.

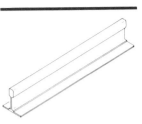

Bulb Tee

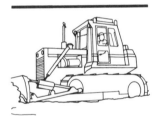

Bulkhead

Bulldozer

Bull Float

built-up rib A rib made of laminations of various sized timbers.

built-up roofing, composition roofing, felt-and-gravel roofing, gravel roofing A continuous roof covering made up of various plies or sheets of saturated or coated felts, cemented together with asphalt. The felt sheets are topped with a cap sheet or a flood coat of asphalt or pitch which may have a surfacing of applied gravel or slag.

built-up string A curved stair string constructed of wood members which are fastened together with counter cramps.

built-up timber A *built-up beam, def. 3*.

bulb angle A hot-rolled angle with a formed bulb on the end of one leg.

bulb bar A rolled steel bar with a formed bulb on one edge.

bulb pile See **pedestal pile**.

bulb pressure See **pressure bulb**.

bulb tee A rolled-steel tee with a formed bulb on the edge of the web.

bulk density The weight of a material per unit volume, including voids, solid particles, and any contained water.

bulkhead (1) A horizontal or inclined door providing outside access to a cellar or shaft. (2) A partition in concrete forms to separate placings. (3) A structure on the roof of a building to provide headroom over a stairwell or other opening. (4) A low structure on a roof covering a shaft or protruding service equipment. (5) A retaining structure which protects a dredged area from earth movement.

bulking Increase in the volume of a quantity of granular material when moist over the volume of the same quantity when dry. Also called "moisture expansion".

bulking factor Ratio of the volume of a quantity of moist granular material to the volume of the same quantity when dry.

bulking value The specific gravity of a pigment, usually expressed as gallons per 100 pounds or liters per kilograms.

bulk modulus of elasticity, modulus of volume The volumetric relationship between stress and strain within the elastic limit of a material. This quantity is measured as the ratio between a pressure acting on a material and the fractional change in volume due to that pressure.

bulk strain, volume strain The ratio of the change in volume of a mass, due to an applied pressure, to the original volume.

bulldog grip A U-bolt threaded at each end.

bulldozer (1) A tractor with a large, blunt blade attached to its front end by hydraulic controlled arms. A bulldozer is used to move earth short distances. (2) A machine used to bend reinforcing bars into U-shapes.

bullet catch A type of door latch consisting of a spring loaded steel ball. The loaded ball holds the door closed but rolls free when the door is pulled.

bulletproof glass See **bullet-resisting glass**.

bullet-resisting glass A laminated assembly of glass, usually at least four sheets, alternated with transparent resin sheets, all bonded under heat and pressure. Bullet-resisting glass is also made of laminations of special plastics.

bull float (1) (OSHA) A tool used to spread out and smooth the concrete. (2) A board of wood, aluminum, or magnesium mounted on a pole and used to spread and smooth freshly placed, horizontal concrete surfaces.

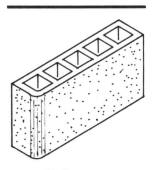

Bullnose

Burner

bull gear (slang) The largest or strongest toothed driving gear.

bull header, bull head A brick with one rounded corner used to form a corner at a door jamb or laid on edge to form a window sill.

bullhead tee, bullheaded tee A piping tee in which the outlet opening on the branch is larger than the openings on the run.

bullnose, bull's-nose (1) A rounded outside corner or edge. (2) A metal bead used in forming a rounded corner on plaster walls.

bullnose block A concrete masonry unit with one or more rounded corners.

bullnose plane A small, hand-held carpenters plane with the blade set forward.

bullnose step, bull stretcher A step, usually the bottom in a flight, with a semicircular end which projects beyond the railing.

bullnose stretcher (1) A masonry stretcher with one of the corners rounded. (2) Any stretcher which is laid on edge, as at a window sill.

bullnose trim A structural member or piece of trim which has a rounded edge, such as a stair tread, window sill, or door sill.

bull pin A tapered steel pin used to align holes in steel members so bolts can be inserted.

bull-point A pointed steel hand drill which is struck with a hammer to chip off small pieces of rock or other masonry.

bulwark A low wall used as a defensive shield.

bumper (1) A device, other than a spring or oil buffer, used to absorb impact from an elevator or counterweight. (2) The rubber silencer on a door frame.

bundle of lath A quantity of lath for plastering, usually 50 pieces of wood lath, 5/16 by 1-1/2 by 48 inches (0.79 by 3.8 by 122 centimeters) or 6 sheets of gypsum lath, 16 x 48 inches (41 by 122 centimeters).

bundler bars An assembly of up to four parallel reinforcing bars in contact with each other and enclosed in stirrups or ties, used as a unit in reinforced concrete, especially columns.

bungalow siding Clapboard siding having a minimum width of 8 inches (20 centimeters).

bunker (1) A compartmented storage container for aggregate or ore. (2) Space in a refrigerator for the ice storage and for the cooling element. (3) A low protective shelter usually constructed of reinforced concrete and partially or fully below ground. (4) A metal shield in a crushing or screening operation which is used to direct raw material to the feed belt.

buoyant foundation A reinforced concrete foundation which is designed and located such that its weight and superimposed permanent load is approximately equal to the weight of displaced soil and/or ground water.

burden (1) The loose material that overlays bedrock. (2) The depth of material to be moved or loosened in a blast.

burner That part of a boiler, furnace, etc., where combustion takes place.

burning Flame-cutting metal plates to a desired shape.

burning-brand test A standard test to determine the resistance of a roof covering to exposure from flying brands.

burnish To polish to a smooth, glossy finish.

burnt lime See **lime**.

burnt sienna Sienna which has been

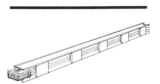

Bus Duct, Busway

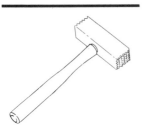

Bush Hammer

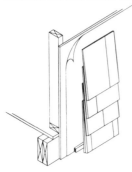

Butt

calcined to obtain a rich brown pigment.

burnt umber Umber which has been calcined to obtain a red to reddish-brown pigment.

burr (1) An uneven or jagged edge left on metal by some cutting tools. (2) Partially fused brick. (3) A batch of bricks which were accidently fused together. (4) A curly figure in lumber that was cut from an enlarged trunk of certain trees, such as walnut.

bursting strength The ability of sheet material to resist rupture under pressure, as measured by one of several tests.

bus, bus bar An electric conductor, often a metal bar, that serves as a common connection for two or more circuits. A bus usually carries a large current.

bus duct, busway A prefabricated unit containing one or more protected busses.

bush hammer A hand or power hammer with a serrated face with many pyramidal points. A bush hammer is used to dress concrete or stone.

bush-hammered concrete Concrete with an exposed aggregate finish that has been obtained by removing the surface cement with a percussive hammer with a serrated face.

bush-hammer finish A stone or concrete finish obtained through use of a percussive hammer with a serrated face used decoratively or to provide a rough surface for better traction.

bushing (1) A threaded or smooth (for soldered tubing) pipe or tube fitting used to connect pipes or tubes of different diameters. (2) A metal sleeve screwed or fitted into an opening to protect and/or support a shaft, rod, or cable passing through the opening. Usually the inside surface of the bushing is machined to close tolerance to reduce friction and

abrasion. (3) An insulating structure for a conductor with provision for an insulated mounting.

business agent An official of a trade union who represents the union in negotiations and disputes and checks jobs for compliance with union regulations and union contracts.

bus wire The wire(s) to which are connected the leg wires of blasting caps.

butment Same as **abutment**.

butt (1) A short length of roofing material. (2) The thick end of a shingle.

butt and break The staggering of joints in lath or on alternating studs in order to reduce cracking of plaster.

butt casement hinge A type of tow-plate hinge intended for use on casement sashes.

butt chisel A carpenter's chisel with a short blade used to shape recesses for hinges, etc.

butted frame A doorframe with a depth which is equal to or less than the thickness of the wall in which the frame is mounted.

butt end The thicker end of a handle, tapered pole, pile, etc.

butt-end treatment The process of protecting that part of a timber or post which will be exposed to soil and/or water by treating it with a preservative chemical.

butter (1) To apply mortar to a masonry unit with a trowel. (2) To spread roofing cement smoothly with a trowel. (3) To apply putty or compound sealant to a flat surface of a member before setting the member, such as buttering a stop before installing it.

butterfly hinge A decorative hinge.

butterfly nut Same as **wing nut**.

Butterfly Valve

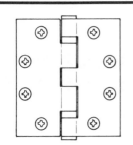

Butt Hinge

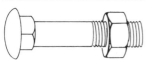

Buttonhead

butterfly roof A roof shape in which two surfaces rise from a control valley to the eaves.

butterfly spring A formed piece of spring metal, set over the pin of a hinge to serve as a door closer.

butterfly valve A valve with excellent throttling characteristics that has a disc that rotates 90 degrees within the valve body.

butterfly wall tie A masonry wall tie made from heavy wire in the shape of a figure 8.

buttering Applying mortar to a masonry unit with a trowel.

buttering trowel A small trowel used to apply mortar to a brick before laying the brick.

butternut, white walnut A moderately soft, medium-textured wood of light to pale brown color. Butternut is used particularly for decorative veneer.

butt fusion A method of joining pipe or sheet, made of a thermoplastic resin, where the ends to be joined are heated to the molten state pressed together and held until the material sets.

butt hinge The common form of hinge consisting of two plates each with one meshing knuckle edge and connected by means of a removable or fixed pin through the knuckles.

butt-hung door A swing door hung with butt hinges.

butt joint (1) A square joint between two members at right angles to each other. The contact surface of the outstanding member is cut square and fits flush to the surface of the other member. (2) A joint in which the ends of two members butt each other so only tensile or compressive loads are transferred.

buttonhead The head of a bolt, rivet, or screw that is shaped like a segment of a sphere and has a flat bearing surface.

button punching Crimping the interlocking lap of metal ducting panels with a dull punching tool. Button punching has been largely replaced by spot welding overlapping edges.

buttress (1) An exterior pier of masonry, often sloped, which is used to strengthen or support a wall or absorb lateral thrusts from roof vaults. (2) An A-shaped formwork of timber or steel used to strengthen or support a wall.

butt splice A butt joint secured by fastening a short piece of wood or steel to each side of the butted members.

butt strap The short pieces used in butt splice.

butt weld A welded joint made by welding two butted piece or sheets together.

butyl stearate A colorless, oily liquid used as a dampproofing for concrete.

buzz saw A power saw with a circular blade.

bypass (1) A pipe used to divert flow around another pipe or piping system. (2) A pipe or duct used to divert flow around an element.

bypass valve A valve which is located to control a bypass.

bypass vent A vent stack parallel to a soil or waste stack and connected together at branch intervals.

byre A stable or barn for livestock.

ABBREVIATIONS

The abbreviations listed below are those most commonly used in the construction industry. Alternative forms (usually nonstandard) are shown in parentheses.

1/C single conductor
2/C two conductors
c candle, cathode, cycle, channel
C carbon, centigrade, Celsius
C&Btr. grade C and better (used in lumber industry)
CAB cement-asbestos board, cabinet
cal calorie
cap. capacity
CAT. catalog, catalogue
CATW catwalk
CB catch basin
CB1S center beam one side
CB2S center beam two sides
CBR California bearing ratio
cc cubic centimeter
CCW counterclockwise
cd candela
ceil ceiling
cem. fin. cement finish
cem. m cement mortar
cent. central
cer ceramic
CF centrifugal force, cost and freight, cooling fan
cfm, CFM cubic feet per minute
CFS cubic feet per second
CG center of gravity, coarse grain, ceiling grille, corner guard
CG2E center groove two edges
CHIM chimney
CHU centigrade heat unit
CI cast iron, certificate of insurance
CIP cast iron pipe
CIR circle, circuit
CIRC circumference
CL center line
cm centimeter
CM construction management, center matched
CMP corrugated metal pipe
CMPA corrugated metal pipe arch

CND conduit
CO change order, certificate of occupancy, cleanout, cutout
coef coefficient
col column
com common
COMB. combination
comp compensate, component, composition
COMPF composition floor
COMPR composition roof, compress, compressor
conc concrete
cond conductivity
const constant, construction
constr construction
CONTR contractor
conv convector
cop. coping
corb corbeled
corn. cornice
corr corrugated
CP cesspool
CPFF cost plus fixed fee
CPM Critical Path Method, cycles per minute
crib. cribbing
CRN cost of reproduction/replacement new
CRP controlled rate of penetration
CRT cathode ray tube
CS cast stone
CSI Construction Specifications Institute
CSK countersink
c/s cycles per second
ct coat, coats
CTB cement treated base
c to c center to center
ctr center
cu cubic
cu. ft. cubic feet
cu. in. cubic inch

cu. yd. cubic yard
cw clockwise
CV1S center vee one side
CV2S center vee two sides
C.W. pt. cold water point
cwt hundred weight
cyl cylinder
CYL L cylinder lock
cyp cypress

C DEFINITIONS

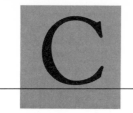

Cabin

Cabinet

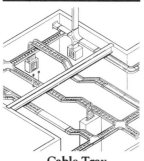

Cable Tray

Cabba (1) A small cube-shaped building. (2) The small stone building in the court of the Great Mosque at Mecca, the sacred shrine of the Muslims.

cabana A small shelter of open, tent, or wood-frame construction placed near swimming pools or the shoreline.

cabin (1) A small simple hut or house. (2) A rustic shelter often made of logs.

cabinet (1) A small room or private apartment especially for study or consultations. (2) A suite of rooms for exhibiting articles and curiosities. (3) A case, box, or piece of furniture with sets of drawers or shelves with doors, primarily used for storage. (4) An enclosure with doors for housing electrical devices and wiring connections.

cabinet file A single-cut hand file that is half round on one side and flat on the other.

cabinet filler The wood piece that closes the space between cabinets and adjacent walls or ceilings.

cabinet finish A varnished, oiled, or polished wood surface finish, as distinguished from a painted surface finish.

cabinet heater A metal housing, enclosing a heating element, with openings to facilitate air flow. The heater frequently contains a fan for controlling the air flow.

cabinet lock A spring bolt or magnetic latch.

cabinet scraper A flat steel blade used for smoothing wood or for removal of a finish, such as paint or varnish, from a surface.

cabinet window A projecting window or type of bay window popular during the nineteenth century for the display of shop goods.

cabinet work Wood joinery as used in the construction of built-in cabinets and shelves.

cable (1) A rope or wire comprised of many smaller fibers or strands wound or twisted together. (2) In electric cable, a conductor consisting of a group of smaller diameter conductor strands twisted together. (3) A group of electric conductors insulated from each other but contained in a common protective cover.

cable conduit See **conduit**. Also see **cable duct**.

cable duct A rigid, metal protective enclosure through which electric conductors are run. For underground installations, concrete or plastic pipes are usually used.

cable molding See **cabling**.

cable pulling compound A material used as a lubricant to facilitate the pulling of wires through a cable duct or conduit.

cable roof A system comprised of a roof deck and covering that are supported by cables.

cable sheath The protective cover that surrounds a cable.

cable support box An electrical box that is mounted on a wall and provides support for the weight of cables within a vertically installed conduit.

cable-supported construction A structure that is supported by a system of cables. This system is used for long-span roofs and also for suspension bridges.

cable tray An open, metal framework used to support electric conductors in a manner similar to cable duct. The primary difference is that the cable tray has a lattice type construction and an

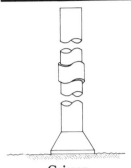

Caisson

Caisson Pile

open top.

cable vault An underground structure utilized in the pulling and joining of underground electric cables.

cableway A cable that bridges a gap between two points and permits materials to be pulled across the gap between the points.

cabling, cable molding An ornament or molding of spiral form similar to the twists of a cable.

cadastral survey A large-scale land survey to define boundaries, areas, and ownership of real estate for purposes of subdivision or apportioning taxes.

cadmium plating A coating of cadmium metal applied over a base metal, usually for corrosion protection.

cadmium yellow A yellow pigment of cadmium sulfide valued in the manufacture of paint for its quality and permanence.

cage (1) The box or enclosed platform of an elevator or lift. (2) An enclosure for electrical lights or signs. (3) Any rigid open box or enclosure.

caisson (1) A drilled, cylindrical foundation shaft used to transfer a load through a soft strata to a firm strata or bed rock. The shaft is filled either with reinforced or unreinforced concrete. (2) A watertight box or chamber used for construction work below water level.

caisson drill A piece of boring equipment used to excavate a shaft, usually vertical, in the earth for construction of a building footing.

caisson pile A cast-in-place pile formed by driving a hollow tube into the ground and filling it with concrete.

caking In construction, the undesirable mass of pigment that collects in the bottom of paint cans.

calcareous A term referring to any material containing calcite or calcium carbonate, or containing calcium, as in ''calcareous soils.''

calcimine, kalsomine A low-cost wash, white or colored, used on ceilings, interior plaster, or other masonry-type surfaces.

calcine The process of heating a substance to a point just below its melting temperature in order to remove chemically combined water.

calcined gypsum A gypsum that has been partially dehydrated. In this form it is used as a base material to which mineral aggregate, fiber, or other material may be added to produce the desired plaster.

calcite The main raw material used in the manufacture of portland cement. Calcite is a crystalized form of calcium carbonate and is the principal constituent of limestone, chalk, and marble.

calcite streak A break or fault in a mineral limestone deposit that has been filled by calcite sedimentation.

calcium aluminate cement A combination of calcium carbonate and aluminates that have been thermally fused or sintered and ground to make cement.

calcium silicate brick See **brick**.

calcium silicate insulation Hydrated calcium silicate that has been molded into rigid shapes and forms. The material is commonly used for pipe insulation where service temperatures reach 1,200 degrees Fahrenheit. It is particularly suited to pipe insulation because it is not appreciably affected by moisture.

calculated live load The useful load-supporting capability above the weight of the structural members, usually specified in the applicable building code.

calculon See **brick**.

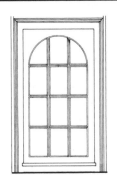

Camber Window

calf's-tongue molding, calves-tongue molding A molding with repeating shapes of a tongue or semi-oval, all pointing in the same direction or, when surrounding an arch, to a common center.

caliber The internal diameter of a pipe. Most pipe in construction is specified by outside diameter, and a further designation for wall thickness.

caliche A term used to describe gravel or sand loosely cemented together by calcium carbonate or other salts.

caliduct A conduit for conveying hot air, hot water, or steam for heating.

California bearing ratio A ratio used to determine the bearing capacity of a foundation. It is a ratio, of the force per unit area per minute required to penetrate a soil mass, to the force required for a similar penetration of a standard crushed rock material.

caliper An instrument with two hinged legs used to determine the thickness or diameter of objects.

calked rivet A rivet that has not been properly driven so as to fill its hole, but has been tightened by having the edge of its head driven under with a cold chisel.

calking See **caulking**.

calliper See **caliper**.

call loan A loan that is payable on formal notice of demand by the lender.

calorie The amount of heat required, at a pressure of one atmosphere, to raise the temperature of one gram of water one degree centigrade.

calorific value The amount of heat liberated by the oxidation or combustion of the unit weight of a solid material, or the unit volume of a gas or liquid.

cam (1) An eccentric wheel mounted on a rotating shaft and used to produce reciprocal or variable motion in an engaged or contacted part. (2) In a lock, the rotating piece attached to the cylinder that actuates the locking mechanism.

camber A slightly convex curvature built into a beam or truss to compensate for deflection under load. Camber is also built into structural components, roadways, or bridges to provide run-off for water.

camber arch A flat arch with a very small upward curve of the under surface. The upper surface may or may not be curved.

camber beam A beam constructed with a slight upward curve in the center.

camber diagram A construction drawing that shows the desired design camber along the beam, truss, or structural component.

camber piece The temporary wood support or template used to lay a slightly curved brick arch.

camber window A window with the top arched.

came A flexible, cast lead rod used in stained glass windows to hold together panes or pieces of glass.

camelback truss A truss whose upper chord is comprised of a series of straight segments so that the assembly looks like the hump of a camel's back.

camelhair mop A high-quality, soft-haired brush used for varnishing, gilding, or painting.

cam handle, locking handle On a hinged window, the handle that rotates against its keeper plate, thus pulling the window tight and locking it.

campaniform A term describing a structure in the shape of a bell.

canal, canalis (1) Any watercourse or channel. (2) A channel or groove fluting in carved ornamentation.

candela A unit of luminous intensity; formerly called "candle."

Canopy

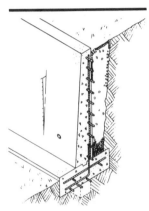

Cantilever Retaining Wall

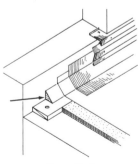

Cant Strip

candlepower The luminous intensity of a light source expressed in candelas.

canopy (1) An overhanging shelter or shade covering. (2) An ornamental rooflike structure over a pulpit.

cant To slope, tilt, or angle from the vertical or horizontal.

cant brick See **splay brick**.

cantilever A beam, girder, or supporting member which projects beyond its support at one end.

cantilever footing (1) A footing with a tie beam connected to another footing to counterbalance an asymmetrical load. (2) In retaining wall construction, a wide reinforced footing with reinforcing steel extending into the retaining wall to resist the overturning moment.

cantilever form A form, used in concrete construction, that is jacked slowly upward and supported by the hardened concrete of the wall previously poured. Also see **slip form**.

cantilever retaining wall A retaining wall that has a wide footing to resist its overturning moment.

cantilever steps Steps supported at one end by a wall and at the other end by the next lower step.

cantilever truss A truss that is anchored at one end and overhangs a support for the other end.

cantilever wall A wall that resists its overturning moment with a cantilever footing.

canting strip A horizontal ledge near the bottom of an exterior wall sloped to conduct water away from the face of a building or its foundation.

cant molding A molding with one face beveled.

cant strip A three-sided piece of wood, one angle of which is square, used under the roofing on a flat roof where the horizontal surface abuts a vertical wall or parapet. The sloped transition facilitates roofing and waterproofing.

canvas A closely woven cloth, usually made of cotton, used for tarpaulins, awnings, sun covers, and temporary canopies.

canvas wall A plaster wall, with a canvas covering, that may be painted or wallpapered.

cap The top piece, often overhanging, of any vertical architectural feature or wall. A cap may be external, as on an outside wall or doorway, or internal as on the top of a column, pilaster, molding, or trim.

capacitance The property of a system that permits the storage of electronically separated charges when potential differences exist between the conductors. Its value is expressed as a ratio of a quantity of electricity to a potential difference, and is usually measured in farads or microfarads.

capacitor A device used to introduce capacitance into an electric circuit. The device consists of conducting plates insulated from each other by a layer of dielectric material.

capacitor motor A single-phase induction motor with a main winding arranged for direct connection to a source of power and an auxiliary winding connected in series with a capacitor. The capacitor may be directly in the auxiliary circuit or connected to it through a transformer.

capillary action In subsurface soil conditions, the rising of water in the soil above the horizontal plane of the water table.

capillary break A space that is intentionally made large enough to prevent moisture flow by capillary action.

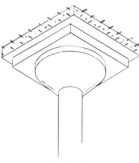

Capital

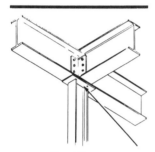

Cap Plate

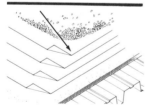

Cap Sheet

capillary flow The flow of moisture through any form, shape, or porous solid by means of capillary action.

capillary joint A pipe or fitting connection made by melting a solder material at the edge of a slip fit so that it is drawn into the joint by capillary action and, on cooling, makes a tight joint.

capillary space A term describing air bubbles that have become embedded in cement paste.

capillary tube (1) A small-diameter tube used in refrigeration as a control to restrict the flow of refrigerant from the condenser to the evaporator. (2) The small-diameter tubing used to connect temperature- and pressure-sensing bulbs to a control mechanism.

capillary water Water, which by virtue of capillary action, is either above the surrounding water level or has penetrated where a body of water would normally not go.

capital The uppermost member of a column, pilaster, etc., crowning the shaft and taking the weight of an entablature.

cap molding, cap trim The molding or trim above a door or window casing.

capping The top component of an assembly after covering and weatherproofing a joint.

capping piece, cap piece, cap plate The timber or masonry used to cover the tops of a series of uprights or other vertical members.

cap plate The plate on top of a column or post that supports the load.

cap screw A threaded fastener that screws into a threaded hole rather than into a nut. Cap screws are specified by diameter, length, and thread pitch.

cap sheet The top ply of mineral-coated felt sheet used on a built-up roof.

capstone A stone segment in a coping.

captain's walk A walk or platform, with a view of the sea, on the roof of early New England style houses.

caracole (1) A zig-zag course or path. (2) A spiral staircase.

car annunciator The floor indicator panel or dial in an elevator.

carbonaceous Material containing or composed of carbon, i.e., organic matter.

carbon-arc cutting A metal cutting process in which arc heat melts a path through metal. The arc is established between an electrode, which forms one terminal of an electric circuit, and the workpiece, which forms the other terminal.

carbon-arc lamp An electric-discharge lamp employing an arc discharge between carbon electrodes. One or more of the electrodes may have a core of special chemicals to enhance the discharge.

carbon-arc welding An arc-welding process in which the heat for fusion is generated by an electric arc between a carbon electrode and the workpiece.

carbonation The process of burning or converting a substance through chemical reaction into a carbonate. For example, the reaction between carbon dioxide and the calcium compounds in cement paste or mortar to produce calcium carbonate.

carbon black A fine carbon obtained as soot by the direct impingement of a flame on a metal surface. Carbon black is used as a pigment to color paints and other materials.

carbon steel Steel that owes its distinctive properties, primarily its strength, to the carbon it contains, as opposed to the characteristics

Carcass

Carcass Roofing

imparted by other alloying elements.

carcass, carcase (1) A body or shell without adornment or life. (2) The structural framework of a building without walls, trim carpentry, masonry, etc.

carcass flooring The joists and associated structures that support the floorboards above and the ceiling below.

carcass roofing The roofing framework before decking, membrane, shingles, etc., have been applied.

card frame, card plate The small frame that attaches to a door surface for insertion of a name card or plate.

care, custody and control A feature of construction-liability insurance policies that excludes damage to property in the care, custody, or control of the insured.

carillon A set of fixed bells sounded by striking with hammers operated either from a keyboard or mechanically. By extension, a bell tower or campanile.

carnauba wax A wax made from the leaves of the Brazillian wax palm. The wax is hard and is used in wood polishes and protective coatings.

Carpenter Gothic Gothic style architecture and ornamentation constructed of wood.

carpenter's brace A crank-shaped tool with a chuck on one end to hold a drill bit or auger. Rotating the crank bores a hole. This type of brace is also called a "bit brace."

carpenter's bracket scaffold A scaffold supporting a platform using wood or metal brackets.

carpenter's level A hand tool used by carpenters to determine a horizontal or vertical plane or line. The level is a wood or metal bar about 2 feet long with four spirit levels set into it.

carpenter's punch A hand tool similar to a nail set having a long slender neck. The punch is used by carpenters to remove nails by driving them through a piece of wood, rather than pulling them out.

carpenter's square, framing square A flat, metal L-shaped tool that constitutes an accurate right angle and is engraved with divisions and markings useful to a carpenter laying out and erecting framing.

carpet backing The base material on the back of a carpet. The backing is usually made of jute, cotton, or carpet rayon and may have a coating of latex.

carpet bedding Beds of low-growing plants which may be used in ornamental design or as an erosion preventing ground cover.

carpet cushion A padding placed on the floor before a carpet is laid. The cushion helps to provide resilience and extends the life of a carpet. It is usually made of hair, felt, jute, foam, or sponge rubber.

carpet density The number of pile tuft rows, per inch, for the length of the carpet.

carpet float A tool used by plasterers to give texture to a sand finish. The tool consists of a wood float covered with a piece of carpeting. The density of the carpet determines the type of finish.

carpet pitch The number of yarns across the width of the carpet expressed in yarn ends per 27 inches of width.

carpet strip (1) A flat strip or molding used to fasten the edge of carpeting. (2) A piece of wood or metal, approximately the thickness of a carpet, installed at the edge of a carpet, as at a threshold.

carpet underlayment See **carpet cushion**.

Car Platform

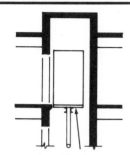

Carport

Cartridge Fuse

**Cased Opening
Trimmed Opening**

car platform The structural floor of an elevator car that supports the load.

carport A roofed shelter for automobiles, usually attached along the side of a dwelling, with one or more sides open.

carreau A square or diamond-shaped piece of glass or tile used in ornamental glazing.

carrel, cubicle A small alcove, as in a cloister or a library, for individual study.

carrelage Decorative tiling that is used particularly in terra-cotta flooring.

carriage The sloping beam that supports the steps of a wooden staircase and is installed between the strings.

carriage bolt A threaded bolt with a round smooth head. The bolt is prevented from rotating in its hole by a square neck directly under the head.

carriage clamp A clamp used in carpentry and cabinetry.

carriage porch A canopy structure that extends from the doorway of a building over a driveway.

carrier angle A metal angle attached to stair carriers or stringers to support the tread or riser.

carrier bar A flat metal bar used to support stair treads or risers in a manner similar to carrier angles.

carrying channel A three-sided metal molding used in the construction of a suspended ceiling.

cartridge fuse A low-voltage fuse consisting of a current-responsive element inside a cylindrical tube with a terminal on each end.

cartridge heater An electric heater in the shape of a cylindrical cartridge with the heating coils at one end and a fan at the other.

carvel joint A flush, longitudinal joint between boards or planks with no tongues, grooves, or laps.

cascade refrigerating system Two or more refrigerating systems connected in series. The cooling coil of one system is used to cool the condenser of the next system. This type of system is an efficient way to produce ultra-low temperatures.

case (1) A box, sheath, or covering. (2) The process of covering one material with another; to encase. (3) A product or food display counter, i.e., a refrigerated case displaying ice cream or frozen produce. (4) A lock housing.

case bay An area of a building floor or roof between two beams or girders.

cased beam An exposed interior beam encased in finished millwork.

cased frame , boxed frame, box frame The hollow frame for double-hung windows with sash weights. The weights travel in the hollow frame.

cased glass, case glass, overlay glass Ornamental glass formed of two or more layers with cuts so that sub layers may show.

cased opening, trimmed opening An interior doorway or opening with all the trim and molding installed, but without a door or closure.

cased post An exposed interior post or column encased in finished millwork.

cased sash-frame See **cased frame**.

case-hardened (1) Steel or iron alloy with a hard surface developed by a special heat-treatment process. (2) Timber whose outer fibers have dried too rapidly, thus causing checking and cracking.

case-hardened glass Glass that has been heated and quenched, thus giving it a hardness and strength several times greater than it originally possessed.

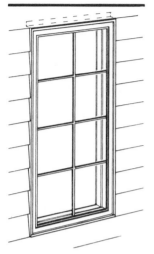

Casement Window

Casework

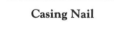

Casing Nail

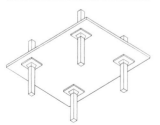

**Cast-in-Place
Concrete**

casein glue Glue made from the white amorphous phosphoprotein occurring naturally in milk. The glue is frequently used in carpentry and joinery.

casein paint A paint using the white amorphous phosphoprotein emulsion from milk as a base adhesive and binder.

casement (1) A window sash that opens on hinges that are fixed along either side. (2) A ventilation panel that opens on vertical hinges like a door. (3) A casing.

casement door (1) A door consisting of a wooden frame around glass panels that make up a major portion of the area of the door. (2) A French door.

casement stay The brace or bar used to hold a casement window open in any of several positions.

casement ventilator A ventilation panel that opens on hinges along one side.

casement window A window assembly having at least one casement or vertically hinged sash.

case mold A frame support, usually made of plaster, that holds smaller plaster pieces in position in a mold.

case steel See **case hardened**.

casework A term for assembled cabinetry or millwork.

cash allowance An amount of money established in the contract documents for inclusion in the contract sum to cover the cost of prescribed items not specified in detail, with provision that variations between such amount and the finally determined cost of the prescribed items will be reflected in change orders appropriately adjusting the contract sum.

casing (1) The exposed millwork enclosure of cased beams, posts, pipes, etc. (2) The exposed trim molding or lining around doors and windows. (3) The pipe liner of a hole in the ground as for a

well, caisson, or pile.

casing bead A metal molding or wood strip which acts as a stop, an edge, or a separation between different materials.

casing-bead doorframe A metal doorframe used in plaster walls that includes a strip or molding along its outer edges as a ground or guide for the plaster.

casing knife A knife used to trim wallpaper around casing, moldings, or other edges.

casing nail (1) A nail with a very small head. (2) A finishing nail.

cassoon A deep panel in a ceiling or soffit.

castable refractory A dry prepared mixture which when reconstituted with water becomes a concrete or mortar suitable for refractory use.

casting An object that has been cast in a mold. The object may be made of iron, steel, plaster, concrete, plastic, or any other castable material.

casting plaster A mixture of plaster with additives to provide properties desirable for casting work.

cast-in-place concrete Concrete placed in forms at its final location.

cast-in-place pile A concrete pile cast or formed at its permanent location. Such a pile can be cast with or without a casing.

cast iron An iron alloy cast in sand molds and machined to make many building products, such as ornamentation, pipe and pipe fittings, and fencing.

cast-iron boiler A boiler made of cast iron sections. Large cast iron boilers may be transported in sections and assembled in place. Such a boiler provides long-lasting, heavy-duty service.

cast molding A molding or molding component that has been cast, removed from the mold, and then fastened in place. The molding may be made of plaster,

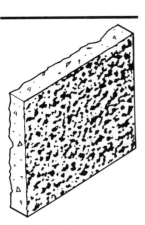

Cast Stone

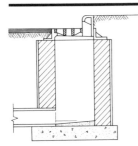

Catch Basin

Caulking Gun

concrete, plastic, or some similar material.

cast stone A mixture of paste or mortar, with an aggregate of stone chips or fragments, which has the appearance of stone when cast into the desired form or structural shape.

catalyst (1) A substance that accelerates chemical reactions. (2) The hardener that accelerates the curing of adhesives, such as synthetic resins.

catalytically blown asphalt Asphalt that has had air with a catalyst blown through it while hot to give it the desired characteristics for a special use.

catch The fitting that a latch or cam mates with to lock a door, gate, window, etc.

catch basin A receptor or reservoir that receives surface water run off or drainage. Typically, a catch basin is made of precast concrete, brick, or concrete masonry units and has a cast iron frame and grate on top.

catch drain A long drain installed across or along a slope to collect and convey surface water.

catch pit See **catch basin**.

catenary A term referring to the curve formed by a flexible cord hanging between two points of support, as in power lines between poles.

catenary arch An arch constructed on the curve of an inverted catenary.

catface A rough flaw in a finish coat of plaster.

cathead The top piece of a hoist tower to which pulleys or sheaves are attached.

Catherine-wheel window (1) A round window with mullions like wheel spokes. (2) A rose window.

cathodic corrosion See **electrochemical corrosion**.

cathodic protection A form of protection against electrolytic

corrosion of fuel tanks and water pipes submerged in water or embedded in earth. Protection is obtained by providing a sacrificial anode which will corrode in lieu of the structural component, or by introducing a counteracting current into the water or soil. This type of protection is also used for aluminum swimming pools, cast iron mains, and metal storage tanks.

cat ladder, gang boarding roof ladder A board with a series of cross pieces nailed to it. This type of ladder is used to provide footing on steep surfaces like the slope of a roof.

cat's eye In a piece of lumber, a small knot 1/4" in diameter or less.

catwalk A small permanent walkway, usually elevated, to provide access to a work area or process, or to provide service/maintenance access to lighting units, stage draperies, etc.

caul A sheet of metal or plywood used to protect the surface of a sheet of plywood, particle board, or fiberboard from the presses during molding.

caulk To fill a joint, crack, or opening with a sealer material. The filling of joints in bell and spigot pipe with lead and oakum.

calk See **caulk**.

caulking, calking The soft puttylike material used to fill joints, seams, and cracks in masonry and carpentry. Lead and oakum materials are used in plumbing.

caulking ferrule A metal ring, usually brass, that is sealed with caulk.

caulking gun A tool, either hand or machine-powered, that extrudes caulking material through a nozzle.

caulking recess In plumbing the space between the bell or hub of one piece of pipe and the spigot

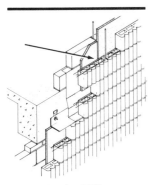

Cavity Fill

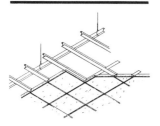

Ceiling

that is inserted into it. The space is filled with caulking.

causeway A passage or roadway that has been raised above wet ground.

caustic dip A strong cleaning solution into which materials are immersed.

caustic etch, frosted finish A finish or decorative design produced on glass or other materials by a corrosive chemical.

caustic lime A material, white when pure, that is obtained by calcining limestone, shells, or other forms of calcium carbonate. Caustic lime is also called "quicklime" or "burnt lime" and is used in mortars and cements.

cavitation The formation of a cavity or hollow in a fluid, as is the case for the partial vacuum in water about a rapidly revolving propeller.

cavitation damage (1) A pitting damage done to propellers, turbine wheels, or other rotating equipment. (2) The pitting of concrete due to the collapse of bubbles in flowing water.

cavity batten A length of wood placed in a cavity wall to catch mortar droppings during construction.

cavity fill Material placed in the hollow of a wall ceiling or floor to provide insulation or sound deadening.

cavity flashing The continuous waterproof barrier that covers the longitudinal opening in a cavity wall.

cavity wall, hollow masonry wall, hollow wall An exterior masonry wall in which the inner and outer wyths are separated by an air space, but tied together with wires or metal stays.

C/B ratio saturation coefficient The ratio of the weight of water absorbed by a masonry brick or block immersed in cold water to the weight

absorbed when immersed in boiling water. This is a measure of the resistance of the masonry to spalling due to freezing and thawing.

C-clamp A type of clamp in the shape of the letter C that is frequently used in carpentry and joinery. Force is applied by rotating a threaded shaft through one jaw of the C to force the work against the other jaw.

cedar A soft wood with reddish heartwood and white sapwood noted for decay resistance and used for shingles, shakes, and gutters.

ceiling The overhead inside lining or finish of a room or area. By extension, any overhanging surface viewed from below.

ceiling area lighting A lighting system in which the entire ceiling functions as a single large luminaire. This lighting includes luminous and louvered ceilings.

ceiling cornice The molding installed at the intersection of the wall and ceiling planes.

ceiling diffuser Any air diffuser located in a ceiling through which warm or cold air is blown into an enclosure. The diffuser is designed to distribute the conditioned air over a given area.

ceiling fitting A fitting, usually electrical, that is attached to a ceiling, i.e., the attaching part of a ceiling light fixture.

ceiling flange An escutcheon or cover placed over the hole where pipes go through the ceiling, and also around sprinkler heads that protrude through ceilings.

ceiling floor Framing intended to support a ceiling but not a floor.

ceiling outlet In a ceiling, a metal or plastic junction box which supports a light fixture and encloses the connecting fixture wires.

ceiling plenum In air conditioning systems, the air space between the

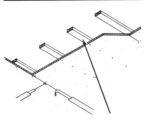

Ceiling Strap

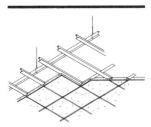

Ceiling Suspension System

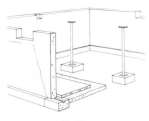

Cellar

hung ceiling and the underside of the floor or roof above acting as a return to the air handling unit, as opposed to a ducted air return.

ceiling sound transmission The transmission of sound between adjacent rooms through a ceiling plenum common to them both.

ceiling strap Wood strips nailed to the bottom of roof rafters or floor joists for the support or suspension of a ceiling.

ceiling strut A temporary brace suspended from an overhead structure or framing and used to hold door frames in place while the adjacent walls are being constructed.

ceiling suspension system A gridwork of metal rails and hangers erected for the support of a suspended ceiling and ceiling-mounted items, e.g., air diffusers, lights, fire detectors, etc.

ceiling switch A chain pull switch.

cellar A room or set of rooms below or predominately below grade, usually under a building.

cellar hole The excavation for a cellar.

cellular brick See **brick**.

cellular concrete, aerated concrete Concrete that has had gas-forming chemicals mixed with the basic ingredients, so that the final set material is lighter than ordinary concrete, due to its porosity. This type of concrete is often used as insulation when it is placed over a roof slab.

cellular construction Construction with concrete components that have been cast with voids for sound and thermal insulation and to decrease weight.

cellular-core door A hollow-core door that has the hollow filled with a lightweight, honeycomb-shaped, expanded material to provide rigidity and support.

cellular framing A method of framing in which the walls are composed of cells with the cross walls transmitting the bearing loads to the foundation.

cellular material A material that contains voids or air pockets.

cellular plastic Plastic molding or extrusion with a pattern of cells that give it rigidity and reduce weight.

cellular raceway A channel in a modular floor or wall that may be used as a raceway for electric conductors.

cellular striation A pattern in a material that contains an integral layer of cells or voids.

cellulose acetate Any of several compounds, insoluble in water, formed by the action of acetic acid, anhydride of acetic acid, and sulfuric acid on cellulose and used for making lacquers, varnishes, rayon, and packaging sheets.

celluloid A thermoplastic material often used in thin sheets because of its good molding properties.

cellulose A naturally occurring substance made up of glucose units and constituting the main ingredient in wood, hemp, and cotton. Cellulose is used in many construction products.

cellulose enamel A lacquer paint made with a cellulose base.

cellulose fiber tile A sound-absorbing acoustical tile made from cellulose fiber.

Celsius scale A thermometer in which the interval between the freezing point and boiling point of water is divided into one hundred parts or degrees. Named after Aners Celsius a Swedish astronomer. Also called a centigrade thermometer.

Celtic cross A cross formed from a long vertical column, a shorter horizontal bar, and a circle about their intersection.

cement Any chemical binder that makes bodies adhere to it or to each other, such as glue, paste, or portland cement.

Cement Mixer

Cement Mortar

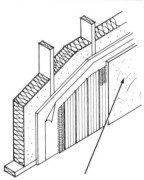

Cement Stucco

cement-aggregate ratio The ratio of cement to aggregate in a mixture, as determined by weight or volume.

cement-asbestos board A stiff, hard, non-combustible board fabricated from asbestos fibers impregnated and bonded with cement and used as roofing slates and wall shingles.

cement block An inaccurate term, usually an erroneous reference to concrete block or to a hollow or sold-cast masonry unit.

cement clinker A lump or ball of fused material, usually 1/8 to 1 inch in diameter, formed by heating cement slurry in a kiln. When cool, the clinker is interground with gypsum to form cement.

cement-coated nail A nail coated with cement to increase its friction and holding power.

cement content The quantity of cement per unit volume of concrete or mortar, expressed in pounds or bags per cubic yard.

cement factor Same as **cement content**.

cement fillet, weather fillet A mortar bead or caulk used to provide a watertight seal in a joint between roofing and a wall and used in place of flashing.

cement gravel Gravel bound together in nature by clay or some other natural binding agent.

cement grout (1) A thin, watery mortar or plaster that is capable of being pumped or forced into joints, cracks, and spaces as an adhesive sealer. (2) A mixture that is pumped into soils around foundations to firm them up and provide better load-bearing characteristics.

cement gun A tool used for spraying or injecting cement grout or mortar material.

cement mixer (1) A concrete mixer. (2) A container with the capability of mixing its contents (concrete ingredients) by means of paddles or a rotary motion. The container may be manually or power operated.

cement mortar A plastic building material made by mixing lime, cement, sand, and water. Cement mortar is used to bind masonry blocks together or to plaster over masonry.

cement paint, concrete paint A paint containing white portland cement and usually applied over masonry surfaces as a waterproofing.

cement paste A mixture of cement and water.

cement plaster Plaster containing portland cement as the binder and commonly used on exterior surfaces or in damp areas.

cement rendering A wash of portland cement and sand applied over a surface.

cement rock, cement stone A natural impure limestone containing the ingredients for the manufacture of portland cement.

cement screed See **screed**.

cement slurry A thin, watery mixture of cement for pumping or for use as a wash over a surface.

cement stucco (1) A mixture of portland cement, sand, and a small percentage of lime. The material thus produced is used to form a hard covering for exterior walls that are usually textured. (2) A fine plaster used for interior decorations and moldings.

cement temper The addition of portland cement to lime plaster to improve its strength and durability characteristics.

cement-wood floor The addition of sawdust to a mixture of portland cement and sand for

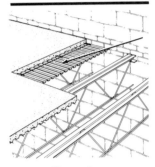

Centering

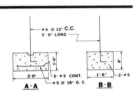

Center-to-Center

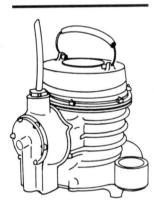

Centrifugal Pump

use as a poured floor.

cementation The firming and hardening of a cementicious material.

cementitious Capable of setting like a cement.

center (1) A point laying exactly halfway between two other points or surfaces. (2) A point equidistant from all points on the circumference of a circle. (3) The internal core of a built-up construction.

center bit A cutter used for boring holes in wood. The cutting end consists of a sharp threaded screw for guidance and pulling the tool deeper into the work piece. The bit also has a scorer for marking the outline of the hole and a lip for cutting away the wood inside the hole. The other end of the bit is usually a tapered rectangular block that is clamped in the chuck of a brace.

center-hung door A door that is supported by, and swings about, pivot pins fixed above and below it and inserted into or attached to the door on the center line of its thickness.

center-hung sash A window sash hung so that it rotates horizontally about hinge pins at its center. Upon opening, usually the top part of the window swings into the room and the bottom part projects outside the building.

centering (1) The temporary support for a masonry arch while it is being built. (2) The temporary forms for all supported concrete work.

center of gravity The location of a point in a body or shape about which all the parts of the body balance each other.

center of mass See **center of gravity**.

center pivot (1) The pins of a window sash that are centrally located in the frame so that the window rotates horizontally about

its center line. (2) The pivot pins of a door that support the door on the center line of its thickness.

center-pivoted door See **center-hung door**.

center punch A punch for making starter indentations in metal where a hole is to be drilled. The punch has a sharp conical point on one end and is hit by a hammer on its other end.

center rail The horizontal portion or member that separates the upper and lower sections of a recessed panel door.

center stringer The structural member of a stair system that supports the treads and risers at the midpoint of the treads.

center-to-center, on center The distance from the center line of one beam, part, or component of a structure to the corresponding center line of the next similar unit, e.g., center-to-center of rafters or joists.

centimeter (1) A metric unit of length equal to a hundredth part of a meter. (2) One inch equals 2.54 centimeters.

central-mixed concrete Concrete that is mixed in a stationary mixer and then transported to the location of its use, in contrast to transit mixed concrete.

centric load, concentric load A load applied perpendicular to the neutral axis of a beam or column. The load acts through the geometric center of gravity (centroid).

centrifugal compressor An air or gas compressor in which the compression is obtained by means of centrifugal force, i.e., the force away from the center of a rapidly spinning impeller.

centrifugal pump A pump that draws fluid into the center of a rapidly spinning impeller and uses the force of the spin (centrifugal force) to impart pressure and velocity to the fluid as it leaves the

Chain

periphery.

ceramic Items made of clay or similar materials that are baked in a kiln to a permanent hardness. Products include pottery, tiles, stoneware, etc.

ceramic coating A coating made of a nonmetallic mineral, such as clay, and compounded to provide the desired properties when baked or cured. A coating is usually applied to metals to provide heat- or abrasive-wear protection.

ceramic color glaze, ceramic glaze A material that is applied over a clay product and then baked to produce a hard, vitreous, glassy surface. The glaze may be transparent or colored.

ceramic-faced glass Glass that has had ceramic material fused to it during the heating and tempering process.

ceramic mosaic tile Small unglazed tiles commonly used for flooring and walls. Patterns of various colors and shapes are placed on sheets for ease in laying.

certificate for insurance A document issued by an insurance company or its agent stating the dates of coverage of insurance in effect along with the types and amounts of coverage for the insured.

certificate of occupancy A certificate issued by a government authority stating that a structure or a portion of a structure has been approved as complying with applicable laws, regulations, and codes and may be occupied and put to its intended use.

certified ballast Fluorescent lighting manufactured to meet standards set by the Certified Ballast Manufacturers Association.

Certified Ballast Manufacturers Association A trade organization of fluorescent lighting ballast manufacturers.

certified output rating The output capacity of a piece of equipment or machinery that is accepted as meeting the standards of a government regulatory body or trade organization.

cesspool A lined excavation in the ground which receives the discharge of a drainage system, especially from sinks and water closets, and is so designed as to retain the organic matter and solids, but to permit the liquids to seep through the bottom and sides.

chain (1) A unit of length used by land surveyors (Gunter's Chain): 7.92 inches = 1 link; 100 links = 66 feet = 4 rods = 1 chain; 80 chains = 1 mile. (2) An engineer's chain is equal to 100 feet. (3) A series of rings or links of metal connected to each other to form useful lengths.

chain block A grooved pulley or sheave suitable for chain. Typically, a chain hook is provided with a hook, eye, or strap by which it may be attached for hoisting or hauling.

chain bolt A spring bolt at the top of a door that is operated by pulling an attached chain.

chain course A bond course formed by masonry headers and held in place by cramps.

chain door fastener A chain that can be secured by a slide bolt between a door stile and a door jamb to allow the door to be opened slightly while remaining securely fastened.

chain fall An assembly of chain with blocks or pulleys for hoisting or pulling a heavy load.

chain hoist See **chain block**.

chaining The process of measuring a distance using a surveyors chain.

chaining pin A metal pin used for marking measurements on the ground.

chain intermittent fillet weld An interrupted fillet weld

Chain Link Fence

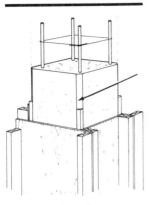

Chamfer

on both sides of a joint; the weld segments are opposite each other.

chain link fence A fence made of steel wire fabric that has been woven together and is supported by metal posts.

chain molding A molding carved to simulate the appearance of a chain.

chain-pipe vise A pipe vise that uses a chain to restrain a pipe.

chain pipe wrench, chain tongs A wrench for turning or restraining pipe. The wrench consists of a lever handle with a serrated jaw that is held in contact with the pipe by a roller chain that wraps around the pipe.

chain-pull switch An electrical switch that is operated by pulling a bead chain. Frequently, this type of switch is an integral part of a light bulb socket.

chain pump A pump in which an endless chain with disks mounted on it is pulled through a pipe to move viscous material.

chain riveting Riveting in which the rivets are placed in parallel adjacent rows. This type of riveting is not as strong as the staggered pattern, which has more tear material between rivets.

chain saw A hand-held saw powered by gasoline or electricity. With this type of saw, an endless chain, with cutting teeth, is driven around a bar extending from the power head.

chain scale An engineer's scale on which the inch divisions are subdivided into 10 parts or multiples of 10, e.g., 20, 30, 40, 50, etc.

chain timber Large beams built horizontally into a masonry wall to provide support and a tie-in during construction.

chair (1) A frame built into a wall to provide support for a sink, lavatory, urinal, or toilet. Also called a "carrier." (2) In concrete construction, a small metal

support for reinforcing steel. The support is used to maintain proper positioning during concrete placement.

chair rail A horizontal piece of wood attached to walls to prevent damage to plaster or paneling from the backs of chairs.

chalk A soft limestone, white, gray, or buff colored, composed of the remains of small marine organisms. In construction, chalk is commonly dyed blue and used for marking dimensions.

chalkboard A flat, slightly abrasive surface used as a marking surface for chalk.

chalkboard trim The frame and mounting hardware for a chalkboard.

chalked A term for porcelain enamel that has lost its hard glossy surface and is powdery when rubbed.

chalking A surface condition in which the binder of a coating breaks down so that the pigment comes off on anything touching it or is washed down by rain, i.e., a self-cleaning white paint.

chalk line (1) A light cord that has been rubbed with chalk (usually blue) for marking. (2) The line left by a chalked string.

chamber test A test developed and conducted by Underwriters' Laboratories to determine the fire-hazard classification of building materials.

chamfer The beveled edge formed at the right-angle corner of a construction member. See also **cant strip**.

chamfer bit A tool bit for beveling the upper edge of a hole, especially in wood, so that a flat-head wood screw can be driven flush with the wood surface.

chamfered rustication Masonry work in which the corner edge around the exposed face of each block is deeply chamfered, in addition to the joint mortar being

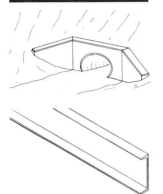

Channel

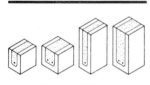

Channel Block

recessed, so that between two adjacent blocks an internal right angle is formed.

chamfer plane A carpenter's plane that has been designed for cutting chamfers.

chamfer stop Any ornamentation or construction piece that ends or blocks the continuation of a chamfer.

chamfer strip See **cant strip**.

chancel An area in the front of a church that is reserved for the use of the clergy and choir.

chancel aisle The side aisle of a chancel.

chancel arch An arch employed in some church designs to indicate the separation of the church body (nave) from the chancel or sanctuary.

chancel rail A railing or barrier (in place of a screen) that separates the church body (nave) from the chancel.

chancel screen The screen that separates the church body from the chancel.

chancery A room, suite of rooms, or building used as a court of record or office of records. A chancellor's court or office, or a secretariat.

chandelier A light fixture, often ornate, having several branches and hanging from a ceiling.

chandlery, chandry A place where lamps, candles, and other lighting supplies are kept.

change In construction, a deviation in the design or scope of the work as defined in the plans and specifications, which were the basis for the original contract.

change order A written order to a contractor with the necessary signatures to make it a legal document and authorizing a change from the original plans, specifications, or other contract documents, as well as a change in the cost.

changeover point That temperature and time at which a building requires neither heating nor cooling.

changes in the work Changes to the original design, specifications, or scope of work requested by the owner. All changes in the work should be documented on change orders.

channel (1) A structural steel member shaped like a U. (2) In glazing, a U-shaped member used to hold a pane or panel. (3) A watercourse, usually man-made. (4) The suspension system for a suspended ceiling.

channel bar Same as **channel iron**.

channel beam A construction member with a U-shaped cross section.

channel block A hollow concrete masonry block with a groove or furrow in it to provide a continuous recess for reinforcing steel and grout.

channel clip A metal spring clip used to support perforated metal pans in place in a suspended ceiling system.

channel glazing A window glazing system in which glass panels are set in a U-shaped channel using removable beads or glazing stops.

channeling Decorative fluting or grooving, as in the grooves in a vertical column.

channel iron, channel bar A U shaped rolled structural steel member, the web of which, is always deeper than the width of the flanges. The web depth is the size of the channel.

channel pipe A section of pipe with the top quarter or half removed. Channel pipe is used for directing flow in a manhole.

channel runner A primary horizontal support in a suspended ceiling system.

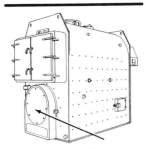

Charging Door

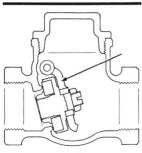

Check

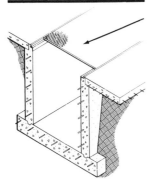

Checkered Plate

channel section A U-shaped section. See **channel**.

chantlate The piece of wood across the rafters at the eaves and projecting so as to prevent rain water from running straight down the wall.

charcoal filter (1) A filter containing pulverized charcoal for removing odors, smoke, and vapors from the air. (2) A filter for purifying and cleaning water.

charette The push or effort to complete an architectural problem within a specified time.

charge (1) A quantity of explosives set in place. (2) A load of refrigerant in a refrigeration or air conditioning system.

charging (1) Inserting refrigerant into a refrigeration or air conditioning system. (2) A predetermined mixture or quantity of materials being inserted into a concrete mixer, furnace, or other piece of processing equipment.

charging chute An enclosed slide in which waste material is dumped and subsequently fed into an incinerator.

charging door The door into a furnace or incinerator through which loads or charges are inserted.

Charpy test A measurement of material strength at impact. A specimen, usually notched, is struck by a swinging pendulum at a velocity of 17.5 feet per second. The energy absorbed in the fracture is then measured.

chase A continuous enclosure in a structure that acts as a housing for pipes, wiring conduits, ducts, etc. A chase is usually located in a wall or adjacent to a column, which provides some physical protection.

chase bonding The continuous vertical joint bond by which new masonry is joined to an older existing wall.

chase mortise, pulley mortise A mortise or recess cut into a beam or structural member that has one side elongated and sloped to allow the tenon to be rotated in place from the side, when space will not permit a direct perpendicular insertion.

chase wedge A hand tool that is wedge-shaped and is used for bossing and forming soft materials, such as lead and copper.

chattel Any item of property, moveable or immovable, except real estate or a freehold.

chattel mortgage A mortgage or security interest in personal property as collateral for a loan.

chatter marks The marks left on a work piece by a vibrating tool.

cheapener An additive to paint that may be for the purpose of imparting desirable characteristics or extending its coverage. Cheapener does not necessarily make the paint cheaper in cost.

check (1) A flapper or valve that permits fluid to flow in one direction only. (2) An attachment that regulates movement, i.e., a door check. (3) A small split running parallel to the grain in wood caused by shrinking during drying.

check cracks See **check def. 3**.

checkered plate A metal plate cast or rolled with a pattern of squared projections like a waffle.

check fillet A water-control barrier or curb used on roofs.

checking, alligatoring, check cracks, map cracks, shelling (1) Small cracks that appear in paint or varnish, but usually do not expose the base material. (2) Small, closely spaced irregular surface cracks in concrete, mortar, or plaster. Also called "crazing."

checking floor hinge A mechanism fixed on or in the floor that acts both as a door hinge and closing speed controller.

Check Rail

check lock A small supplementary lock whose purpose is to control the bolt or action of a larger lock.

check nut A second nut, usually thin, that when tightened on a bolt or shaft against the first nut prevents its vibrating or shaking loose. A locknut.

check rail The horizontal meeting rail of a double-hung window.

check stop A molding strip or bead used to restrain a sliding unit such as a window sash.

check strip A bead or molding used for parting.

check throat The groove cut longitudinally along the underside of window and door sills to prevent rainwater from running back to the building wall.

check valve, back-pressure valve, reflux valve A valve designed to limit the flow of fluid through it to one direction only.

cheek In general, the side of any feature, e.g., the side of an opening.

cheek boards The vertical stops or bulkheads placed at the ends of a concrete wall form.

cheek cut, side cut The oblique angular cut made in a rafter or jack rafter to permit a tight fit against a hip or valley rafter or to square the lower end of a jack rafter.

cheesiness The stage at which partially dried paint can be torn and shredded.

chemical bond A bond between similar materials caused by the interlocking of their crystalline structures. Once formed, there is no defined parting line in the bond that would permit a future clean separation.

chemical brown stain The brown-colored stain that appears during kiln or air drying of lumber due to changes in the concentrations of the wood's moisture components.

chemical closet A toilet that contains or recirculates a chemical that breaks wastes down while serving as a disinfectant and deodorant. It does not use water to flush and carry away sanitary waste to a public septic system.

chemical flux cutting The use of a chemical compound or flux to expedite oxygen torch cutting.

chemically foamed plastic Spongy or cellular plastic material that is formed by the gas producing action of its component chemical compounds.

chemical stabilization The forcing of chemical materials under pressure into soil to increase its bearing strength and other characteristics.

chemical toilet See **chemical closet.**

cherry picker A powered lift for raising men or materials, in a basket or on a lift, at the end of a telescoping boom. It may be either self- powered or mounted on a truck.

chert An impure flintlike rock, usually dark in color. Some of its common impurities react with cement making. Its use as an aggregate is undesirable for certain applications.

chevron (1) A V-shaped stripe that may point up or down and is used as a distinguishing mark to indicate rank or service. (2) Architecturally, it may be used as an ornamental repeating pattern on molding or trim.

chevron slat A V-shaped configuration used in fences and other architectural locations to provide privacy while permitting ventilation.

chicken ladder A board with wood strips or cleats nailed across it at regular intervals to provide safe footing across a void or on an inclined surface like a roof.

chilled-water refrigeration system A cooling system in which cold water is circulated to

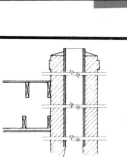

Chimney

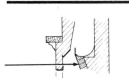

**Chimney Arch,
Chimney Bar**

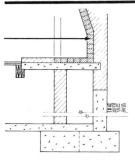

Chimney Back

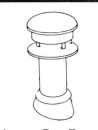

Chimney Cap, Bonnet

remote blower units.

chilling (1) The process of applying refrigeration moderately as to achieve cooling without freezing, i.e., the chilling of produce or meats. (2) In painting, the clouding of a surface or loss of shine due to cold air blowing over the drying surface.

chimney A vertical non-combustible structure with one or more flues to carry smoke and other gases of combustion into the atmosphere.

chimney apron The metal flashing built into the chimney and roof at the point of penetration as a moisture barrier.

chimney arch The upper structure over a fireplace.

chimney back The inside backwall of a fireplace, which may be made of fire brick masonry or ornamental metal, and is intended to reflect heat out into the room.

chimney bar, turning bar A lintel bar of iron or steel that bridges the fireplace opening, supported by the side masonry, and in turn supporting the masonry above the opening.

chimney block Solid concrete blocks cast as segments of a circle for laying up or building round flues.

chimney board A board or screen to cover or block the opening of a fireplace when it is not being used.

chimney bond A pattern of stretcher bond used in laying up the internal brickwork in masonry chimneys.

chimney breast, chimney piece That part of the front of a fireplace that projects out into the room.

chimney can A cylindrical extension that may be placed on top of a chimney to improve the draft. The can may be masonry or metal.

chimney cap, bonnet (1) The slab or masonry that is the top piece

on a chimney. (2) A metal top piece that is designed to minimize the entry of rain down the flue while encouraging good draft.

chimney cheek The masonry on either side of a fireplace opening that supports the mantel and the upper chimney construction.

chimney connector The smokepipe or metal breeching that connects the top of a stove, furnace, water heater, or boiler to a chimney flue.

chimney corner, inglenook, roofed ingle A small area or alcove beside a fireplace, frequently with built in benches.

chimney cowl An elaborate chimney pot to improve drafts.

chimney crane A swinging iron arm attached to the inside of a fireplace to support cooking pots over the fire.

chimney cricket A small false roof built behind a chimney on the main roof to divert rain water away from the chimney.

chimney crook, chimney hook An extension bar that hangs on the chimney crane and supports cooking pots, letting smaller pots hang lower and closer to the fire. It usually has adjustments for height.

chimney effect, flue effect, stack effect The process by which air and gases, when heated, become less dense and rise. The rising gases in a chimney create a draft that draws in cooler gases or air from below. Stairwells, elevator shafts, and chases in a building often draw in cold air from lower floors or outside through this same process.

chimney flue The vertical passageway in a chimney through which the hot gases flow. A chimney may contain one or several flues. Flues are typically lined with fired clay pipes to resist corrosion and facilitate cleaning.

chimney gutter The flashing used around a chimney for

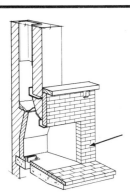

Chimney Jamb

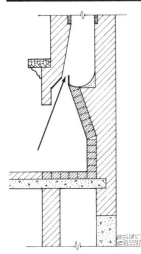

Chimney Throat

waterproofing the roof penetration.

chimney head See **chimney cap**.

chimney hood A covering top or cap that blocks rain from entering the flue.

chimney jamb The vertical sides of a fireplace opening.

chimney lining The non-combustible heat resisting material that lines the flue inside the chimney. Typically, the lining consists of round or rectangular fired clay material.

chimney piece Ornamentation around a fireplace, i.e., a mantelpiece.

chimney pot, chimney can A metal, masonry, or ceramic extension placed on top of a chimney to improve draft, appearance, or to block rain.

chimney shaft The segment of chimney that projects above the roof.

chimney stack (1) A very large and tall chimney used for utilities, commercial and industrial installations, and for large apartment houses. (2) There can be several flues in a single chimney.

chimney stalk See **chimney stack**.

chimney throat, chimney waist (1) The narrowest portion of a flue. (2) The area above a fireplace and below the smoke shelf where the damper is usually located.

chimney waist See **chimney throat**.

chimney wing The sides of a fireplace chimney above the opening that close in to the throat, or damper opening.

china sanitary ware Water closets, lavatories, or other items made of vitrified china with a glazed finish.

China white, Chinese white A

white paint pigment containing white lead.

china wood oil A drying oil used in paints and varnishes.

chinbeak molding A molding with a concave section followed by a convex section and forming an S curve.

Chinese blue A blue pigment for paint that is created from a cyanogen compound of iron.

Chinese lacquer, Japanese lacquer, lacquer A natural varnish, originating in China or Japan, and extracted from the sap of a sumac tree. This lacquer forms a high-quality surface.

Chinese white A dense white pigment of zinc oxide.

chink A small elongated cleft, rent, or fissure, e.g., the openings between logs in a log cabin wall.

chinking Any material used to fill in voids, such as those occurring in stone walls or between the timbers of a log cabin.

chip A small fragment of metal, wood, stone, or marble, etc. A chip may have been chopped, cut, or broken from its parent piece.

chip ax A hand tool used for chipping wood or stone to shape.

chipboard A flat panel manufactured to various thicknesses by bonding flakes of wood with a binder. Chipboard is an economical, strong material used for sheathing, sub-flooring, and cabinetry.

chip carving Ornamental work created by cutting away chips of material in various patterns.

chip cracks, eggshelling A process, similar to alligatoring or checking, except that the edges of the cracks are raised up from the base surface.

chipped grain On a wood surface, those depressions or voids occurring where the wood has been torn out as the result of dull planing or machining against the direction of growth.

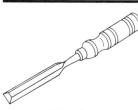

Chisel

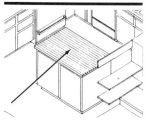

Chopping Block

Chord

chipper A small pneumatic or electric reciprocating tool for cutting or breaking concrete.

chipping The use of a hammer and cold chisel to remove weld spatter, rust, or old paint from iron work.

chisel A metal tool, with a cutting edge at one end, used in dressing, shaping, or working wood, stone, or metal. A chisel is usually tapped with a hammer or mallet.

chisel bar A heavy metal bar with a sharp edge on one end.

chisel knife A sharp, narrow square-edged knife used to remove paint or wallpaper.

chlorinated paraffin wax A thick viscous material used in flame-retardant paints.

chlorinated rubber A rubber-based product used in paints, plastics, and adhesives.

chock A wedge-shaped block used to prevent the movement of a wheeled vehicle or other object. Chocks are usually used in pairs.

chopping block A single slab of hardwood, or a composite of rectangular pieces that have been bonded together to form a kitchen work surface.

chord (1) The top or bottom members of a truss, typically horizontal, as distinguished from the web members. (2) A straight line between two points on a curve.

chrismatory A recess in a church that is used to hold a basin of consecrated oil for baptism, confirmation, ordination, etc.

chromate A greenish, yellow primer used to coat steel members in order to inhibit corrosion. Chromate is a salt of chromic acid, frequently lead or zinc chromate.

chrome green A green pigment produced by blending chromate salts and used to color paint. The pigment has rust-inhibiting properties.

chrome steel Carbon steel to which chromium, in proportions of 0.5% to 2.0%, has been added to increase the steel's hardness and yield strength.

chrome yellow, Leipzig yellow A yellow pigment consisting essentially of lead chromate and used to color paint. The pigment also has rust-inhibiting properties.

chromium A grayish-white metallic element that is hard, brittle, and resistant to corrosion. It is added to steel to enhance strength and hardness. It is also used as a plating medium because of the rust-resisting surface that results. A chromium-plated material produces a highly polished surface.

chromium plating The application, by electroplating, of a thin coat of chromium to provide a hard, protective wearing surface or a surface that can take a decorative high polish.

church stile An archaic term for "pulpit."

churn molding A molding with a repeating wavy or chevron pattern.

chute An inclined plane or trough down which various items or bulk materials may pass or slide to a lower level, propelled by gravity.

cima See **cyma.**

cimbia A decorating band or fillet encircling the shaft of a column.

cinder block A masonry block that is made of crushed cinders and portland cement. This type of block is lighter and has a higher insulating value than concrete. Because moisture causes deterioration of cinder block, they are used primarily for interior walls, rather than exterior walls.

cinder concrete A lightweight concrete that uses cinders for coarse aggregate.

cinders (1) A partly burned combustible material. (2) Slag from a metal furnace. (3) Ashes,

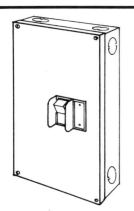

Circuit Breaker

Circular Stair, Spiral Stair

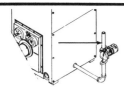

Circulation Pipe

especially from the burning of soft coal. (4) A form of volcanic lava.

circle trowel A trowel with a curved blade that is used for plastering a curved surface.

circline lamp A fluorescent lamp with a circular tube.

circuit (1) A path of conductors for an electric current. (2) A piping loop for a liquid or gas.

circuit breaker An electrical device for discontinuing current flow during an abnormal condition. By resetting a switch, a circuit breaker becomes reusable, unlike a fuse.

circuit vent A vent to prevent pressure buildup or to eliminate air in a closed circuit system, as in a heating system or a solar heat-exchanger system.

circular arch An arch whose inner surface, and occasionally a portion of its outer surface, describe a circle.

circular cutting and waste The cutting of material such as flooring, roofing, or decking to provide a curved surface or to go around a curved intersection. Also, a measure of the resulting waste material.

circular face A surface that in plan view is the outside or convex side of a circle.

circular mil An electrical term for the cross-sectional area of a wire. A conductor with a diameter of 1 mil (0.001 inch) has a cross-sectional area of 1 circular mil.

circular mil-foot A unit of measure for an electrical conductor that is 1 circular mil in cross-sectional area and 1 foot long.

circular plane A carpenter's wood planing tool with a curved baseplate for finishing curved surfaces.

circular saw A thin steel disk, with teeth on its periphery, that rotates on a power-driven spindle and is used either as a hand tool or is table mounted.

circular spike A metal timber connector with teeth set in a ring. The teeth grip the wood as a central axial bolt is tightened.

circular stair, spiral stair A stairway having a cylindrical staircase as a segment of a corkscrew.

circular sunk face A surface that in plan view is the inside or concave side of a circle; the opposite of a circular face.

circulating head, circulating pressure The pump-induced pressure in a piping system, such as one for hydronic heat, chilled water, or domestic hot water.

circulation (1) The flow of a liquid or a gas within a closed series of pipes. (2) The flow of air (fresh, heated, or cooled) in a building. (3) The flow of people through a building.

circulation pipe Any pipe forming a part of a liquid or gas circuit.

cissing, sissing A term describing a mild form of alligatoring, i.e., random surface cracks.

cistern A catch basin on a roof or an enlargement at the end of a gutter with an outlet into a cistern.

city plan A large-scale map of a city showing the streets and all important buildings, parks, and other features.

city planning, town planning, urban planning The conscious control of growth or change to be allowed to occur in a city, town, or community, taking into account esthetics, industry, utilities, transportation, and many other considerations that affect the quality of life.

civic center In an urban area, a group of government or public service buildings which could include such structures as a city hall, a library, a museum, an art

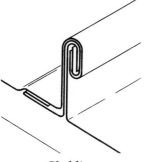

Cladding

Clapboard

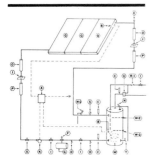

Clarification Drawing

center, a parking garage, a court house, etc.

civil engineer An engineer specializing in the design and construction of roads, buildings, dams, bridges, and other structures, as well as water distribution, drainage, and sanitary sewer systems.

C-label A door certified by Underwriters' Laboratories as meeting their class C level requirements.

clack valve A little-used term referring to a check valve, especially a swing check valve, that permits fluid flow in one direction only.

clad alloy A metal with a molecular-bonded surface coat to provide an improved wearing surface, a better appearance, or corrosion protection.

clad brazing sheet A sheet of metal, intended for use in a brazing operation, that has the brazing filler material pre-bonded to one or both sides.

cladding A covering or sheathing usually applied to provide desirable surface properties, such as durability, weathering, corrosion, or impact resistance. See **clad alloy**.

clamp A mechanical device commonly used to hold items together or firmly in place while other operations are being performed. The clamping force may be applied by screws, wedges, cams, or a pneumatic/hydraulic piston.

clamp brick A brick that was retained in a clamp while being fired in a kiln.

clamping plate A metal splice plate that is bolted across a joint in a wooden frame to increase its strength.

clamping screw A screw that in connection with wood or metal jaws may be used to provide a clamping force.

clamping time The time that a bonded joint must remain immobile in a clamp for the glue to set.

clamp nail A mechanical fastener designed to bridge a joint and pull the two pieces tightly together as it is driven in place.

clamshell (1) A bucket used on a derrick or crane for handling loose granular material. The bucket's two halves are hinged at the top, thus resembling a clam shell. (2) A wood molding with a cross section that resembles a clamshell.

clapboard, bevel siding, lap siding Narrow wood boards, thicker at one edge, used for the exterior covering of frame buildings. The boards are applied horizontally and overlapped.

clapboard gauge, siding gauge A spacing tool used to maintain a uniform overlap when applying clapboards.

clapper valve A swing check valve that permits fluid flow in one direction only.

clarification drawing An illustration, provided by an architect or engineer, to explain in more detail some area or item on the contract documents, or as a part of a job change order or modification.

Clarke beam A composite beam consisting of planks bolted with their flat sides together and then strengthened with a board nailed across the joint.

clasp nail See **cut nail**.

class (1) A group of items ranked together on the basis of common characteristics or requirements. (2) A division grouping or distinction based on grade or quality.

class A, B, C, D, E, Fire-resistance ratings applied to building components such as doors or windows. The term ''class'' also

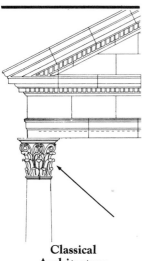

Classical Architecture

Clay Pipe

refers to the opening into which the door or window will be fitted.

class-A Underwriters' Laboratories classification for a component having a 3-hour fire endurance rating.

class-B Underwriters' Laboratories classification for an interior component having a 1-hour or a 1-1/2-hour fire endurance rating.

class-C Underwriters' Laboratories classification for an interior component having a 3/4-hour fire endurance rating.

class-D Underwriters' Laboratories classification for an exterior component having a 1-1/2-hour fire endurance rating.

class-E Underwriters' Laboratories classification for an exterior component having a 3/4-hour fire endurance rating.

Classical architecture Characteristic of or pertaining to the styles, types, and modes of structural construction and decoration practiced by the Greeks and Romans.

classicism A term referring to architecture that is derived from the basic principles and styles of Greco-Roman or Italian Renaissance buildings.

classroom window A wide window, the upper portion of which is a fixed light, while the lower portion has two or more tilt-in panels positioned side by side.

clause A subdivision of a paragraph or subparagraph, particularly within a legal document such as a contract document. A clause is usually numbered or lettered for easy reference.

clavel, clavis A voussoir or keystone of a arch.

clavis See **clavel**.

claw bar A steel bar that is straight at one end, with a chisel point, and bent at the other end, which has a notch for pulling nails. A claw bar is used for general demolition work.

claw chisel A hand tool with a notched edge for cutting stone.

claw hammer A hammer with a hardened face on one end of the head for driving nails, and curved forked tines on the other end for pulling nails out.

claw hatchet A hatchet on which one end of the head is cleft.

claw plate A round timber connector that has raised prongs to bite into the wood.

clay A fine-grained material, consisting mainly of hydrated silicates of aluminum, that is soft and cohesive when moist, but becomes hard when baked or fired. Clay is used to make bricks, tiles, pipe, earthenware, etc.

clay binder A material, consisting of fine clay particles, that has good holding or fixing properties in a soil mass.

clay cable cover A fired-clay half pipe that is placed over underground electric cables as a protective cover.

clay content In any mixture of soil or earth, the percentage of clay by weight.

clay-mortar mix A mixture consisting of pulverized clay that has been added to masonry mortar to act as a plasticizer.

clay pipe Pipe that is made of earthenware and glazed to eliminate porosity. Clay pipe is used for drainage systems and sanitary sewers.

clay puddle A mixture of clay and water used to prevent or retard the passage of water.

clay spade A wide, flat chisel-like cutter tool that is used in a pneumatic hammer to cut into tight, hard materials like clays.

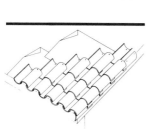

Clay Tile

Cleanout

clay tile A fired earthenware tile used on roofs. Also called "quarry tile" when used for flooring.

clean aggregate Sand or gravel aggregate that is free of clay, dirt, or organic material.

clean back The end of a bondstone that shows in masonry work.

cleaning eye An access plug, in a cleanout pipe fitting, used for inspection and cleaning purposes.

cleaning sash A window sash that is normally inoperable, but may be opened with a special tool.

cleanout (1) A pipe fitting with a removable threaded plug that permits inspection and cleaning of the run. (2) A door in the base of a chimney that permits access for cleaning. (3) A small door in a ventilation duct used to permit removal of grease, dust, and dirt blockages.

cleanout door The door in a cleanout frame. See **cleanout**.

clean room A special-purpose room that meets requirements for the absence of lint, dust, or other particulate matter. In a clean room, the filter systems are high efficiency and the air exchange is one-directional laminar flow.

clean timber Wood that is free of knots or other defects.

clear A distance between any two points without interruptions, obstructions, or discontinuities.

clearage See **clearance**.

clearance (1) The distance by which one item is separated or clear from another; the empty space between them. (2) An intentional gap left to allow for minor dimensional variations of a component or a part.

clearcole, clairecolle (1) A primer and sealer composed of water, glue, and a whitening pigment. (2) A clear coating used in the application of gold leaf.

clear glaze A transparent glaze for ceramic applications.

clearing The removal of trees, vegetation, or other obstructions from an area of land.

clearing arm A branch on a pipe to provide access for cleaning.

clear lumber, clean timber, clears, clear stuff, clear timber, free stuff Wood that is without knots, splits, blemishes, or other defects.

clear span The distance between the inside surfaces of the two supports of a structural member.

clearstory That part of a church or building which rises above the roofs of the other parts of the building with windows for admitting light to the interior central area.

cleat A small block of wood nailed on the surface of a wood member to stop or support another member.

cleat wiring Electrical wiring that is exposed and supported on standoff insulators.

cleavage A natural parting or splitting plane evidenced in slate, mica, and some types of wood.

cleavage plane In crystalline material, the plane along which splitting can be easily performed.

cleft timber A wooden beam that has been split to the desired size or shape.

clench See **clinch**.

clench bolt A bolt designed to have one end bent over to retain it in place.

clench nail A nail made to have the protruding point bent over after being driven.

clerestory See **clearstory**.

clerk of the works A representative of the architect or owner who oversees construction, handles administrative matters, and ensures that construction is in accordance with the contract documents.

Clevis

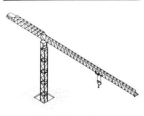

Climbing Crane

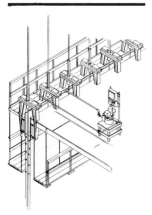

Climbing Form

clevis A U-shaped bar of metal with the ends drilled to receive a pin or bolt. A type of shackle.

climbing crane A hoisting device with a vertical portion that is attached to the structure of a building so that as the construction is built higher and higher, the crane rises along with it.

climbing form Vertical concrete formwork that is successively raised after each pour has hardened. The formwork is anchored to the concrete below. See also **slip form**.

clinch The process of securing a nail in place by bending the protruding point at a right angle.

clinch bolt See **clench bolt**.

clinch nail See **clench nail**.

clink (1) The sealed edge between sheets of metal roofing. (2) A pointed steel bar used to break up concrete.

clinker A vitrified residue of burnt coal sometimes used as the aggregate in cinder block. See also **cement clinker**.

clinker block Masonry block that contains ground cinders or clinkers as a lightweight aggregate. Same as **cinder block**.

clinker brick A hard brick that is frequently formed with a distorted shape and is used for paving.

clinometer An instrument for measuring angles of elevation or inclination.

clip (1) To cut off as with shears. (2) A short piece of brick that has been cut. (3) A small metal fastening device.

clip angle, lug angle A small piece of angle iron attached to a structural member to accept or retain some structural load.

clip bond A style of decorative masonry for walls in which the corners of stretchers are cut forming a V-notch to receive the corner of a diagonal header.

clip joint A thicker than usual mortar joint used to raise the height of a masonry course.

clipped gable A gable roof with a shortened ridge pole and the end peaks cut off to about midway down the gable end. The gable roof thus resembles a hip roof from the gable end intersection up to the ridge pole.

clipped header, false header In masonry, a half-brick placed in a wall to look like a header.

clipped lintel A lintel connected to another structural member that can help carry the load.

cloak rail A board attached to a wall containing pegs or hooks for hanging clothes.

cloakroom A room, usually near a main entrance, for the purpose of checking or leaving coats, hats, and other items of outer clothing.

cloister (1) A monastic establishment such as a monastery or convent. (2) A covered passage on the side of an open courtyard, usually having one side walled and the other an open arcade or colonnade.

cloistered arch A structure whose sides have the appearance of four quarter rounds fitted together so that the corner joints appear as an X when viewed from above.

cloistered vault See **cloistered arch**.

cloister garth An open courtyard surrounded by a cloister.

cloistral A term that refers to a cloister.

clone The reproduction of plant varieties by vegetative means, as by cuttings.

close (1) An enclosed place such as the area around a cathedral. (2) A passage from a street to a court

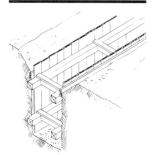

Close-coupled Tank and Bowl

Closed Sheathing

Closer

and the houses facing it. (3) The common stairway of an apartment building.

close-boarded, close-sheeted (1) Sheathing with boards tight against each other. (2) Privacy fencing in which adjacent boards are tight against each other.

close-contact glue A glue that requires a tight contact between the surfaces to be bonded.

close couple Opposite rafters fastened at the peak and strengthened by a horizontal tie part way down their lengths.

close-coupled tank and bowl A water closet that has a bowl and water tank that are manufactured separately and then bolted tightly together with only a gasket at the joint.

close-cut The trimming of a roof covering (shingle, tile, or slate) where adjacent surfaces of a hip or valley meet.

closed cell A term descriptive of a foamed material in which each cell is totally enclosed and separate so that the bulk of the material will not soak up liquid like a sponge. Closed cell material is characteristic of certain types of insulation board and flexible glazing gaskets.

closed-cell foam See **closed cell**.

closed-circuit grouting A system of injection grouting that fills all voids to be grouted at each hole and returns excess grout to the grouting apparatus.

closed cornice A cornice that projects only slightly beyond a vertical wall so that there is no soffit, the intersection being filled by a crown molding.

closed eaves Projecting roof rafters that are boxed in.

closed list of bidders A list of contractors that have been approved by the architect and owner as the only ones from whom bid prices will be accepted.

closed newel The continuous central or inside wall of a circular staircase.

closed shaft A vertical opening in a building that is covered over or roofed at the top.

closed sheathing A continuous structure with boards or planks placed side by side to support or retain the sides of an excavation or trench.

closed sheeting, closed sheathing, tight sheeting Same as **closed sheathing**.

closed shelving Shelving that can be concealed by closing a door or panel.

closed specifications Specifications stipulating the use of specific or proprietary products or processes without provision for substitution. See **base bid specifications.**

closed stair A stairway with covered strings on both sides so that the ends of the treads and risers are not visible. See also **box stair.**

closed stair string See **closed stair.**

closed string See **closed stair.**

closed system Any heating or cooling system in which the circulating medium is not expended, but is recirculated through the system.

closed valley A roof covering, in a roof valley, laid so that the flashing is not visible.

close-grained Wood that has its annual growth rings close together, thus indicating slow growth.

close nipple A short piece of pipe threaded from both ends and leaving no smooth outside surface.

closer (1) A mechanism for automatically closing a door. (2) The final masonry unit (brick, stone, or block) that completes a

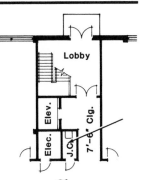

Closet

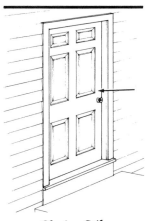

Closing Stile

horizontal course. The unit may be whole or trimmed to size.

closer mold A wooden guide or jig for cutting closer bricks. See **closer**, def. 2.

closer reinforcement A metal plate attached to a door and/or a frame as reinforcement where a door closer is installed.

closer reinforcing sleeve A plate to reinforce the corner joints of a metal doorframe.

close-sheeted See **close-boarded**.

closed sheeting See **closed sheathing**.

close string, close stringer, closed stringer, curb string, housed string A stair string that has concealed tread and riser supports on its inside surface. The top and bottom edges of the string are straight and parallel.

closet (1) A small room or recess for storing utensils or clothing. (2) A water closet.

closet bolt A special bolt for securing a floor-mounted water closet to the floor and soil pipe flange.

closet lining Thin cedar boards, with tongue-and-groove or lap edges, used to cover the inner surface of a closet in order to repel moths.

close tolerance A term describing smaller allowable dimension deviation than would normally be allowed.

closet pole, closet rod A strong rod mounted horizontally in a closet to support clothes hangers.

closet screw A screw used to hold a water closet floor flange to the floor.

closet stop valve The valve underneath a tank-type water closet. The valve is used to shut the water off so that repairs can be made to the flushing mechanism inside the tank.

close-up casement hinge A casement-window support hinge

with its pivot point placed so that when the window is open, there is a minimum space between the sash and the frame.

closing costs The costs associated with the sale/purchase of real estate property, such as legal fees, recording fees, and title search and insurance, but not including the cost of the property itself.

closing device, automatic closing device, self-closing device (1) A mechanism that closes a door automatically and slowly after it has been opened. (2) A mechanism, usually fused, that releases and closes a fire door, damper, or shutter in the event of a fire.

closing stile The vertical side of a door or casement sash that seats against the jamb or frame when the door or window is closed.

closure bar A flat metal bar that is attached to a stair string next to a wall. The bar bridges or covers the opening between the string and the wall.

clothes chute A vertical shaft used to drop dirty clothes and linens from one floor to another, especially in a hospital. Also called a "dirty linen chute."

cloudiness A haziness or lack of transparency in a protective film, as in a varnish or urethane.

clout (1) A metal sheathing or protection plate on a moving wood member subject to wear. (2) A type of nail.

clout nail A nail with a large flat head, round shank, and long flattened point used on sheet metal and dry wallboard.

clustered column Two or more columns, joined together at the top and/or bottom, which share the structural load equally.

clustered pier A pier consisting of several shafts grouped together as a single unit, frequently surrounding a heavier central member.

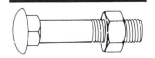

Coach Bolt

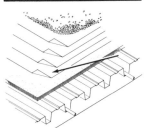

**Coated Base Sheet
Coated Base Felt**

Coat Rack

cluster housing A planned residential development with dwelling units closely spaced and common open space for the use of all residents.

clutch A friction or hydraulic coupling between a power source and a drive train.

coach bolt A bolt having a smooth circular head. Rotation of the bolt in its hole during tightening is prevented by the square shoulder or neck under the head.

coach house, carriage house A structure like a garage for the protection and storage of carriages.

coach screw A screw with a square head and a coarse pitched thread. See also **lag bolt**.

coak (1) A projection from the end of a piece of wood as a tenon for inserting into a mating hole in another piece to make a joint. (2) A wooden pin used to hold beams together.

coal-tar pitch, tar A black bituminous material made from the distillation of coal and used as a water proofing material on built-up roofing and around elements that protrude through a roof.

coaming A raised frame or curb around a roof hatchway or skylight to prevent water from flowing in.

coarse aggregate Aggregate that will not pass through a 1/4-inch sieve screen.

coarse filter A prefilter used in air conditioning systems to remove all large particles such as dust and lint. This type of filter is usually reusable after being vacuumed or washed.

coarse-grained (1) A term used to describe wood with the annual growth rings widely separated, thus indicating fast growth. (2) A pebbly effect on a photographic enlargement.

coarse stuff A thick plaster intended for use as a base coat and usually containing lime, sand, putty, and a fibrous material.

coarse-textured, coarse-grained, open-grained A term referring to a porous open-grained wood that usually requires priming and filling to produce a smooth surface.

coat A single covering layer of any type of material such as paint plaster, stain, or surface sealant.

coated base sheet, coated base felt The underlaying sheet of asphalt-impregnated felt used in built-up roofing.

coated nail A nail with a coating on it. The coating may be of cement, copper, zinc, or enamel.

coat rack A piece of furniture that supports several coat hangers and that may include a stand for umbrellas and a tray for galoshes.

coatroom (1) A room for checking or storing outer garments, such as coats and hats, and usually located near the main entrance of a building. (2) A cloakroom.

coaxial cable A cable consisting of a tube of conducting material surrounding a central conductor. The tube is separated from contact with the cable by insulation. Coaxial cable is used to transmit telephone, telegraph, television, and computer signals.

cob A lump or piece of anything, such as coal, ore, or stone.

cobble, cobblestone (1) A stone usually between 6 and 10 inches in diameter used for paving roads and for constructing foundations and walls. (2) A rough-cut, rectangular-shaped stone used for paving.

cob wall A wall formed of cob material, i.e., a mixture of clay, straw, and gravel.

cobwebbing In painting, the expulsion of a mass of fiber strands of dried or partially dried paint from a spray gun.

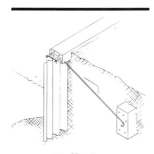

Cofferdam

cochlea (1) A spiral staircase. (2) The enclosure or tower for a spiral staircase.

cochleary, cochleated A spiral or helix twisting pattern as in a spiral- shelled snail.

cock (1) A valve having a hole in a tapered plug. The plug is rotated to provide a passageway for fluid. (2) Any control for the flow of liquid or gas through a pipe. See also **sill cock, bibcock**.

cock bead A bead that projects higher than the adjacent surface.

cocking The fitting together of timbers, beams, or pieces of wood that have been notched to form a joint.

cocking piece A board attached to rafters at the eaves of a roof to give a slight upward tilt in order to break the straight roof line.

cockle stair A circular or spiral staircase.

cockloft An upper attic or loft.

cockscomb A piece of metal with a serrated edge used to scratch plaster and create a rough gripping surface for the next coat.

cockspur fastener A casement window catch.

coctile Building units made in an oven, such as brick or porcelain.

code Regulations, ordinances, or statutory requirements of a governmental agency relating to building construction and occupancy. Codes are adopted and administered for the protection of the public health, safety, and welfare.

code of practice A publication of an agreement by any group of trades, or manufacturers of similar products, that lists minimum quality standards for material, workmanship, or conduct to be followed by group members.

coefficient of expansion For building materials, the increase in dimension per unit of length caused by a rise in temperature of 1 degree Fahrenheit, expressed in decimals.

coefficient of friction A ratio of the force causing an object to slide to the total force perpendicular to the sliding plane. The ratio is a measure of the resistance to sliding.

coefficient of heat transmission See **coefficient of thermal transmission**.

coefficient of performance (1) In a heat pump, the ratio of heat produced to the energy expended. (2) In a refrigeration system, the ratio of the heat removed to the energy expanded.

coefficient of static friction The coefficient of friction required to start an object sliding.

coefficient of thermal expansion See **coefficient of heat transmission**.

coefficient of thermal transmission The amount of heat passing through a partition per unit of area per degree difference in air temperature between the two sides of the partition. For example: BTU per square foot per hour per degree Fahrenheit.

coffer, lacunar (1) An ornamental recessed panel in the ceiling of a vault or dome. (2) A cofferdam, a caisson.

cofferdam A watertight enclosure from which water is pumped to expose an area or formation in order to permit access for construction or repairs.

coffering A ceiling with ornamental recessed panels.

cog (1) In a cogged joint, the portion of wood left where a beam has been notched. (2) A nib on roofing tiles.

cogged joint A joining of two crossing pieces of wood, each of which has been notched to fit into the other.

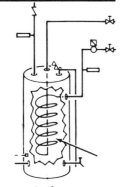

Coil

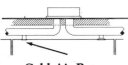

Cold-Air Return

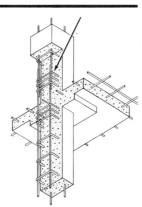

Cold Bending

cogging, cocking The assembly of two pieces of wood that have been notched or cogged.

cohesion (1) In soils, the sticking together of particles. (2) In adhesives or sealants, the ability of sticking together without cracking.

cohensionless soil A soil, such as clean sand, with no cohesive properties.

cohesive failure The internal shearing of a joint sealant that is weaker cohesively than the adhesion bond to the adjacent joint surfaces. The possible causes for failure include insufficient curing, excessive joint movement, improper joint dimensions, or improper selection of the sealant itself.

cohesive soil Soil that tends to hold together in a comparatively stable clump or mass, such as clay.

coil A term applied to a heat exchanger using connected pipes or tubing in rows, layers, or windings, as in steam heating, water heating, and refrigeration condensers and evaporators.

coiled expansion loop A gradual bend, loop, or coil of pipe or tube intended to absorb expansion and contraction caused by temperature changes. By flexing like a spring, undesireable stresses on the piping system are relieved or minimized.

cold-air return In a heating system, the return air duct that transports cool air back into the system to be heated.

cold bending, cold gagging The bending of any material without the application of heat to soften it. Some ductile materials can be bent in this manner without fracturing.

cold-cathode lamp An electric-discharge lamp that operates with low current, high voltage, and low temperature. The color of the light depends upon the vapor used in the lamp.

cold cellar An underground storage area where fruit and vegetables are stored during the winter. The constant ground temperature keeps the contents from freezing and retards rotting.

cold check Fine cracks that appear in some wood finishes, such as paint and varnish, after being subjected to alternating warm and cold temperatures.

cold chisel A steel chisel with a tempered edge for cutting cold steel.

cold cut, cold cutter A cold chisel with an offset handle like a hammer for holding it during striking with a maul.

cold-drawn A process of metal forming in which metal is formed in shape and size by being pulled through dies while cold. This type of metal forming has an effect on final surface finish and hardness and is often used in the manufacture of wire, tubing, and rods.

cold-driven rivet A rivet that is driven and formed without being preheated.

cold-finished bar A metal bar that has been drawn or rolled to its final dimensions while cold. The process of cold finishing produces a hard smooth surface finish.

cold-finished steel Any steel shape that is rolled, drawn, or forged to its final shape and size when cold. The resulting finish is hard and smooth.

cold flow The continuing permanent change in shape of a structural system under constant load.

cold glue A glue that flows and bonds without being heated.

cold joint The joint between two consecutive pours of concrete if the time elapsed between the first and second pours is such that the first pour has started to harden.

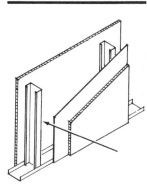

Cold-rolled

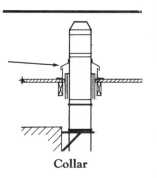

Collar

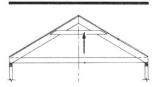

Collar Tie

cold-laid mixture Any mixture that may spread at ambient temperatures without being pre-heated.

cold mix An asphaltic concrete made with slow curing asphalts and used primarily as a temporary patching material when hot mix plants are closed. Cold mix is used to repair pot holes, but is less durable than hot mix.

cold molding (1) Material that is shaped at ambient temperature and hardened by baking. (2) The material that is used in the cold-molding process.

cold patch See **cold mix**.

cold pie Mortar left after laying a masonry unit.

cold-process roofing Built-up roofing consisting of layers of asphalt- impregnated felts that are bonded and sealed with a cold application of asphalt-based roof cement.

cold riveting Riveting done without the rivets being preheated.

cold-rolled Descriptive of a metal shape that has been formed by rolling at room temperature, thus giving a dense smooth surface finish and high tensile strength.

cold room A walk-in refrigerator for storage of perishable materials.

cold saw A saw for cutting metal at room temperature. A cold saw may be either circular or reciprocating.

cold set A hardened hand tool used to flatten sheet metal seams.

cold-setting adhesive A glue that sets at ambient room temperature.

cold-solder A soft-metal, air-hardening material used for bonding two metals at ambient temperatures. Cold solder is not a true solder, which would require a hot bond.

cold-solder joint An electrical wiring joint in which the metals to be joined have not been heated enough for the solder to bond.

cold-start lamp A fluorescent lamp that is started with high voltage without having to preheat the electrodes.

cold-water paint Paint with the pigment and binder mixed in cold water.

cold welding The process of fusing together two pieces of metal that have had their common faces well cleaned and then held in contact under high pressure. No heat is applied in cold welding.

cold-worked steel Steel that has been rolled, drawn, or formed to achieve permanent deformation at room temperature.

collar (1) A flashing for a metal vent or chimney where it passes through a roof. (2) A trim piece to cover the hole where a vent goes through a wall or ceiling. (3) A metal band that encircles a metal or wooden shaft.

collar beam The horizontal board that joins the approximate midpoints of two opposite rafters in order to increase rigidity.

collar beam roof, collar roof A roof with rafters connected by collar beams.

collaring The pointing of masonry or tile joints under overhangs.

collar joint See **collar beam**.

collar tie A beam that prevents wood roof framing from shifting.

collected plants Plantings that have been gathered from locations other than nursery stock.

colloid Any substance divided into fine particles that range in diameter from 0.2 to .005 microns. When the substance is mixed in a liquid, it remains in suspension.

colloidal concrete Concrete that is a mix of aggregate and colloidal grout.

Colonial Architecture

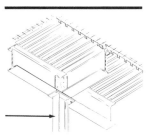

Column

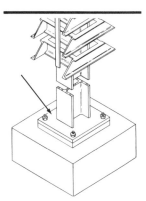

Column Baseplate

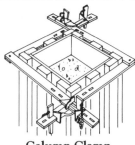

Column Clamp

colloidal grout A mortar mix grout whose solid particles are in a fixed suspension.

colloidal mixer A mixing machine for making colloidal grout.

collusion A secret agreement or action of a fraudulent, illegal, or deceitful nature.

colombage Timber construction using half timbers, i.e., members measuring not less than 5 inches by 10 inches in cross section.

colonial architecture (1) An architectural style from a mother country that has been incorporated into the buildings of settlements or colonies in distant locations. (2) Eighteenth century English Georgian architecture as reproduced in the original colonies of the United States.

colonial casing A style of door and window trim molding.

colonial panel door A type of door with recessed panels.

colonial revival An architectural style popular during the end of the nineteenth century in the United States. The style copied designs and details from eighteenth century English Georgian colonial buildings.

colonial siding Exterior siding material consisting of wide, square-edged boards applied as clapboards.

colonnade A row of evenly spaced columns that usually support one side or end of a roof.

color chart A standardized card or folder showing a display of colors or finishes.

colored aggregate Aggregate material that has been selected for use because of its natural color.

colored cement Cement containing a colored pigment.

colored concrete Concrete mixed with colored cement or to which pigment has been added during mixing.

colored finishes The addition of a pigment or colored aggregate to plaster for a final coat so that the color goes through the full depth of the coat.

color pigment Any substance that can be mixed with a suitable liquid, in which it is relatively insoluble, to color paints or stains.

color retention The capability of a surface finish to maintain its original shade and intensity over a period of time and or exposure to sunlight or weather.

column A type of supporting pillar that is long and relatively slender. A column is usually loaded axially in compression.

column baseplate A bearing plate beneath a column that distributes the load coming down through the column over a wider area.

column capital The enlarged cross sectional area at the top of a column.

column clamp A latching device for holding the sections of a concrete- column form together while the concrete is being placed.

column footing The foundation under a column that spreads the loads out to an area large enough so that the bearing capacity of the soil is not exceeded and differential settling does not occur.

column head Same as **column capital**.

comb A tool used to produce a rough surface or pattern in plaster or concrete.

comb board A board, with notches on its upper edge, that is used to cover the joint at the ridge of a pitched roof.

comb cut A vertical cut similar to the cut on the end of a rafter where it is joined to a ridgeboard.

combed Descriptive of a surface that has been raked with a comb to produce a rough or patterned finish.

Combination Door

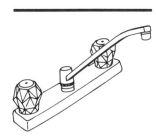

Combination Faucet

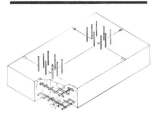

Combined Footing

combed-finish tile Tile with a serrated rear face to provide a stronger grip with the bonding mastic.

combed joint A joint made by fitting the ends of two pieces of wood together that have been cut with a deep V-pattern, thus giving the joint a zig-zag appearance.

comb-grained A piece of wood that has been sawn so that the growth rings are at an angle of 45 degrees with the wide face viewed from the end.

combination column A column of structural steel encased in concrete and designed so that each material carries a portion of the total load.

combination door An exterior door that has interchangeable panels of glazing and screening, one for summer and the other for winter use.

combination faucet A faucet that is connected to both hot- and cold-water supply pipes so that the water dispensed can be of the desired temperature.

combination fixture (1) A combination kitchen and laundry sink in a single unit, with both deep and shallow bowls side by side. (2) An institutional or prison fixture that combines the functions of a lavatory and a water closet into a single fixture.

combination plane A carpenter's or joiner's hand tool that has interchangeable blades or guides so that it can be used for various shaping operations.

combination pliers Any of several styles of pliers that can be used for cutting, swaging, or hammering, as well as for gripping.

combination square A carpenter's hand tool with a head that slides along a metal rule and may be locked at any location. The head has 90- and 45-degree angle surfaces, a bubble level, and a scribe pin.

combination window A window with replaceable screen and glass inserts for summer and winter use.

combined aggregate A mixture of both fine and coarse aggregate for concrete.

combined-aggregate grading The proportions of fine and coarse aggregate in a mix design.

combined footing A footing that receives loading from more than one column or load-supporting element.

combined frame A doorway with a fixed panels of glass on either or both sides of the door.

combined load The total of all the loads (live, dead, wind, snow, etc.) on a structure.

combined sewer A sewer that receives both storm water and sewage.

combined stresses A condition of stress loading that cannot be represented by a single resultant stress.

combing (1) The act of dressing a stone surface. (2) The use of a serrated comb or brush to create a surface pattern or roughness in paint, plaster, or concrete. (3) Those shingles on a roof that project above the ridge line. (4) The top ridge on a roof.

combplate The fixed floor plate at either end of an escalator or moving walkway that has fingers that project into the serrations of the moving segments to prevent an object from wedging into the seam.

combustible Descriptive of a substance that will burn in the air, pressure, and temperature conditions of a burning building.

come-along A portable ratcheting winch consisting of a fastening device on one end and a hook attached to a cable at the other

Common Bond

Common Joist

Common Nail

end. Operating a lever turns a drum which takes up the cable and exerts a pulling force on the hook.

comfort chart A chart showing the limits of human comfort for various combinations of temperature, relative humidity, and circulating air velocity.

comfort station A building or area with toilet and lavatory fixtures accessible for public use.

comfort zone The range of effective temperatures over which a majority of adults will feel comfortable.

commercial bronze A corrosion-resistant weatherstripping material made from an alloy of copper and zinc.

commercial projected window A type of steel window used in commercial and industrial applications where decorative trims and moldings are not used. A projected window has one or more panels that rotate either inward or outward on a horizontal axis.

commercial tolerances Allowable variations in the dimensions of items mass produced for the commercial market.

commode step On a staircase, the bottom step or steps that curve out around the newel post and are wider than the other steps.

common area An area in a building, or adjacent to a building or group of buildings, for the use of tennants, such as a courtyard, entry area, lounge, etc.

common ashlar A rough dressed block of stone masonry.

common bond, American bond A brick masonry wall pattern in which a header course of brick is laid after every five or six stretchers courses.

common brass A copper-zinc alloy of which 30 to 40 percent is zinc. It is relatively inexpensive, very ductile, and readily worked.

common brick Brick not selected for color or texture, and thus useful as filler or backing. Though usually not less durable or of lower quality than face brick, common brick typically costs less. Greater dimensional variations are also permitted.

common dovetail, box dovetail, through dovetail A corner joint of interlacing wedge-shaped fingers of wood. The ends of each finger show through at the edge.

common ground A strip of wood recessed in a masonry or concrete wall to provide a surface to which something else can be nailed.

common joist, bridging joist One of a series of parallel beams laid on edge to support a floor or ceiling.

common lap In roofing, an offset of one-half a shingle width for alternate rows of shingles.

common lime Hydrated lime or quicklime as used in plaster or mortar.

common nail A general-use headed nail with a diamond shaped point that is used where appearance is not important, as in framing.

common rafter, intermediate rafter, spar A wood framing member that reaches from the rafter plate at the eaves to the ridgeboard.

common vent A plumbing vent that connects at the junction of two fixture drains and serves as a vent for both fixtures. Also known as a "dual vent."

common wall (1) A wall built on the boundary between separate lots of land so that it does not belong completely to either owner. (2) A party wall between dwellings in a condominium or row house.

communicating frame A frame for two side-by-side, single-swing doors that open in opposite directions.

Compaction - Concrete Vibrator

Compactor

Compass

compacted volume The volume of any mass of material after it has been compressed, particularly soil in an embankment or fill area.

compacting factor The ratio of the weight of a concrete that has been gravity filled in a cylinder to the weight of compacted concrete in a similar cylinder.

compaction (1) The compression of any material into a smaller volume; for example, waste compaction. (2) The elimination of voids in construction materials, as in concrete, plaster, or soil by vibration, tamping, rolling, or some other method or combination of methods. In specifying compaction of embankment or fill, a percent compaction at optimum moisture content is often used. Another method is to specify the equipment, height of lift, and number of passes.

compactor A machine that compresses or compacts materials, usually using hydraulic force, weight, or vibration.

companion flange A flat, plate-like pipe connection device that is drilled and machined to match other flanges on pipe of fittings.

compartment One of the parts, divisions, or sections into which an enclosed space is divided.

compartment ceiling A ceiling that has been divided into decorative panels, usually surrounded by moldings.

compass (1) An instrument for locating the magnetic north pole of the earth. (2) A mechanical drawing device used to draw circles.

compass brick A curved brick used in circular patterns or to form an arch.

compass-headed arch A semicircular arch.

compass plane A carpenter's plane with a curved baseplate and/or blade for finishing curved woodpieces.

compass rafter A rafter with one or both sides curved.

compass roof A roof built of curved rafters.

compass saw A small tapering handsaw used to cut rounded shapes or circles. Sometimes called a "keyhole saw."

compass survey A surveyor's traverse using bearings determined by a compass, as opposed to bearings determined by recording the angles turned by the instrument.

compass window A window that projects outward from the face of a wall. Similar to a bow window.

compass work, circular work Carpentry or joinery based on a circular or curved motif.

compensation (1) Payment for services rendered or products or materials furnished or delivered. (2) Payment in satisfaction of claims for damages suffered.

completed operations insurance Liability insurance coverage for injuries to persons or damage to property occurring after an operation or work is completed. This type of coverage does not apply to damage to the completed work itself.

complete fusion In welding, a total merging of the welded surfaces, including all the weld filler material.

completion bond, construction bond, contract bond The guarantee by a bonding company that a job contract will be completed and will be clear of all liens and encumbrances.

completion date The date certified by the architect when the work in a building, or a designated portion thereof, is sufficiently complete, in accordance with the contract documents, so that the owner can occupy the work or designated

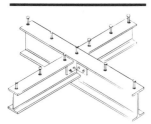

Composite Beam
Composite Girder

Composite
Construction

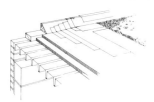

Composition Roofing

portion thereof for the use for which it is intended.

completion list The final list of items of work to be completed or corrected by the contractor. Sometimes called a "punch list".

compo A slang term referring to any type of composition material.

composite arch An arch formed from three or four centers of curvature.

composite beam A beam combining different materials to work as a single unit such as structural steel and concrete or cast-in-place and precast concrete.

composite board A fabricated board that can be manufactured from many different materials. Composite board may be used for sheathing, insulation, or as an acoustical barrier.

composite column A concrete compression member-reinforced longitudinally with structural steel shapes, pipe or tubing with or without longitudinal reinforcing bars.

composite construction A building constructed of various different building materials, or by using more than one construction method, such as a structural steel frame with a precast roof, or a masonry structure with a laminated wood beam roof.

composite door A door with a wooden or metal shell over a lightweight core that is usually made of foam or of a material that is shaped like a honeycomb cell.

composite fire door A door made of metal or chemically treated wood surrounding a core material. The door is rated as resistant to fire for some specified period of time.

composite girder A girder assembled from component parts of similar or differing materials; for example, a plate girder or a bar joist with wooden flanges.

composite joint (1) A joint made by using more than one method of securing it, such as riveting and welding. The loads through each method are indeterminate. (2) In plumbing, a general reference to joining bell and spigot pipe with packing materials such as rope and rosin, or cement and hemp.

composite pile A pile consisting of more than one member or material, but designed to act as one unit.

composite truss A load-supporting framework of several members. Those members in compression are made of wood; the tension members are usually made of steel.

composition board A manufactured board consisting of any of several materials usually pressed together with a binder. Composition board is frequently used as sheathing, wall board, or as an insulation or acoustical barrier.

composition roofing A roof consisting of several layers, thicknesses, or pieces. Also called "built-up roofing."

composition shingles Shingles made from felts impregnated and covered with asphalt, and then coated with colored granules on the exposed side. Also called "asphalt shingles."

compost A mixture of organic materials that have decomposed into a humus or earthy type material that is beneficial as a lightener and fertilizer when added to soil.

compound arch A built-up arch, or one having the appearance of several concentric arches.

compound beam, built-up beam A beam consisting of smaller components that have been assembled and fastened together to function as a single unit.

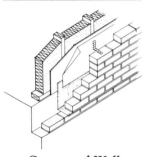

Compound Wall

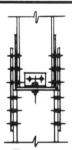

Compression Bearing Joint

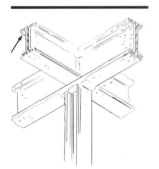

Compression Flange

compound pier, compound pillar A column usually consisting of one central shaft surrounded by secondary shafts, or by two or more equal shafts working in conjunction.

compound wall A wall made of more than one material.

compregnated wood, resin-treated wood A process in which wood is impregnated with a resin under pressure and then cured. The process improves the strength and rot resistance of the cured wood.

comprehensive general liability insurance A broad form of liability insurance covering claims for bodily injury and property damage which combines under one policy coverage for all liability exposures (except those specifically excluded) on a blanket basis and automatically covers new and unknown hazards that may develop. Comprehensive general liability insurance automatically includes contractual liability coverage for certain types of contracts. Products liability, completed operations liability and broader contractual liability coverages are available on an optional basis. This policy may also be written to include automobile liability.

comprehensive services Professional services performed by an architect, including traditional services and such other services as project analysis, programming, land use studies, feasibility investigations, financing, construction management, and special consulting services.

compressed cork Cork that has been ground up and then pressed with a binder into the desired shape or sheet. Compressed cork is used as a vibration or sound absorber, as a wall covering, and for many other commercial and industrial applications.

compressed fiberboard Wood fiber that has been compressed into sheets and is usually used as a paneling material.

compressed straw slab A board manufactured from straw compressed with a binder and cured.

compressed wood, densified wood Wood that has been impregnated with resin or other materials and then compressed under high pressure to increase its strength, density, and rot resistance. Also called compregnated wood and resin-treated wood.

compression Structurally, the force that pushes together or crushes, as opposed to the force that pulls apart, as in tension.

compression bearing joint A joint consisting of a structural member in compression, pressing or bearing against another member also in compression.

compression flange The upper flange of a beam. If loaded as a simple span, the top flange will be in compression, and the bottom flange will be in tension.

compression molding A method of molding plastic in which the plastic material is injected into a metal mold under heat and pressure.

compression set (1) The permanent shortened or flattened condition of any material that has been compressed beyond its yield point, and thus cannot return to its original size and shape. (2) The permanent imprint in a compression gasket.

compression test (1) A test of any material by compression to determine its useful strength characteristics. The test is most often performed in accordance with standards published by the American Society for Testing

115

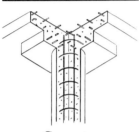

Compressor

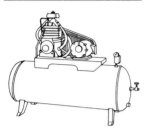

Concrete

and Materials, or some other professional or industrial organization. (2) A determination of the compression developed in each cylinder of an internal combustion engine.

compressive strength The resistance capacity of any material, but especially structural members, to crushing force. Compressive strength is usually expressed as the maximum number of pounds per square inch that can be resisted without failure.

compressive stress The resistance of a material to an external pushing or shortening force.

compressor (1) A machine that compresses air or gases. (2) In refrigeration/air conditioning, a machine that compresses a refrigerant gas, which then goes to an evaporator.

concave joint A masonry joint formed by a curved pointing tool and used particularly on exterior masonry walls.

concentric tendon One of several steel cables or rods that run through the center of gravity of a prestressed concrete member, thus placing it in compression.

concourse An open area for the circulation of large crowds within a building, as in an airport terminal or shopping mall.

concrete A composite material which consists essentially of a binding medium within which are embedded particles or fragments of aggregate; in portland cement concrete, the binder is a mixture of portland cement and water.

admixture A special substance or chemical added to a concrete mix. Typically, an admixture is used to control setting, entrain air, impart color, control workability, or to waterproof.

aerated Concrete that has had a gas forming chemical added to it

so that when it sets it contains many air holes and is light weight. See *concrete, foamed*.

aggregate Granular mineral material that is mixed with cement and water to form concrete or mortar.

air entrained Concrete containing minute air bubbles that improve its workability and frost resistance.

block A masonry building unit of concrete that has been cast into a standard shape, size, and style.

bond, bond plaster A thin gypsum plaster coat applied to a concrete face as a base for subsequent coats of plaster.

collar, doughnut A ring of reinforced concrete cast around an existing column and used to jack up the column.

contraction The shrinkage of concrete that occurs as it cures and dries.

curing compound A chemical applied to the surface of fresh concrete to minimize the loss of moisture during the first stages of setting and hardening.

cyclopean A mass concrete placement into which large stones weighing 100 pounds or more are placed.

cylinder test A compression test for concrete strength. Wet samples of concrete are carefully placed in especially made containers, 6 inches in diameter and 12 inches high. The cylinders are sent to a laboratory where a compression test is performed. This is done by putting the concrete in a hydraulic machine that measures the pressure(s) needed to crush it. Cylinder tests are usually performed for each pour on a project that requires concrete strength control.

dry-packed The forceful injection of a damp concrete mixture into

Finishing Machine

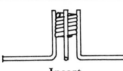

Insert

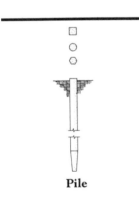

Pile

Pump

a confined area to provide a solid low-shrinkage fill.

fibrous Concrete in which fibrous material, such as glass fibers, has been mixed to improve tensile strength while reducing weight.

finish The smoothness, texture, or hardness of a concrete surface. Floors are trowelled with steel blades to compress the surface into a dense protective coat. Walls that are exposed to the weather are often ground with a carborundum stone or wheel, cement being added to fill the small voids. A smooth surface is desired in order that water cannot enter the small holes, freeze, and deteriorate the surface.

finishing machine (1) A portable machine with large paddles like fan blades used to float and finish concrete floors and slabs. (2) A large power-driven machine mounted on wheels that ride on steel pavement forms used to finish concrete pavements.

floor hardener An additive used to impart extra wear and chip resistance to concrete floors. The additive may be placed in the mix before the floor is cast, or it may be applied to the surface in liquid or granular form.

foamed Concrete to which a chemical foaming agent has been added. The result is a light, porous material, full of air holes, with low strength and good thermal properties.

gun A hand-controlled tool at the end of a hose for injecting or spraying concrete by pneumatic pressure.

insert A device such as a pipe sleeve, a threaded bolt, or a nailing block that is attached to a concrete form before placing concrete. When the concrete forms are removed, the insert

remains embedded in the concrete.

in-situ, cast in place Concrete cast in its final location in the finished structure.

masonry (1) Concrete blocks laid with mortar or grout in a manner similar to bricks. (2) Concrete that may be poured in place or as special tilt-up building walls.

mixer, cement mixer A machine that mixes cement, aggregate, and water to make concrete. The components are loaded into a rotating drum.

nail A steel nail with a diamond point that has been hardened so that it can be driven into concrete without bending.

normal weight Concrete made with the standard mix of component materials and weighing approximately 148 pounds per cubic feet.

pile A slender, precast concrete-reinforced member, either prestressed or not prestressed, that is embedded in the soil by driving, jetting, or insertion into a predrilled hole.

planer A machine with cutters or grinders used to level and refinish old concrete pavement.

plank A hollow-core or solid flat beam used for floor or roof decking. Concrete planks are usually precast and prestressed.

preplaced-aggregate Coarse aggregate placed in a form, with portland cement grout injected later.

prestressed Concrete members with internal tendons that have been tensioned to put a compressive load on the members. When a load is applied to a prestressed member, compression is decreased where tension would normally occur.

pump A pump that forces premixed concrete through a

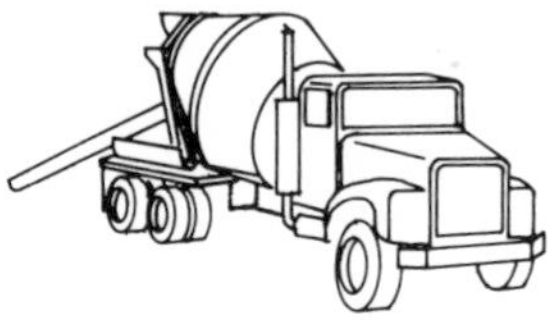

Ready-mixed

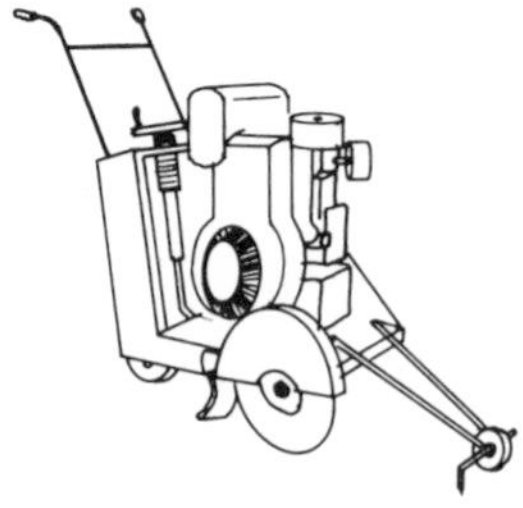

Saw

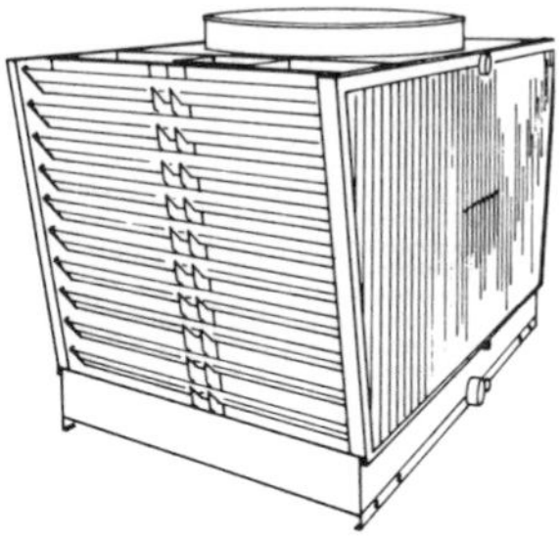

Condenser

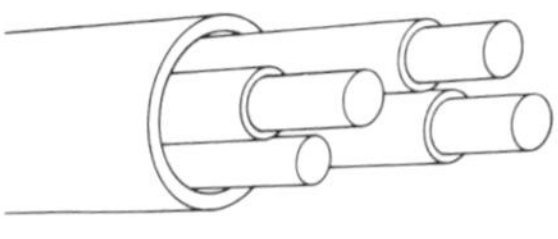

Conductor

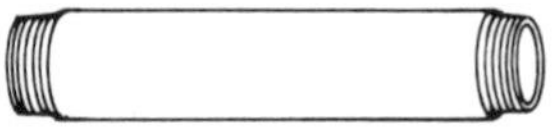

Conduit

hose to a desired location.

ready-mixed Concrete delivered to a site all mixed and ready for placement. Also called "transit-mixed".

reinforcement Metal bars, rods, or wires placed within formwork before concrete is added. The concrete and the reinforcement are designed to act as a single unit in resisting forces.

saw A power saw used, for example, to cut concrete in order to remove damaged sections of pavement or to groove the surface to create a control joint.

slump test A test to determine the plasticity of concrete. A sample of wet concrete is placed in a cone-shaped container 12 inches high. The cone is removed by slowly pulling it upward. If the concrete flattens out into a pile only 4 inches high, it is said to have an 8-inch slump. This test is done on the job site. If more water is added to the concrete mix, the strength of the concrete decreases and the slump increases.

condemnation (1) A declaration by the appropriate local governing authority that a structure is no longer fit or safe to use. (2) The seizure of private property by a governmental authority in order to use that property for public purposes, as in the exercise of eminent domain.

condensation The conversion of moisture in the air to water, as in the formation of water droplets on the outside of cold pipes.

condenser In a refrigeration system, the heat exchanger that removes heat from the refrigerant and transforms a hot gas under pressure into a cool liquid.

condominium A multifamily dwelling in which each apartment is individually owned and ownership of common areas and facilities is shared.

conduction See **thermal conduction.**

conductive flooring Flooring designed to prevent electrostatic buildup and sparking. Typical uses include computer areas and hospital operating rooms.

conductive mortar A tile mortar to which electric-conductive material has been added.

conductor (1) A pipe that leads rainwater to a drain. (2) A material that transfers heat. (3) A wire, bar, or device that has a low resistance to the flow of an electric current.

conduit (1) A pipe, tube, or channel used to direct the flow of a fluid. (2) A pipe or tube used to enclose electric wires to protect them from damage.

configurated glass, figured glass Glass that has a surface pattern impressed into it during manufacture to diffuse light or to obscure vision, e.g., bathroom window glazing.

conglomerate A rock deposit with stones of varying sizes naturally cemented together as a natural concretion.

conifer Any of an order of mostly evergreen trees and shrubs, including those with true cones, such as pines, and with arillate fruit, such as yews.

connected load The total load, as measured in watts, that is connected to an electric supply system if all lights, equipment, etc., are turned on.

connector A device that will hold two or more electrical wires in close contact and is quickly and easily removable.

consistency A reference to the relative mobility or plasticity of freshly mixed concrete or mortar, as measured by a slump test for concrete, and by a flow test for mortar and grout.

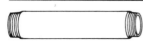

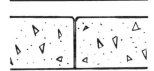

Construction Joint

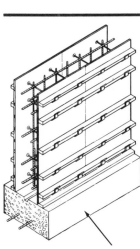

Contact Ceiling

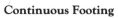

Continuous Beam

Continuous Footing

consistometer A device for measuring the firmness or consistency of concrete, mortar, grout, plaster, etc.

console (1) A control station with switches and gages to govern the operation of mechanical, electrical, or electro/mechanical equipment. (2) An ornamental bracket-like member used to support a cornice.

consolidation (1) Compaction of freshly poured concrete by tamping, rodding, or vibration to eliminate voids and to assure total envelopment of aggregate and reinforcement. (2) Compaction of soil in an embankment to achieve a higher bearing strength.

constant-voltage transformer A transformer designed to minimize or eliminate the variations in standard line voltage and to produce an unchanging voltage, as required by computers and some instrumentation.

construction bolt Any of several standard bolts used in construction to temporarily hold things together or in place, e.g., forms.

construction documents Drawings and specifications setting forth in detail the requirements for the construction of a project.

construction joint The interface or meeting surface between two successive pours of concrete.

constructive eviction An unlawful act by a landlord that makes a rented property uninhabitable.

consultant A person, or organization, with an area of expertise or professional training who contracts to perform a service.

contact adhesive, dry-bond adhesive A bonding agent that is applied to two surfaces to be joined and allowed to dry prior to their being pressed together.

contact ceiling A ceiling that is attached directly to the construction above, with no intermediate furring strips.

contact pressure The pressure or force that a footing and all the structure that is resting on it exerts on the soil below.

contact splice A connection between concrete reinforcing bars where the bars are lapped and in direct contact with each other.

continental seating Seats placed in an auditorium in continuous rows having no aisles or crossovers. Access is from either a wall aisle or though the doors in the side walls.

contingent agreement An agreement, generally between an owner and an architect, in which some portion of the architect's compensation is contingent upon the owner's obtaining funds for the project (such as by successful referendum, sale of bonds, or securing of other financing), or upon some other specially prescribed condition.

continuous beam A beam supported at three or more points, and thus having two or more spans.

continuous footing A concrete footing supporting a wall or two or more columns. The footing may vary in depth and width.

continuous girder A girder supported at three or more points.

continuous header Wood frame construction in which the top plate is replaced by a double member, such as a 2 x 8 on edge, that acts as a lintel over all wall openings.

continuous hinge, piano hinge A hinge that extends the full length of the moving part to which it is attached.

continuously reinforced pavement A longitudinally reinforced concrete pavement

Cooling Tower

with no intermediate transverse expansion or contraction joints.

continuous span A rigid span over three or more supports so that bending moments are transmitted from one segment to the next.

continuous vent A vertical plumbing vent that is a continuation of the drain to which it connects.

continuous waste and vent A single pipe, the upper portion of which is a vent, while the lower portion functions as a waste drain.

contour line A line on a map or drawing indicating a horizontal plane of constant elevation. All points on a contour line are at the same elevation.

contract documents (1) The owner-contractor agreement. (2) The conditions of the contract (general, supplementary, and other conditions). (3) The drawings. (4) The specifications. (5) All addenda issued prior to, and all modifications issued after, execution of the contract. (6) Any items that may be specifically stipulated as being included in the contract documents.

contraction joint A formed, sawed, or tooled groove in a concrete structure. The purpose of the joint is to create a weakened plane and to regulate the location of cracking resulting from the dimensional change of different parts of the structure.

contraflexure point See **point of inflection**.

control factor The ratio of the minimum to the average compressive strength of a material.

control joint See **contraction joint**.

controlled fill A backfilling or embankment operation in which the moisture content, depth of lift, and compaction equipment are closely regulated by specification and inspection.

controller An electrical, electronic, pneumatic, or mechanical device designed to regulate an operation or function.

convection The movement of a gas or liquid upward as it is heated and downward as it is cooled. The movement is caused by the change in density. The warm, less dense material rises, and the cooler, denser material falls.

conversion (1) The change in use of a building, as from a warehouse to residential units, that may require changes in the mechanical, electrical, and structural systems. (2) The sawing or milling of lumber into smaller units.

conversion factor Any factor used to change from one unit of measurement to another. For example, multiply yards by three (3) to obtain feet.

conveyance An instrument or deed by which a title of property is transferred.

coolant A fluid used to transfer heat from a heat source to a heat exchanger, where the heat is removed and the coolant is usually recycled.

cooling tower An outdoor structure, frequently placed on roofs, over which warm water is circulated to cool it by evaporation and exposure to the air.

cooperative, co-op A type of participant ownership in a building or complex. Actual ownership is by a non-profit corporation. Individuals are part owners of the corporation and pay monthly fees for use or occupancy of part of the building.

coopered joint A joint made on a curved surface similar to the joints made between the staves of a barrel.

coordinator The device on a pair of double doors that permits the doors to close in the correct sequence. If the door with the

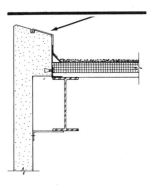

Coping

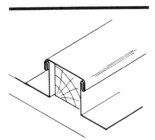

Copper Roofing

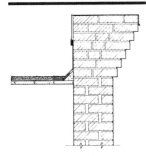

Corbel

overlapping astragal closed first, the other door would strike the astragal and not close properly.

cope (1) To cut the end of a molding so as to match the contour of the adjacent piece. (2) To cut structural steel beams so that they fit tightly together.

coped joint, scribed joint The intersection of two pieces of molding where one is cut to fit the contour of the other.

coping The protective top member of any vertical construction, such as a wall or chimney. A coping may be masonry, metal, or wood and is usually sloping or beveled to shed water in such a way that it does not run down the vertical face of the wall. Copings often project out from a wall with a drip groove on the under side.

coping saw A light hand saw with a narrow blade mounted in a U-shaped frame with a handle. A coping saw is used to cut curves in wood. The blade may be rotated to keep the frame away from the work.

copper fitting In piping, an elbow, tee, reducer, or other fitting made of wrought copper, cast brass, or bronze.

copper roofing A roof covering made of copper sheets that are joined by weatherproof seams.

corbel A course or unit of masonry that projects out beyond the course below. A corbel may be used entirely for decoration or for a ledge to support a load from above.

corded door A suspended door or divider made of plastic, vinyl, or wood slats interconnected with cord or tapes. When opened, the door slides in tracks folding flat back in accordion pleats.

core (1) The interior structure of a hollow-core door. (2) A cylindrical sample of concrete or rock extracted by a core drill. (3) The void in a concrete masonry unit. (4) The center layer of a sheet of plywood. (5) The vertical stack of service areas in a multistory building. (6) The central part of an electrical winding. (7) The rubble filling in a thick masonry wall.

coreboard A manufactured board of wood fiber or wood chip center with bonded veneer faces on both sides.

cored beam (1) A precast concrete beam with longitudinal holes through it. (2) A beam that has had core samples removed for testing.

core drill (1) A drill used to remove a core of rock or earth material for analysis. (2) The drilling of a hole through a concrete floor, wall, or ceiling for running pipe or conduit.

core frame See **buck frame**.

core gap An open joint extending through, or partly through, a plywood panel, that occurs when core veneers are not tightly butted. When center veneers are involved, the condition is referred to as a "center gap".

core test A compression test on a sample of hardened concrete cut out with a core drill.

coring The core drilling of a sample of rock, soil, or concrete to obtain a test sample.

corkboard Compressed and baked granulated cork used in flooring and sound conditioning and as a vibration absorber.

cork tile Tiles made from cork particles bound and pressed into sheets and covered with a protective wearing finish.

corner board The vertical boards that are butted together to form the outside corner of a wood frame building. The siding or shingles butt up against it.

corner brace The diagonal braces let into the studs of wood frame structures for reinforcement.

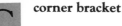

Cornice

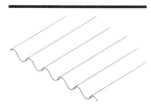

Corrugated Roofing

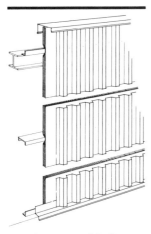

Corrugated Siding

Counter (3)

corner bracket A fitting mounted on a door frame at the upper hinge corner for mounting an exposed overhead door closer.

corner post (1) A glazing mullion in the corner of a structure which retains glazing in both walls. (2) A vertical member located at the corner of a timber structure.

corner reinforcement (1) The reinforcement in the upper corners of a metal door frame. (2) The metal angle strip used at the corners of plaster or gypsum board construction. (3) The bent reinforcing rods embedded at the corners of a cast in place concrete wall.

corner stud Same as **corner post**.

cornice (1) An ornamental molding of wood or plaster that circles a room just below the ceiling. (2) An ornamental topping that crowns the structure it is on. (3) An exterior ornamental trim at the meeting of the roof and wall. This type of cornice usually includes a bed molding, a soffit, a fascia, and a crown molding.

corridor A long, interior passageway usually with doors opening to rooms, apartments, or offices and leading to an exit.

corrosion The oxidation or eating away of a metal or other material by chemical or electrochemical action after prolonged exposure.

corrosion inhibitor A protective layer or coating of material, paint, or other surface finish applied to prevent oxidation of, or chemical attack on, the base material.

corrugated aluminum Sheet aluminum that has been rolled into parallel waves to impart stiffness.

corrugated fastener A small, wavy steel fastener with one edge sharpened. The fastener is driven into two pieces of wood, bridging the joint in order to hold them together.

corrugated metal Sheet metal that has been rolled into a parallel wave pattern for stiffness and rigidity.

corrugated roofing Corrugated metal, fiberglass, or cement asbestos mounted on rafters as sheet roofing.

corrugated siding Siding made of sheet metal or asbestos-cement composition board and used for siding on factory and other nondecorative buildings. By corrugating the material, the structural strength is increased.

cost breakdown See **schedule of values**.

cost-plus-fee agreement An agreement under which the contractor (in a contractor-owner agreement) or the architect (in an owner-architect agreement) is reimbursed for the direct and indirect costs of performance of the agreement and, in addition, is paid a fee for services. The fee is usually stated as a stipulated sum or as a percentage of cost.

cotter pin A split metal pin, which when pushed through a hole in a shaft or a bolt (and having the ends bent), prevents parts or nuts from vibrating loose.

counter (1) A person or device that keeps a tally of the occurrences of some event, such as the number of loads of fill. (2) A long flat surface over which business or sales are transacted at a store or business. (3) A flat work surface in a kitchen.

counterbalanced window A double-hung window in which weights on pulleys balance the weight of the sash, thus making it relatively easy to raise and lower.

counterbatten A strip of wood attached to the backs of boards to stiffen them.

counterbrace A diagonal brace that balances the force from another brace as part of a web network in a truss.

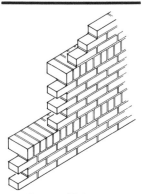

Coursed Masonry

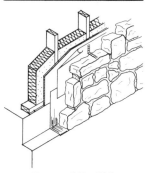

Coursed Rubble

counterflashing A thin strip of metal frequently inserted into masonry construction and bent down over other flashing to prevent water from running down the masonry and behind the upturned edge of the base flashing.

counterfort In masonry construction a pier, buttress, or pilaster on the inner side of a wall to resist thrust.

countersink (1) A conical depression cut to receive the head of a flat head bolt or screw for driving it flush with the surrounding surface. (2) A bit used to cut a conical depression.

countersunk bolt A flat head bolt that tapers to the shank diameter so that it will be flush with the top of a countersunk hole.

countersunk rivet A rivet used in countersunk holes.

counterweight A weight that balances another weight; for example, a sash weight to balance a window sash or a large weight that balances a lift bridge.

counterweight system A permanent system of weights mounted, for example, on a theatre stage to balance curtains, scenery, or lighting equipment.

couple Two equal, opposite, and parallel forces that tend to produce rotation. The moment equals the product of one force and the perpendicular distance between the two.

coupler A metal clamp-like device used to join and extend tubular frames and braces used for barricades or scaffolding.

couple roof, coupled roof A narrow-span, double-pitched roof in which opposite rafters are not tied together, thus leaving the outward forces to be resisted by the walls.

coupling pin A pin used to connect scaffolding sections or lifts vertically.

course (1) A horizontal layer of bricks or blocks in a masonry wall. (2) A row or layer of any type of building material, such as siding, shingles, etc.

coursed masonry Any masonry units laid in regular courses, as opposed to rough or random rubble.

coursed rubble Masonry construction consisting of roughly dressed stones of mixed sizes and shapes that are used with small stones to fill the irregular voids.

coursing joint The horizontal mortar joint between two courses of masonry. In an arch it is the arched joint between two curving courses.

coved vault An enclosure whose sides are like four quarter cylinders intersecting so that the joints look like an "x" in plain view.

cove lighting Indirect lighting in which the fixtures are behind a molding or valance, and thus out of sight.

cover (1) That which envelopes or hides, as a protecting cover of paint. (2) That part of a tile or shingle that is overlapped by the next course. (3) The minimum thickness of concrete between the reinforcing steel and the outer surface of the concrete.

coverage (1) The nominal square feet of area that a can of stain or paint can be expected to cover. (2) The amount of surface that may be covered by a unit, such as a bundle, square, or ton of building material.

coverstone A flat stone laid on a steel beam or girder as the foundation for the masonry above.

coving (1) Concave molding, as used at the intersection of a wall with a ceiling. (2) The outward curve of an exterior wall to meet the eaves. (3) The curving sides of

Crawler Tractor

Crib

a fireplace that narrows at the back.

cowl (1) An elaborate chimney pot to improve drafts. (2) A covering over vent pipes to block rain and snow.

crack-control reinforcement The use of steel reinforcing rods in concrete and masonry to minimize crack size and occurrence.

cracking load The load that imposes a tensile stress in concrete in excess of its tensile strength, causing it to crack.

cradle (1) A scaffold that is suspended outside the roof or top of a structure. (2) A U-shaped support for pipe or conduit.

cramp (1) A device for supporting a frame in place during construction. (2) A U-shaped metal bar used to lock adjacent blocks of masonry together, as in a parapet wall.

crandall A hand tool, for dressing stone, consisting of several sharp pointed bars fixed to the end of a handle in a hammerlike configuration.

crane A machine for raising, shifting, or lowering heavy weights, commonly by means of a projecting swinging arm.

crawler tractor A powered earthmoving vehicle that moves on continuous, segmented cleated treads or tracks that provide low ground-bearing pressure and a high level of mobility and power.

crawl space (1) In a building or portion of a building without a basement, the accessible space between the surface of the ground and the bottom of the first floor joists, with less than normal headroom. (2) Any interior space of limited height designed to permit access to components such as ductwork, wiring, and pipe fittings.

crazing Surface cracks that appear randomly, usually due to

shrinkage in paints, plaster, concrete, varnish, etc.

creasing One or more courses of masonry units laid on the top of a wall, chimney, or other vertical surface and overhanging the course below so water will drip off and not run down the wall below. Coping is placed above the creasing if it is used.

creep (1) The continuing, but slow, permanent deformation of a material under continual sustained stress. (2) The very slow movement of rock or soil under pressure.

creep strength The maximum stress that can be applied to a material at a specified temperature without causing more than a specified percentage increase in dimension in a specified time.

creosote An oily liquid obtained from tar and used to prevent wood from decaying. Creosote is extensively used for railroad ties, wood piles, posts, and wood foundations.

crescent truss A truss in which the upper and lower chords are both curved in the same direction, but with different radii of curvature, so that they meet at the ends, thus giving the assembly a crescent-shaped appearance.

crib (1) A boxed in area, the sides of which may be an open lattice, often filled with stone, and used as a retaining wall or support for construction above. (2) A retaining framework lining in a shaft or tunnel.

cribbing A system of open boxes built with heave timbers and filled with rocks or previous fill and used as a retaining wall.

cribwork An open construction of beams, at the face of an embankment, the alternate layers of which project to provide lateral stability, prevent erosion, and resist thrust or overturning.

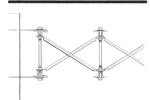

Critical Path Method, CPM

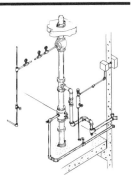

Cross Bracing

Cross Connection

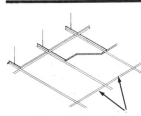

Cross Runner

crimp A sharp bend in a metal sheet, as the joint in metal roofing.

crimped wire Wire that has been deformed or bent to improve its bonding effectiveness when used to reinforce concrete.

cripple (1) In construction framing, members that are less than full length; for example, studs above a door or below a window. (2) In roofing, a bracket secured at the ridge of a pitched roof to carry the scaffold for roofers.

critical angle The maximum angle or rise of a stair or ramp before it becomes unsafe for public use. For stairs the angle is 50 degrees; for ramps, it is 20 degrees.

critical path method, CPM A charting of all events and operations to be encountered in completing a given process. The method is rendered in a form permitting determination of the relative significance of each event, and establishing the optimum sequence and duration of operations.

crook The warp in a board, as the deviation of an edge from a straight line drawn from end to end.

crossband, crossbanding (1) Any ornamental strip or band whose grain pattern is perpendicular to the main surface. (2) In plywood, a layer whose grain direction is perpendicular to the surface veneer.

cross bar In a grating or grill, one of the bars perpendicular to the primary bar.

cross beam (1) A large beam that spans between the two walls or sides of a structure. (2) A brace between opposite walings or sheeting in an excavation.

cross bracing Diagonal braces placed in pairs that cross each other.

cross connection A piping connection between two otherwise separate piping systems, one of which contains potable water, and the other water of unknown or questionable safety. (2) In a fire protection system, a connection from a siamese fitting to a standpipe or sprinkler system.

crosscut saw A saw with its teeth filed and set to cut across the grain of a piece of wood.

cross-furring, brandering, counterlathing Strips or slats attached to the lower edge of joists to which plastering lathing is nailed.

cross-garnet hinge A hinge that is shaped like the letter T, with the head attached to the door frame and the long part fastened to the door. Also called a "T-hinge."

cross-laminated Laminated wood members in which the grain of some layers is at right angles to the grain of other members.

crosslap joint A wood joint in which two pieces are each cut to half their thickness at the overlap, so that the total thickness of the system does not change.

crossover (1) In an auditorium, a walkway that is parallel to the stage and connects with the aisles, leading from back to front. (2) A pipe fitting used to by-pass a section of pipe. (3) A connection between two piping systems.

cross runner In a suspended or T-Bar ceiling, one of a series of short pieces that span between the long runners.

cross section A diagram or illustration showing the internal construction of a part or assembly if the front portion were removed, exposing the rest.

cross tongue A piece of wood, usually plywood, glued between two wood members at a joint for reinforcement.

cross ventilation The circulation or flow of air through openings, such as doors, windows, or grilles,

Crown

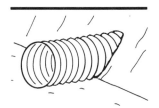

Culvert

Curb Cock, Curb Stop

that are on opposite sides of a room.

cross welt, transverse seam On sheet metal roofing, a seam or joint that runs parallel to the ridge or across the flow of storm water down the roof.

crown (1) An ornamental architectural topping. (2) The central top section of an arch or vault. (3) The high point in the center of a road that causes water to flow to the edges. (4) The top of a tree or flowering plant. (5) The convex curvature or camber in a beam.

crown course (1) The top row of a roofing material. (2) The row that actually covers the ridge.

crown saw A hole saw with a hollow rotating cylinder that has teeth around the edge for drilling round holes in wood.

crushed gravel Gravel that has been crushed and screened so that substantially one face of each particle is a fractured face.

crushed stone Stone crushed and screened so that substantially all faces result from fracturing.

crusher A machine for reducing gravel, rocks, boulders, or blasted rock to smaller sizes.

crusher-run aggregate Aggregate that has been ground in a crusher, but has not been sorted for particle size.

crushing strength The greatest compressive load a material can withstand without fracturing.

crush plate A replaceable strip of wood attached to the edge of a concrete form to take the abuse of stripping or prying operations.

crystallized finish A random pattern of wrinkles in the finish of paint or varnish.

cube strength An analytical strength test of portland cement in which a standard size concrete cube is loaded to failure.

cul-de-sac A dead-end street that

ends in an enlarged turn-around area.

culvert A transverse drain under a roadway, canal, or embankment other than a bridge. Most culverts are fabricated with materials such as corrugated metal and precast concrete pipe.

cumulative batching The weighing of a batch of material by adding the components successively to the same container and balancing the scale at each new total weight.

cup joint A straight joint used in copper tubing. A swaging tool is driven into one of the ends to be joined, enlarging it so that it will fit over the other tube for a soldered connection.

curb cock, curb stop A control valve that is placed in the water supply pipe that runs from a water main in a street to a building.

curb form (1) A form used with a curb machine for extruding concrete curbs to produce a desired shape and finish on a concrete curb. (2) A reusable metal form for cast-in-place concrete curbs.

curb level The elevation of the curb grade in front of a building as measured at the center of its street frontage.

curb machine A self-propelled, hand-operated machine for extruding bituminous or concrete curbs.

curb plate (1) The wall plate at the top of a circular or eliptical structure. (2) The plate on which the upper rafters of a curb roof rest.

curb roof A roof with the slope divided into two pitches on each side. Also called a gambrel or mansard roof.

cure (1) A change in the physical and/or chemical properties of an adhesive or sealant by chemical reaction when mixed with a catalyst or subjected to heat or

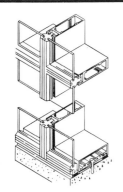

Curtain Wall

Cut

Cut and Fill

Cut Nail

pressure. (2) The process of maintaining the proper moisture and temperature after placing or finishing concrete to assure proper hydration and hardening.

curing blanket A covering of straw, burlap, sawdust, or other suitable material placed over fresh concrete and moistened to help maintain humidity and temperature for hydration.

curing membrane Any of several kinds of sheet material or spray-on coatings used to temporarily retard the evaporation of water from the exposed surface of fresh concrete, thus ensuring a proper cure.

curling The change in shape of wood, such as in straightness or flatness, due to drying or temperature differences.

current-carrying capacity The maximum rated current, measured in amperes, that an electrical device is allowed to carry. Exceeding this limit could lead to early failure of the device or create a fire hazard. Also called "ampacity".

curtain board, draft curtain A heavy fireproof fabric hung from a roof or ceiling to isolate a hazardous area and to act as a shield against the spread of a fire, by containing heat and smoke for direct venting.

curtain grouting Injection of grout into the ground in a broad line to act as a barrier to sub surface water flow.

curtain wall The exterior closure or skin of a building. A curtain wall is non-bearing and is not supported by beams or girders.

cushion (1) Wood placed so as to absorb a force by acting as a buffer or by transmitting it over a larger area. (2) A stone placed to accept and spread out a vertical load. (3) An isolating pad against shock and vibration for glass, machinery, equipment, etc.

cut (1) Material excavated from a construction site. (2) A term for the area after the excavated material has been removed. (3) The depth of material to be removed, as in a 5-foot cut.

cut-and-cover An operation for the laying of pipe or conduit, or for constructing a tunnel, in which a trench is dug, the pipe, conduit, or tunnel liner is placed, and the trench is covered by the excavated material.

cut and fill An operation common to road building and other rock and earthmoving operations in which the material excavated and removed from one location is used as fill material at another location.

cutaway drawing A drawing of an area or object that shows what would be seen if a slice could be made into the area or object and a piece removed. See also **cross section**.

cutback asphalt A bituminous roof coating or cement that has been thinned with a solvent so that it may be applied without heat to roofs or other areas needing sealing or cementing. Also used for damp proofing concrete and masonry.

cut nail A hard wedge-shaped nail used for nailing hardwoods such as oak flooring.

cutoff A wall or barrier placed to minimize underground water percolation or flow. A cutoff is often used at the inlet and outlet of culverts. (2) The design elevation at which the tops of driven piles are cut.

cutout (1) A mechanical or electrical device used to stop a machine when its safe limits have been reached. (2) An opening in a wall or surface for access or other reasons. (3) A piece stamped out of sheet metal or other sheet material.

cutout box A metal box used in electrical wiring to house circuit breakers, fuses, or a disconnect switch.

cyclopean concrete See **concrete, cyclopean**.

cylinder lock A lock in which the keyhole and tumbler mechanism are contained in a cylinder or escutcheon separate from the lock case.

cyma A molding having an S-curve section.

cypress A wood of average strength and very good decay resistance and durability.

D

ABBREVIATIONS

The abbreviations listed below are those most commonly used in the construction industry. Alternative forms (usually nonstandard) are shown in parentheses.

d degree, density, penny (nail size)

D diameter, dimensional, deep, depth, discharge

D&CM dressed and center matched

D&M dressed and matched

D&MB dressed and matched beaded

D&SM dressed and standard matched

D1s dressed one side

D2s dressed two sides

D2S&CM dressed two sides and center matched

D2S&SM dressed two sides and standard matched

D4S dressed on four sides

DAD. double acting door

dB decibel

dBA a unit of sound level (as from the A-scale of a sound-level meter)

DB. Clg double-headed ceiling

DBL double

DBT dry-bulb temperature

DEC decimal

DEG degree, degrees

DEL delineation

DEPT department

DET detail, detached, double end trimmed

DF drinking fountain, drainage free, direction finder

dflct deflection

d.f.u. drainage fixture unit

DHW double-hung window

DIA diameter

DIAG diagonal

DIM. dimension

DIN Dutch Industry Normal (German industry standard)

DIV division

DL dead load, deadlight

DN down

DO ditto

DOZ dozen

DP dew point, double pitched, degree of polymerization

DPC dampproof course

DR drain, dressing room, dining room, driver

DRG drawing

drn drain, drainage

drwl drywall

DS downspout

DSGN design

DT drum trap

DT&G double tongue and groove

DU disposal unit

DUP duplicate

DVTL dovetail

DWG, dwg drawing

DWV drain, waste, and vent

D DEFINITIONS

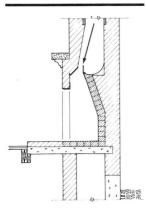

Damper

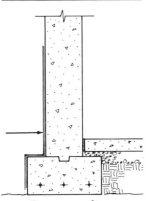

Dampproofing

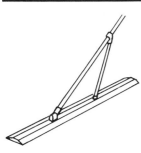

Darby

dabber A soft round tipped brush for applying varnish or for polishing and finishing gilding.

dabbing, daubing Dressing a stone face with a sharp tool to give it a pitted appearance.

dado (1) The flat bottomed groove cut into one board to receive the end of another; usually cut across the grain. If it is cut at the edge of a board, it is called a " rabbet." (2) An ornamental paneling above the baseboard of a room finish. (3) Part of a column between the base and the cap or cornice.

dado cap A cornice or chair rail along the highest part of a framed dado.

dado head A power saw blade designed specifically to produce dados or flat bottomed grooves.

dado joint A joint created by the end of one piece of wood fitting, usually at right angles, into a groove cut across the width of another piece to a depth of half its thickness.

dado rail see **chair rail**.

dais A raised platform at one end of a hall usually reserved for officers, dignitaries, or speakers.

damages Usually, per diem amounts specified in a contract and payable only when incurred loss can be proved to have resulted from a contractor's delays or breach of contract. Also see **liquidated damages**.

damp check See **damp course**.

damp course, damp check, dampproof course In masonry construction, an impervious course or any material preventing moisture from passing through to the next course. Usually, a horizontal layer of material such as metal, tile, or dense limestone.

damper (1) A blade or louver within an air duct, inlet, or outlet which can be adjusted to regulate the flow of air. (2) A pivotal cast-iron plate positioned just below the smoke chamber of a fireplace and used to regulate drafts.

damping The force that acts to reduce vibrations in the same way that friction acts to reduce ordinary motion. Also, the gradual dissipation of energy over of period of time.

damping material Viscous material applied to the surface of a vibrating plate to subdue the vibrations and noise.

dampproof course See **damp course**.

dampproofing An application of a water resisting treatment or material to the surface of a concrete or masonry wall to prevent passage or absorption of water or moisture. Can also be accomplished by an admixture to the concrete mix.

dancers Another word for **stairs**.

dancing step See **balanced step**.

dancing winder See **balanced step**.

dao, paldao A moderately hard, heavy wood that is variegated in color and is indigenous to the Philippines and New Guinea. The wood is used mostly for cabinets, plywood, and interior finishes.

dap A notch made in one timber to fit the end of another.

darby, derby slicker A hand float or trowel, about 4-feet long, used by concrete finishers and plasterers in preliminary floating and leveling operations. The float is usually made of wood or aluminum and has either one or two handles.

dash-bond coat A thick slurry of portland cement, sand, and water which is applied to a concrete wall with a brush or wiskbroom as a

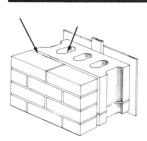

Dead Air Space

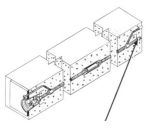

Dead End

bond for a subsequent plaster coat.

date of agreement The date shown on the face of an agreement, or the date the agreement is signed. It is usually the date of the award.

date of commencement of the work The date established in the notice to proceed or, in the absence of such notice, the date of the agreement or such other date as may be established therein or by the parties thereto.

date of substantial completion The date certified by the architect when the work or a designated portion thereof is sufficiently complete, in accordance with the contract documents, so that the owner may occupy the work or designated portion thereof for the use for which it is intended.

datum A base elevation to which other elevations are referred.

datum line An elevation reference line.

daub To rough coat with plaster. To rough coat a stone surface by striking it with a special hammer.

day In glazing, one division of a window, particularly in large church windows.

day gate In a bank, the metal grille door used when the the main vault door is open.

daylight glass A blue colored glass used with incandescent lamps to produce the optical characteristics of daylight.

daylight lamp A lamp manufactured with daylight glass. The output is generally 35% less than that from an incandescent lamp.

daylight width, sight size width The width of that portion of a window which admits light.

deactivator A tank which contains iron filings and through which hot water is passed to be purged of its active oxygen and other corrosive elements.

D-cracking The fine hairline cracks in concrete surfaces. In highway pavement, those running parallel to joints and edges and cutting diagonally across corners.

dead A conductor not connected to an electrical source.

dead-air space Unventilated air space between structural elements. The space is used for thermal and sound insulation.

dead bolt The bolt of a type of door lock that must be operated positively in both directions by turning a key or a thumb bolt.

dead-burnt gypsum Anhydrous calcium sulfate (gypsum from which all the water of crystallization has been removed).

dead end (1) In stressing of a tendon, the end opposite to the one at which stress is applied. (2) In plumbing, a drain line or vent which has been purposefully terminated by a cap, plug, or other fitting.

dead-end anchorage The anchorage for the stressing steel in a tendon opposite the jacking end.

deadening The use of insulating and damping materials to restrict the passage of sound.

dead-front An electrical device in which the front part is insulated from voltage and can be touched without receiving an electric shock.

dead knot A knot in cut lumber which is no longer connected to the surrounding lumber.

deadlatch See **night latch**.

dead level The state of being absolutely level, with no pitch or slope. The term also refers to a grade of asphalt used on a level or nearly level roof.

deadlight A fixed window sash.

dead load A calculation of the weight of the structural components, the fixtures, and the

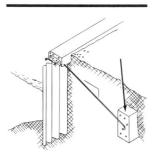

Deadman

Deck

Deck Curb

Decking

permanently attached equipment used in designing a building and its foundations.

deadlock A type of door lock employing a dead bolt.

deadman A heavy anchor, such as a concrete block or a log, usually buried underground for securing the end of a tie or guy. This device is frequently used with sheeting and retaining walls.

dead parking Long-term, unattended parking of vehicles.

dead-piled Freshly pressed plywood piled flat without spacers and weighted down while it attains normal temperature and moisture levels.

dead room A room designed for high sound-absorption qualities.

dead sand An unwashed, ungraded sand used for fill.

dead shore An upright timber used as a temporary support for the dead load of a building during structural alterations. This type of shore is commonly used in pairs to support a needle.

dead-soft temper A term describing the temper of sheet copper used for flashing.

deadwood Dead tree limbs or the timber cut from dead trees.

deal A term used in the lumber industry referring to boards or planks usually more than 9 inches wide and 3 to 5 inches thick.

debt service The cost of money. The periodic repayment of loans including interest and a portion of the principal.

decal A design that can be transferred from a special paper onto several different surfaces.

decay The disintegration, particularly of wood, caused by the action of fungi.

decayed knot A knot that is softer than its surrounding wood.

decay rate (1) The rate at which

sound reverberations decrease in an enclosure expressed in decibels per second. (2) The rate at which vibrations in a mechanism decrease over an elapsed time.

decenter The act of lowering or removing shoring or centering.

decibel The standard unit of loudness measurement for sound.

deciduous A term used to describe those trees that shed their leaves annually. The term includes most hardwoods and some softwoods.

deck (1) An uncovered wood platform usually attached to or on the roof of a structure. (2) The flooring of a building. (3) The structural system to which a roof covering is applied.

deck clip (1) The H-shaped accessory placed between plywood sheets used as roof decking to prevent uneven deflection. This allows a thinner sheet of plywood to be used with no blocking beneath the joints. (2) A metal fastener used to attach roof deck units to the structural frame. (3) An accessory made for attaching rigid insulation material to a roof deck.

deck curb A curb around the edge of a roof deck or around roof mounted equipment.

decking (1) Light-gauge, corrugated metal sheets used in constructing roofs or floors. (2) Heavy planking used on roofs or floors. (3) Another name for slab forms that are left in place to save stripping costs.

deck-on-hip A flat roof constructed over a hip roof.

deck paint A special hard-surface paint which is resistant to abrasive wear and is used particularly on decks and porches.

decor A combination of materials and furnishings which creates a particular style of interior decorating.

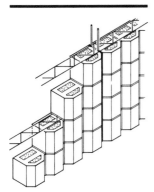

Decorative Block

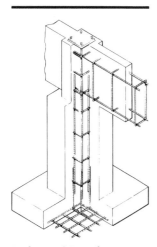

Deformed Reinforcement

decorative block A concrete masonry unit manufactured or treated to have a desired architectural effect. The particular effect may be either in color or texture, or both.

dedicated street A street whose ownership has been relinquished and which has been accepted by a governmental agency for public use and maintenance.

deduction The amount of money deducted from the contract sum by a change order.

deductive alternate An alternative bid that is lower than that bidder's base bid.

deed (1) A legal document giving rights or ownership to a property. (2) A document that forms part of a contract and which, when signed by both parties, legally commits the contractor to perform the work according to the contract documents, and the client to pay for the work.

deed restriction Specified limitations on the use of property legally documented in the deed.

deep bead See **draft bead**.

deep cutting, deeping (1) The resawing of timber parallel to its face. (2) The cutting out of a structural member to a depth relatively far below its surface.

deep-seal trap, antisiphon trap A U-shaped plumbing trap having a seal of 4 inches or more.

defect Any condition or characteristic which detracts from the appearance, strength, or durability of an object.

defective work Work that does not comply with the requirements of a contract.

deficiencies See **defective work**.

deflected tendons Tendons, occurring in a concrete member, that have a curved trajectory with respect to the gravity axis of the member.

deflection The bending of a structural member as a result of

its own weight or an applied load. Also, the amount of displacement resulting from this bending.

deflection angle In surveying, a term describing a horizontal angle as measured from the prolongation of the preceding transit line to the next line.

deflectometer An instrument for measuring the degree to which a transverse load causes a beam or other structure to bend.

deformation A change in the shape or form of a structural member that does not cause failure or rupture.

deformed bar, deformed reinforcing bar A reinforcing bar manufactured with surface deformations to provide bonding strength when embedded in concrete.

deformed metal plate A corrugated or otherwise horizontally deformed metal plate used at vertical joints to provide a mechanical interlock between sections.

deformed reinforcement In reinforced concrete, reinforcement comprising bars, rods, deformed wire, welded wire fabric, and welded deformed wire fabric. See also **deformed bar**.

degradation Deterioration of a painted surface by heat, light, moisture, or other elements.

degree-day A unit of measure for heating-fuel consumption. The unit is used to specify the nominal heating load of a building in winter. One degree-day is equal to the number of degrees, during a 24-hour day, that the mean temperature is below 65 degrees Fahrenheit, which is the "base temperature" in the United States.

degree of saturation The ratio of the volume of water in a given soil mass to the volume of intergranular voids, expressed as a percentage.

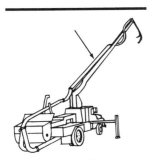

Delivery Hose

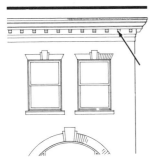

Demountable Partition

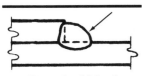

Dentils

Deposited Metal

dehumidification The removal of moisture from the ambient air by physical or chemical means.

dehumidifier A device for removing water vapor from the air.

dehydration The removal of water vapor from the air by means of absorption or adsorption.

delamination Separation of plies of materials usually through failure of the adhesive, such as in veneers, roofing, and laminated wood beams.

deliquescence The absorption of moisture from the air by certain salts in plaster or brick, resulting in damp spots that appear darker than the surrounding material.

delivery hose The hose through which concrete, mortar, grout, or shotcrete is pumped.

deluge sprinkler system A dry-pipe sprinkler system particularly good for areas that may experience temperatures below freezing. The system is actuated by a heat- or smoke-detection device, which then turns on a valve admitting the water.

demand The electric load integrated over a specific interval of time and usually expressed in watts or kilowatts.

demand factor In an electrical system, the ratio of the maximum demand to the connected load.

demand mortgage loan A mortgage loan which may be called at any time by the mortgagee.

demolition The intentional destruction of all or part of a structure.

demountable partition, relocatable partition A non-loadbearing wall made of prefabricated sections that can be readily disassembled and relocated. These partitions can be full height or partial height. Often the expense is outweighed by their versatility.

den A small room in a home for work or leisure.

dense concrete Concrete containing a minimum number of voids.

dense-graded aggregate Aggregate sized so as to contain a minimum of voids, and therefore the maximum weight when compacted.

densified impregnated wood Laminated wood that has been impregnated with resins and compressed to increase greatly its density and strength.

density A measure of the number of units per area or volume, as in housing units per acre.

density control The testing of concrete used in a structure to maintain the specified density.

dentil band A wood molding creating the effect of a series of dentils.

dentils Square toothlike blocks used as ornaments under a cornice.

depolished glass Glass whose surface has been diffused by sand blasting, etching, or some other surface treatment.

deposited metal The filler metal placed during welding.

deposit for bidding documents A deposit required for each set of the plans, specifications, and other bidding documents for a contract. This deposit is returned to the bidder upon return of the documents in good condition within a specified time.

depository A location where bids are received by an awarding authority. Also see **bank depository**.

depot (1) A storehouse, warehouse, or transfer station. (2) A railroad or bus station for ticketing, sheltering, transferring, and shipping passengers and freight.

Derrick

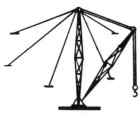

Detached Dwelling

Detached Garage

depreciation factor The ratio of initial illumination on an area to the present illumination of the same area, used in lighting calculations to account for depreciation of lamp intensity and reflective surfaces.

depressed arch An arch having less pitch than an equilateral triangle.

depreter A form of stucco, with rock dash finish, that is made by imbedding aggregate in the plaster.

depth gauge An instrument for measuring the depth of tire tread, dados, grooves, drilled holes, and various other recesses by spanning the edges with a cross piece and measuring the depth with a scale that slides through the middle.

derby, derby float See **darby.**

derrick A device consisting of a vertical mast and a horizontal or sloping boom operated by cables to a separate engine or motor. The device is used for hoisting and moving heavy loads or objects.

desiccant An absorbent material that removes water or vapors from a material or from the air. The material is used in the glazing and refrigeration industries.

design (1) To create a graphic representation of a structure. (2) The graphic architectural concept of a structure. (3) To make a preliminary sketch, drawing, or outline.

design-build, design-construct A term designating a contractor who provides both design and construction services to an owner.

design development phase The second phase of an architect's basic services, which include developing structural, mechanical, and electrical drawings and specifying materials and the probable cost of construction.

design load (1) In structural analysis, the total load on a structural system under the worst possible loading conditions. (2) In air conditioning, the maximum heat load a system is designed to withstand.

design strength (1) The assumed value for the strength of concrete and yield stress of steel used to develop the ultimate strength of a section. (2) The load bearing capacity of a structural member determined by the allowable stresses assumed in the design.

design ultimate load, factored load The working load multiplied by the load factor.

detachable bit In rock drilling, the cutting bit which is screwed into the end of the drill steel. The bit is provided with holes for blowing out the rock dust.

detached dwelling A structure, intended for habitation, that is surrounded by open space.

detached garage A parking structure whose exterior walls are surrounded by open space.

detail A large-scale drawing, engineering or architectural related, indicating specific configurations and dimensions of any construction elements. If the large-scale drawing differs from the general drawing, it is the architect's or engineer's intention that the large-scale drawing be used to clarify the general drawing.

detailed estimate of construction cost A forecast of the cost to construct a project based on unit prices of materials, labor, and equipment as contrasted to a parameter estimate or square-foot estimate.

detailer A draftsman who prepares shop drawings and lists of materials.

detector See **sensor.**

detention door A special steel door with fixed lights and steel bars used in prisons and mental institutions.

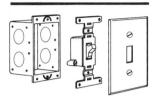

Device

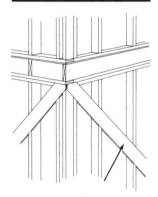

Diagonal Brace

detention window A narrow, metal awning window manufactured especially for the security of prisons and mental institutions.

detonating cord A flexible cord attached to a highly explosive core and used for setting off the explosive to which it is attached.

detonator Any of a number of devices such as blasting caps for detonating explosives.

detritus Loose material formed from disintegrating rock.

detritus tank A settling tank in a sewage-treatment system used for collecting sediment without interrupting the flow of sewage.

detrusion The shearing of wood fibers along the grain.

developed area An area or site upon which improvements have been made.

developed length The length of a pipe line, including fittings, measured along its centerline.

development A tract of land which has been subdivided for housing and/or commercial usage and includes streets and all necessary utilities.

development bond stress See **anchorage bond stress**.

device In an electrical system, a component that carries but does not consume electricity, such as a switch or a receptacle.

devil A fire grate used to heat asphalt spreading tools.

devil float, devil, nail float A hand-held flat tool, with nails in each edge, that is used to scratch a coat of plaster prior to applying the next coat.

deviling The act of scratching a plaster coat prior to applying another coat.

dewater To remove water from a job site by pumps, wellpoints, or drainage systems.

dewpoint The temperature at which air of a given moisture content becomes saturated with water vapor. Also, the temperature at which the relative humidity of the air is 100%.

dextrin, amylin, starch gum A starch compound used for wall sizing and wallpaper adhesive.

diabase An igneous rock used extensively in construction materials, such as asphaltic concrete, concrete, and pavement base courses, and commonly called "traprock." The rock crushes into good sharp angular fragments that interlock well with one another.

diaglyph A relief sculptured below the general surface.

diagonal A straight structural member forming the hypotenuse of a right triangle as in the diagonal bracing of a stud wall.

diagonal bond A pattern of brick laying in which every sixth course is a header course, with the bricks placed diagonally to the face of the wall.

diagonal brace A sloping member installed at an angle to make a rectangular frame more rigid.

diagonal bridging Diagonal bracing, in crossing pairs, between the top of one floor joist and the bottom of the adjacent joist. The bracing is used to distribute the load and to decrease deflection.

diagonal buttress A support structure constructed at 45 degrees to the right angle corner extension of two exterior walls.

diagonal crack A crack forming at about 45 degrees from the longitudinal dimension of a concrete member, beginning at the tension surface of the member.

diagonal grain In lumber, a deviation of the grain from a line parallel to the longitudinal edges caused by improper sawing.

diagonal pitch The distance between one rivet or bolt in one row to the nearest rivet or bolt in the next row of a structural

Diamond Lath

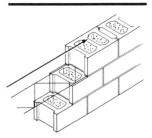

Diaphragm

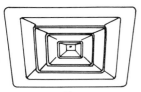

Diffuser

member with two or more rows of bolts.

diagonal rib A structural member crossing a bay of a vault in a diagonal direction.

diagonal sheathing A covering of wood boards over an exterior stud wall. The boards are placed diagonally and, although slightly more expensive to install, this method provides a more rigid frame than horizontally installed boards. The method is used primarily in residential construction.

diagonal slating, drop-point slating A method of applying diamond-shaped roofing slates with one diagonal laid horizontally.

diagonal tension In concrete structural members, the tensile stress resulting from shearing stresses within the member.

diamond fret, lozenge fret, lozenge molding A molding with a diamond pattern.

diamond lath, diamond mesh A type of expanded-metal strip used as a base for plaster and manufactured by slitting and expanding metal sheets.

diamond matching A method of laying up veneers in which four pieces are used, each with the grain running perpendicular to a line drawn diagonally across the whole panel. The resulting design is diamond shaped.

diamond slate An asbestos-cement roofing shingle shaped like a diamond and used in diagonal slating.

diamondwork Masonry laid up so as to incorporate diamond shapes, usually every sixth course.

diaphragm (1) A stiffening member between two structural steel members used to increase rigidity. (2) The web across a hollow masonry unit.

diaphragm pump, mud sucker A reciprocating water pump with a flexible diaphragm good for continuous dewatering of excavations containing mud and small stones.

diaphragm valve A valve that is actuated by fluid pressure on a diaphragm.

dichroic reflector lamp An incandescent lamp with a filter for correcting the lamp's light color to make it approximate daylight.

die A tool for cutting threads on pipe and bolts.

die-cast A method of casting by forcing molten metal into a mold.

differential leveling A method for determining the difference in elevation between two points using a level and rod.

differential settlement Uneven, downward movement of the foundation of a structure, usually caused by varying soil or loading conditions and resulting in cracks and distortions in the foundation.

diffused light Light reflecting from a surface rather than radiating directly from a light source.

diffuser Any device or surface which scatters light or sound from a source. See **air diffuser**.

diffusing glass A glass with an irregular surface, such as ground glass or sand blasted glass. Irregular surfaces can also be produced by rolling during manufacturing. This type of glass is used for providing a softer light for eye comfort.

diffusing panel A translucent material, usually plastic, used as a cover over a light source such as a luminaire to distribute the light more evenly.

diffusing surface A reflecting surface, usually rough textured, used to scatter light or sound.

dike, dyke An earth embankment for retaining water.

diluent Any agent used as a thinner.

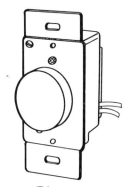

Dimmer

Dipper Stick

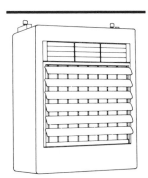

**Direct-fired
Air Heater**

dimensionally stable A term applied to any building material that does not alter shape appreciably with changes in temperature, moisture, and loading conditions.

dimension lumber, dimension stuff Lumber that is cut from 2 inches up to, but not including, 5 inches thick and 2 or more inches wide.

dimension shingles Shingles cut to a uniform size, usually 5 or 6 inches wide, and used for special architectural effects.

dimension stock Hardwood plywood that has been manufactured to specific dimensions for a particular user.

dimension stone Stone that has been trimmed or cut to specifications for a particular use, such as curbs, building stone, paving blocks, and the like.

dimension stuff See **dimension lumber**.

diminished arch, skeen arch, skene arch An arch with less rise than a semicircle.

diminished stile, diminishing stile, gunstock stile A door stile that has a narrower dimension at a panel, particularly at a glazed panel.

diminishing course See **graduated course**.

diminishing pipe, taper pipe A pipe reducer.

dimmer An electrical device that varies the light given off by an electrical lamp by varying the current.

dinging A rough stucco coat applied to a wall and sometimes scored to imitate masonry joints.

diorama A three-dimensional scene constructed to scale, usually with a painted background.

diorite A granular, igneous rock composed of feldspar and hornblende.

dip (1) In geology, the slope of a fault or vein. (2) In plumbing, the lowest point of the inside top of a trap.

dipcoat A covering of paint or other coating applied by immersing an object in a liquid and usually applied as an anticorrosive.

dipper The digging bucket attached to the stick or arm of a power shovel or backhoe.

dipper stick The straight arm that connects the dipper with the boom of a power shovel.

dipper trip The mechanism that releases the latch to the bucket door of a power shovel to discharge the load.

dip solution The solution used to produce a specific finish or color on copper or copper-alloy products.

direct-acting thermostat A device that activates a control circuit within a predetermined temperature range.

direct cross-connection A connection between two systems for maintaining equal water pressure.

direct current A term applied to an electrical circuit in which the current flows in one direction only.

direct dumping The placing of concrete directly into open forms from a chute or bucket at or near its final position.

direct expense All items of expense directly incurred by or attributable to a specific project, assignment, or task.

direct-fired air heater An air heating system in which the heat of combustion is applied directly to the intake air with no heat exchanger.

direct heating The heating of a space by a heat source such as a stove, radiator, or fire.

Disappearing Stairway

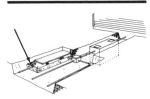

Disposal Field

direct-indirect lighting A lighting system employing diffused light luminares that emit little or no light horizontal to the lighting unit.

directional lighting The lighting of an object or work place from a desirable source location.

direct leveling The determination of the difference in elevation of two or more points by use of a surveyor's level and a graduated rod.

direct lighting A lighting system in which all or most of the light is directed toward the surface to be illuminated.

direct luminaire A luminare that directs 90% to 100% of its light downward.

direct nailing See **face nailing**.

directory board An informational panel with changeable letters such as those located in an office building lobby.

direct personnel expense Salaries and wages, including fringes, of all principals and employees attributable to a particular project or task.

direct return system A heating or cooling system in which the heating or cooling fluid is returned from the heat exchanger to the boiler or evaporator by the shortest most direct route.

direct stress Tensil or compressive force applied without shear or bending.

direct system A heating, air conditioning, or refrigerating system which employs no intermediate heat exchanger to heat or cool a space.

dirt-depreciation factor A measurement of the reduction of light emitted by a luminare as a result of accumulated dirt.

disability glare Objectionable effect of brightness that hampers visual performance and causes discomfort.

disappearing stair A hinged stair, usually folding, that is attached to a trap door in a ceiling and can be raised out of sight when not in use.

discharging arch An arch built above the lintel of a door or window to distribute the weight to each side of the wall above.

discoloration Any change in hue, shade, or color from the original or from the adjacent color.

discomfort glare An objectionable effect of brightness which does not necessarily hamper visibility.

discontinuous construction A method of construction whereby structural members are staggered to reduce sound transmission through a wall or floor.

discontinuous easement A right of accommodation (for a specific purpose) requiring action by one of the parties, such as a right-of-way.

disintegration Deterioration or crumbling caused by exposure, oxidation, freezing, or other elements. The term is often applied to the deleterious effects suffered by exposed concrete.

disk sander A power hand tool that has a rotating, circular abrasive disk used for smoothing or polishing a surface.

dispersant A material capable of holding finely ground particles in suspension and used as a slurry thinner or grinding compound.

dispersing agent An additive that increases the workability or fluidity of a paste, mortar, or concrete.

dispersion In painting, a paint or varnish holding small particles of pigment or latex in suspension.

disposal field See **absorption field**.

disposal unit An electric motor-driven unit installed in a sink for grinding food waste for disposal through the sanitary sewer system.

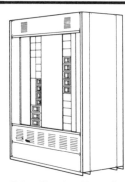

Distribution Board

Ditch

Divider Strips

distance separation The distance, specified by fire-protection codes, from an exterior wall to an adjacent building, property line, or the centerline of an adjacent street.

distemper An interior wall paint in which colors are tempered in a glutinous base.

distemper brush A wide, flat long-bristled brush used to apply distemper.

distributed load A load distributed evenly over the length of a structural member or the surface of a floor or roof, expressed in weight per length or weight per area.

distribution The movement of heated or conditioned air to desired locations. Also, the placement of concrete from where it is discharged to its final location.

distribution board, distribution box An electric switchboard or panel used to distribute electricity within a building. The switchboard is enclosed in a box and contains circuit breakers, fuses, and switches.

distribution box A container, located at the outlet of a septic tank, that distributes the effluent evenly to the drain tiles in the absorption field. See also **distribution board**.

distribution cutout In a primary electrical circuit, a safety device which disconnects a circuit to prevent overload.

distribution line The main electrical feed line to which other circuits are connected.

distribution panel See **distribution board**.

distribution reinforcement The placement of small-diameter reinforcing steel at right angles to the main reinforcing steel in a slab. This method of reinforcement is used to distribute the load and prevent cracking.

distribution steel See **distribution reinforcement**.

distribution switchboard See **distribution board**.

distribution tile The tile laid in an absorption field to distribute effluent from a septic tank.

distributor truck A truck with an insulated tank, heating units, and spray bars for applying liquid asphalt to a pavement area.

ditch A long, open trench used for drainage, irrigation, or for burying underground utilities.

ditcher, ditching machine A device used for digging temporary ditches or trenches with vertical slopes. The device operates by means of buckets or teeth mounted on a continuously rotating wheel or chain.

diver's paralysis Caisson disease; also called the "bends."

diversion A pipe, trench, or channel, usually temporary, used to divert the flow of water from its usual course.

diversity factor In electrical systems, the sum of the individual demands of the subsystems divided by the maximum demand of the whole system.

divided light door A French door with the glass divided into small panes.

divided tenon Two tenons side by side in the end of a member.

dividers A compass, with both legs terminating in points, used for transferring measurements from a plan or map to a scale, or vice versa.

divider strips The nonferrous metal or plastic strips used in terrazzo work as screeds and also used to divide the panels.

division One of the standard sixteen major Uniform Construction Index classifications used in specifying, pricing, and filing construction data.

Dog

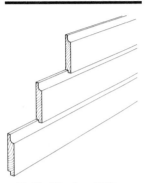

Dolly Varden Siding

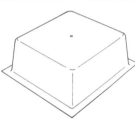

Dome

division bar See **muntin**.

division wall, fire wall A fire-resistant wall, satisfying fire code regulations, extending from the lowest floor of a building through the roof to prevent the spread of fire.

D-line crack (1) One of several fine, closely spaced, randomly patterned cracks in concrete surfaces. (2) Similar cracks parallel to the edges or joints in a slab. (3) Larger cracks in a highway slab running diagonally across its corners.

dobying Blasting a boulder by mud-capping rather than drilling.

dock A raised platform for the loading and unloading of trucks. Often, the elevation of a portion of the dock is adjustable to accommodate the differing heights of truck beds.

dock bumper A recoiling or resilient device attached to a dock and used to absorb the impact of a truck.

dog (1) In concrete forming, the hardware which holds the end of a snap tie. (2) Any device used for holding, gripping or fastening an object. (3) See also **dog anchor**.

dog anchor A short piece of iron, with its two pointed ends bent at right angles, driven into two adjoining pieces of timber to hold them together.

dog bars Intermediate vertical members in the lower part of a gate used to prevent animals from passing through.

dog-ear In roofing or flashing, the forming of an external corner by folding rather than cutting the metal.

dog iron See **dog anchor**.

dogleg stair A stair that forms a right angle turn at a landing.

dog nail A large nail or spike whose head projects over one side.

dolly (1) One of a variety of small, wheeled devices used to move heavy loads. (2) A block of wood placed on the upper end of a pile to cushion the blow of a hammer. (3) A steel bar with a shaped head used to back-up a rivet while the rivet is being driven.

Dolly Varden siding Beveled wood siding that is rabbeted along the bottom edge, such as novelty siding.

dolomite A mineral containing calcium and magnesium carbonate used primarily as an aggregate in making concrete. See also **dolomitic limestone**.

dolomitic lime A name commonly used for high-magnesium lime.

dolomitic limestone A form of cut limestone that contains dolomite and is used in building construction.

dome (1) A hemispherical roof such as those commonly used on government structures. (2) A rectangular pan form used in two-way joist or waffle concrete-floor construction.

dome light A dome-shaped skylight made of glass or plastic.

domestic hot-water heater A packaged unit that heats water for household purposes.

domestic sewage Household waste, including that originating from single- or multi-family housing.

dominant estate Pertaining to legal rights and benefits in property deeds. The property receiving the right or benefit is the dominant estate, whereas the property granting the right or benefit is the servient estate.

dook A wooden plug used for attaching trim to a wall.

door A movable member used to close the opening in a wall.

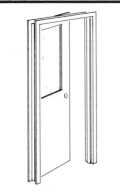

Door

Door Closer

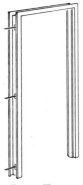

Door Frame

bolt A sliding bar or rod for locking a door.

buck A rough wood or metal frame in which the finished door frame is aligned and attached.

bumper A device placed on the wall or floor behind a door to limit the swing of the door.

casing The finished visible frame into which a door fits.

catch A device for fastening a door, usually attached to only one side of the door.

check An automatic door closer.

cheek See *doorjamb*.

class A fire-rating classification for doors.

clearance The distance or open space between the bottom of a door and a finished floor, or between two double doors.

closer A device that closes as well as controls the speed and force of closure of a door.

contact A switch, for an electrical circuit, that is opened or closed by opening or closing a door.

frame The surrounding assembly, into which a door fits, consisting of two uprights, jambs, and a head over the top.

frame anchor A device used to align and attach the doorframe to the wall. The type of device depends upon the type of wall and the material of the frame.

hand The direction of the swing of a door, expressed as either left-hand or right-hand.

head The horizontal upper member of a door frame; a lintel.

holder A device for holding a door open.

jack A device for holding a wooden door while the hinges are being set or while the door is being planed to fit.

jamb The vertical member on each side of a door frame.

latch A device for holding a door closed. Usually, a beveled bolt self-activated by a spring or gravity when hitting the strike.

light The glass area of a door.

louver The slats, usually in the bottom panel of a door, that permit the passage of air.

mullion The vertical member in the middle of a double-door frame that contains the strike of the two leaves.

nail A large-headed nail against which a knocker strikes. Also, any large-headed exposed nails in a wooden door.

opening The size of a door-frame opening between jambs and from finished floor or threshold to head.

operator A power-operated mechanism for opening and closing an elevator door.

pivot The pin on which a swinging door rotates.

plate A nameplate on the outside of a door.

pocket The opening in a wall or partition that receives a door that slides into the opening. The door is installed on tracks during the rough framing stage of construction.

post See *door jamb*.

rail Any of the horizontal members that connect the stiles of a door. In a flush door, the members are hidden. In a panel door, the members are exposed.

roller The operational hardware used primarily on heavy doors such as refrigeration doors in order to allow them to roll.

saddle A threshold or doorsill.

schedule A table in the contract documents listing all the doors by size, specifications, and location.

sill See *threshold*.

stile The outside upright members of a door.

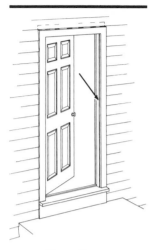

Door Stop

Dormer

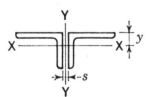

Double Angle

stop The strip on the door frame against which the door closes.

strip A weatherproofing device used around a door.

swing See *door hand.*

switch See *door contact.*

trim The casing or molding around a door frame that covers the joint between the frame and the wall.

way Any opening in a wall that has a door.

dope (1) In plumbing, a material placed on pipe threads to make them waterproof. (2) An additive to mortar or plaster used to accelerate or retard the set.

dormer A structure projecting from a sloped roof, usually with a window.

dormer cheek The side wall of a dormer.

dormer window A window on the front vertical face of a dormer.

dormitory A residential building including sleeping facilities most commonly found at an institution such as a school.

dosing tank A holding tank, for raw or partially treated sewage, that discharges automatically when a desired quantity is accumulated.

dot A small amount of plaster placed on a surface between grounds to aid the plasterer in maintaining a constant thickness.

dote An early form of decay in wood. Usually, the decay can be cut out and the remainder of a piece salvaged.

doty A term used to describe wood that shows an early form of decay.

double-acting butt A butt hinge that swings both ways.

double-acting door, swinging door A door with double-acting hinges that swings both ways.

double-acting frame A door frame without stops which allows a double-acting door to swing both ways.

double-acting hinge A hinge used on a swinging door which swings both ways.

double angle A structural member formed by using two L-shaped angle irons back to back.

double-bitted ax An ax with two cutting edges.

double back A method of applying two coats of plaster in which the second coat is applied before the first coat has set.

double-bend fitting An S-shaped pipe fitting.

double-beveled edge The edge of a swinging door stile that is beveled from the center of the edge toward the two faces.

double-break switch An electrical switch that opens and closes a conductor in two places.

double breasted A firm that operates or has ownership in both open shop and union companies.

double bridging Bridging that breaks the span of joists into three sections.

double-center theodolite A precision instrument used in surveying and with which repeated readings of the same angle can be made. A standard procedure is to take the sum of six readings and divide by six to minimize human error.

double-cleat ladder A wide, site-constructed ladder with a center rail to allow both ascending and descending traffic.

double corner block A concrete masonry unit having four flat rectangular faces.

double course Two thicknesses of roofing or siding shingles laid one over the other so that there is complete coverage of at least two layers overall.

double-cut file A file having two sets of diagonal cutting ridges

Double Door

crossing each other on the face of the file.

double-cut saw A saw with teeth which cut both on the push and the pull stroke.

double door Two single doors or leaves hung in the same door frame.

double-dovetail key A hardwood wedge used to join two timbers. The wedge has a dovetail on each end which is driven into a corresponding recess in each timber.

double drum hoist A hoisting engine for a material or personnel hoist that has two drums which can be separately driven.

double eaves course A double course of roofing installed at the eaves of a sloping roof.

double egress frame A door frame designed for two doors, both of the same hand, swinging in opposite directions. The door that swings toward the viewer is called a "reverse-hand door."

double-end-trimmed Passed through saws to be smoothly trimmed at both ends, commonly in length increments of 2 feet.

double-extra-strong pipe A specification for steel pipe that indicates the wall thickness has been increased to give the pipe double strength.

double-faced hammer A hammer whose head has two striking faces.

double-faced stock AA and AB grades of sanded plywood, both interior and exterior, having a good appearance on both sides.

double Flemish bond Brickwork laid up so that it has a pattern of headers and stretchers laid alternately on both faces.

double floor A floor constructed with a sub-floor under a finished wood floor.

double-framed floor A floor system constructed with both binding joists and common joists.

The ceiling is attached to the binding joists and the floor is laid on the common joists. The binding joists are framed by girders.

double-framed roof A roof framing system in which both longitudinal and lateral members are used, such as one that employs purlins.

double-fronted lot A lot bounded by a street on both the front and the back.

double-gable roof An M-shaped roof with a valley between two gables.

double glazing Fenestration with two panes of glass with an air space between for increased thermal and sound insulation.

doublehanded saw A crosscut saw with a handle on each end, designed to be used by two people.

double-headed nail, duplex nail, form nail, scaffold nail A nail with two heads, one above the other, used for temporary construction such as concrete forms and scaffolds. The nail can secure the work but still be easily extracted.

double header A header joist for an opening made stronger by using two pieces of lumber.

double-hung window A window having two vertically sliding sashes, one above the other, on separate guides so that either or both can be opened at one time. This is the type most commonly used on wood frame houses.

double jack rafter A rafter that joins the roof hip to the valley.

double layer Two layers of gypsum board installed over each other to improve fire resistance or sound-suppression characteristics.

double-lock seam A seam used in metal roofing and ductwork formed by double folding the edges of adjacent sheets and laying them down.

Double T-Beam

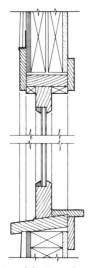

Double Window

double L stair A stair with two landings turning 90 degrees each, one near the top and one near the bottom.

double-margin door An extra-wide door constructed to appear like two doors.

double nailing A method of applying drywall using two nails in cases where one nail would ordinarily be used.

double offset In piping, a dogleg or two changes of direction near to each other.

double partition A partition constructed with two rows of studs to form a pocket door or a cavity for sound proofing.

double-pitched A roof having a second pitch or slope, such as a gambrel roof.

double-pole scaffold A scaffold erected with two rows of uprights independent of an adjacent wall.

double-pole switch An electrical switch having two blades and contacts that open or close both sides of a circuit simultaneously.

double pour In built-up roofing, a second final application of bitumen without another layer of felt.

double-rabbeted frame A door frame with a longitudinal groove cut on each edge so that a door can be hung on each side.

double-rebated frame See **double rabbeted frame**.

double return stair A stair with one flight to a landing, separating to two flights to the next floor.

double roof A roof-framing system in which the rafters rest on purlins.

double-shell tile A ceramic tile with two finished faces separated by a web.

double step In timber framing a double notch, cut into a tie beam supporting the rafters, that reduces horizontal shear.

double-strength glass Plate or float glass which is 1/8-inch thick.

double T-beam A precast concrete member composed of two beams with a flat slab cast monolithically across and projecting beyond the top of the two beams.

double tenons Two tenons side by side in the end of a wood beam. See **tenon**.

double-throw switch An electrical switch that can make connections in either one of two circuits by changing its position.

double-tier partition A partition that is continuous through two stories of a building.

double up See **double back**.

double-wall cofferdam A cofferdam erected with two rows of sheet piling with fill between them and used principally for high cofferdams that require greater stability.

double-walled heat exchanger A heat exchanger with two walls between the collector fluid and potable water.

double waste and vent In plumbing, a single vent serving two fixtures.

double-welded joint A joint welded on both sides.

double window A double-glazed window.

doubling course See **double course**.

doubling piece A cant strip or fillet.

doubly prestressed concrete A concrete member prestressed in two directions perpendicular to each other.

doubly reinforced concrete A concrete member cast with compression reinforcing steel as well as tension reinforcing steel.

doughnut (1) A large washer used as a concrete form accessory to increase bearing in area on the forms. (2) A small, round concrete casting with a hole in the middle used to keep reinforcing

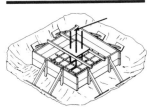

**Dowel-Bar
Reinforcement**

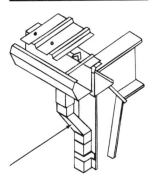

Downspout

Dozer Shovel

steel the desired distance from the forms during the placement of concrete. (3) A concrete collar cast around a column and used to jack up the column.

Douglas Fir *Pseudotsuga menziesii.* A softwood found throughout the Western U.S. and Canada and grown abundantly on the western slopes of the Cascade Mountains. The lumber from this tree is widely used in general construction, as well as in finish applications.

dovetail In finish carpentry, an interlocking joint, shaped like a dove's tail, that is wider at its end than at its base.

dovetail cramp A dovetail-shaped metal fastening used for lifting masonry units.

dovetail cutter A rotary power tool used to shape dovetail joints and mortises.

dovetail half-lap joint A dovetail joint and mortise which are cut halfway through adjoining pieces, thus giving the appearance of a butt or miter joint on the opposite side.

dovetail hinge A strap hinge with both leaves shaped like a dove's tail.

dovetail joint See **dovetail.**

dovetail lath See **rib lath.**

dovetail miter A joint having a concealed dovetail so that it appears like a miter joint.

dovetail plane A wood-finishing tool used for preparing dovetail joints.

dovetail saw A small tenon saw or back saw having a stiffening strip along its back and fine teeth for preparing dovetails.

dowel A cylindrical piece of stock inserted into holes in adjacent pieces of material at a joint. The stock serves to align and/or attach the two pieces.

dowel-bar reinforcement Short sections of reinforcing steel extending from one concrete pour into the next and used to increase strength in the joint.

dowel bit, spoon bit A carpenter's bit used with a brace for drilling dowel holes. The bit is shaped like a half cylinder.

dowel lubricant A lubricant applied to dowels placed in adjoining concrete slabs to allow longitudinal movement in expansion joints.

dowel pin A special nail pointed at both ends and used to fasten mortise-and-tenon joints.

dowel plate A steel plate with holes through which dowels are driven to trim them to size.

dowel screw A threaded dowel.

downdraft A downward current of air such as in a chimney flue or a stairwell.

down-feed system A piping system in which the distribution lines are above the systems being supplied.

downlight A small, direct-lighting unit usually recessed in the ceiling and directing light downward only.

downspout, conductor, leader A vertical pipe used to carry rainwater from a roof to the ground or into a drainage system.

downtime The amount of time a piece of equipment cannot be used for reasons of repair or maintenance.

doze See **dote.**

dozer See **bulldozer.**

dozer shovel A tractor mounted unit which can be used for pushing, digging, or loading earth.

d.p.c. brick A brick whose average water absorption does not exceed 4.5% by weight.

draft The drawing power of air and gasses through a chimney flue. Also, a smooth strip worked on a stone surface to be used as a guide for finishing the surface.

draft bead, draft stop A fillet or bead on a window sill of a double-

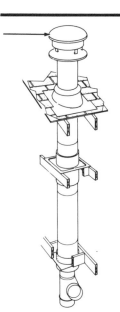

Draft Hood

Drain

Drapery Track

hung window that permits ventilation of the joint, but prevents a draft from entering.

draft curtain A non-combustible curtain, surrounding a high-hazard area, that contains and directs flame, smoke, and fumes.

draft fillet A narrow strip of material on which glass rests in glazing.

draft hood A cap that fits over a chimney flue to prevent downdrafts.

drafting machine A device attached to a drafting table which combines the functions of a T square, triangle, scale, and protractor.

drafting pen One of a variety of drawing instruments used for drafting.

draft stop See **draft bead**.

drag (1) A long serrated plate used to level and score plaster in preparation for the next coat. (2) Any combination of logs, chains, or other materials dragged behind a tractor used to fine-grade earth.

dragline A bucket attachment for a crane that digs by drawing the bucket towards itself using a cable. The attachment is most commonly used in soft, wet materials that must be excavated at some distance from the crane. This device is used extensively in marsh or marine work.

dragon beam A short horizontal member that bisects the angle of the wall plate at the corner of a wood-frame building and bisects the angle brace at its other end. The member supports a hip rafter at one end and is supported by the angle brace at the other end.

dragon tie An angle brace that supports a dragon beam.

drag shovel See **backhoe**.

drain A pipe, ditch, or trench designed to carry away waste water.

drainage Waste water, or the system by which it is carried away.

drainage piping The piping comprising the plumbing of a drainage system.

drainage system All the components that convey the sewage and other waste water to a point of disposal.

drain cock A cock or spigot at the lowest point of a water system where the system can be drained.

drain field See **absorption field**.

drainpipe See **downspout**.

drain tile Tile in short-length sections laid with open joints, usually surrounded with aggregate and covered with asphaltic paper or straw, and used to drain the water from an area.

drain trap See **trap**.

draped tenon See **deflected tendon**.

drapery track A system of tracks or bars that support draperies and allows them to be drawn.

drawbar The bar attached to the back of a tractor for pulling equipment.

drawbore A hole in a tenon which is out of line with the holes in the mortise of a mortise-and-tenon joint, so that the joint is tightened when a pin is driven into the hole.

drawdown The distance by which the groundwater level is lowered by excavation, dewatering, or pumping.

drawer dovetail A dovetail joint in which the tenons of one member do not pass completely through the other member, such as the joint used on the front of a drawer.

drawer kicker A block that prevents a drawer from tilting downward when open.

drawer roller Prefabricated drawer-slide hardware employing rollers to facilitate the opening of a drawer.

drawer runner, drawer slip The strips on which a drawer slides.

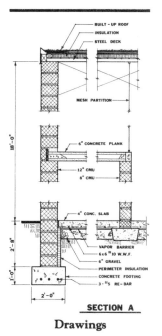

SECTION A

Drawings

Dressed & Matched

drawer slide The hardware on which a drawer slides.

drawer stop A device that prevents a drawer from sliding all the way out.

drawings Graphic and pictorial documents showing the design, location, and dimensions of the elements of a project. Drawings generally include plans, elevations, sections, details, schedules, and diagrams. When capitalized, the term refers to the graphic and pictorial portions of the Contract Documents.

draw-in system An electrical wiring system in which all cables are pulled through conduits, ducts, and raceways.

drawknife A two-handled curved knife used in woodworking which is utilized by pulling it toward the user.

drawn finish A smooth, bright finish on longitudinal stock obtained by drawing the stock through a die.

draw pin A pin that attaches a tongue to a drawbar.

drawshave See **drawknife**.

draw tongue The bar attached to a machine that is used to pull the machine.

dredge (1) To excavate underwater. (2) The equipment for excavating underwater, such as a clamshell, dragline, or suction line.

drencher system A fire-protection sprinkler system that protects the outside of a building.

dressed Building material that has been prepared, shaped, or finished on at least one side or edge.

dressed lumber Lumber that has been processed through a planing machine for the purpose of attaining a smooth surface and uniformity of size on at least one side or edge.

dressed and matched Another term for "tongued and grooved."

The term refers to boards or planks that have been machined so that each piece has a tongue on one edge and a groove on the other.

dressed size The true dimensions of lumber after sawing and planing, as opposed to the nominal size.

dressed stuff See **dressed lumber**.

dresser A plumber's tool for flattening sheet lead and straightening lead pipe.

Dresser coupling A device for connecting unthreaded pipe.

dressing compound Liquid bituminous material used to cover and protect the exposed surfaces of roofing felt.

dress up To fasten and fabricate structural members on the ground before erection.

drier (1) A unit containing a desiccant which is installed in a refrigeration circuit to collect excess water in the system. (2) An additive to paints and varnishes which speeds the drying process.

drift (1) In a water spraying system, the entrained unevaporated water picked up by the air movement through it. (2) In aerial surveying, the angle a plane must crab, or turn its nose into the wind, in order to fly a predetermined line.

driftbolt A short rod or bar, 1 to 2 feet long, driven into prebored holes in timber to secure them to each other. The driftbolt is usually of square stock and of slightly larger section than the hole.

drifter A pneumatic drill used for drilling horizontal or sloped blastholes.

driftpin A tapered pin used in steel erection to align holes for bolting or riveting.

drift plug A hardwood plug that is driven into a lead pipe to straighten it or to flare the end.

drift punch See **driftpin**.

Drill

Drilled-In Caisson

Drop Ceiling

drill (1) A hand-held, manually operated or power-driven rotary tool for making holes in construction materials. (2) A large machine capable of drilling 4-inch diameter blast holes 100 feet deep in rock cuts or quarries. (3) A machine capable of taking core samples in rock or earth.

drill bit The part of a drill which does the cutting.

drilled-in caisson A caisson that is socketed into rock. See also **caisson**.

drill press A rotary drill, mounted on a permanent stand, that operates along a vertical shaft.

drill steel Hollow steel sections that connect the percussion drill with the bit. Air is blown through steel to clean out the drill hole.

drip (1) A groove in the underside of a projection, such as a window sill, that prevents water from running back into the building wall. (2) A condensation drain in a steam heating system.

drip cap A horizontal molding placed over exterior door or window frames to divert rainwater.

drip channel See **drip**.

drip edge The edge of a roof which drips into a gutter or into the open. Also, the metal or wood strip that stiffens and protects this edge.

drip mold, drip molding See **drip cap**.

dripstone A drip cap made of stone.

dripstone course A course of stone projecting out beyond the face of a wall to shed water away from the wall.

drive band A steel band placed around the top of a wood pile to protect the pile during driving.

driven pile A pile that is driven into place by striking it with a pile driving hammer.

driven well A pipe fitted with a

well point driven into water-bearing soil.

drivepoint See **well point**.

drivescrew, screw nail A screw that is driven with a hammer and removed with a screwdriver.

drive shoe A cap placed over the bottom of a pile to protect it during driving.

driveway A private access for vehicular traffic.

driving band See **drive band**.

driving cap A steel cap placed over the top of a pile during driving to protect the pile.

driving resistance The number of blows required to drive a pile a given distance.

drop In air conditioning, the vertical distance that a horizontally projected air stream has fallen when it reaches the end of its throw.

drop apron A metal strip that is bent vertically downward at the edge of a metal roof to act as a drip.

drop bottom bucket A concrete bucket whose bottom opens when it touches the surface where the concrete is to be deposited.

drop bottom-seal See **automatic door bottom**.

drop box An electrical outlet box hung from above, as from a ceiling.

drop ceiling, dropped ceiling A non-structural ceiling suspended below the structural system, usually in a modular grid. A drop ceiling usually contains lighting systems.

drop chute A device used to place concrete with a minimum amount of segregation, such as an elephant's trunk.

drop hammer A pile driving hammer which gains impact by free fall after being lifted by a cable in the leads.

drop panel The part of a cast-in-place flat slab that has been

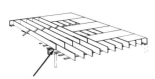

Dropped Girder

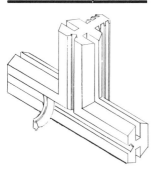

Dry Glazing

thickened on the underside at the location of the columns.

dropped girder A girder occurring below the floor joists and supporting them.

dropped girt A horizontal member occurring below the floor joists and supporting them.

drop-point slating See **diagonal slating**.

drop siding A type of siding that is tongued and grooved or ship lapped so that the edge of each board fits into the edge of the adjoining board.

drop tee A pipe tee with lugs in the sides for attachment to a support.

drop vent In plumbing, a vent that is connected at a point below the fixture.

drop window A vertically sliding window that slides into a recess below the window stool so that the entire opening can be used for ventilation.

drop wire An electrical conductor dropped from a pole to a building in order to supply electric power.

drove A mason's broad, blunt chisel used for facing stone.

drum (1) One of the stone cylinders forming a column. (2) The wall supporting a dome. (3) The rotating cylinder used to wind up cable on a hoist. (4) The cylinder in which transit-mixed concrete is transported, mixed, and agitated.

drum paneling A door panel that is covered with leather.

drum trap A cylindrical plumbing trap usually set flush with a floor so that easy access can be gained by unscrewing the top.

drunken saw, wobble saw A type of circular saw designed to operate with a built-in wobble so that the kerf it makes is greater than the thickness of the saw. This type of saw is used for grooving and for other special purposes in carpentry.

dry area A narrow roofed area, between a basement wall and a retaining wall, constructed to keep the basement wall dry.

dry batch weight The weight of the materials, excluding water, used to make a batch of concrete.

dry-bond adhesive An adhesive that seems dry but which adheres on contact.

dry-bulb temperature The temperature indicated by an ordinary thermometer with an unmoistened bulb.

dry-bulb thermometer An ordinary thermometer for determining the temperature of ambient air.

dry concrete Concrete having a low water content, which thus makes it relatively stiff. The effects are a lower water-cement ratio, less pressure on forms, lower heat of hydration, and a consistency which allows for placement on a sloping surface.

dry construction Building construction without the use of plaster or mortar. The use of dry materials speeds the construction and allows earlier occupancy.

dry course The first course of a built-up roof which lies directly on the insulation without bitumen.

dry density The weight of a soil sample which has been heated to 105 degrees Centigrade (Celsius) to remove the moisture.

dry filter A device for removing pollutants from the air in a system by passing it through various screens and dry porous materials. Most dry filters today in HVAC systems are made of spun fiberglass.

dry glazing The installation of glass by using dry, preformed gaskets in place of a glazing compound.

dry hydrate See **hydrated lime**.

drying inhibitor A substance added to paint to prevent the

Dry-Pipe Sprinkler System

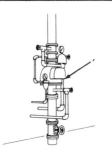

Dry-Pipe Valve

Dry Rubble Construction

surface from drying too rapidly, which could cause a wrinkled or cracked surface.

drying oil An oil used in paints to promote quick, smooth, hard, and elastic drying qualities.

drying shrinkage Contraction by loss of moisture, particularly in concrete, mortar, and plaster.

dry joint A masonry joint without mortar.

dry kiln A chamber in which wood products are seasoned by applying heat and withdrawing moist air.

dry masonry Masonry laid without mortar.

dry mix A concrete or mortar mixture with little water in proportion to its other ingredients.

dry-mix shotcrete Concrete or mortar that is pumped through a hose and to which water is added at the nozzle during application.

dry mortar Mortar mixed to a very dry consistency, but still wet enough for hydration. This type of mortar is used to prevent compaction in the joints of high narrow lifts.

dryout In plywood manufacturing, an unsatisfactory glue bond caused by the glue becoming too dry to adhere properly when pressure is applied in the hot press.

dry-pack To ram a dry mix of mortar or concrete into a space where high strength and little shrinkage is desired, such as under a bearing plate.

dry-packed concrete Concrete with only a small amount of moisture, so that it must be rammed in place in order to be compacted.

dry-pipe sprinkler system A sprinkler system having no water in the pipes until it is activated. This type of system is particularly useful where there is danger of freezing.

dry-pipe valve The valve that activates a dry-pipe sprinkler system. The pipe must be protected from freezing.

dry-powder fire extinguisher A hand-held fire extinguisher that discharges a dry powder by means of a compressed gas. This type of extinguisher is effective for class B and class C fires.

dry press A mechanical device for molding masonry units from a very dry mixture.

dry-press brick Bricks made in a dry press.

dry return In a steam heating system, a return pipe above the water level.

dry-rodded volume A standard method of compacting aggregate when designing a concrete mix.

dry-rodded weight The unit weight of aggregate compacted under standard conditions and used when designing a concrete mix.

dry rodding Compaction of aggregate in a container under standard conditions for determining the unit weight in concrete mix design.

dry rot A type of decomposition in seasoned wood caused by fungi.

dry rubble construction A method of construction using stone rubble but no mortar.

dry shake A concrete surface treatment such as color, hardening, or anti-skid which is applied to a concrete slab by shaking on a dry granular material before the concrete has set, and then troweling it in.

dry sheet The nonbituminous felt or light roofing paper laid between the roof deck and the roofing material to prevent adherence and to allow for independent motion of the two systems.

dry sprinkler See **dry-pipe sprinkler**.

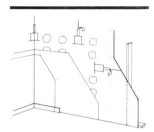

Drywall Construction

Drywall Frame

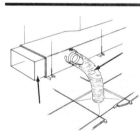

Duct

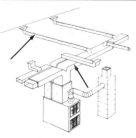

Duct System

dry stone wall A stone wall constructed without mortar.

dry strength The strength of an adhesive joint determined under standard laboratory testing conditions.

dry-tamp process See **dry pack**.

dry tape The application of joint tape to gypsum board using an adhesive other than the standard joint compound.

dry topping See **dry shake**.

dry-type transformer A transformer whose core and coils are not immersed in an oil bath.

dry-volume measurement Measurement of the ingredients of a concrete or mortar mix by using their bulk volume. This method is used principally in mix design.

drywall The term commonly applied to interior finish construction using preformed sheets, such as gypsum wallboard, as opposed to using plaster.

drywall construction Interior construction using drywall rather than plaster.

drywall frame A door frame specifically designed to be used with drywall construction.

dry well A pit or well that is either filled or lined with coarse aggregate or rocks and is designed to contain drainage water until it can be absorbed into the surrounding ground.

dual duct In electrical wiring, a duct having two individual raceways.

dual-duct system An HVAC system using two ducts, one for hot and one for cold air. The air from these ducts is blended in mixing boxes before distribution to each location.

dual-fuel system A heating unit that can use either of two fuels.

dual glazing See **double glazing**.

dual-head nail See **double headed nail**.

dual vent A single plumbing vent which serves two fixtures.

dubbing, dubbing out Filling in irregularities in a plaster wall before the final coat. Also, roughing in a plaster cornice before forming it with the final coat.

duckboard A wooden walkway across a wet area.

duct (1) In electrical systems, an enclosure for wires or cables often embedded in concrete floors or encased in concrete underground. (2) In HVAC systems, the conduit used to distribute the air. (3) In posttensioning, the hole through which the cable is pulled.

duct sheet The metal used in fabricating HVAC ducts.

duct silencer See **sound attenuator**.

duct system The total of connected elements of an air-distribution system through ductwork.

ductwork The ducts of an HVAC system.

due care A legal term defining the standards of performance which can be expected in the execution of a particular task, either by contract or by implication.

dug well A term that refers to a hand-excavated water well that was cased either during or after excavation.

dumbwaiter A small hoisting mechanism or elevator in a building used for hoisting materials only. The dumbwaiter was originally developed by Thomas Jefferson for use in his house at Monticello.

dummy In CPM (Critical Path Method) scheduling, a restraint with no activity and no time.

dummy joint A groove formed on the surface of a concrete slab or wall for appearance and crack control.

Duplex House

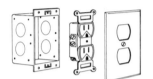

Duplex Receptacle

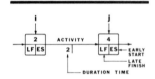

Duration

dumped fill Soil that has been deposited, usually by truck, and which has not been spread or compacted.

dumpling Unexcavated soil, left in the middle of an excavation, which can be used for bracing against sheeting and forms during construction of the subgrade walls.

dumpy level A surveying instrument used to determine relative elevation. The telescope is rigidly attached to the vertical spindle.

dunnage Structural support for a system within a building which is independent of the building's structural frame. An example would be the supports for an air-conditioning cooling tower.

dunter A pneumatic machine for polishing stone surfaces.

duplex A house with two separate dwelling units.

duplex apartment An apartment with rooms on two floors connected by a private staircase.

duplex burner A gas heating system having two burners which can either burn simultaneously for rapid heating or separately during lower heating demand.

duplex cable Two conductors, separately insulated, encased in a common insulated cover.

duplex-head nail See **double headed nail**.

duplex house, two-family house See **duplex**.

duplex outlet See **duplex receptacle**.

duplex receptacle Two electrical receptacles housed in the same outlet box.

duraluminum An aluminum alloy used principally in extrusions and rolled sheets and possessing good anticorrosion properties.

duramen The hard inner portion, or heartwood, of a tree.

duration In CPM (Critical Path Method) scheduling, the estimated time to complete an activity.

Durham fitting A cast iron threaded pitched fitting used on drainage pipes.

Durham system A waste-water system using all threaded pipes and pitched fittings.

durometer An instrument for determining the relative hardness of a material.

dust collector Any device used to collect the dust produced by a machine or tool, such as the dust collectors required at all rock drilling operations and those used with bench saws.

dustfree A term used in paint and varnish drying describing the stage after which dust will no longer stick to the surface.

dust-free time The time required for a paint or varnish application to become dust-free under given environmental conditions.

dusting The development of dust on the surface of a concrete floor. Dusting can be the result of troweling too soon, too much water in the mix, improper mix design, or other reasons.

dust laying oil Usually, a petroleum distilate applied cold for dust control in unpaved areas.

dustproof strike A strike plate for a lock that has a spring plunger which completely fills the bolt hole when the bolt is not in it.

Dutch door A door with two leaves, one above the other.

Dutch door bolt A bolt for locking together the upper and lower halves of a Dutch door.

Dutch lap A method of applying shingles with a head lap and a side lap.

Dutch light A removable sash used in greenhouses.

dutchman In carpentry and joinery, a small piece inserted to cover up a defect or cover a joint.

dwang A crowbar.

Dwelling

dwarf partition A partition that does not extend to the full ceiling height.

dwarf rafter See **jack rafter**.

dwarf wall Any wall of less than one-story height, such as a subwall for a crawl space or a wall that supports sleepers for the floor above.

dwelling Any building designed for permanent living accommodations for one or more persons.

dwelling unit A room or rooms designed for permanent living accommodations for one or more persons.

dynamic analysis Analysis of stresses in a framing system under dynamic loading conditions.

dynamic load A load on a structural system that is not constant, such as a moving live load or wind load.

dynamic loading Loading imposed by a piece of machinery through vibration in addition to its dead load.

dynamic resistance The resistance of a pile to blows from a pile hammer, expressed in blows per unit of penetration.

ABBREVIATIONS

The abbreviations listed below are those most commonly used in the construction industry. Alternative forms (usually nonstandard) are shown in parentheses.

e eccentricity, erg

E Modulus of Elasticity, Engineer

ea. each

EA Exhaust Air

E and OE Errors and Omissions Excepted

EB1S edge bead one side

Econ economy

EDP Electronic Data Processing

EDR Equivalent Direct Radiation

EE eased edges, electrical engineer, errors expected

EEO Equal Opportunity Employer

eff efficiency

EG edge (vertical) grain

ehf extremely high frequency

EHP Effective Horse Power, Electric Horsepower

elec electric or electrical

elev, EL elevation, elevator

EM end matched

EMF Electromotive Force

enam enamelled

encl enclosure

eng engine

engr engineer

EPDM Ethylene Propylene Diene Monomer

eq equal

equip equipment

equiv equivalent

erec erection

ERW Electric Resistance Welding

est estimate

esu electrostatic unit

EV electron volt

evap evaporate

EV1S edge vee one side

EW each way

EWT Entering Water Temperature

ex extra, example

exc excavation, except

excav excavation

exh exhaust

exp expansion

exp bt expansion bolt

ext exterior

extg extracting

extru extrusion

exx examples

DEFINITIONS

Ear

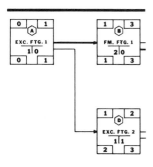

KEY

Early Start
Early Finish

ear (1) A small decorative or structural projecting member or part of a structure or piece. (2) A small metal projection on a pipe by which it can be nailed to a wall. See also **shoulder**.

ear lamp An incandescent lamp, part of whose envelope serves as an ellipsoidal reflector.

earliest event occurrence time In CPM (Critical Path Method) terms, the earliest completion of all the activities that precede the event.

early finish In CPM (Critical Path Method) terms, the first day on which no work is to be done for an activity, assuming work began on its early start time.

early start In CPM (Critical Path Method) terms, the first day of a project on which work on an activity can begin if all preceding activities are completed as early as possible.

early stiffening The premature development of rigidity in freshly mixed portland cement paste, mortar, or concrete. Stiffening occurs without much heat generation from the batch, a condition that can be overcome by further mixing without additional water.

early strength Strength of concrete or mortar developed usually within 72 hours after placement.

earlywood See **springwood**.

earth See **soil**.

earth auger A screw-like drill (usually powered) for cutting cylindrical holes in earth or rock.

earth berm A small, earthen dike-like embankment usually used for diverting run-off water.

earth borer, earth drill Same as *earth auger*.

earth electrode A metal plate, water pipe, or other conductor of electricity, partially buried in the earth so as to constitute and provide a reliable conductive path to the ground.

earthing lead The conductor making the final connection to an earth electrode.

earth pigment Pigment derived from the processing of materials or substances mined directly from the earth.

earth plate A buried metal plate serving as an earth electrode.

earth pressure The horizontal pressure exerted by retained earth.

earthquake load The total force that an earthquake exerts on a given structure.

earthwork A general term encompassing all operations relative to the movement, shaping, compacting, etc. of earth.

eased edge Any slightly rounded edge.

easement (1) The legal right afforded a party to cross or to make limited use of land owned by another. (2) A curve formed at the juncture of two members which otherwise would intersect on an angle.

easing Excavation that allows an allotted space to accommodate a foreign piece or part.

eastern red cedar A distinctively red, fine-textured aromatic wood often used for shingles and closet linings.

eastern S-P-F spruce, pine, fir Softwoods whose primary uses include structural lumber as in posts, beams, framing, and sheathing.

eaves (1) Those portions of a roof that project beyond the outside walls of a building. (2) The bottom edges of a sloping roof.

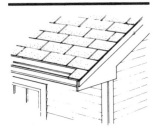

Eaves

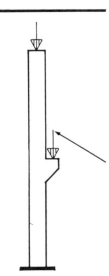

Eccentric Load

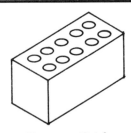

Economy Brick

board A thick, feather-edged board nailed across rafters at the eaves of a building so as to slightly raise the first course of shingles.

channel A long, shallow depression that serves to direct roof drippings along the top of a wall and into downspouts or gargoyles.

course The first course or row of shingles, tiles, etc. at the eaves of a roof.

fascia, fascia board A board secured horizontally to and covering the vertical ends of roof rafters. The board may support a gutter.

flashing A thin metal strip dressed into an eaves gutter from a roof in order to prevent overflowing water from running down the exterior siding.

gutter A long, shallow wood or metal trough fitting under and paralleling the eaves in order to catch and direct water dripping from a roof.

plate A wood beam used horizontally at the eaves, in the absence of a rafter-supporting wall, to support the feet of roof rafters between existing posts or piers.

strut A structural piece spanning columns at the edge of a roof. The term is usually associated with pre-engineered steel buildings.

tile A shorter than usual roofing tile; used specifically in the eaves course.

trough Same as *eaves gutter*.

ebony Dark, durable tropical hardwood usually used for ornamental work or carving.

eccentric Not having the same center or centerline.

fitting A fitting whose opening or centerline is offset from the run of pipe.

load A load or force upon a portion of a column or pile not symmetric with its central axis, thus producing bending.

eccentric tendon In prestressed concrete, a steel element or tendon which does not follow a trajectory coincidental with the gravity axis of the member.

echo The reflection of sound waves with magnitude and time delay sufficient to be perceived by a listener as separate from sound communicated directly to the listener.

economy brick A cored, modular brick with nominal dimensions of 4 inches x 4 inches x 8 inches and intrinsic dimensions of about 3-1/2 inches x 3-1/2 inches x 7-1/2 inches.

economy wall A 4-inch thick, back-mortared brick wall strengthened at intervals with vertical pilasters to support roof framing or walls.

edge beam A stiffening beam at the edge of a slab.

edge form Formwork used to limit the horizontal spread of fresh concrete on flat surfaces such as pavements or floors.

edge grain Lumber sawn so as to reveal the annual rings intersecting the wide face or surface at an angle of at least 45 degrees.

edge joint (1) A joint between two veneers or laminations in the direction of the grain. (2) A joint created by the edges of two boards or surfaces united so as to form an angle or corner.

edge molding Any molding along the edge of a relatively thin member, such as a door or a counter.

edge nailing Nailing through the edges of successive boards, such as floorboards, so that each board conceals the nails in the board

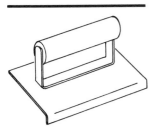

Edger

Edging Strip

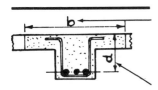

Effective Depth
Effective Flange Width

adjacent to it.

edge plate An angle iron or channel-shaped guard protecting the edge of a door.

edge pull Mortising in the edge of a sliding doors or windows by which they may be pulled open or closed.

edger A tool used to fashion finishing edges or round corners on fresh concrete or plaster.

edge shot Planed on the edges.

edgestone A stone used for curbing.

edge tool Any tool having a sharp cutting edge, such as a chisel, plane, or hatchet.

edge vent An opening at the perimeter of a roof for the relief of water-vapor pressure within the roof system.

edging (1) An edge molding. (2) An edging strip. (3) The process of rounding to reduce the possibility of chipping or spalling of the exposed edges of concrete slabs.

edging strip A plain or molded strip of wood, metal, etc. used protectively on the edges of panels or doors and/or cosmetically to conceal laminations in materials such as plywood.

edging trowel Same as **edger**.

effective area (1) The net area of an air inlet or outlet system through which air can pass. The effective area equals the free area of the device multiplied by the coefficient of discharge. (2) The cross-sectional area of a structural member, calculated to resist applied stress.

effective area of reinforcement In reinforced concrete, the product derived by multiplying the cross-sectional area of the steel reinforcement by the cosine of the angle between its direction and the direction for

which its effectiveness is considered.

effective bond In brickwork, a bond that is completed at the ends with a 2-1/2 inches closer.

effective depth The depth of a beam or slab section as measured from the top of the member to the centroid of the tensile reinforcement.

effective flange width The width of a slab poured monolithically with a stem acting as the compression flange of a T section. This width is usually equal to one-half of the span on each side of the stem.

effective length The distance between inflection points in a column that is bent.

effective opening The minimal cross-sectional area of the opening at the point of water supply discharge, expressed in terms of the diameter of a circle. The diameter of a circle of equivalent cross-sectional area is given in instances where the opening is not circular.

effective prestress The stress remaining in concrete resulting from prestressing after loss of prestress. Effective prestress includes the effect of the weight of the member, but not the effects from any superimposed loads.

effective reinforcement The reinforcement assumed active in resisting applied stress.

effective span The center-to-center distance between the supports of a beam, girder, or other structural member.

effective stress In prestressed concrete, the remaining stress in the tendons after the occurrence of loss of prestress.

effective temperature A single-figure index reflecting the combined effects of temperature, humidity, and air movement on

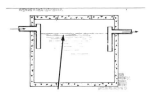

Effluent

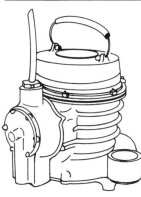

Ejector

the sensation of warmth and/or cold felt by the human body, and numerically equivalent to the temperature of still, saturated air, which produces an identical sensation.

effective thickness of a wall The thickness of a wall to be assumed for calculating the slenderness ratio.

efficiency The ratio of actual performance to theoretical maximum performance.

efflorescence A whitish, powdery deposit of soluble salts carried to the surface of stone, brick, plaster, or mortar by moisture, which itself evaporates but leaves the residue.

effluent In sanitary engineering, the liquid sewage discharged as waste, as in the discharge from a septic tank. Generally, the discharged gas, liquid, or dust by-product of a process.

eggcrate diffuser A metal or plastic eggcrate-shaped diffuser used below a lighting fixture.

eggcrate louver A louver whose rectangular openings resemble the dividers used in some egg containers.

eggshell A semi-matte glaze or porcelain enamel surface that resembles the texture of an eggshell.

eggshell gloss A low-gloss designation for a paint film that is more glossy than flat or satin, but less glossy than semi-gloss or glass. (2) A smooth matte face to building stone.

egress An exit or means of exiting.

ejector (1) A pump for ejecting liquid. (2) In plumbing, a device used to pump sewage from a lower to a higher elevation.

ejector grill A ventilating grill having slots shaped so as to force out air in divergent streams.

elastic A term describing the ability of a material to return to its initial

state after having been deformed by an outside force.

elastic arch An arch designed on the concept of the elastic theory of materials.

elastic constant A constant or coefficient that expresses the degree to which a material possesses elasticity. In an elastic material that has been subjected to strain below it elastic limit, the elastic constant is the ratio of the unit stress to the corresponding unit strain.

elastic curve In structural analysis, the curve showing the deflected shape of a beam with an applied load.

elastic deformation A modification in the shape of an elastic material, but not of its elastic purpose.

elastic design An analytical method in which the design of a structural member is based on a linear stress-strain relationship that assumes that the active stresses are equal to only a fraction of the given material's limit of elasticity.

elasticity See **elastic.**

elastic limit The maximum to which a material can sustain deforming stress and still return to its initial shape without incurring any permanent deformation.

elastic loss In pretensioned concrete, the condition resulting in the reduction of prestressing load due to the elastic shortening of a member.

elastic shortening (1) A linearly proportional decrease in length of a structural member under an imposed load. (2) In prestressed concrete, the shortening of a member caused by, and occurring immediately upon, application of force induced by prestressing.

elastomer A term descriptive of various polymers which, after being temporarily deformed by

Elbow

Electric Baseboard Heater

Electric Drill

stress to a substantial degree, will return to their initial size and shape immediately upon the release of the stress.

elastomeric bearing An expansion bearing made from an elastomer. This type of bearing allows movement of the structure it supports.

elastomeric waterproofing A method of foundation moisture protection using impervious flexible sheet metal.

elbow A rather sharply bent or fabricated angle fitting, usually of pipe, conduit, or sheet metal.

elbow board (1) A window board installed under a window on the interior side. (2) An elbow rail.

elbow catch Usually, a spring-loaded locking device having on one end a hook that engages a strike, and on the other end a right angle bend that provides a means for releasing the catch. This type of catch is commonly used to lock the inactive leaf of a pair of cabinet doors.

elbow rail A length of millwork or metal rail attached to a partition or under a window on the interior side and used as an armrest.

electrical insulator An electronics component which, because of the electrical resistance inherent in the material(s) from which it is fabricated, is itself considered to be a nonconductor of electric current. The component is sometimes designed so as to physically support a conductor while electrically separating it from another conductor or object.

electrically supervised Usually descriptive of a closed-circuit wiring system that utilizes a current-responsive device to indicate a failure within the circuit or an accidental grounding.

electrical resistance The opposition, measured in ohms, to the flow of an electric current as manifested by a device, conductor, substance, or circuit element.

electrical tape A tough fibrous or thermoplastic adhesive tape used primarily to insulate electrical joints or conductors.

electric arc welding The joining of metal pieces by fusion for which the heat is provided by an electric arc or the flow of electric current between the electrode and the base metal. The electrode is a consumable, melting and bead-depositing, or non-consumable, metal rod. The base is the parent metal.

electric baseboard heater A heating system with electric heating elements installed in longitudinal panels usually along the baseboards of exterior walls.

electric blasting cap A blasting cap that is detonated by an electric current.

electric cable (1) An electric conductor comprising several smaller-diameter strands that are twisted together. (2) A group of electric conductors that are insulated from one another.

electric delay blasting caps Electronically controlled blasting caps designed to detonate only after a predetermined length of time has elapsed, after the ignition system has been electronically activated.

electric drill A portable, hand-held, motor-driven tool used to turn a bit for boring holes in a material and powered either by direct or alternating current.

electric eye A photoelectric cell whose resistance varies in response to the light that falls on the cell. Hence, an electric eye is often incorporated into electric circuits where it is used in measuring or control devices that depend on illumination levels or on the interruption of a light beam.

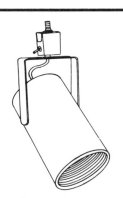

Electric Fixture

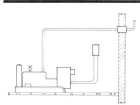

Electric Generator

Electric Panelboard

electric feeder In power distribution, a set of electric conductors that originate at a primary distribution center and proceed to supply one or more secondary distribution centers, one or more branch-circuit distribution centers, or a combination of these. See **feeder**

electric field A particular region or space characterized by the existence of a detectable electric intensity at every point within its perimeters.

electric fixture An electrical device that is fastened to a ceiling or wall and used to hold lamps.

electric generator A mechanism that transforms mechanical power into electrical power.

electric heat Heat resulting from the flow of electric current through resistance elements.

electric heating element A heat-producing device consisting of a length of resistance material, insulated supports, and terminals that connect to an electrical power source.

electric lock A locking mechanism in which the motion of a latch or bolt is controlled by applying voltage to the terminals of the mechanism.

electric meter A device that measures and registers the integral of an electrical quantity with respect to time.

electric operator An electrically powered device used for opening and/or closing casement windows, doors, or dampers.

electric outlet A point in an electric wiring system at which current is taken through receptacles equipped with sockets for plugs, making it available to supply lights, appliances, power tools, and other electrically powered devices.

electric panelboard A panel or group of individual panel units

capable of being assembled as a single panel and designed to include fuses, switches, circuit breakers, etc. and to be housed in a cabinet or cutout box positioned in or against a wall or partition and accessible only from the front.

electric receptacle A contact device, usually installed in an outlet box, which provides the socket for the attachment of a plug to supply electric current to portable power equipment, appliances, and other electrically operated devices.

electric resistance welding A coalescence-producing set of welding processes whereby the required heat is derived from the application of pressure and from the resistance of the work to the flow of electric current in a circuit of which the work itself is an integral part.

electric space heater A portable, self-contained heating device in which electricity supplies the heat energy which is then blown into a given space or room by a powerful electric fan also contained within the unit.

electric squib An electronically activated mechanism used in blasting operations to ignite a charge.

electric strike An electrical mechanism that allows the release of a door at a remote location.

electric switch A device used for opening or closing electric circuits or for changing the connection of a circuit.

electric welding See **electric arc welding** and/or **electric resistance welding**.

electrode (1) A solid electric conductor through which an electric current enters or leaves an electrolyte, vacuum, gas, or other medium. (2) That component in an arc welding circuit through which electric current is

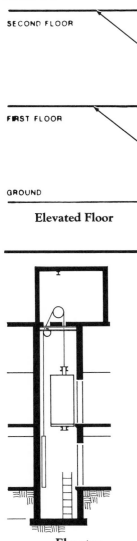

Elevated Floor

Elevator

conducted between the electrode holder and the arc itself. (3) That component in resistance welding through which the electric current in the welding machine passes directly to the work and which is usually accompanied by pressure, such as a clamp.

electrode boiler A boiler containing submerged electrodes that heat the surrounding water by passing electric current through it.

electrogalvanizing A method of coating iron or steel with zinc deposited by electroplating rather than by immersion.

electrogas welding A manner of gas metal-arc welding or flux-cored arc welding that utilizes an externally supplied gas.

electrolier A hanging electric light fixture, such as a chandelier.

electrolysis The process of producing chemical changes or the breakdown of chemical compounds into their constituent parts by passing a current between electrodes and through an electrolyte.

electrolyte (1) A substance that separates into ions when in a solution or when fused, thereby becoming electrically conducting. (2) A conducting medium within which the flow of electric current occurs by the migration of ions.

electrolytic copper Copper refined through electrolytic deposition and used in the manufacturing of tough pitch copper and copper alloys.

electrolytic corrosion The deterioration of metal or concrete by chemical or electrochemical reaction resulting from electrolysis.

electromagnet An iron-based object that becomes magnetic when a current is passed through the coils around it. The object looses its magnetis when the current is stopped.

electronic controls In an HVAC system, the electronically operated sensors and controls.

electroplating The deposition of a thin, sometimes decorative but usually protective metal coating onto another metal. The deposition is achieved by submersing the receiving metal into an electrolyte, where electrons from the anode (the coating metal) flow to and are deposited on the receiving metal (or cathode).

electrostatic air cleaner Same as **electrostatic precipitator**.

electrostatic precipitator A type of filtering device that prevents smoke and dust from escaping into the atmosphere by charging them electrically as they pass through a screen, from which they are attracted to one of two electrically charged plates and subsequently removed.

elephant trunk A long, cylindrical, tubular chute-like device having a hopper-like top through which concrete is placed in deep shafts or forms. When in use the device is kept filled with concrete to eliminate the free-fall material and the possible resultant segregation of its constituents.

elevated conduit An electrical conduit hung from structural floor members.

elevated floor Any floor system not supported by the subgrade.

elevated slab A floor or roof slab supported by structural members.

elevated water tank A domestic water tank supported on an elevated structural framework to obtain the required head of pressure.

elevation (1) A vertical distance relative to a reference point. (2) A view or drawing of the interior or exterior of a structure as if projected onto a vertical plane.

elevator A "car" or platform that moves within a shaft or guides and is used for the vertical hoisting

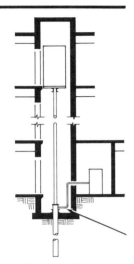

Elevator Pit

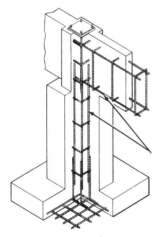

Embedded Reinforcement

and/or lowering of people or material between two or more floors of a structure. An elevator is usually electrically powered, although some short-distance elevators (serving fewer than six or seven floors) are powered hydraulically.

elevator car That part of an elevator that includes the platform, enclosure, car frame, and gate or door.

elevator car safety A mechanism that is attached to the car frame of an elevator or to the frame of the counterweight and whose function it is to slow, stop, and hold the car or counterweight in the event of integral equipment malfunction or failure, such as excessive speed or free-fall of the car.

elevator interlock A device on the door of each elevator landing which prevents movement of the elevator until the door is closed and locked.

elevator machine beam A steel beam that is usually positioned directly over the elevator in the elevator machinery room and is used to support elevator equipment.

elevator pit That part of an elevator shaft that extends from the threshold level of the lowest landing door down to the floor at the very bottom of the shaft.

elevator shaft A hoistway through which one or more elevators may travel.

ell, el (1) An extension, addition, or secondary wing that joins the principal dimension of a building at right angles. (2) Same as **elbow**.

elliptical arch An arch whose form is that of a semi-ellipse and whose intrados is in the form of an ellipse.

elm A strong, tough, dull brown, easily worked hardwood of moderately high density and twisted, interlocking grain. Elm is

used more for ornamental and shade purposes than for veneer, piles, or planks.

elongated piece A particle of aggregate whose length/width ratio exceeds a specified value.

embankment A ridge constructed of earth, fill rocks, or gravel. The length of an embankment exceeds both its width and its height. The usual function of an embankment is to retain water or to carry a roadway.

embedded length That length of embedded steel reinforcement that extends beyond a critical section.

embedded reinforcement In reinforced concrete, any slender members such as rods, bars, or wires which are usually made of steel and are embedded in concrete in such a way as to function together with the concrete in resisting forces.

embossed Raised or indented on the surface of a material.

emergency exit A door, hatch, or other device leading to the outside and usually kept closed and locked. The exit is used chiefly for the emergency evacuation of a building, airplane, etc. when conventional exits fail, are insufficient, or are rendered inaccessible, as by fire.

emergency lighting Temporary illumination provided by battery or generator and essential to safety during the failure or interruption of the conventional electric power supply.

emergency power Electricity temporarily produced and supplied by a standby power generator when the conventional electric power supply fails or is interrupted. Emergency power is essential in operations like hospitals, where even relatively short power outages would be life threatening.

emergency release On a door, a safety device different from a panic exit device, but one which also allows exit during emergency conditions.

emergency stop switch A safety switch in the cab of an elevator which can be manually operated so as to remove electric power from the driving machine motor and brake of an electric elevator or from the electrically operated valves and/or pump motor of a hydraulic elevation.

emery Bluish-grey or black, impure carborundum in granular form used as an abrasive for grinding and polishing the surfaces of materials such as metal, stone, or glass.

emery cloth A finely abrasive cloth manufactured by sprinkling powdered emery or aluminum oxide onto a thin cloth coated with glue and used either wet or dry for polishing metallic surfaces.

emery wheel An abrasive wheel similar to a grinding wheel but composed primarily of emery (or aluminum oxide) and rotated at high speeds for fine grinding or polishing.

eminent domain The legal right or power of a government to take, for public use, hitherto privately owned property, usually with some degree of compensation to its owner.

emissivity The ratio of radiation intensity from a given body or surface to the radiation intensity at the same wave length from a perfect blackbody at the same temperature and in the same ambience.

emittance A percentage of the energy absorbed by a solar energy collector.

employer's liability insurance Insurance that protects an employer from his employees' claims for damages resulting from sickness or injury sustained during their course of work and based on negligence of common law rather than on liability under workmen's compensation.

empty-cell process A procedure in which wood is impregnated with liquid preservatives under pressure.

emulsified asphalt Asphalt cement that has been mixed with water containing a small amount of an emulsifying agent.

emulsifier A substance that serves to modify the surface tension of colloidal droplets, inhibiting their coalescence and maintaining their suspension.

emulsion (1) A mixture of two liquids that are insoluble in one another and in which globules of one are suspended in the other (as oil globules in water). (2) The mixture of solid particles and the liquid in which they are suspended but insoluble, as in a mixture of uniformly dispersed bitumen particles in water, where the cementing action required in roofing and waterproofing would occur as the water evaporates.

emulsion glue A usually cold-setting glue manufactured from emulsified synthetic polymers.

emulsion paint A paint comprising tiny beads of resin binder that, along with pigments, are dispersed in water which, upon evaporation, effects the coalescence of the resin particles, thus forming a film that adheres to the surface and binds the pigment particles. Vinyl, latex, or plastic paints are examples and may be based on polyvinyl acetate emulsion.

emulsion sprayer An apparatus used in highway resurfacing to spray an emulsified asphalt tack coat on the existing surface.

enamel A type of paint composed of particles of finely ground pigments and a resin binder which dries to form a hard glossy film with very little surface texture.

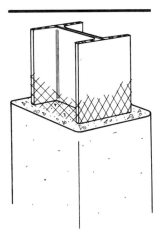

Encase

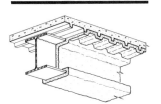

Encased Beam

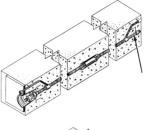

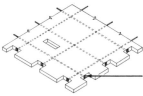

End Anchorage

enameled brick In masonry, a brick or tile with a glazed ceramic finish.

encase To cover with or enclose in a case or lining.

encased beam A metal beam usually enclosed in concrete or some other similar material.

encaustic (1) Descriptive of a process in which a material is covered with a mixture of paint solution and wax, then set by heat after application. (2) Relating to paint or pigment which has been applied to glass, tile, brick, or porcelain and then set or fixed by the application of heat.

encaustic tile A decorative or pavement tile whose pattern is inlaid in clay of one color in a background of a different colored clay.

enchased Describing hammered metalwork whose pattern in relief is effected by hammering down the background or depressed portions of the design.

enclosed knot A wood knot that is invisible from the surface of a wood member because it is completely covered by surrounding wood.

enclosed stair An interior staircase having a closed string on each side, often encased in walls or partitions, and having door openings at various floor levels, thus making it accessible to hallways or living units.

enclosure wall An interior or exterior non-load-bearing wall of skeleton construction and usually anchored to columns, piers, or floors.

encroachment The unauthorized intrusion or extension of a building or structure onto the property of another.

encumbrance (1) A lien or other claim on a piece of real property. (2) A restriction on the use of a piece of property.

end anchorage A mechanism designed to transmit prestressing force to the reinforced concrete of a posttensioned member.

end bearing pile A pile supported mainly by point resistance. The pile's point (toe) rests on or is imbedded in a posttensioned member.

end block A member having an enlarged end section designed to reduce anchorage stresses to allowable values.

end butt joint (1) A joint formed when boards are connected square end to square end. (2) A joint between two veneers, formed perpendicular to the grain. (3) In masonry, a joint in which mortar connects the butt ends of two bricks.

end channel A metal stiffener welded horizontally into the tops and bottoms of hollow metal doors to supply strength and rigidity.

end checks Small cracks that develop in the end grain of drying lumber.

end-construction tile Tile that receives its principal stress parallel to the axes of the cells, and which is laid so that the axes of cells are vertical.

end distance The distance from an end of a member to the center of its nearest bolt hole, nail, or screw.

end grain The grain that is exposed when a piece of wood is cut perpendicular to the grain.

end-grain core Plywood or panel core made from wood blocks that are sawn and glued in such a way that the grain is at right angles to the forces of the panels.

end-grain nailing The driving of nails into the end grain surface of wood so that the resultant position of each nail shank is parallel to the grain.

end joint See **end butt joint**.

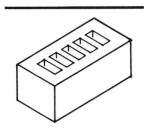

Engineer Brick

end lap The overlap in a lap joint, such as at the end of a ply of roofing felt.

end lap joint An angle joint involving two members, each having been cut to half its thickness and lapped over the other in such a manner as to result in a change of direction.

endless saw A band saw.

end matched Descriptive of boards or strips that are tongued along one end and grooved along the other.

endothermic Descriptive of a reaction occurring with the absorption of heat.

end post A post or other structural member in compression at the end of a truss.

end product specifications Embankment compaction specifications written to define the requirements rather than the method or equipment to obtain those requirements.

end scarf A scarf joint between two timbers made by notching and lapping the ends and then inserting the end of one into an end of the other.

end section In a drainage system, the prefabricated, flared metal end attached at the inlet and outlet to prevent erosion.

end span In a continuous-span floor design, the exterior span, which is often more heavily reinforced.

end stiffener One of the vertical angles connected to and serving to stiffen the web of a beam or girder at its ends and to transfer the end shear to the shoe, baseplate, or supporting member.

end thrust The force that the end of a structural member exerts.

endurance limit The maximum stress that can be applied to a metal during a specified number (usually 10 million) of stress cycles without causing failure of the material being stressed.

energized Electrically connected to a source of voltage.

energy audit A survey of heat loss through the components of a structure.

energy efficiency ratio In HVAC, the ratio of cooling capacity in Btu/hr to the electricity required in watts.

energy head The elevation of the slope of open or underground water measured at any section, plus the head.

enfilade The axial alignment of a series of doors through a sequence of rooms.

engaged Apparently or actually attached to a wall by virtue of being bonded to it or partially embedded in it.

engaged column A column that is attached or apparently attached to a wall by being bonded to it or partially embedded in it. The column is not actually free-standing, but is at least partially built into a wall.

engineer A person trained, experienced in, or licensed to practice the profession of engineering.

engineer brick Brick whose nominal dimensions are 3-1/5 inches x 4 inches x 8 inches.

engineering officer A person designated by a government agency or by a private corporation as having authoritative control over specified engineering operations and/or duties.

engineering survey A survey undertaken for the purpose of obtaining information essential for the planning of an engineering project.

engineer-in-training A person designated by statute as being qualified for professional engineering registration in every respect but the required professional experience.

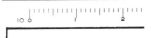

Engineer's Scale

**English Garden
Wall Bond**

Entrance

Equalizing Bed

engineer's chain A chain-like device consisting of a series of one-hundred links, each 1 foot long. The chain is used in land surveying for measuring distances.

engineer's level A precision leveling instrument used for establishing a horizontal line of sight, and hence for determining differences of elevation.

engineer's scale A straightedge on which each inch is divided into uniform multiples of 10, thus enabling drawings to be made with distances, loads, forces, and other calculations expressed in decimal values.

engineeer's transit (1) A instrument used in surveying to measure and to lay out vertical and horizontal angles, distance, directions, and differences in elevation. (2) A theodolite, the directions of whose alidade and telescope are reversible.

English basement A story that is partially below but mostly above the ground level of a residential building, but which does not contain the main entrance to the building.

English bond (1) Brickwork consisting of alternate courses of headers and stretchers. (2) A strong, easily laid bond.

English cross bond Alternate courses of headers and stretchers on which the stretcher course breaks joints with the stretcher courses next to it. Also called "St. Andrew's cross bond."

English garden wall bond Widely used brickwork that can be laid quickly because headers constitute only every fifth or sixth course, with all the other courses being stretchers. Also called "American bond or common bond."

English tile A smooth, flat single-lap clay roofing tile with interlocking sides.

enriched Adorned with ornate and/or decorative elements.

entrained air Nearly spherical microscopic air bubbles which are purposely incorporated into mortar or concrete during mixing. Entrained air is added in order to lessen the effects of the freeze/thaw cycle for concrete that will be exposed to the weather.

entrance An exterior door, lobby, passage, or other designed point of entry into a building.

entrapped air Voids of at least 1 millimeter in diameter in concrete which are the result of air other than entrained air.

entry A passage hallway that affords entrance inside an exterior door.

envelope A term used to denote the extreme outside surface and dimensions of a building.

environmental design The adaption of a building to its surroundings and the consequences of its incorporation into the setting.

environmental impact statement A detailed analysis, as required by the National Environmental Policy Act of 1969, of the probable significant environmental effects of either large-scale, federally funded construction or other proposed construction having potentially significant consequences upon the environment.

epoxy Any of several synthetic, usually thermosetting, resins that provide superior adhesives and hard, tough, chemical, and corrosion-resistant coatings.

epoxy joint In masonry, an unsealed joint in which epoxy resin is used in place of mortar.

epoxy resin A material containing an average of more than one epoxy group per molecule.

equalizing bed A layer of sand, stone, or concrete laid in the

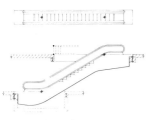

Escalator

Escutcheon

bottom of a trench as a resting place for buried pipes.

equalizer Any device that compensates for an undesired characteristic of a system or component.

equipment ground (1) A connection providing a path to ground from any equipment housing whose exposed metal parts could become energized, as from the failure of insulation around a conductor within the equipment. (2) A ground connection to any metal part of a wiring system or equipment that does not carry current.

equity The residual value of a business or property, often calculated by the subtraction of outstanding liens or mortgages from the total value of the business or property.

equivalent embedment length That longitudinal part of an embedded reinforcement which is capable of developing a stress equivalent to that developed by a hook or mechanical anchorage.

equivalent round A dimensional term used to define the size of an oblong-shaped pipe. The equivalent round of the oblong pipe is equal to the diameter of a circular pipe with an equal circumference.

erecting bill A list of structural components for a building prepared by the fabricator and indicating the tagging and location of each member.

erection The positioning and/or installation of structural components or preassembled structural members of a building, often with the assistance of powered equipment such as a hoist or crane.

erection bracing Usually, temporary bracing used to hold framework in a safe condition during construction until such time as enough permanent

construction is in place to provide complete stability.

erection drawing A shop drawing that illustrates the constructional components of an entire project or part thereof, each being lettered or numbered so as to facilitate erection.

erection stress The stress effected by loads applied during erection.

erosion The gradual deterioration of metal or other material by the abrasive action of liquids, gases, and/or solids.

errantum An error in printing, writing, or editing, especially one included in a list of corrections appended to a book or publication in which the error appeared.

errors and omissions insurance Professional liability insurance protecting architects or engineers from claims for damages which may result from alleged professional negligence.

escalator A continuously moving, power-driven inclined stairway usually used to transport passengers between different floor levels of a building or into and out of underground stations.

escape (1) The curved portion of a column shaft where it springs out from the base.

escutcheon (1) The protective plate, usually metal or plastic, which surrounds a door keyhole or a light switch. (2) A pipe flange covering a floor hole through which the pipe passes.

escutcheon pin A small nail, usually brass and often ornamental, used to fix an escutcheon.

esker A long narrow deposit of coarse gravel left by a stream.

espalier (1) A framework or trellis that serves to direct branches of fruit trees or other decorative plantings to grow in a predesigned shape or to form in a single plane, for improved air circulation and

171

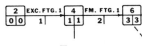

Event

Excavation

exposure to sun as well as for ornamental effect. (2) The trees or plants grown in this manner.

esplanade A flat, unobstructed stretch of grass or pavement, usually for promenading but sometimes for driving and observing a view.

esquisse A rough sketch demonstrating the general features of a project.

Essex board measure A chart appearing on a certain type of carpenter's steel framing square, listing the number of board feet contained in a board 1-inch thick and of several standard sizes.

estimate The anticipated cost of materials, labor, services, or any combination of these for a proposed construction project.

estimated design load The sum of: (1) the useful heat transfer, (2) the heat transfer to or from the connected piping, and (3) the heat transfer that occurs in any auxiliary apparatus connected to a heating or air conditioning system.

estimated maximum load The calculated maximum heat transfer that a heating or air conditioning system might have to provide.

estimating The process of determining the anticipated cost of materials, labor, and equipment of a proposed project.

etching The process of using abrasive action or a strong acid to wear away the surface of glass or metal, often in a decorative pattern.

ethylene glycol (1) Anti-freeze. (2) A water-miscible alcohol used to transfer heat in heating and cooling systems and to provide stability in latex- and water-based paints during freezing conditions.

ettringite A naturally occurring mineral containing a large amount of sulfate calcium sulfoaluminate. Ettringite is also produced synthetically by sulfate attack on mortar and concrete.

eurythmy In architecture, orderliness or harmony of proportion.

evaporable water Water held by the surface forces or in the capillaries of set cement paste, measured as the water that can be removed by drying under specific conditions.

evaporation Loss from a liquid of moisture in vapor form.

evaporative cooling Cooling achieved by the evaporation of water in air, thus increasing humidity and decreasing dry-bulb temperature.

evaporator The part of a refrigeration system in which vaporization of the refrigerant occurs, absorbing heat from the surrounding fluid and producing cooling.

event In a CPM (Critical Path Method) arrow diagram, the starting point for an activity that cannot occur until all work preceding it has been performed.

eviction The lawful or unlawful removal of a tenant from a property.

excavation (1) The removal of earth, usually to allow the construction of a foundation or basement. (2) The hole resulting from such removal.

excavator (1) A company or individual who contracts to perform excavation. (2) Any of several types of power-driven machines used in excavation.

excelsior Thin, curved wood shavings usually for packing and stuffing.

exfiltration The flow of air outward through walls, joints, or other apertures.

exfoliated vermiculite Vermiculite whose original volume has been expanded several times by heat processing so as to render it suitable for lightweight

Exhaust Fan

**Exit Device,
Panic Exit Device**

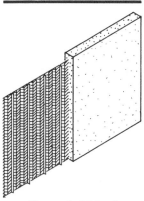

**Expanded Metal
Expanded Metal Partition**

aggregation, such as thermal insulation.

exfoliation (1) The flaking, scaling, or peeling of stone or other mineral surfaces usually caused by physical weathering, but sometimes by heating or chemical weathering. (2) The heat treatment of minerals for the purpose of expanding their original volume many times over.

exhaust fan A device used to draw unwanted contaminated air away from a particular room or area of a building to the outside.

exhaust fume hood A prefabricated hood unit that serves to confine noxious or toxic fumes for subsequent exhausting or filtration.

exhaust grille A grate or louvers through which contaminated air exits to the atmosphere.

exhaust shaft A duct through which exhaust air is carried to the outside atmosphere.

exhaust ventilation A method of ventilation that allows fresh air to enter a space through available or controlled openings and employs mechanical means such as fans to remove foul air from the same space.

existing building In codes and regulations, an already completed building or one that prior laws or regulations allow to be built.

exit That part of an exit system which, because of its separation from the rest of the building by devices such as walls, doors, floors, provides a reasonably safe and protected emergency escape route from a building.

exit access The corridor or door leading to an exterior exit.

exit control alarm An electronic device that activates an alarm when a fire exit door is opened.

exit corridor An enclosed passageway or corridor that connects a required exit with

direct or easy access to a street or alley.

exit device, panic exit device An exit door-locking device, consisting of a bar across the inside of a door, which when pushed releases the door latch.

exit discharge That part of an exit system beginning at the termination of the exit at the exterior of a building and ending at the ground level.

exit light An illuminating sign above an exit, identifying it as an exit.

exothermic Descriptive of a reaction during which heat is released.

expanded aluminum grating An aluminum grating manufactured by cutting and mechanically stretching a single piece of sheet aluminum.

expanded blast-furnace slag A cellular material, used as lightweight aggregate, derived from treating molten blast-furnace slag with water, steam, and/or other agents.

expanded cement See **expansive cement**.

expanded clay Clay expanded to several times its original volume by the formation of internal gas caused by heating it to a semi-plastic state.

expanded glass, foam glass A thermal insulation having a closed-cell structure, manufactured by foaming softened glass so as to produce a myriad of sealed bubbles, and then molded into boards and blocks.

expanded metal A type of open-mesh metal lath made in different patterns and thicknesses by slitting and stretching sheet metal.

expanded metal partition A partition constructed from thin framing or support members covered with heavy expanded

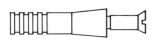

Expansion Anchor
Expansion Bolt

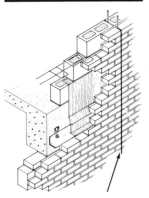

Expansion Joint

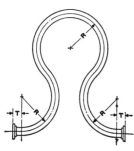

Expansion Loop

metal lath and plastered on both sides to produce a solid finished product 1-1/2 inches to 2-1/2 inches thick.

expanded perlite A glassy, lightweight naturally occurring cellular volcanic material whose properties make it suitable for aggregate in concrete.

expanded plastic, cellular plastic (1) A plastic containing myriads of cells incorporated uniformly throughout its mass. (2) A plastic that has been expanded chemically, mechanically, or thermally so as to form a lightweight, closed-cell structure. The material is commonly used for thermal insulation.

expanded polystyrene A type of foamed styrene plastic having a high resistance to heat flow and a high strength-to-weight ratio.

expanded rubber Closed-cell rubber manufactured from a solid rubber compound.

expanded shale A material suitable as a lightweight aggregate and manufactured by heat treating shale, which dramatically expands its volume.

expanded slate Slate whose subjection to exfoliation results in a porous material suitable as a lightweight aggregate.

expanding bit A drilling bit whose blade can be adjusted to bore holes of different diameters.

expanding cement See **expansive cement**.

expanding plug A drain stopper most often located at the lowest point of a piping system. When inflated, the plug acts to seal a pipe.

expansion The increase in dimensions or volume of a body or material caused by thermal variations, moisture, or other environmental conditions.

expansion anchor See **expansion bolt**.

expansion attic In a completed house, an attic left unfinished to allow for its future conversion into liveable area.

expansion bearing An end support of a span, which allows for the expansion or contraction of a structure.

expansion bend, expansion loop In a pipe run, a usually horseshoe-shaped bend inserted to allow thermal expansion of the pipe.

expansion bit See **expanding bit**.

expansion bolt An anchoring or fastening device used in masonry, which expands within a predrilled hole as a bolt is tightened.

expansion coil An evaporator fabricated from tubing or pipe.

expansion fastener See **expansion bolt**.

expansion joint In a building structure or concrete work, a joint or gap between adjacent parts which allows for safe and inconsequential relative movement of the parts, as caused by thermal variations or other conditions.

expansion joint cover The prefabricated protective cover of an expansion joint, which also remains unaffected by the relative movement of the two joined surfaces.

expansion joint filler Material used to fill an expansion joint and to keep it clean and dry. Common materials used are felt, rubber, and neoprene.

expansion loop See **expansion bend**.

expansion shield Same as **expansion bolt**.

expansion sleeve Usually, a short length of metal or plastic pipe built into a wall or floor to allow

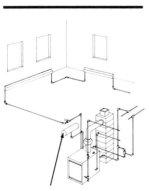

Expansion Tank

Explosion Proof Box

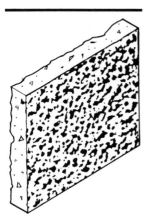

Exposed Aggregate

for the inconsequential expansion and/or contraction of the element (usually another, smaller pipe) which passes through it.

expansion strip (1) The material used in an expansion joint. (2) Resilient insulating material used as a joint filler between a partition and a structural member.

expansion tank In a hot-water system, the tank that provides for the increased volume of the water heated in the heating tank below it.

expansion valve The valve that controls the flow of refrigerant to the cooling element in a refrigeration system.

expansive cement, sulfoaluminate cement A type of cement whose set paste increases in volume substantially more than the set paste of portland cement. Hence, it is often used in applications where the desired results include the compensation for volume decrease due to shrinkage, or the induction of tensile stress in reinforcement.

expansive-cement concrete A concrete made from expansive cement for the purpose of reducing or controlling volume changes that occur during curing.

expert witness In a court case or other legal or arbitration proceeding, an individual whose exceptional knowledge, experience, or skill in a particular field or subject is recognized by the court, or by those presiding over other legal or arbitration proceedings, as being qualified to render an authoritative opinion in matters relating to his field of expertise.

expletive (1) In masonry, the stone used to fill a cavity. (2) That which is used to fill a void of a machine, device, apparatus, or construction.

exploded view An illustration depicting the disassembled individual components in proper relationship to their assembled positions therein.

explosion proof box Electrical-equipment housing designed in compliance with hazardous location requirements to withstand an explosion that could occur within it and to prevent ignition by such explosion of flammable material, liquid, or gas that might externally surround it.

explosion proof fixture Electric lighting that will not explode or cause ignition when in contact with a flammable gas or liquid.

explosion proof lighting An electrical fixture that will not explode or ignite flammable liquids or gasses surrounding it.

explosive Any device or chemical compound whose primary function is to produce an explosion.

explosive actuated gun A stud driver in which the discharge of a blank cartridge provides the impact. Also called a "stud gun."

explosive rivet A rivet whose explosive-filled shank is exploded by striking it with a hammer after the rivet has been inserted.

exposed Descriptive of an unprotected or uninsulated *live part* of an electrical system whose position renders it susceptible to being inadvertently touched or approached by a person at nearer than a safe distance.

exposed aggregate The coarse aggregate in concrete work revealed when the outer skin of mortar is removed, usually before the full hardening of the concrete.

exposed area In roofing, the part of a shingle that is not covered by another shingle.

exposed masonry Any masonry construction whose only surface finish is paint applied to the face of the wall.

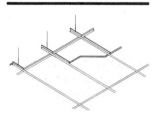

Exposed Suspension System

Exterior Stair

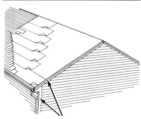

Exterior Trim

exposed nailing A method of nailing that leaves the nails exposed to the weather.

exposed suspension system A method of installing a suspended acoustical ceiling in which the panel-supporting grid is left exposed in a room.

exposure hazard The probability of a building's exposure to fire from an adjoining or nearby property.

extended coverage insurance Work-site insurance (included in property insurance) against loss or damage caused by wind, hail, riot, civil commotion, aircraft, land vehicles, smoke, and explosion (except steam boiler explosion).

extended-service lamp, long-life lamp A lamp designed to have a life substantially longer than others in its general class. Such an incandescent lamp provides an output of fewer lumens than a standard lamp of comparable voltage.

extended surface A fitting, usually composed of metal fins or ribs, that provides additional surface area on a pipe or tube through which heat is transferred.

extender (1) An opaque, white, inert mineral pigment such as calcium carbonate, silica, diatomaceous earth, talc, or clay, that is added to paint to provide texture or lower gloss or, by providing bulk, to reduce paint cost. (2) A volume-increasing, hence cost-reducing, additive in synthetic resin adhesives.

extensibility The capability of a sealant to be stretched in tension.

extension A wing, ell, or other addition to an existing building.

extension bolt See **extension flush bolt.**

extension casement hinge An exterior hinge on a casement window whose sash swings outward, located so as to provide enough clearance to allow the cleaning of the hinged side from the inside when the window is open.

extension device Any device other than an adjustment screw used for accomplishing vertical adjustment.

extension flush bolt A type of flush bolt whose head connects to the operating mechanism via a rod which is inserted through a hole bored in the door.

extension ladder A ladder comprising more than one section, each of which slides within the other and locks on the other, thus allowing lengthening.

extension link That device which provides a long backset in the bored lock of a door.

extension rule A rule containing an extendable, calibrated sliding insert.

exterior glazed Descriptive of a glazing that has been set from the outside of a building.

exterior paint Paint formulated with durable binder and pigment, which make it suitable for exposure to weather.

exterior panel In a concrete slab, a panel with at least one of its edges not adjacent to another panel.

exterior plywood Plywood whose layers of veneer are bonded with a waterproof glue.

exterior stair An often legally required exit consisting of a series of flights of steps exposed to the outdoors.

exterior trim Any visible or exposed finish material on the outside of a building.

external vibration Brisk agitation of freshly mixed concrete by a vibrating device strategically positioned on concrete forms.

extra An item desired or work performed in addition to that specified in the contract and usually involving additional cost.

extra-strong pipe The common designation for steel or wrought

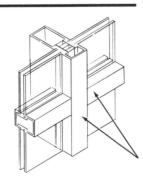

Extruded Section

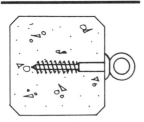

Eye Bolt

iron pipe having wall thickness greater than that of standard-weight pipe.

extra work Any work, desired or performed, but not included in the original contract.

extruded Descriptive of any item manufactured by forcing its hot constituent material through an orifice in a die.

extruded section Structural sections formed by extrusion and used in light construction.

extruded tile A tile formed by pushing clay through a die and cutting it into specified lengths.

extrusion coating Extruded molten resin, a thin film of which is pressed onto a substrate to produce a coating with an adhesive.

eye (1) In architecture, the opening in the uppermost portion of a cupola. (2) The nearly circular center of the roll or volute of an Ionic capital. (3) The middle roundel of a pattern or ornament. (4) A hole through a member providing access, as for the passage of a pin. (5) In tools, the handle-receiving orifice in the head of the implement.

eye bar A bar that functions as a tension member in a steel truss, as either or both of its ends contain an eye which provides for the passage of a pin, thus forming a joint.

eye bolt An anchoring device comprising a threaded shank with a lopped head designed to accept a hook, cable, or rope.

eyebrow A window or ventilation opening though the surface of a roof, but unlike a dormer in that it forms no sharp angles with the roof, but rather is incorporated into the general horizontal line of the continuous roof, which is carried over it in a wave line.

eye of a dome The opening at the very top of a dome.

F ABBREVIATIONS

The abbreviations listed below are those most commonly used in the construction industry. Alternative forms (usually nonstandard) are shown in parentheses.

f fine, focal length, force, frequency
F Fahrenheit, fluorine
FA fresh air, fire alarm
fab fabricate
fac facsimile
FAI fresh air intake
F.A.I.A. Fellow of the American Institute of Architects
FAO finish all over
FAR floor-area ratio
FAS free alongside ship, firsts and seconds
FBM foot board measure
f.c./F.C. footcandle
f.c.c. face-centered cubic
FDB forced-draft blower
FDC fire-department connection
fdn/fdtn/fds/FDN foundation, foundations
fdry foundry
Fe ferrum (iron)
FE fire escape
FEA Federal Energy Administration
FFA full freight allowance
FG fine grain, flat grain
F.G. finished grade
FHA Federal Housing Administration
FHC fire-hose cabinet
Fig. figure
fill. filling
Fin. finish
Fixt fixture
fl floor, fluid
FL floorline, floor, flashing
flash. flashing
FLG flooring
fl oz fluid ounce
Flr floor
FLUOR fluorescent
fm fathom
FM frequency modulation, Factory Mutual

FMT flush metal threshold
FMV fair market value
fndtn. foundation
FOB free on board
FOC free of charge
FOHC free of heart centers
FOK free of knots
Fount. fountain
fp fireplace, freezing point
f.pfg. fireproofing
fpm feet per minute
FPRF fireproof
fps feet per second
fr frame
Fr. fire rating
F.R. fire rating
frmg framing, forming
FRP fiber reinforced plastic
frt freight
frwy freeway
FS Federal specifications
FST flat seam tin
ft foot, feet
ftc footcandle
ftg footing
fth fathom
ft lb foot pound
Furn furnish(ed)
fus fusible
fv face velocity
FW flash welding
fwd forward

F DEFINITIONS

Face Block

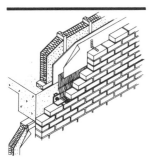

Face Brick

face (1) The surface of a wall, masonry unit, or sheet of material that is exposed to view or designed to be exposed in finish work. (2) To cover the surface layer of one material with another, as to "face" a wall with brick or fieldstone.

face block, faced block A unit of concrete masonry with a plastic or ceramic face surface, often glazed or polished for special architectural uses.

face brick See **facing brick**.

face cord A measurement of firewood. A face cord is 4 feet high x 8 feet long, but only as deep as the length of the pieces. Thus, a face cord may be 4 x 8 feet, but only 16 inches deep, and contains only one-third of the wood volume of a plywood, or standard, cord.

faced plywood Plywood that is faced with plastic, metal, or any material other than wood.

faced wall A wall with facing and backing bonded so that both act together under load.

face edge (1) The best of the two narrow faces on a rectangular piece of lumber. (2) The edge from which other edges are measured or "trued".

face glazing The compound applied with a glazing knife after a light has been bedded, set, and fastened in a rabbeted sash.

face grain The grain on the face of a plywood panel. The face grain should always be placed at right angles to the supports when applying it to a roof or subfloor.

face joint A visible joint on the surface of a masonry wall.

face mark A mark made on the face of a piece of lumber, usually in pencil or crayon, and near the edge or end of the piece that

identifies the work face of a planed timber.

face measure The measurement of the area of a board or panel. Face measure is not the same as board measure, except when the piece being measured is 1-inch thick.

face milling Cutting done on the wide, exposed surface of a piece of lumber in order to produce a design.

face mix A concrete mix bonded to the exposed surface of a cast stone building unit.

face mold (1) A template used in ornamental woodworking to mark the boards from which pieces will be cut. (2) A device used to examine the shape of wood and stone faces and surfaces.

face nailing Nailing perpendicular to the face of wood.

face oiling A light coating of oil on the face of concrete panels to aid in removing the forms after the concrete sets.

faceplate Any protective and/or decorative plate such as an escutcheon.

face putty, front putty Putty used on the side of window glass that is exposed to view.

face side The better wide side of a rectangular piece of lumber.

face stock See **face veneer**.

face string, finish string An exterior string, usually of better material or finish, placed over the rough string in the construction of a staircase.

face veneer Those veneers of higher grade and quality that are used for the faces of plywood panels, especially in the sanded grades. Face veneer is selected for decorative qualities rather than for strength.

Face Width

Face Brick

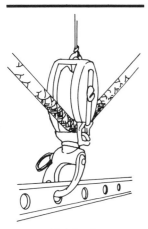

Fairlead

face width The width of the face of a piece of dressed lumber. In tongued or lapped lumber, the face width does not include the width of the tongue or lap.

facing, facework Any material used to cover a rough or inferior surface, in order to provide a finished and more attractive appearance.

facing brick, face brick Brick that is especially made or selected for color, texture, or other properties. Facing brick is intended for use in the exposed face(s) of a masonry wall.

factored load See **design ultimate load**.

factor of safety (1) Stress factor of safety: the ratio of the ultimate strength, or yield point, of a material to the design working stress. (2) Load factor of safety: the ratio of ultimate load, moment, or shear of a structural member to the working load, moment, or shear, respectively, assumed in design.

factory built A reference to a construction, usually a dwelling, that is built, or at least partially preassembled, in a factory rather than "on site." Most factory-built houses are constructed in two or more modules, often complete with plumbing, wiring, etc. The modules are delivered to a building site and assembled there. Finish work is then performed. Factory-built houses are generally less expensive to build, because of savings gained through mass production and factory efficiencies, and are quicker to erect.

factory-filled particleboard Particleboard that has had its edges filled with plastic during manufacture.

factory finished A product that has been coated or stained as part of the manufacturing process.

factory lumber A broad category of lumber that includes stock of various grades and species intended for remanufacturing into items such as furniture, doors, windows, moldings, and boxes.

factory primed A product to which an initial undercoat of paint has been applied.

factory select A grade of shop lumber containing 70 percent or more of No. 1 door stock, but having other defects, such as pitch or bark pockets, beyond the restrictions allowed in the No. 1 grade of lumber.

failure A deficiency of a structural element that makes it unable to continue the load-bearing function for which it was originally designed.

fair face concrete A concrete surface which, on completion of the forming process, requires no further (concrete) treatment other than curing. See also **architectural concrete**.

fairlead A device such as a ring or a block of wood with a hole in it, through which cable or rope is led for alignment.

fair market value A price that is fair and reasonable in light of current conditions in the marketplace.

fall The slope in a channel, conduit, or pipe, stated either as a percentage or in inches per feet.

falldown (1) Those lumber or plywood items of a lesser grade or quality that are produced as an adjunct to the processing of higher quality stock. (2) A term used in waterborne shipments to indicate that a cargo was not ready for loading when a vessel called for it.

falling wedge A V-shaped piece of steel, aluminum, or plastic used by driving it into the back cut when felling a tree. The purpose is to keep the saw from binding, and to help control the direction of the fall.

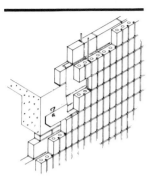

False Joint

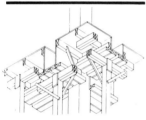

Falsework

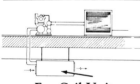

Fan-Coil Unit

Fascia

fallout shelter A room or structure intended for shelter from the effects of radiation caused by nuclear explosions.

false attic An architectural addition, built above the main cornice of a structure, that conceals the roof rafters but has no rooms or windows.

false ceiling A ceiling suspended a foot or more below the actual ceiling to provide space for and easy access to wiring and ducts, or to alter the dimensions of a room. See **suspended ceiling**.

false door A non-operable door placed in a wall for architectural appearance.

false front A front wall that extends beyond the sidewalls and/or above the roof of a building to create a more imposing facade.

false heartwood Dark innerwood that has been colored by disease or fungi so that it resembles heartwood in color.

false joint A groove in a solid masonry block or stone, simulating the appearance of a joint.

false set The rapid development of rigidity in a freshly mixed portland cement paste, mortar, or concrete without the evolution of much heat. Rigidity can be dispelled and plasticity regained by further mixing without addition of water. "Premature stiffening", "hesitation set", "early stiffening", and "rubber set" are terms referring to the same phenomenon, but "false set" is the preferred designation.

falsework The temporary structure erected to support work in the process of construction. Falsework consists of shoring or vertical posting formwork for beams and slabs, and lateral bracing. See also **centering**.

fan-coil unit An air conditioning unit that houses an air filter, heating or cooling coils, and a centrifugal fan, and operates by moving air through an opening in the unit and across the coils.

fancy butt A shingle with the butt end machined in some pattern. Fancy butt shingles are usually used on sidewalls to form geometric designs in the shingle pattern.

fan Fink truss A form of Fink truss with subdiagonals that extend outward from a central point.

fanlight A semi-circular or semi-elliptical window built above the opening of a door. A fanlight often has triangular panes of glass, in between radiating bars or leads, and resembles the form of an open fan.

fanlight catch The spring catch on a fanlight window or other hinged windows. The catch often has a cord attached for opening.

fan truss A truss with struts that radiate like the ribs of a fan, supported at their base by a common suspension member.

fascia, facia (1) A board used on the outside vertical face of a cornice. (2) The board connecting the top of the siding with the bottom of a soffit. (3) A board nailed across the ends of the rafters at the eaves. (4) The edge beam of a bridge. (5) A flat member or band at the surface of a building.

fascia board See **eaves fascia**.

fastener Any mechanical device used to hold together two or more pieces, parts, members, etc.

fast-pin hinge A hinge with a pin fixed permanently in place.

fat (1) Material accumulating on a trowel during smoothing. Fat is used to fill in small imperfections. (2) Descriptive of a mortar containing a relatively high proportion of cement.

fat area See **fat spot**.

fat board A bricklayers mortar board.

Featheredge

fat concrete Concrete with a relatively large proportion of mortar.

fat edge A ridge of paint at the bottom edge of a surface when too much paint has been applied or because the paint runs too freely.

fatigue The weakening of a material caused by repeated or alternating loads. Fatigue may result in cracks or complete failure.

fatigue failure The phenomenon of rupture that occurs when a material is subjected to repeated loadings at a stress substantially less than the ultimate tensile strength.

fatigue life (1) The number of cycles of a specified loading that a given specimen can be subjected to before failure occurs. (2) A measure of the useful life of similar specimens.

fatigue strength The maximum stress that can be sustained for a given number of stress cycles without failure.

fat lime, rich lime A quicklime or hydrated lime used in plastering and masonry and obtained by burning a pure or nearly pure limestone source.

fat mix, rich mix A mortar or concrete mix with a relatively high cement content. Fat mix is more easily spread and worked than a mix with the minimum amount of cement required for strength.

fat mortar, rich mortar Mortar with a high percentage of cement. Fat mortar is sticky and adheres to a trowel.

fat spot, fat area A thick spot in an area paved with bituminous concrete.

fattening The thickening of paint in a partially filled can that has been left standing over a period of time.

faucet ear The projection on a pipe or fitting bell, used to bolt it to another pipe or fitting.

fault (1) A defect in an electrical system caused by poor insulation, imperfect connections, grounding, or shorting. (2) A shifting along a plane in a rock formation that causes differential displacement.

faulting Differential vertical displacement of a slab or other member adjacent to a joint or crack.

faying surface The surface of a metal or wooden member joined closely to another member or in near proximity to the member to which it will be joined.

feasibility study A thorough study of a proposed construction project to evaluate its economic, financial, technical, functional, environmental, and cultural advisability.

feather To blend the edge or finish of new material smoothly into an older surface.

feather edge, featheredge The beveling or tapering of the edge of one surface or surface coating where it meets another.

featheredge rule A straight-edged device used to shape and define angles in plastering.

feather joint A joint between two squared, closely butted boards made by plowing a groove along the length of each and fitting them with a common tongue.

feather tip A shingle or shake with an extra-thin end.

Federal Home Loan Mortgage Corp. A government-sponsored organization, nicknamed "Freddie Mac", that provides a secondary market for mortgages.

Federal Housing Administration (FHA) A division of the Department of Housing and Urban Development. The FHA works through lending agencies to provide mortgage insurance on private residences that meet the agency's minimum property standards. The FHA has also been

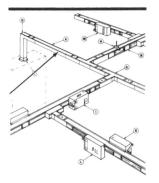

Feeder

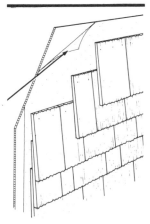

Felt

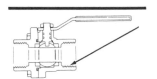

Female Thread

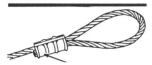

Ferrule

charged with administering a number of special housing programs.

Federal National Mortgage Association A corporation that provides a secondary mortgage market for FHA-insured and VA-guaranteed home loans.

Federal style The Classic Revival style in the U.S.A., from 1790 to 1830.

fee Remuneration for professional services.

feeder (1) An electrical cable or group of electrical conductors that runs power from a larger central source to one or more secondary or branch-circuit distribution centers. (2) A tributary to a reservoir or canal. (3) A bolt or device that pushes material on to a crusher or conveyor.

feed rate The rate at which material can be fed to a machine, such as an edger or planer, without causing a jam.

feed wheel A material distributor or regulator in certain types of shotcrete equipment.

feeler gauge A gauge used to measure the thickness of a gap, consisting of a series of blades of graduated thicknesses.

fee owned timber Timber that is presently owned free and clear. The term "fee" comes from the legal phrases "fee simple" and "fee simple absolute". A company's fee owned timber includes timber on land owned by the firm and also may include timber that is owned by the firm but is on land owned by another party.

fee simple An enduring, inheritable interest in land which may be legally honored until the death of all potential heirs of the original owner, and which the owner is free to convey at any time.

feldspar Any of a group of igneous minerals chemically composed of calcium silicates, potassium silicates, or sodium-aluminum silicates and having a consistency softer than quartz.

fell To cut down a tree. A faller fells a tree with a falling saw.

felt A fabric of matted compressed fibers, usually manufactured from cellulose fibers from wood, paper, or rags, or from asbestos or glass fibers.

felt-and-gravel roofing See **built-up roofing**.

felt nail See **clout nail**.

felt paper A type of building paper.

female coupling A coupling with internal threads on both ends.

female thread See **inside thread**.

femerell A ventilator, often louvered, used to draw smoke through a roof when a chimney is not provided.

fence A straight-edge guide mounted parallel to a saw blade to guide a cant as it passes through the saw.

fender A protective bumper often of wood or old tires.

fenestration The layout and design of windows in a building.

ferrocement A composite hydraulic structural material having thin sections consisting of cement mortar reinforced by a number of very closely spaced layers of steel wire mesh.

ferrous metal Metal in which the principal element is iron.

ferrule (1) A tube or metallic sleeve, fitted with a screwed cap to the side of a pipe to provide access for maintenance. (2) A metal sleeve or collar attached to the end of a short cable in making a choker. The ferrule fits into the bell of the choker. Also called a "nubbin".

fertilizer An organic or inorganic material that acts as a nutrient.

Fiberglass

Field Gluing

Fertilizer may be natural or manufactured.

fiber A thread-like structure of a plant that contributes to stiffness or strength.

fiberboard A general term referring to any of various panel products, such as particleboard, hardboard, chipboard, or other type formed by bonding wood fibers by heat and pressure.

fibered plaster Gypsum plaster reinforced with fibers of hair, glass, nylon, or sisal.

fiberglass, fibrous glass, glass fiber Filaments of glass formed by pulling molten glass into random lengths that are either gathered in a wool-like mass or formed as continuous threads. The wool-like form is used as thermal and acoustical insulation. The thread-like form is used as reinforcing material and in textiles, glass fabrics, and electrical insulation.

fiberglass reinforced plastic (FRP) A coating of glass fibers and resins applied as a protective layer to plywood. The resulting composite is tough and scuff resistant. It is used in construction of containers and truck bodies, and as concrete form.

fiber-reinforced concrete See **concrete, fibrous**.

fiber stress The longitudinal compressive or tensile stress in a structural member.

fibrous concrete Concrete with asbestos, glass, or other fibers added to increase tensile strength.

fibrous plaster Cast plaster reinforced canvas, excelsior, or other fibrous material.

fiddleback grain A rippling or undulating grain common to certain hardwoods, such as Maple and Sycamore. Pieces containing this grain pattern are often used on the backs of violins.

fiducial mark A standard reference point or line used as a base in surveying.

field (1) In masonry, an expanse of brick work between two openings or corners. (2) A term used to designate a construction project site.

field applied (1) The application of a material, such as paint, at a job site, as opposed to being applied at a factory. (2) The construction or assembly of components in the field.

field bending Bending of reinforcing bars on the job rather than in a fabricating shop.

field chipping The reduction to pulp chips of slash or small trees using a portable chipper in the woods.

field concrete See **concrete, field**.

field-cured cylinders Test cylinders cured nearly in the same manner as the concrete in the structure to indicate when supporting forms may be removed, additional construction loads may be imposed, or the structure may be placed in service.

field drain See **agricultural drain pipe**.

field engineer A representative of certain government agencies who oversees projects at the site. Also called a "project representative" or "field representative".

field gluing A method of gluing plywood floors in which specially developed glues are applied to the top edges of floor joists. Plywood is then laid on the joists and nailed in place. The combination of gluing and nailing results in a stiffer floor construction and tends to minimize squeaks and nail-popping.

field inspection The reinspection of lumber or plywood "in the field", usually at the buyer's location. A field inspection is called by a buyer when he believes the product he receives to be inferior to that specified. See also **reinspection**.

Fieldstone

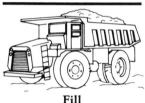

Fill

field man A representative of any of various trade associations. He represents the association "in the field" and is available to answer questions regarding grades, usage, and characteristics of various species, and the products produced by his association's members. He also conducts clinics and demonstrations and otherwise promotes his association's products.

field-molded sealant A joint sealant in liquid or semisolid form that can be molded to the desired shape as it is installed in the joint.

field order In construction, a written order passed to the contractor from the architect to effect a minor change in work, requiring no further adjustment to the contract sum or expected date of completion.

field painting The painting of structural steel or other metals after they have been erected and fastened.

field representative See **project representative**.

field rivet A rivet driven in a field connection of structural steel.

fieldstone (1) Loose stone removed from the soil. (2) Flat slabs of stone suitable for use in dry wall masonry.

field supervision The site supervisory work performed by an architect or engineer.

field work Any work performed at a job site.

fifths The lowest of the common lumber grades normally exported from Scandinavia and Eastern Europe.

fifth wheel (1) The swivel connection between highway-type tractors and semi-trailers. (2) (slang) Any extra or unnecessary person or thing.

figured glass Translucent sheet glass rolled with a bas-relief pattern on one face, providing high light transmission, but obscure, with the degree of obscurity depending on the pattern.

filament A fine wire used in an incandescent lamp. The form is designated by a letter: S, straight wire; E, coil; CC, coiled coil.

file A hand-held steel tool with teeth or raised oblique ridges, used for scraping, redressing, or smoothing metal or wood.

filer The person in a sawmill or logging operation who keeps the saws sharp. Also called a "saw filer".

fill (1) The soil or other material used to raise the grade of a site area. (2) A sub-floor leveling material.

filled particleboard Particleboard to which plastic material has been applied to fill gaps between particles on the edges or faces.

filler (1) Finely divided inert material, such as pulverized limestone, silica, or colloidal substances sometimes added to portland cement paint or other materials to reduce shrinkage, improve workability, or act as an extender. (2) Material used to fill an opening in a form.

filler block Concrete masonry units placed between joists or beams under a cast-in-place floor.

filler coat A coat of paint, varnish, etc., used as a primer.

filler metal Metal, added by a welding process, that has a melting point either approximately the same as or below the base metal.

filler plate A steel plate or shim used to fill any gap between columns or between a column and a roof member.

fillet (1) A narrow band of wood between two flutes in a wood member. (2) A flat, square molding separating other moldings. (3) A concave junction

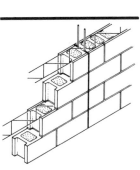

Fill Insulation

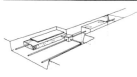

Filter Bed

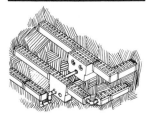

Filter Block

formed where two surfaces meet. See also **Chamfer strip**.

fillet weld A weld of approximately triangular cross section, used to join two surfaces at approximately right angles to each other. The weld metal may be deposited in one pass or built up with multiple passes.

filling piece A piece inserted into another to form a continuous surface.

fill insulation (1) Thermal insulation placed in prepared or natural cavities of a building. (2) Any variety of loose insulation that is poured into place.

fill-type insulation See **fill insulation**.

film glue A thin sheet of paper or scrim which has been impregnated with a thermosetting resin, used to bond expensive decorative veneers without bleed-through or in hot-pressing nonporous laminates.

film surfaced hardboard siding Hardboard to which a thin, dry sheet of paper, coated on both sides with a phenol-formaldehyde resin adhesive, has been applied. Such coatings are often printed with grain patterns in imitation of actual veneers, and are used as an inexpensive substitute for the higher-priced veneer panels.

filter (1) A device to separate solids from air or liquids, such as a filter that removes dust from the air or impurities from water. (2) Granular matter placed on an area to provide drainage while preventing the entry and flow of sediment and silt.

filter bed A bed of granular material used to filter water or sewage.

filter block A hollow, vitrified clay masonry unit used for trickling filter floors in sewage treatment plants.

fin A narrow, linear projection on a formed concrete surface, resulting from mortar flowing out between spaces in the formwork.

final acceptance The formal acceptance of a contractor's completed construction project by the owner, upon notification from an architect that the job fulfills the contract requirements. Final acceptance is often accompanied by a final payment agreed upon in the contract.

final completion A term denoting that the work has been completed in accordance with the terms and conditions of the Contract Documents.

final inspection An architect's last review of a completed project before issuance of the final certificate for payment.

final payment The payment an owner awards to the contractor upon receipt of the final certificate for payment from the architect. Final payment usually covers the whole unpaid balance agreed to in the contract sum, plus or minus amounts altered by change orders.

final prestress See **final stress**.

final set A degree of stiffening of a mixture of cement and water greater than initial set, generally stated as an empirical value indicating the time in hours and minutes required for a cement paste to stiffen sufficiently to resist to an established penetration of a weighted test needle. See also **initial set**.

final setting time The time required for a freshly mixed cement paste, mortar, or concrete to achieve final set. See also **initial setting time**.

final stress (1) In prestressed concrete, the stress that remains in a member after substantially all losses in stress have occurred. (2) The stress in a member after all loads are applied.

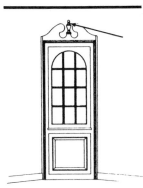

Finial

Finish Carpentry

fine aggregate (1) Aggregate passing the 3/8-in. (9.5-mm) sieve and almost entirely passing the No. 4 (4.75-mm) sieve and predominantly retained on the No. 200 (75-micrometer) sieve. (2) That portion of an aggregate passing the No.4 (4.75-mm) sieve and predominantly retained on the No. 200 (75-micrometer) sieve. See also **aggregate** and **sand**.

fine grading The final grading of ground to prepare for seeding, planting, or paving.

fine grained soil Soil in which the smaller grain sizes predominate, such as fine sand, silt, and clay.

fineness A measure of particle size.

fineness modulus A factor obtained by adding the total percentages by weight of an aggregate sample retained on each of a specified series of sieves, and dividing the sum by 100. In the United States the standard sieve sizes are No. 100 (150 micrometer), No. 50 (300 micrometer), No. 30 (600 micrometer), No. 16 (1.18 mm), No. 8 (2.36 mm) and No. 4 (4.75mm), and 3/8 in (9.5 mm), 3/4 in. (19mm), 1-1/2 in. (38.1mm), 3 in. (75 mm), and 6 in. (150 mm).

fines (1) Soil which passes through a No. 200 sieve. (2) Fine-milled chips used in the production of particleboard. Fines are larger than sander dust or wood flour and are used on the faces of particleboard panels, with coarser chips used as the core of the board. See **discrete**. (3) An undesirable byproduct of cutting wafers and strands for waferboard and oriented strand board.

finger joint A method of joining two pieces of lumber end-to-end by sawing into the end of each piece a set of projecting "fingers" that interlock. When the pieces are pushed together, these form a strong glue joint.

finial A decoration on the point of a spire, pinnacle, or conical roof.

finish (1) The texture of a surface after compacting and finishing operations have been performed. (2) A high-quality piece of lumber graded for appearance and often used for interior trim or cabinet work.

finish builders' hardware Visible, functional hardware that has a finished appearance, including such features as hinges, locks, catches, pulls, and knobs.

finish carpentry The wood finish to a building, such as moldings, doors, and windows. Also called "joinery".

finish casing The finish trim around a casing.

finish coat (1) Final thin coat of shotcrete preparatory to hand finishing. (2) An exposed coat of plaster and stucco.

finished grade The top surface of an area after construction is completed, such as the top of a road, lawn, or walk.

finished size The final size of any completed object including trim.

finisher A tradesman who applies the final treatment to a concrete surface, including patching voids and smoothing.

finish floor The wearing surface of a floor.

finish flooring The material used to make the wearing surface of a floor, such as hardwood, tile, or terrazzo.

finish grade See **finished grade**.

finish grinding (1) The final grinding of clinker into cement, with calcium sulfate in the form of gypsum or anhydrite generally being added. (2) The final grinding operation required for a finished concrete surface, such as bump cutting of pavement, fin removal from structural concrete, or terrazzo floor grinding.

finish hardware See **finish builders' hardware**.

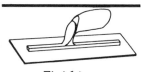

Finishing

Finishing Machine

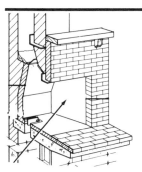

Fire Brick

finishing Leveling, smoothing, compacting, and otherwise treating surfaces of fresh or recently placed concrete or mortar to produce the desired appearance and service. See also **float** and **trowel**.

finishing machine A power-operated machine used to give the desired surface texture to a concrete slab.

finishing nail A slender nail, with a narrow head, that may be driven entirely into a wood surface leaving a small hole that can be filled with putty or a similar compound.

finish sag A defect in hardboard.

finish size See **finished size**.

finish skip A defect in hardboard.

finish string See **face string**.

Fink truss A symmetrical truss, formed by three triangles, commonly used in supporting large, sloping roofs. Also called "Belgian truss" or "French truss".

fir (1) A form of softwood indigenous to temperate zones, used principally for interior trim and framing. Varieties include Douglas fir, silver fir, balsam fir, and white fir. (2) Although used most often to refer to Douglas Fir, which is also a pseudo-fir, a general term for any of a number of species of conifers, including the true firs.

fir & larch A mixture of Douglas Fir and Western Larch, sold together as one species grouping. The two species are intermixed in the inland regions of the Western U.S. and British Columbia. Because they have similar characteristics and are used in similar ways, the two woods are usually mixed together.

fire alarm system An electrical system, installed within a home, industrial plant, or office building that sounds a loud blast or bell when smoke and flames are detected. Certain alarms are engineered to trigger sprinkler systems for added protection.

fire area An area in a building enclosed by fire resistant walls, fire resistant floor-ceiling assemblies, and/or exterior walls with all penetrations properly protected.

fireback, chimney back (1) The heat-resistant back wall of a fireplace, built of masonry, cast metal, or wrought metal, that withstands flames and radiates heat into the area served. (2) A fire-resistant attachment to the back wall of a fireplace that keeps the masonry from absorbing too much heat.

fireboard, chimney board, summer piece A device used to close or cover the opening of a fireplace when it is not in service.

firebreak (1) A strategic space between buildings, clusters of buildings, or sections of a city that helps to keep fires from leaping or spreading to surrounding areas before they can be contained. (2) Any doors, walls, floors, or other interior structures engineered to go inside a building.

fire brick A flame-resistant, refractory ceramic brick used in fireplaces, chimneys, and incinerators.

fire cement A cement, such as calcium aluminate cement, used in mortars for laying refractory brick.

fire clay An earthy or stony mineral aggregate having as the essential constituent hydrous silicates of aluminum with or without free silica. Fire clay is plastic when sufficiently pulverized and wetted, rigid when subsequently dried, and of suitable refractoriness for use in commercial refractory products.

fire control damper An automatic damper used to close a duct if a fire is detected.

Fire Extinguisher

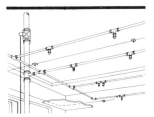

Fire-Extinguishing System

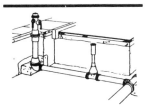

Fire Hydrant

fire cut A sloping cut on the end of a wood beam or joist supported on a masonry wall. If, somewhere along its length, the joist or beam burns through, injury to the wall is minimized.

fire damper See **fire control damper**.

fire detection system A series of sensors and interconnected monitoring equipment which detect the effects of a fire and activate an alarm system.

fire division wall A wall that subdivides one or more floors of a building to discourage the spread of fire.

fire door (1) A highly fire-resistant door system, usually equipped with an automatic closing mechanism, that provides a certain designated degree of fire protection when it is closed. (2) The opening in a furnace or a boiler through which fuel is added.

fire-door rating A system of evaluating the endurance and fire resistance of door, window, or shutter assemblies, according to standards set by Underwriters' Laboratories, Inc., or other recognized safety authorities. Ratings of A through F are assigned in descending order of their effectiveness against fire.

fired strength The compressive or flexural strength of refractory concrete, determined upon cooling after first firing to a specified temperature for a specified time.

fired unit weight The unit weight of refractory concrete, upon cooling, after having been exposed to a specified firing temperature for a specified time.

fire endurance (1) The length of time a wall, floor-ceiling assembly, or roof-ceiling assembly will resist a standard fire without exceeding specified heat transmission, fire penetration, or limitations on structural components. (2) The length of time a structural member, such as a column or beam will resist a standard fire without exceeding specified temperature limits or collapse. (3) The length of time a door, window or shutter assembly will resist a standard fire without exceeding specified deformation limits.

fire escape A continuous, unobstructed route of escape from a building in case of fire, sometimes located on the outside of an exterior wall.

fire-exit bolt See **panic exit device**.

fire extinguisher A portable device for immediate use in suppressing a fire. There are four classes of fires: A, B, C or D. A single device is designed for use on one or more classes.

fire-extinguishing system An installation of automatic sprinklers, foam distribution system, fire hoses, and/or portable fire extinguishers designed for extinguishing a fire in an area.

fire frame A cast-iron housing installed in a fireplace opening to reduce its size.

fire hazard The relative danger that a fire will start and spread, that smoke or toxic gases will be generated, or that an explosion will occur, endangering the occupants of a building or area.

fire-hazard classification A designation of high, ordinary, or low assigned to a building, based on its potential susceptibility to fire, judged by its contents, functions, and the flame-spread rating of its inner furnishings and finishes.

fire hydrant A supply outlet from a water main, for use in fire fighting.

fire limits The boundary of an area where the relative danger

191

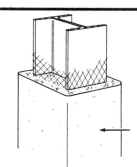

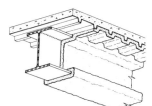

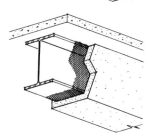

Fireproofing

exists that a fire will start and spread or that an explosion will occur, endangering the lives of the occupants of the building.

fire load, fire loading The amount of flammable contents or finishes within a building per unit of floor area, stated either in pounds per square foot or BTUs per square foot.

fire partition An interior partition that does not fully qualify as a fire wall, but has a fire-endurance rating of not less than two hours.

fireproof (1) Descriptive of materials, devices, or structures with such high resistance to flame that they are practically unburnable. (2) To treat a material with chemicals in order to make it fire resistant.

fireproofing Material applied to structural members to increase their resistance to the effects of fire.

fireproof wood Chemically treated wood that is fire resistive. No wood can be completely fireproof, but wood can be treated so that it is highly resistant to heat and flame and very difficult to ignite under ordinary circumstances.

fire protection The practice of minimizing the probable loss of life or property resulting from fire, by fire-safe design and construction, the use of detection and suppression systems, establishment of adequate fire fighting services, and training building occupants in fire safety and emergency evacuation.

fire-protection rating See **fire endurance**.

fire-protection sprinkler system An automatic fire-suppression system, commonly heat-activated, that sounds an alarm and deluges an area with water from overhead sprinklers when the heat of a fire melts a fusible link.

fire-protection sprinkler valve A

valve, normally open, that is used to control the flow of water in a sprinkler system.

fire-rated system Wall, floor, or roof construction using specific materials and designs that have been tested and rated for conformance to fire safety criteria, such as the flame spread rate.

fire resistance (1) The property of a material or assembly to withstand fire, characterized by the ability to confine a fire and/or to continue to perform a given structural function. (2) According to OSHA, the property of a material or assembly that makes it so resistant to fire that, for a specified time and under conditions of a standard heat intensity, it will not fail structurally and will not permit the side away from the fire to become hotter than a specified temperature.

fire-resistance rating See **fire endurance**.

fire-resistant door A door designed to confine fire to one part of a structure, keeping it from spreading through an entire building. It may be a solid-core wooden door, or one sheathed in metal, depending on the intended location. Doors are rated for the projected time they could be expected to perform their function during a fire. Most building codes require that the door between living quarters and a garage be fire resistant.

fire-resistive Capable of resisting the effects of fire to some degree.

fire-resistive construction Construction in which the floors, walls, roof, and other components are built exclusively of noncombustible materials, with fire-endurance ratings equal to or greater than those mandated by law.

fire-resistive wall A wall having the fire-resistance rating required by a building code or in

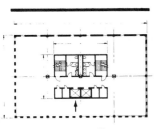

Fire Wall

accordance with underwriters' recommendation.

fire retardant (1) A chemical applied to lumber or other wood products to slow combustion and flame spread. (2) A chemical used to fight forest fires, often by dropping it from an airplane or helicopter.

fire-retardant finish Paint composed in part of noncombustible substances such as silicones, chlorinated waxes and resins, and antimony oxide, used to protect combustible surfaces, inhibiting the rapid spread of flames.

fire-retardant wood Wood chemically impregnated or otherwise treated such that the fire hazard has been reduced.

fire-retarding glazing See **wire glass.**

fire road A road built to provide access to an area in case of fire. Such roads are usually built as part of a fire-protection system, rather than at the time a fire is actually burning.

fire screen A screen set in front of a fireplace to retain flying sparks or embers.

fire separation A floor or wall, with any penetrations properly protected, having the fire endurance required by authorities and acting to retard fire spread in a building.

fire shutter A complete metal shutter assembly with the fire endurance required by code. The rating depends on the location and nature of the opening.

firestat A fixed thermostat in an air-conditioning system, preset at a temperature required by code or other authorities, usually 125 degrees F (52 degrees C).

fire stop A short piece of wood, usually a 2x4 or 2x6, placed horizontally between the studs of a wall. Firestops are equal in width to the studs. They are usually placed halfway up the height of the wall to slow the spread of fire by limiting drafts in the space between the sheathing on the two sides of the wall. Building codes usually do not require fire stops in walls that are less than eight feet high.

fire tower A vertical enclosure, with a stairway, having the fire endurance rating required by code and used for egress and as a base for fire-fighting.

fire wall An interior or exterior wall that runs from the foundation of a building to the roof or above, constructed to stop the spread of fire.

fire window A window assembly with the fire-endurance rating required by code for the location of the opening.

fir fixed Unplaned lumber held in place by nails only.

firing The controlled-heat treatment, in a kiln or furnace, of ceramic or brick ware.

firmer chisel A long, narrow carpenter's chisel used for mortising.

firmer gouge A long, narrow carpenter's gouge used for cutting grooves.

firmwood The solid wood in a log suitable for manufacture into wood products or chips.

first coat The initial application of mortar, called "base coat" in two-coat work, and "scratch coat" in three-coat work.

first floor (1) In the U.S.A., the floor of a building at or closest to grade. (2) In Europe, the floor of a building above the "ground floor".

first in, first out (FIFO) A type of accounting for inventory in which items purchased first are assumed to be the first to be sold. An accountant will compute the cost and profit on the oldest or

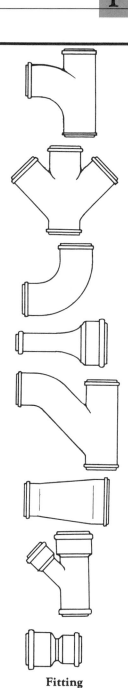

Fitting

first in the inventory. See also **last in, first out (LIFO)**.

first mortgage The mortgage or security in a property that takes precedence over all other mortgages on the property.

firsts and seconds The highest standard grades of hardwood lumber.

fish beam Two beams attached end-to-end to form a single built-up beam with fishplates securing the opposite sides of the joint.

fish-bellied Descriptive of a beam or truss that has its bottom flange or chord convex and sagging downward.

fished joint A butted timber joint made using fishplates.

fisheye A spot, approximately 1/4 in. (6.4mm) in diameter, in a finish plaster coat caused by a piece of lumpy lime.

fishing wire See **snake**.

fishplate A wood or metal plate fastened to the sides of two beams or rails to splice the two ends together.

fish scale A fancy-butt shingle pattern with the exposed portion of the shingle rounded. When used as siding, the overall effect is the pattern of fish scales.

fishtail A wedge-shaped piece of wood used as part of the support form between tapered pans in concrete joist construction.

fishtail bolt A bolt with a split end embedded in concrete as an anchor.

fish tape See **snake**.

fitch (1) A thin, long-handled paintbrush used especially for touching up in recessed areas. (2) A veneer, or a bundle of veneers stacked in the order in which they are cut from a log. (3) A board constituting part of a flitch beam.

fitting (1) A standardized part of a piping system used for attaching sections of pipe together, such as a coupling, elbow, bend, cross, or tee. (2) The process of installing floor coverings around walks, doors, and other obstacles or projections. (3) In electrical wiring, a component, such as a bushing or locknut, that serves a mechanical purpose rather than an electrical function.

fitting-up Erecting a structure by partially bolting the members prior to aligning and completing the connections.

fitting-up bolt A temporary bolt used to secure structural members before they are permanently connected.

five-quarter (5/4) See **quarter measure**.

fixed assets Assets, such as real property, that are not readily converted to cash.

fixed-bar grille A grille with preset, nonadjustable bars, used to cover return and exhaust openings in air-conditioning systems.

fixed beam A structural beam with fixed connections, as opposed to hinged connections.

fixed fee A specific amount specified in a contract to be paid for all materials and services required to complete the contract.

fixed light A window or portion of a window that does not open.

fixed limit of construction cost A maximum amount to be paid for a construction job as specified in the agreement between an owner and the architect.

fixed rate mortgage A mortgage on which the interest rate remains the same through the period of the loan.

fixed retaining wall A retaining wall supported both at its top and bottom.

fixed transom A glass light or panel that is built above a door and cannot be opened.

fixing The installation of glass panels in a ceiling, wall, or

Fixture

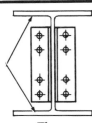

Flange

partition. Fixing is distinct from glazing, which covers most other uses, such as the installation of glass in windows, doors, and showcases.

fixture (1) An electrical device, such as a luminaire, attached to a wall or ceiling. (2) An article that becomes part of real property by virtue of being merged with or affixed to it, although it was originally a distinct item of personal property.

fixture branch A pipe connecting and servicing two or more plumbing fixtures.

fixture drain A drain that leads from the trap of a fixture to another drain pipe.

fixture joint An electrical connection formed by twisting two conductors together and then folding them over.

fixture supply The pipe connecting a plumbing fixture to a supply line.

fixture unit A measure of the rate at which various plumbing fixtures discharge into a drainage system, stated in units of cubic volume per minute.

fixture-unit flow rate The total number of gallons discharged per minute from a single plumbing fixture, divided by 7.5 to provide the flow rate of that fixture in cubic feet per minute.

flagging (1) A walk or patio paved with flagstones. (2) The process of setting flagstones. See **flagstone**.

flagman A person who directs vehicular traffic by means of flags or signs.

flagstone A flat, irregular-shaped stone, usually 1 to 4 inches thick, used as pedestrian paving or flooring. Flagstones are set in mortar or aggregate outdoors and in mortar indoors.

flake board, flakeboard A building board composed of wood particles or flakes solidified with a binder. See **particleboard** and **waferboard**.

flaking A peeling of paint caused by loss of adhesion and/or cohesion.

flame cleaning A method of cleaning mill scale, moisture, paint, and dirt from steel by applying a hot flame.

flame cutting Cutting metal with an oxyacetylene torch.

flameproof Able to resist the spread of flame and not easily ignited, but having less fire resistance than fire-retardant substances.

flame-retardant Treated to resist flame spread and ignition.

flame-spread rating A numerical value given to a material designating its resistance to flaming combustion. Materials can have a rating of 100, as in untreated lumber, to 0, as in asbestos sheets.

flame treating A process by which inert thermoplastic materials and objects are prepared to receive paints, adhesives, lacquers, and inks by immersion in open flame, causing surface oxidation.

flammability A material's ability to burn.

flammable (OSHA) Capable of being ignited easily, burning intensely, or having a rapid rate of flame spread.

flanch, flaunch To broaden the top of a chimney stack and to cant the edges to direct water away from the flue.

flange (1) A projecting ring, ridge, or collar on a pipe or shaft to strengthen, prevent sliding, or to accommodate attachments. (2) The longitudinal part of a beam, or other structural member, that resists tension and compression.

flange angle An angle shape used as one of the parts making up the flange of a built-up girder.

flange cut A cut-out in a flange of a beam or girder to accommodate

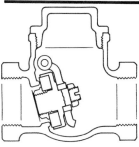

Flap Trap

Flashing

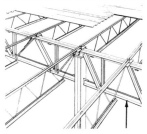

Flat-chord Truss

attachment or passage of another structural element.

flange union A method of connecting threaded pipe using a pair of flanges screwed onto the ends of the pipe and bolted together.

flanking transmission Transmission of sound by paths around a building component rather than through it.

flap trap, flap valve A hinged flap within a plumbing system that permits water to flow in only one direction, preventing backflow.

flare fitting A soft-tube connector made by flaring the end of the tube to provide a mechanical seal.

flash (1) To make a joint weathertight using flashing. (2) An intentional or accidental color variation on the surface of a brick. (3) A variation in paint color resulting from variable wall absorption.

flash chamber A tank placed between the expansion valve and evaporator in a refrigeration system to separate and bypass any flash gas formed in the valve.

flash coat A light coat of shotcrete used to cover minor blemishes on a concrete surface.

flashing A thin, impervious sheet of material placed in construction to prevent water penetration or direct the flow of water. Flashing is used especially at roof hips and valleys, roof penetrations, joints between a roof and a vertical wall, and in masonry walls to direct the flow of water and moisture.

flashing block A concrete masonry unit with a slot cast in one face to receive and hold one edge of a flashing.

flashing board A board to which one edge of a flashing is fastened.

flashing cement A mixture of solvent and bitumen, reinforced with inorganic glass or asbestos fibers, and applied with a trowel.

flashing ring A collar, around a pipe where it passes through a wall or floor, used to seal the opening.

flash point The minimum temperature at which a combustible liquid will give off sufficient vapor to produce a combustible mixture when mixed with air and ignited. Defined flash points are for specific enclosure conditions and ignition energy.

flash set The rapid development of rigidity in a freshly mixed portland cement paste, mortar, or concrete, usually with the evolution of considerable heat. This rigidity cannot be dispelled, nor can the plasticity be regained, by further mixing without addition of water. Also referred to as "quick set" or "grab set".

flash welding A welding process that joins metals with the heat produced by the resistance to an electric current between the surfaces of the metals, followed by application of pressure.

flat (1) Descriptive of a structural element having no slope, such as a flat roof. (2) One floor of a multi-level building, or an apartment occupying one floor. (3) Descriptive of low-gloss paint, used either as an undercoat or as a final coat. (4) A thin iron or steel bar with a rectangular cross section.

flat arch An arch with slight curvature.

flat-chord truss A truss with the top and bottom chords nearly flat and parallel.

flat coat An intermediate coat of paint applied between the primer and finish coat.

flat enamel brush A brush used to apply smooth films of enamel on wood work, about 2 to 3 in. (5 to 7.5 cm) wide and having flagged and tapered bristles.

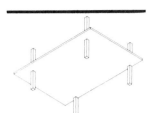

Flat Plate

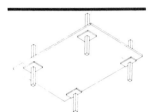

Flat Slab

Flemish bond

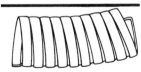

Flexible conduit

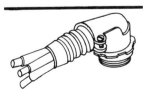

Flexible Connector

flat glass See **window glass, plate glass, float glass,** and **rolled glass**.

flathead (1) A screw or bolt with a flat top surface and a conical bearing surface. (2) A rivet with a flattened head.

flathead rivet See **flathead, def. 2**.

flat jack A hydraulic jack made of light-gauge metal welded into a flat rectangular shape that expands under hydraulic pressure.

flat-joint jointed pointing A flush masonry joint with a narrow groove in the center of the joint for ornamentation.

flat paint A paint that dries to a low gloss or flat finish.

flat piece (of aggregate) One in which the ratio of the width to thickness of its circumscribing rectangular prism is greater than a specified value. See also **elongated piece (of aggregate)**.

flat plate A flat slab without column capitals or drop panels. See also **flat slab**.

flat plate collector A panel of metal or other suitable material used to convert sunlight into heat. Flat plate collectors are usually a flat black color and transfer the collected heat to circulating air or water.

flat pointing, flat-joint pointing The process of finishing mortar in the joints of brickwork by applying it flush with the masonry surface, using a flat trowel.

flat rolled Steel plates, sheets, or strips manufactured by rolling the steel through flat rollers.

flat roof A roof with only enough pitch to allow drainage.

flat seam A seam between two joined sheets of metal, formed by turning up, folding, flattening, and finally soldering the edges.

flat skylight A skylight installed in a horizontal position, slanted only enough to permit rainwater runoff.

flat slab A concrete slab reinforced in two or more directions, generally without beams or girders but with drop panels at supports. See also **flat plate**.

flat spot A spot on a glossy painted surface that fails to absorb the paint properly, usually because of a porous place on the undercoat, resulting in a flat finish that flaws the appearance of the surface.

flat varnish, matte varnish A varnish that dries to a finish with low gloss or no gloss.

flat wall brush A paint brush, usually 4 to 6 in. (10 to 15cm) wide, with long stiff bristles that are often synthetic.

fleck A small spot or mark in wood, usually caused by wood rays or other irregular growth characteristics.

Flemish bond A brick wall laid up with alternate headers and stretchers in each course. Headers in the next course are centered over the stretchers in the course below.

fleur-de-lys The French royal lily adapted to ornamentation in Late Gothic architecture.

flexible conduit See **flexible metal conduit, flexible metallic hose, flexible nonmetallic tubing,** and **flexible seamless tubing**.

flexible connector (1) In ductwork, an airtight connection of nonmetallic materials installed between ducts or between a duct and fan to isolate vibration and noise. (2) A connector in a piping system that reduces vibration along the pipes and compensates for misalignment. (3) An electrical connection that permits movement from expansion, contraction, vibration, and/or rotation.

flexible coupling A mechanical connection between rotating parts that adapts to misalignment, such as a universal joint.

flexible metal conduit A flexible raceway of circular cross section for pulling electric cables through.

flexible metallic hose Hose made from a continuous interlocking coil of metal with packing wound in the grooves, used for fluids at low pressure.

flexible-metal roofing Roof coverings of flat sheet metal, such as copper, galvanized iron, or aluminum.

flexible mounting A flexible support for machinery to reduce vibration between the machinery its foundation or slab. Flexible mountings are usually made of rubber, neoprene, steel springs, or a combination of these.

flexible nonmetallic tubing, loom A flexible tubing with a smooth interior and a wall of nonconducting, fibrous material, used as a mechanical protection for electric conductors.

flexible pavement A pavement structure that maintains intimate contact with and distributes loads to the subgrade and depends on aggregate interlock, particle friction, and cohesion for stability. Cementing agents, where used, are generally bituminous materials, as contrasted to portland cement in the case of rigid pavement. See also **rigid pavement**.

flexible seamless tubing A seamless, welded, or soldered metal tube widely used for volatile gases and gases under pressure.

flexural bond In prestressed concrete, the stress between the concrete and the tendon that results from external loads.

flexural rigidity A measure of stiffness of a member, indicated by the product of the modulus of elasticity and moment of inertia divided by the length of the member.

flexural strength A property of a material or structural member that indicates its ability to resist failure in bending. See also **modulus of rupture**.

flight A run of steps without intermediate landings.

flight header A horizontal structural member used to support stair strings at a floor or platform.

flight rise The vertical distance between two levels connected by a flight of stairs.

flight run The horizontal distance between top and bottom risers in a flight of stairs.

flint A variety of chert. See also **chert**.

flitch (1) A log, sawn on two or more sides, from which veneer is sliced. (2) Thin layers of veneer sliced from a cross-section of a log, as opposed to turning the log on a lathe and peeling from the outer edge in a continuous ribbon. Flitch veneers are often kept in order as they are sliced from a log. This provides a pattern to the veneer after it is laid up in panels. Panels that are laid up with matching flitches are said to have a flitch pattern. (3) A product cut from a log by sawing on two sides and leaving two rounded sides, often used for joinery.

flitch beam, flitch girder sandwich beam A beam built up by two or more pieces bolted together, sometimes with a steel plate in the middle. The pieces, or flitches, are cut from a squared log sawn up the middle. The outer faces of the log are placed together, with their ends reversed to equalize their strength.

flitch plate The steel plate pressed between components of a flitch beam.

float A tool (not a darby), usually of wood, aluminum, or magnesium, used in concrete finishing operations to impart a relatively even but still open

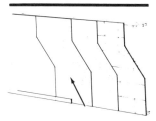

Floated Coat, Topping Coat

Float Finish

Float Valve, Float-Controlled Valve

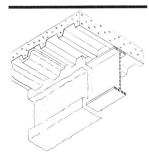

Floor Beam

texture to an unformed fresh concrete surface.

float coat A finish coat of cement paste applied with a float.

floated coat, topping coat A layer of plaster applied to a surface with a float, usually the intermediate coat between the scratch coat and finish coat.

float finish A rather rough concrete surface texture obtained by finishing with a float.

float glass A high quality, flat glass sheet with smooth surfaces, manufactured by floating molten glass on a bed of molten metal at a temperature high enough for the glass to flow smoothly.

floating The operation of finishing a fresh concrete or mortar surface by use of a float, preceding troweling when that is the final finish.

floating floor A floor used in sound-insulating construction. The finish floor assembly is isolated from the structural floor by a resilient underlayment or by resilient mounting devices for machinery. The construction isolates machinery from the building frame.

floating rule A long straightedge used as a float.

floating wood floor A floating floor constructed of wood flooring on a layer of resilient material that isolates it from the building structure.

float scaffold A large platform supported by ropes from overhead, used as a scaffold.

floatstone A stone used in bricklaying to rub the rough spots and ax marks from curved work.

float trap A steam trap regulated by a float in the condensate chamber.

float valve, float-controlled valve A valve controlled by a float in a chamber, such as the float valve in a tank for a water closet.

flood coat (1) The top layer of bitumen poured over a built-up roof, which may have gravel or slag added as a protective layer. (2) See **flow coat**.

flooding The stratification of different colored pigments in a film of paint.

flood-level rim The edge of a plumbing fixture or receptacle over which water will flow if the device is full.

floodlight (1) A projector type of luminaire designed to light an area or object to a level of illumination higher than the surrounding illumination. (2) A metal housing containing one or more lamps used to illuminate a stage uniformly.

floor (1) The surface within a room upon which one walks. (2) The horizontal division between two stories of a building, formed by assembled structural components or a continuous mass, such as a flat concrete slab.

floor arch A flat arch used as a floor.

floor-area ratio The ratio of the total floor area of a building to the area of the building lot.

floor batten A batten attached to a concrete subfloor, to be used as a nailer for a wood-finish floor.

floor beam Any beam that supports a floor of a building or a deck of a bridge.

floorboard One of the boards or planks used to form the finished floor.

floor box A metal electrical outlet box providing outlets from conduits in or under the floor.

floor brick A brick selected as a floor finish because of its smooth, dense texture and resistance to abrasion.

floor clamp A clamp used to force floor boards together while nailing them to joists.

floor clearance The distance between the bottom of a door and

Floor Drain

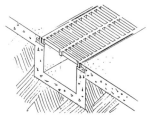

Floor Framing

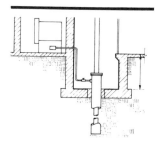

Floor Pit

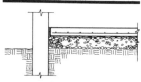

Floor Slab

the finished floor or threshold.

floor clip See **sleeper clip**.

floor closer A device installed in a recess in the floor below a door used to regulate the opening and closing swing of the door.

floor drain A fixture set into a floor, used to drain water into a plumbing drainage system.

floor framing Structural supports for floor systems, including joists, bridging, and subflooring.

floor furring Wood furring used to raise a finished floor above the subfloor and provide a space for piping or conduit.

floor guide A groove in a floor or hardware mounted on a floor, used as a guide for a sliding door.

floor hanger See **stirrup, def. 2**.

floor hatch A unit with a hinged panel, providing access through a floor.

floor hole An opening in a floor, roof, or platform measuring less than twelve inches in its least dimension, so that a person cannot fall through. See also **floor opening**.

flooring A tongued and grooved piece of lumber used in constructing a floor. The basic size of flooring is 1x4 inches, although other sizes are used. Flooring is sold mostly in Superior and Prime grades, and is produced either as vertical grain or flat grain.

flooring cement See **Keene's cement**.

flooring nail A steel nail with a deformed shank, often helically threaded, having a countersunk head and a short diamond-shaped point.

flooring saw A handsaw with teeth on both edges and a blade that tapers to a narrow point, used to cut holes in wood floors.

floor joist Any light beam that supports a floor.

floor light A heavy glass pane in a floor that transmits light to the room below.

floor load The live load for which a floor has been designed, selected from a building code, or developed from an estimate of expected storage, equipment weights, and/or activity.

floor opening A hole or opening in a floor, roof, or platform that measures 12 inches or more in its least dimension, wide enough to permit a person to fall through. See also **floor hole**.

floor panel A prefabricated section of floor, consisting of finish floor, subfloor, and joists.

floor pit Any deep recess in a floor used to provide access to machinery, such as an elevator pit.

floor plan A drawing showing the outline of a floor, or part of a floor, interior and exterior walls, doors, windows, and details such as floor openings and curbs. Each floor of a building has its own floor plan.

floor plate (1) A metal plate set into a floor, sometimes fitted with slots to which equipment may be fastened. (2) A steel plate with a raised pattern on the surface that helps prevent accidents from slipping.

floor plug See **floor receptacle**.

floor receptacle An electric outlet set flush with the floor or mounted in a short pedestal at floor level.

floor register A register set flush with the floor that allows the passage of air.

floor sealer A liquid sealer applied with a brush, sprayer, or squeegee to seal floor surfaces, such as concrete or wood.

floor slab A structural slab, usually concrete, used as a floor or a subfloor.

floor sleeve A short tube set into and penetrating a floor.

floor span (1) The distance, in inches, between the centers of

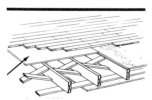

Floor Underlayment

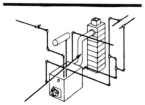

Flue Pipe

floor joists in a floor system. (2) The span covered by floor joists between supports.

floor stilt A device attached to the bottom of a doorframe to keep it above the finished floor.

floor stop A door stop mounted on the floor.

floor tile (1) Modular units used as finish flooring. Floor tile may be resilient material, such as asphalt, vinyl, rubber, or cork, or it may be ceramic or masonry units. (2) Structural units used for floor or roof slab construction.

floor-type heater, floor furnace An air-heating unit mounted below a floor grille or grate. Warm air rises from the center of the grille and cold air returns at the perimeter of the grille.

floor underlayment Particleboard plywood, waterboard, or similar products used on a sub-floor to provide a smooth surface on which to lay the finish floor. See also **carpet underlayment.**

floor varnish A tough, durable varnish used on floors.

flow (1) Time-dependent irrecoverable deformation. See **rheology.** (2) A measure of the consistency of freshly mixed concrete, mortar, or cement paste in terms of the increase in diameter of a molded, truncated cone specimen after jigging a specified number of times.

flow coat A paint coating achieved by immersing an object in streams of paint and draining off the excess.

flow cone (1) A device for measurement of grout consistency in which a predetermined volume of grout is permitted to escape through a precisely sized orifice. The time of efflux (flow factor) is used to determine the consistency. (2) The mold used to prepare a specimen for the flow test.

flow factor See **flow cone.**

flow pressure The pressure at or near an orifice during full fluid flow.

flow promoter A substance added to a surface coating, such as paint, to enhance the brushability, flow, and leveling properties.

flow table A jigging device used in making flow tests for consistency of cement paste, mortar, or concrete. See also **flow, def. 2.**

flow trough A sloping trough used to convey concrete by gravity flow from a transit mix truck, or a receiving hopper to the point of placement. See also **chute.**

flue (1) A noncombustible and heat-resistant passage in a chimney used to convey products of combustion from a furnace fireplace or boiler to the atmosphere. (2) The chimney itself, if there is a single passage.

flue effect See **chimney effect.**

flue grouping The consolidation of multiple flues in one chimney or stack to limit the number of vertical shafts through a structure.

flue lining, chimney lining The lining of a chimney flue composed of heat-resistant firebrick or other fireclay materials, which prevents fire, smoke, and gases from escaping the flue to contaminate surroundings.

flue pipe A tight duct that conveys products of combustion to a chimney or flue.

fluid-filled column A hollow column, usually rectangular, filled with liquid to increase the fire endurance of the column. A series of such columns may be connected to an expansion tank to increase circulation and endurance.

fluidifier An admixture employed in grout to decrease the flow factor without changing water content.

fluing Expanding or splaying, as in the construction of a window, by

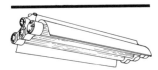

Fluorescent Strip

Flush Door

building the face of a jamb at an oblique angle to the wall in which it is set, enlarging the opening on one or both sides.

fluorescence Visible light generated and emitted from a substance such as a phosphor by absorbing radiation of shorter wave lengths.

fluorescent lamp A low-pressure mercury electric-discharge lamp in which a phosphor coating on the inside of the tube transforms some of the ultraviolet energy generated by the discharge into visible light.

fluorescent lighting fixture A luminaire designed for fluorescent lamps.

fluorescent pigments Pigments that absorb ultraviolet radiant energy and convert it to brilliant, visible light.

fluorescent reflector lamp A fluorescent lamp with the coating on only a part of its cross section so that the light produced will be directed.

fluorescent snaking The visible twisting of the arc in a new fluorescent lamp, usually corrected by turning the appliance on and off several times.

fluorescent strip A luminaire in which fluorescent lamps are mounted on a wiring channel containing the ballast and lamp sockets. Usually reflectors or lenses are not used.

fluorescent U-lamp The tubular fluorescent lamp bent at 180 degrees through its center, forming a U-shape.

fluosilicate Magnesium or zinc silico-fluoride used to prepare aqueous solutions sometimes applied to concrete as surface-hardening agents.

flush Having a surface or face even with the adjacent surface.

flush bolt A door bolt or other bolt mounted on a surface.

flush bolt backset The distance between the center line of a door bolt and the vertical center line of the door's leading edge.

flush bushing In plumbing, a bushing without a shoulder and engineered to fit flush into the fitting with which it connects.

flush-cup pull A door pull for a sliding door. The pull is mortised flush into the door and has a curved recess serving as a finger grip.

flush door A door with flush surfaces and concealed structural parts.

flush glazing Glazing in which glass is set in a channel, which may be formed by a rabbet and stops, in a frame. The glazing, formed or a compound, is flush with the frame at the top of the channel.

flush-head rivet A rivet with a countersunk head.

flush joint Any joint with its surface flush with the adjacent surfaces.

flushometer, flushometer valve A flushing valve, designed for use without a flush tank or cistern, that is activated by direct water pressure to deliver a certain quantity of water for flushing needs.

flush panel A panel in which the exposed surface is in the same plane as the exposed surfaces of the surrounding frame.

flush paneled door A door with one or both surfaces in the same planes as the surfaces of the rails and stiles.

flush pipe A straight pipe that carries flushing water from a cistern or other main source to plumbing fixtures, such as toilets, equipped with a flushing function.

flush plate In electricity, the metal or plastic cover that shields the flush wiring device in a wiring box and provides covering for an outlet or switch, with holes cut

Flush Tank

Flush Wall Box

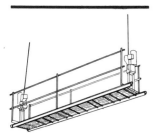

Flying Scaffold

into its face to accommodate switch handles and plugs.

flush ring A door pull mortised into a door, having a pulling ring that folds flat into a recess when not in use.

flush soffit The smooth, visible underside of a flight of spandrel steps.

flush switch An electrical switch installed in a flush wall box in such a way that only its front face is exposed to view.

flush tank A tank that holds water for flushing one or more plumbing fixtures.

flush valve A valve installed in the bottom of a toilet tank to discharge the water needed to flush the fixture.

flush wall box A wall box that houses an electric device and is embedded in a wall, floor, or ceiling with its exposed face in the same plane as the surrounding surface.

flush water See **wash water**.

flute In architecture, one of multiple grooves or channels of semi-circular to semi-elliptical section, used to decorate and to embellish members, such as the shafts of columns.

flux (1) A substance that facilitates the fusion of metals and helps prevent surface oxidation during welding, brazing, and soldering. (2) A liquified bituminous substance used to soften other bituminous materials.

flux-cored arc welding A welding process that produces coalescence by the heating of an arc between the materials being welded and a continuous electrode of filler metal.

fly ash The finely divided residue resulting from the combustion of coal and which is transported from the fire box through the boiler by flue gases. Fly ash is a common additive to concrete to improve strength, workability,

and waterproof qualities.

flying bond A masonry bond formed by laying occasional headers at random intervals.

flying buttress A masonry substructure that transfers the horizontal thrust of a roof or vault to a detached pier or buttress, usually sloping slightly towards the pier or buttress.

flying scaffold Staging suspended by ropes or cables from outrigger beams attached at the top of a structure.

flying shore A horizontal member that provides temporary support between two walls.

foam concrete A lightweight, cellular concrete made by infusing an unhardened concrete mixture with prepared foam or by generating gases within the mixture.

foam core The center of a plywood "sandwich" panel, consisting of plastic foam between wood veneers. The foam may be introduced in a liquid form that is forced under pressure into a space between the wood veneer skins, or the skins may be applied to a rigid plastic foam board.

foamed concrete, foam concrete See **concrete, foamed**.

foamed-in-place insulation A plastic foam employed for thermal insulation, prepared by mixing insulation substances with a foaming agent just before it is poured or sprayed with a gun into the enclosed receptacle cavities.

foamed polystyrene A foamed plastic, weighing about 1 lb. per cu. ft. (0.016 g/cc), which has high insulation value, is low in cost, and is grease-resistant.

foam glass, cellular glass, expanded glass A thermal insulation made by foaming glass with hydrogen sulfide. Foam glass has a closed cell structure and is a low fire-hazard material. It is manufactured in the form of

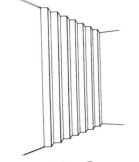

Folding Door

block or board with a density of 9 to 10 lb. per cu. ft. (14 to 16 kg per cu m).

foaming agent A substance incorporated into plastic mixtures of concrete, rubber, gypsum, or other materials to produce a light, foamy consistency in the mixture by releasing gases into it.

fog curing (1) The storage of concrete in a moist room in which the desired high humidity is achieved by the atomization of fresh water. See also **moist room**. (2) The application of atomized fresh water to concrete, stucco, mortar, or plaster.

fog room See **moist room**.

fog sealed Descriptive of surfaces lightly treated with asphalt without a mineral cover.

foil (1) In masonry, one of multiple circular or nearly circular holes, set tangent to the inside of a larger arc, that meet each other in pointed cusps around the arc's inner perimeter. (2) A metal formed into thin sheets by rolling.

foil-backed gypsum board A gypsum board with aluminum foil on one face. The foil acts as a vapor barrier.

folded plate (1) A framing assembly composed of sloping slabs in a hipped or gabled arrangement. (2) A prismatic shell with an open polygonal section.

folded-plate construction A type of construction used with span roofs. Thin, flat elements of concrete, steel or timber are connected rigidly at angles to each other, similar to accordion folds, to form members with deep cross sections.

folded plate roof A roofing system consisting of sloping "plates" in a repetitive gable arrangement, usually a combination of plywood skins and lumber trusses. A folded plate roof permits a large, clear floor area.

folding casement (1) One of two

casements with rabbeted masting stiles, installed in a single frame with a mullion. (2) One of several casements hinged together in a way that permits them to open and fold in a limited space.

folding door An assembly of two or more vertical panels hinged together so they can open or close in a confined space. A floor- or ceiling-mounted track is usually provided as a guide.

folding partition (1) Large panels hung from a ceiling track, sometimes supported also by a floor track, which form a solid partition when closed but stack together when the partition is maneuvered into an open position. (2) A partition, faced with fabric and hung from a ceiling track, that folds up flexibly when opened like the pleated balloons of an accordion.

folding rule A rule made of lengths that are joined by pivots so it can be folded when not in use.

folding stair See **disappearing stair**.

foliated joint A joint made between two boards by fitting their rabbeted edge together, forming a continuous surface on either face.

foot (1) The bottom or base on an object. (2) A unit of measurement of length in the English System. (3) A projection on a cylindrical roller used to compact a layer of earth fill.

foot block A mat of concrete, steel, or timber used to distribute a vertical load from a post or column over an area of supporting soil.

foot bolt A spring-operated bolt fastened to the lower part of a door and controlled with pressure of the foot.

footcandle A unit of illumination equal to 1 lumen per sq. ft.

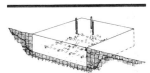

Footing

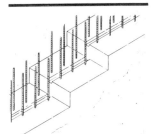

Footing Step

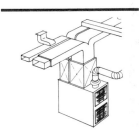

Forced-Air Furnace

footing That portion of the foundation of a structure that spreads and transmits load directly to the soil.

footing beam (1) A reinforced concrete beam connecting pile caps or spread footings to distribute horizontal loads caused by eccentric loading. (2) The tie beam in a roof system.

footing course A broad coarse of masonry at the foot of a wall, which helps prevent the wall from settling.

footing step A change in elevation of a strip footing.

foot-meter rod A surveyor's rod graduated in feet and tenths on one side, and meters and hundredths on the other.

footpiece An item of ductwork in a heating, air-conditioning, or ventilating system that serves to change the direction of the air flow.

footplate A timber used in wood-frame construction to disperse heavily concentrated structural loads among a number of supporting members, as a plate installed below a row of wall studs.

foot-pound A unit measure of work or energy in the English System.

footstall The base of a pier or pillar, characterized by a particular architectural order or treatment. Also called a "plinth".

force account A term for any work ordered on a construction project without an earlier agreement on its lump sum or unit price cost and performed with the understanding that the contractor will bill the owner according to the cost of labor, materials, equipment, insurance, and taxes plus a certain percentage for overhead and profit.

forced-air furnace A warm-air furnace, outfitted with a blower, that heats an area by transmitting air through the furnace and connecting ducts.

forced circulation The circulation of a fluid by mechanical means, as by a pump or fan.

forced convection Heat transfer produced by the forced circulation of air, water, etc.

forced draft A draft of air that is mixed with fuel before the mixture is fed into the combustion chamber of a furnace or boiler.

forced-draft boiler A boiler equipped with a power-driven fan that feeds air into the burner and boiler and expels smoke, gases, and ash through the chimney.

foreclosure The legal transfer of a property deed or title to a bank or other creditor because of the owner's failure to pay the mortgage, whereupon the owner loses the right to the property.

foreclosure sale The optional right of the mortgagee or lending institution to sell mortgaged property if the mortgagor fails to make payment, applying proceeds from the sale toward the outstanding debt.

forest land As defined by the U.S. Forest Service, forest land is land at least 10% stocked with live trees, or land formerly having such a tree cover and not currently developed for non-forest use. The minimum area of forest land recognized is one acre.

forging A piece of metal worked into a desired shape by one or more processes, including pressing, rolling, hammering, and upsetting.

forked tenon A joint formed by a tenon cut into a long rail and inserted into an open mortise.

forklift truck A self-powered vehicle equipped with strong prongs, or forks, that can be raised or lowered. A forklift truck is used to move objects, especially material on pallets, from one location and/or level to another.

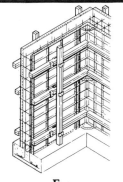

Form

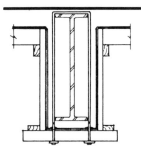

Form Hanger

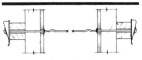

Form Tie

Formwork

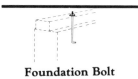

Foundation Bolt

form A temporary structure or mold for the support of concrete while it is setting and gaining sufficient strength to be self-supporting. See also **formwork**.

form anchor A device used to secure formwork to previously placed concrete of adequate strength. The device is normally embedded in the concrete during placement.

format In construction, the standard arrangement of a project manual as set by the A.I.A. with bidding information, contract forms, conditions of the contract, and sixteen subdivisions of specifications.

form board A board or sheet of wood used in formwork.

form coating A liquid applied to interior formwork surfaces for a specific purpose, usually to promote easy release from the concrete, to preserve the form material, or to retard the set of the near-surface matrix for preparation of exposed-aggregate finishes.

formed plywood Curved plywood, prepared by a special pressing technique using rigid forming dies. See **molded plywood**.

form hanger A device used to support formwork from a structural framework. The dead load of forms, weight of concrete, and construction and impact must be supported.

forming A process of shaping metal by a mechanical process other than machining, forging, or casting.

form insulation Thermal insulation, equipped with an airtight seal, that is applied to the exterior of concrete forms to preserve the heat of hydration at required levels so that concrete can set properly in cold weather.

form lining Selected materials used to line the face of formwork in order to impart a smooth or patterned finish to the concrete surface, to absorb moisture from the concrete, or to apply a set-retarding chemical to the formed surface.

form nail See **double-headed nail**.

form oil Oil applied to the interior surface of formwork to promote easy release from the concrete when forms are removed.

form pressure Lateral pressure acting on vertical or inclined formed surfaces, resulting from the fluid-like behavior of the unhardened concrete.

form release agent See **release agent**.

form scabbing The inadvertent removal of the surface of concrete as a result of adhesion to the form.

form spreader See **spreader**.

form stop A wooden piece used in concrete formwork to regulate or limit the flow of concrete at the end of a work day.

form stripping agent See **release agent**.

form tie A tensile unit adapted to prevent concrete forms from spreading due to the fluid pressure of freshly placed, unhardened concrete.

formwork The total system of support for freshly placed concrete, including the mold or sheathing which contacts the concrete, as well as all supporting members, hardware, and necessary bracing.

forty A standard unit of measurement used in surveying and in describing units of forest land, equivalent to forty acres, or 1/16th of a square mile.

foundation The material or materials through which the load of a structure is transmitted to the earth.

foundation bolt See **anchor bolt**.

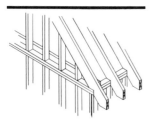

Frame Construction

Frame Wall

foundation drainage tile Tile or piping, either porous or set with open joints, used to collect subsurface water or for the dispersion of effluent.

foundation engineering The category of engineering concerned with evaluating the ability of a locus to support a given structural load, and with designing the substructure or transition member needed to support the construction.

foundation wall That part of the foundation of a building forming a retaining wall for the portion of the building that is below grade.

fourths A common grade of lumber exported from Scandinavia and Eastern Europe.

four-way reinforcement See **reinforcement, fourway**.

four-way switch A switch used in a wiring layout to allow a circuit to be turned on or off from more than two places. Two three-way switches are used, and the remainder are four-way.

foxtail See **foxtail wedge**.

foxtail saw See **dovetail saw**.

foxtail wedge A small wedge used to spread the split end of a bolt in a hole or the split end of a tenon in a mortise to secure the bolt or tenon.

foyer A subordinate space between an entrance and the main interior of a theater, hotel, house or apartment.

fracture load See **breaking load**.

frame An assembly of vertical and horizontal structural members.

frame building See **framed building**.

frame clearance The clearance between a door and the door frame.

frame construction A construction system in which the structural parts are wood or dependent on a wood framework for support. The balloon system consists of vertical members running from the foundation to the roof plate, and to which floor joists are attached. In platform construction, floor joists of each floor rest on the plates of the floor below.

framed building A type of building construction in which vertical loads are carried to the ground on a frame.

framed door Any door with a stiff frame made up of a top rail, lock rail, bottom rail, lock stile, and hanging stile.

framed floor See **double floor**.

framed ground One of the vertical wood members fastened around an opening to which a door casing is attached, usually built with a tenon joint between the head and doorjambs.

framed joist A joist that is specially cut or notched in order to be joined securely with other joists or timbers.

framed partition, trussed partition A partition made by covering framing of studs, braces, and struts that form a truss.

frame gasket A strip of flexible material applied to the stop of a doorframe to ensure tight closure.

frame-high A term used in masonry to denote a height equal to the top of a door or window frame or to the lintel of an opening.

frame house A house of frame construction, usually with exterior walls sheathed and covered with wood siding.

frame pulley A pulley installed in a window frame to carry a sash chord.

frame wall A wall of frame construction.

framework A network of structural members or components joined to form a structure, such as a truss or multi-level building.

framing

Freestanding

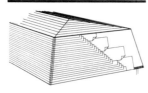

French Roof

framing (1) Structural timbers assembled into a given construction system. (2) Any construction work involving and incorporating a frame, as around a window or door opening. (3) The unfinished structure, or underlying rough timbers of a building, including walls, roofs, and floors.

framing anchor A sheet metal device used to join light wood members such as studs, joists and rafters.

framing plan A drawing of each floor of a building showing the framing members and their connections.

framing square See **carpenter's square**.

framing table See **rafter table**.

f rating The measurement of stress (symbolized by the letter "f") in a piece of lumber. Generally, the higher the "f rating", the stronger the piece of lumber.

freeboard The distance between the design or normal water level and the top or overflow of a tank, dam, or vessel.

free fall Descent of freshly mixed concrete into forms without dropchutes or other means of confinement, or the distance the concrete falls.

free float A term used in project management, planning, and scheduling methods such as PERT, and CPM. The free float of an activity is the amount by which the completion of that activity can be deferred without delaying the start of the following activities or affecting any other activity in the network.

freehold A tenure of property that an owner may hold in fee simple, fee tail, or for life. The term also refers to property held under freehold.

free lime Calcium oxide (CaO) as in clinker and cement that has not combined with SiO_2, Al_2O_3, or Fe_2O_3 during the burning process, because of underburning, insufficient grinding of the raw mix, or the presence of traces of inhibitors.

free moisture Moisture having essentially the properties of pure water not absorbed by aggregate in a test sample or a stock pile. (See also **surface moisture**.)

free on board (FOB) Refers to the point to which the seller will deliver goods without charge to the buyer. Additional freight or other charges connected with transporting or handling the product become the responsibility of the buyer.

freestanding (1) Said of a structural element that is fixed at its base and not braced at any upper level. (2) Cantilevered.

free water See **free moisture**. (See also **surface moisture**.)

freezer A room or cabinet mechanically refrigerated to maintain a temperature of about 10 degrees F (-12 degrees C) used for food storage.

freeze-thaw The cycle from freezing to thawing often detrimental to construction materials such as concrete. Air entrainment of concrete to be exposed to the elements is helpful in preventing damage from freeze-thaw cycles.

French door, casement door, door window A door, or pair of doors, with glass panes constituting all or nearly all of its surface area.

French drain A drainage ditch containing loose stone covered with earth.

Frenchman A kitchen knife with the end bent over, used with a jointing rule for pointing mortar joints.

French roof A mansard roof with nearly perpendicular sides.

French window A doorway with a single full-glazed door or a pair of

208

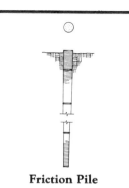

Friction Pile

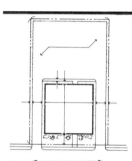

Frontage

Front-End Loader

such doors. See also **French door.**

frequency A measure of oscillations per second, applied to the current or voltage of AC electrical circuits, sound waves, or vibrating solid objects, and stated in hertz (HZ) or cycles per second (CPS).

fresh-air inlet (1) A connection to a building drain, located above the drain trap and leading to the atmosphere. (2) An outside air vent for an HVAC system.

fret A form of architectural ornamentation, consisting of elongated rectangles that are painted, curved, or raised in an elaborate pattern of fillets, bands, and ringlets; frequently made up of continuous lines crafted in repeating rectangular shapes.

fret-work Ornamental openwork or relief consisting of bands of interlocking geometric patterns of contrasting light and dark elements.

friable Said of a material that can be easily pulverized or crumbled.

friction The resistance to relative motion, sliding, or rolling, between two surfaces in contact.

friction catch Any catch held in position by friction when engaged with its strike.

friction hinge A hinge that uses internal friction to hold a door or window in a selected position.

friction loss The stress loss in a prestressing tendon resulting from friction between the tendon and duct or other device during stressing.

friction pile, floating pile foundation A load-bearing pile that receives its principal vertical support from skin friction between the surface of the buried pile and the surrounding soil.

friction shoe A friction device used to hold a sash in any open position. The shoe position may be adjustable or preset.

friction tape An insulating tape used by electricians, made of a fibrous base impregnated with a moisture resisting compound that will stick to itself but not to most other materials.

friction welding A process by which thermoplastic materials are softened and welded together with heat produced by friction.

frog A depression in the bed surface of a masonry unit that is sometimes called a panel.

frontage The length of a property line or building line along the street or beachfront that forms its boundary line on one side.

front-end loader A machine with a bucket fixed to its front end, having a lift-arm assembly that raises and lowers the bucket. A front-end loader is used in earth moving and loading operations and in rehandling stockpiled materials.

front hearth, outer hearth The part of a hearth that occupies the room served by a fireplace, skirting the front of the fireplace opening.

frontispiece The architecturally adorned front wall or bay of a building or edifice. Also, an ornamental porch or pediment.

frost action The loosening, spalling, and/or lifting process caused by alternate freezing and thawing of moisture in a material.

frost boil (1) An imperfection on a concrete surface caused by the freezing of trapped moisture that swells and crumbles the affected concrete. (2) The softening of soil caused by melting and release of subsurface water during a thaw.

frost cracks Splits or cracks in the trunk of a tree, caused by extreme cold. Such cracks are defects in lumber manufactured from this timber.

frost heave The lifting of a soil surface or pavement due to the

Fuel-Fired Boiler

Full-Penetration Butt Weld

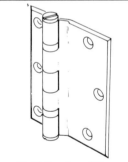

Full-Surface Hinge

freezing of moisture in the underlying soil.

frostproof closet A water-closet bowl that retains no standing water. The trap and the valve for its water supply are located below the frost line.

fudge (slang) To depart from the design drawing for sake of appearance.

fuel-fired boiler An automatic, self-contained mechanical unit or system that produces heat by burning solid, liquid, or gaseous fuels.

fugitive dye A dye whose color fades in a few days to neutral on exposure, usually to ultraviolet rays in sunlight. Fugitive dye is often used to temporarily color membrane-curing compounds so that coverage of the concrete surface can be observed.

full A reference to lumber that is slightly oversize.

full bond A masonry bond in which all bricks are laid as headers.

Fuller faucet A faucet in which a rubber ball is forced into the inlet to stop the flow.

Fuller's curve An empirical curve for gradation of aggregates, also known as the Fuller-Thompson ideal grading curve. The curve is designed by fitting either a parabola or an ellipse to a tangent at the point where the aggregate fraction is one-tenth of the maximum size fraction. See also **grading curve**.

full-flush door A door made of two sheets of steel assembled in hollow-metal construction with a top and bottom either flush or closed off with end panels, and seams that are visible only on the edge of the door.

full glass door A door with glass in the area between rails and stiles except for dividing muntins. The glass is usually heat treated or tempered.

full-louvered door A door with louvers filling the entire area between rails and stiles.

full-penetration butt weld A butt weld with a depth equal to the thickness of the smaller of the two members between which it lies.

full size Said of a drawing depicting an object at true size: not scaled.

full splice A splice that will transmit the full strength of the joined members.

full-surface hinge A hinge that may be installed on the surface of a door and jamb without need of mortising.

fully welded seamless door A door having all joints on its faces and edges fully welded and ground smooth, so as to conceal the joints.

fungicide A substance poisonous to fungi, used to prevent or retard the growth of fungi.

furnace (1) That part of a boiler or warm air heating plant in which the combustion takes place. (2) A complete heating unit that transfers heat from burning fuel to a heating system.

furnish The raw material used to make reconstituted wood-based non-veneer panel products.

furniture grade Lumber of a quality and size suitable for the manufacture of furniture.

furred Provided with furring to leave an air space as between a structural wall and plaster or between a subfloor and wood flooring.

furring (1) Strips of wood or metal fastened to a wall or other surface to even it, to form an air space, to give appearance of greater thickness, or for the application of an interior finish such as plaster. (2) Lumber one inch in thickness (nominal) and less than four inches in width, frequently the product of resawing a wider piece. The most common

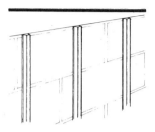

Furring Strip

Fuse

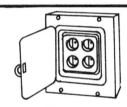

Fuse Box

sizes of furring are 1x2 and 1x3. See **strip**.

furring brick A hollow brick carrying no superimposed load and used as lining, furring, or as a key for plastering on an interior wall.

furring nail A galvanized, low-carbon steel nail, having a flat head and a diamond point. The shank is provided with a spacer for fastening wire lath and spacing it from the nailer.

furring strip A wood strip used as furring.

furring tile Tile designed for lining the inside of exterior masonry walls. It is not intended to support superimposed loads and has a scored face for the application of plaster.

furrowing The process of making furrows in the mortar bed with the tip of a trowel to speed up bricklaying.

fuse A protective device, made of a metal strip, wire or ribbon that guards against overcurrent in an electrical system by melting if too much current is generated and breaking the circuit.

fuse box A metal box with a hinged cover which houses fuses for electric circuits.

fusetron A special fuse that will carry an overload of current for a short time without "blowing" or opening the circuit. A fusetron is used where a heavy load, such as the starting of a motor, may overload a circuit momentarily.

fusible link A metal link made of two parts held together by a low-melting-point alloy. When exposed to fire-condition temperatures, the link separates allowing a door, damper or device to be closed.

fusible metal An alloy with a low melting point used in high temperature detecting devices.

fusible solder An alloy, usually containing bismuth, having a melting point below that of tin-lead solder: 361 degrees F (183 degrees C).

fusible tape See **joint tape**.

fusiform rust A type of blister rust disease, caused by the fungus, Cronartium fusiform. The disease attacks certain southern pine species. Similar rusts infect western pines.

fusion In welding, the melting and coalescence of a filler metal and base metal or two base metals.

fuzzy grain In surfacing lumber, a condition of the board surface in which some fibers are not completely severed in the surfacing process, giving it a fuzzy appearance.

G

 # ABBREVIATIONS

The abbreviations listed below are those most commonly used in the construction industry. Alternative forms (usually nonstandard) are shown in parentheses.

g gram, gravity, guage, girth, gain
G gas
ga gauge
gal. gallon
galv galvanized
gar garage
GB glass block
GC general contractor
GCF greatest common factor
gen general
GI galvanized iron
gl glass, glazing
GM grade marked
GMV gram molecular volume
Goth Gothic
gov/govt government
gpd gallons per day
gph gallons per hour
gpm gallons per minute
gr grade, gravity, gross, grains
G/R grooved roofing
gran granular
gr.fl. ground floor
gr.fl.ar ground floor area
grnd ground
gr.wt gross weight
GT gross ton
gtd guaranteed
g.u.p. grading under pavement
GYP gypsum

G DEFINITIONS

Gabion

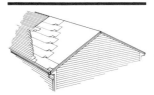

Gable

Gable Dormer

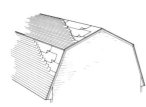

Gambrel Roof

gabion A large compartmentalized container, usually cylindrical or rectangular, often fabricated from galvanized steel hexagonal wire mesh and, when filled with stone, used in the construction of foundations, dams, erosion breaks, etc.

gable The portion of the end of a building that extends from the eaves upward to the peak or ridge of the roof. The gable's shape is determined by the type of building on which it is used: triangular in a building with a simple ridged roof, or semi-octagonal in a building with a gambrel roof.

gableboard A board covering the timbers that extend over the gable end of a gable roof.

gable dormer A dormer that protrudes horizontally outward from a sloping roof and has its own gabled end, whose base meets that of the sloping roof.

gable end An end wall with a gable.

gable molding The molding used as finish for a gable end of a roof.

gable post A short post into which the gableboards are fitted at the peak of a gable.

gable roof A ridged roof having one or two gabled ends.

gable shoulder The projection created by the gable springer at the foot of a gable.

gable springer The projecting stone under the loosest kneeler at the foot of a gable coping.

gable wall A wall whose upper portion is a gable.

gable window A window built into or shaped like a gable.

gage See **gauge**.

gag process The use of a gag press to bend structural shapes.

gain The mortise or notch in a piece of wood into which another piece of wood fits.

gaine A pedestal, square in section but with one end tapered downward, used primarily for decoration.

gallery (1) A long enclosure that functions as a corridor inside or outside a building or between different buildings. (2) The elevated (usually the highest) designated seating section in an auditorium, theater, church, etc. (3) A room or building in which artistic works are exhibited.

gallery apartment house An apartment house whose individual units can be entered from an exterior corridor on each floor.

gallet A chip of rock, stone, or masonry.

galleting, garreting The use of chips of rock, stone, or masonry to fill the joints of rough masonry, either for appearance or to minimize the required amount of mortar.

gallows bracket A triangular wall bracket often used for the support of shelving.

galvanic corrosion The electrochemical action that occurs as a result of the contact of dissimilar metals in the presence of an electrolyte.

galvanize The process of protectively coating iron or steel with zinc, either by immersion or electroplating.

galvanized iron Zinc-coated iron sheet metal.

galvanized pipe Zinc-coated steel or wrought-iron pipe.

gambrel roof A roof whose slope on each side is interrupted by an obtuse angle that forms two pitches on each side, the lower

Gang Form

Garden Apartment

slope being steeper than the upper.

gandy dancer A slang term for a "railroad worker" who performs earth and tracklaying work.

gandy stick Railroad slang for a "tamping bar."

gang A working technique in which several machines and/or apparatuses are controlled by a single force and combined in such a way as to function as a single unit.

gang form Prefabricated form panels connected together to produce large reusable units. Gang forms are usually lifted by crane or rolled to the next location.

gang nail A metal plate having sharp spikes perpendicular to it and functioning as a timber connector.

gang saw A powered mechanical cutting device with two or more saw blades mounted parallel to each other on the same arbor or on the same sash. (2) In stone cutting, a powered mechanical cutting device with parallel-mounted reciprocating saw blades.

gangway A temporary planked path that runs up to a building or over an unfinished portion thereof before the exterior steps are constructed. The gangway provides a passageway for men and materials on the construction site.

ganister A mixture of ground quartz and a bonding material, such as fire clay, used primarily for fire-proofing, as around the hearths of furnaces.

gantry A framework or overhead structure of beams or timbers used as a working platform or as a means to support equipment.

gantry crane A revolving crane situated atop a heavy framework which moves along the ground on tracks.

gap-filling glue A type of glue used in applications where close-

contact glue is inadequate because of the poor fit between the surfaces to be joined.

gap grading A particle-size distribution of aggregate which is practically or completely void of certain intermediate sizes.

garbage disposal unit An electrically motorized device that grinds waste food and mixes it with water before disposing of it through standard plumbing drainage pipes.

garden apartment (1) A two- or three-story suburban, multiple dwelling whose emphasis on patios, balconies, walks, or even communal gardens serves to create or promote the illusion of privacy. (2) A ground-floor apartment having access to an adjacent outdoor space or garden.

garden city A well-treed and planted residential development composed largely of single-family, detached houses and providing vehicular parking.

garden house In a garden, a small gazebo or other shelter.

garden tie Usually, a railroad tie or similar timber used in landscaping.

garden tile Molded structural ceramic units used in a garden or on a patio as stepping stones.

garden wall bond Usually, ornamental brickwork that is only a single brick thick and is composed mostly of stretchers showing a fair face on each side.

gargoyle A waterspout, often with carvings of grotesque human or animal features, that projects from a roof gutter so as to drop rainwater clear of a wall, often through its open mouth.

garnet A silicate mineral with an isometric crystal structure. Garnet occurs in a variety of colors and is used as gemstones, abrasives, or gallets.

garnet hinge A T-shaped hinge whose longer leg attaches to the

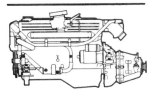

Gas Engine

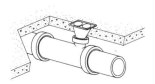

Gas Vent

door leaf and whose shorter leg attaches to the frame.

garnet paper A finishing and polishing paper whose abrasive is powdered garnet.

garret The space immediately below a roof structure, such as an attic.

garrison house A style of house whose second story extends outward beyond the first-story facade on one or more walls.

gas concrete A concrete whose light weight and insulating qualities are the result of cement alkalies and an aluminum powder admixture that reacts to generate hydrogen gas, and hence to produce voids in the unhardened mix.

gas engine An internal-combustion engine that uses gas rather than gasoline as fuel.

gas filled lamp A type of incandescent lamp whose bulb atmosphere consists of an inert gas within which the filament operates.

gas-fired Heated by the burning of a gaseous fuel.

gas flow meter A device that measures the volume and/or velocity of flowing gases.

gas furnace A furnace that burns gas to produce heat.

gasket Any of a variety of seals made from resilient materials and placed between two joining parts (as between a door and its frame, an oil filter and its seat, pipe threads and their fitting, etc.) to prevent the leakage of air, water, gas, or fluid.

gaskin (1) A ring of rope placed into the socket of a stoneware pipe joint before it is sealed with cement mortar. (2) A type of inflatable ring for temporarily sealing a pipe joint.

gas main The public's gas supply as piped from the providing utility company into a community.

gas metal-arc welding A method of welding that achieves coalescence by arc-heating between the work and a consumable electrode.

gas meter A mechanical device that measures and records in cubic feet the volume of gas passing a given point.

gas pliers Rugged pliers whose concave jaws have serrated faces for the efficient gripping of pipes.

gas pocket In a casting, a hole or void caused by the entrapment of air or gas that is produced during the solidification of the metal.

gas refrigeration A refrigeration system in which a gas flame is used to heat the refrigerant.

gas vent An exhaust pipe through which undesirable gaseous by-products of combustion are vented to the outside.

gas welding Any of several welding methods in which a gas flame provides the heat necessary for coalescence and in which pressure and/or a filler metal may or may not be employed.

gate A usually hinged device of solid or open construction that is installed as part of a fence, wall, or similar barrier and which, when opened (either by swinging, sliding, or lifting), provides access through that barrier.

gate hook, gudgeon A metal bar (usually one of a pair, spaced one directly above the other) whose pointed or threaded end is driven or screwed into a wooden post or secured in a masonry or brick gate pier, and whose exposed upright pin engages the gate hinge, thus hanging the gate.

gate house (1) A building that accommodates the gateway to a castle, manor house, estate, or other important building or place.

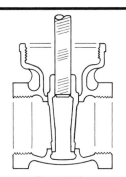

Gate Valve

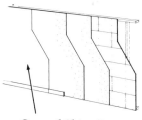

Gauged Skim Coat

gate operator An electronically operated mechanism which, when activated, serves to open or close a gate.

gate pier, gate post A post of wood, brick, stone, or masonry on which a gate is hung or hinged, or against which it latches.

gate valve A piping device consisting of a housed wedge or disc positioned perpendicular to the flow, which it regulates by being raised or lowered.

gathering The bringing together of flues within a single stack.

gauge, gage (1) The numerically designated thickness of sheet metal. (2) A metal tubing or the similarly designated diameter of a screw or wire. (3) A measuring device for pressure or liquid level. (4) The distance between rows of bolt or rivet holes in the same member. (5) A wood or metal strip used as a thickness-control guide in bituminous or concrete paving operations. (6) In plastering, a screed. (7) The act of adding or the amount of gauging plaster added to hasten the setting of common plaster. (8) In laid roofing, the exposed length of a shingle, slate, or tile.

gauge board, gauging board (1) A 3-foot square board on which plaster and tools are carried. (2) A board on whose surface plaster, cement, or mortar is mixed. (3) A carpenter's pitch board.

gauge box, gauging box A container in which a batch of concrete, mortar, or plaster is measured and mixed.

gauged A term that refers to a material that has been ground so as to produce particles of uniform shape and/or thickness.

gauged arch An arch constructed from bricks whose wedge shape causes the joints to radiate from a common center.

gauged brick Brick that has been sawn, ground, rubbed, or otherwise shaped to accurate dimensions for special applications.

gauged mortar (1) Mortar made of cement, sand, and lime in specific proportions. (2) Any plastering mortar that is mixed with plaster of paris to hasten setting.

gauged skim coat A mixture of gauging plaster and lime putty applied very thinly as a final coat in plastering. The mixture is troweled to produce a smooth, hard finish.

gauged stuff (1) A mixture of lime putty and gypsum plaster that is used as a finish coat in plastering. (2) Gauged mortar.

gauged work (1) Any precision brickwork using gauged brick. (2) Plastering that uses gauged plaster as for molding or ornamental work.

gauge glass A glass tubular device or vertical cylindrical device, often graduated, that indicates the level or amount of liquid in a tank or vessel.

gauge pile See **guide pile**.

gauge pressure The inherent pressure of a gas or liquid, exclusive of the value of atmospheric pressure.

gauge rod A stick used for measuring the gauge in brickwork. If used to mark floor and sill levels, the rod is termed a "story rod."

gauge stick (1) A dimensionally specific length of wood, metal, etc., used in measuring repetitive dimensions. (2) In roofing, a scantle used to measure tiles for accurate cutting.

gauging plaster Plaster of paris (gypsum plaster) that is usually mixed with lime putty to produce a quick-drying finish coat.

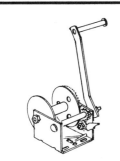

Gear

gaul A hollow place in a coat of plaster or mortar, usually the finish coat, caused by poor troweling.

gazebo A small, round, octagonal, or similarly shaped structure, usually roofed but open-sided, built in parks or large gardens to provide shelter or place to view the surrounding area.

G-clamp, G-cramp A large G-shaped screw clamp often used in joining pieces of glued wood.

gear A toothed wheel, cone, cylinder, or other machined element that is designed to mesh with another similarly toothed element for purposes which include the transmission of power and the change of speed or direction.

gear jammer A slang term for a "truck driver."

gefle standard A measure of roundwood volume equal to 100 cubic feet. The measure is used principally in the export trade.

gel The colloidal state of a material.

gelatin blasting The strongest and highest velocity, commercially available explosive. The material is manufactured by dissolving nitrocotton in nitroglycerin.

gelatin mold A semi-rigid mold composed primarily of gelatin and used in manufacturing complicated undercut fibrous plaster coatings.

gel coat A relatively thin, often pigmented layer of resin which is usually sprayed onto a waxed mold to form the attractive exterior surface of a reinforced plastic molding, such as a fiberglass boat.

gemel, gimmer, gymmer, jimmer Any two structural elements that go together as a pair.

gemel window A two-bayed window.

general conditions The portion of the contract document in which the rights, responsibilities, and relationships of the involved parties are itemized.

general contract In a single-contract system, the documented agreement between the owner and the general contractor for all the construction for the entire job.

general contractor For an inclusive construction project, the primary contractor who oversees and is responsible for all the work performed on the site, and to whom any subcontractors on the same job are responsible.

general diffuse lighting Lighting from units that direct 40-60 percent of their emitted light upward, and the remainder downward.

general drawing A drawing that illustrates structural cross-section, main dimension, elevation plan, substructural borings, and other basic details of a construction project.

general foreman The general contractor's on-site representative, often referred to as the "superintendent" on large construction projects. It is the responsibility of the general foreman to coordinate the work of various trades and to oversee all labor performed at the site.

general industrial occupancy The designation of a conventionally designed building that can be used for all but high-hazard types of manufacturing or production operations.

general lighting Lighting designed to produce a fairly consistent level of illumination over an entire area.

general requirements The designation or title of Division I (the first of 16) in the Construction Specifications Institute's Uniform System. General requirements usually include overhead items and equipment rentals.

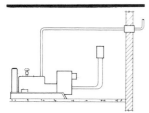

Generator

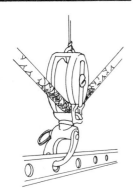

Gin Block

generator A mechanical or electromechanical device that converts mechanical energy into electrical power, as an alternator producing alternating current or a dynamo producing direct current.

geodesic dome A stable, dome-shaped structure fabricated from similar lightweight members connected to form a grid of interlocking polygons.

geodetic survey A land survey, usually of vast areas, whose calculations incorporate consideration of the earth's curvature and can accurately mark specific points from which other surveys can be controlled.

geological map A map showing the character and distribution of outcrops of strata or igneous rocks, including faults, antidives, and other sizeable formations. Some such maps show only the solid outcrops, excluding the overlying drift.

geometrical stairs A stair, either winding or spiral, that is built around a stairwell and returns on itself as it ascends, but does not employ newels at the turns and has no landings between floors.

Georgia buggy A slang term for an on-site concrete-carrying hopper with heavy rubber wheels.

Georgia marble A general descriptive term applied to sparkling, crystalline, light grey, or white marble.

Georgian glass Thick glass into which steel wire mesh is incorporated as reinforcement.

German siding A type of drop siding installed so that the grooved bottom edge of the board above fits over the concave top edge of the board below.

gesso A plaster base coat, for gilding or decorative painting, consisting of gypsum plaster, calcium carbonate, and glue.

ghosting Descriptive of a thin coat of paint that leaves a skimpy appearance.

gib A metal strap used to fasten two members together.

gib-and-cotter joint A joint in which a gib, drawn tightly and wedged with cotters, secures the joined members together.

gig stick See **radius rod**.

gilding (1) The application of gilt (as gold leaf or flakes) to a surface. (2) A surface ornamented with gilt.

Gilmore needle A device used to determine the setting time of hydraulic cement.

gilsonite A naturally occurring, hard, brittle form of asphalt that is mined and used in the manufacturing of floor tile, paving, roofing, and paint.

gimlet A small hand tool, of ancient origin, whose handle is perpendicular to a screw point, and which is used for boring small holes (less than 1/4 inch) in wood.

gin A simple lifting device consisting of a vertical pole, tripod, or other frame.

gin block A simple tackle block, with a single pulley in a frame, having a hook on top for hanging.

gingerbread An ornate style of architecture, usually employed in exterior house trim, and common in the United States during the 19th century.

gin pole A cable-supported vertical pole used in conjunction with blocks and tackle for hoisting.

ginnywink A slang term for an A-frame derrick that has a fixed rear leg.

girandole A branched fixture or bracket for holding lamps or candles, either free standing or protruding from a wall, often having a mirror behind it.

girder A large principal beam of steel, reinforced concrete, wood,

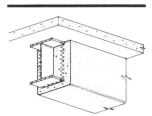

Girder Casing

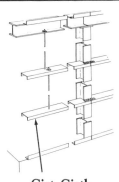

Girt, Girth

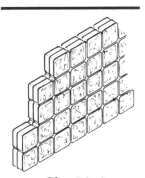

Glass Block

or a combination of these, used to support other structural members at isolated points along its length.

girder casing Material that encloses and protects (as from fire) the part of a girder extending below ceiling level.

girder post Any support member, such as a post or column, for a girder.

girdle A horizontal band around the shaft of a column.

girt, girth (1) A horizontal member used as a stiffener between studs, columns, or posts at intermediate level. (2) A rail or intermediate beam that receives the ends of floor joists on an outside wall.

girt strip A board secured horizontally to studding to carry floor joists. See **ledger board**.

give-and-take lines Straight equalizing lines used in calculating band areas and their boundaries.

glacial till Glacial deposits of sand, gravel, clay, silt, and boulders found in unstratified pockets. Glacial till usually has good load-bearing properties.

gland (1) In plumbing, a brass or copper ring that is compressed by a screwed fitting over a copper tube and that deforms to become a seal between the two. (2) In electrical work, a seal used to prevent water from entering the end of a cable.

gland bolt A bolt used to tighten or loosen an unthreaded gland.

gland joint A joint on a metal soil pipe or hot water pipe which allows for thermal expansion.

glass A hard, brittle, inorganic product, ordinarily transparent or translucent, made by the fusion of silica, flux, and a stabilizer, and cooled without crystallizing. Glass can be rolled, blown, cast, or pressed for a variety of uses.

glass bead See *glazing bead*.

glass block A hollow, translucent block of glass, often with molded patterns on either or both faces, which affords pleasantly diffused light when used in non-load-bearing walls or partitions.

glass cement Any glue or other adhesive material which serves to bind glass to glass, or glass to another material.

glass concrete A concrete panel or slab into which are set a pattern of translucent glass lenses that allow the passage of light.

glass cutter A small hand tool having a pointed diamond tip or a sharp, small, hardened-steel wheel that scores glass.

glass door A door fabricated without stiles or rails and consisting entirely of thick, heat-strengthened or tempered glass.

glass fiber, glass fibre See *fiberglass*.

glasspaper A fine sandpaper or polishing paper whose abrasive is powdered glass.

glass pipe A term defining both glass and glass-lined pipe used in process piping.

glass seam In limestone, a crack that has been cemented, annealled, and rendered structurally sound by the deposition of transparent calcite.

glass size The size to which a piece of glass is cut so as to glaze a given opening. Its length and width should be 1/8 inch less than the distance between the outside edges of the rebates.

glass stop (1) A glazing bead. (2) A fitting at the lower end of a patent glazing bar to prevent the pane from sliding down.

glass tile Transparent or translucent units installed in a roof surface which allow light to enter the room below.

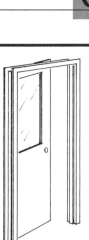

Glazed Door

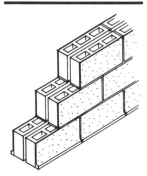

Glazed Structural Unit

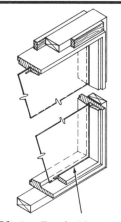

Glazing Bead, Glass Stop

glass wool Spun fiberglass used mostly for thermal and acoustical insulation and as a filtering medium in air and water filters.

glaze (1) To install glass panes in a window, door, or another part of a structure. (2) A hard, thin, glossy ceramic coating on the surface of pottery, earthenware, ceramics, and similar goods.

glaze coat (1) The smooth top layer of asphalt in built-up roofing. (2) A temporary, protective coat of bitumen applied to built-up roofing that is awaiting top-pouring and surfacing. (3) In painting, the application of a nearly transparent coat that enhances and protects the coat below it.

glazed brick Brick or tile with a glassy surface produced by fusing it with a glazing material.

glazed door Any door with glass panes or panels and top and bottom rails.

glazed structural unit A hollow or solid unit to whose surface a smooth, glassy covering, such as glazed tile, has been applied.

glazed tile Ceramic or masonry tile having an impervious, glossy finish.

glazier A person whose trade is to install glass in structures. The glazier removes old putty, cuts glass to fit the openings requiring it, and secures it there by whatever means are necessary or appropriate.

glazier's chisel A chisel-shaped putty knife used in setting glass.

glazier's point A small, thin, flat, rigid, three- or four-cornered piece of metal which is sometimes pushed partway into the outside edges of the rebate, and buried in the face putty around a pane of glass to hold it in place.

glazier's putty A glazing compound often made from a mixture of linseed oil and plaster of paris, and sometimes including white lead.

glazing (1) Fixing glass in an opening. (2) The glass surface of an opening which has been glazed.

glazing bar A wood or metal, vertical or horizontal bar that subdivides a window and holds panes of glass.

glazing bead, glass stop (1) A narrow strip of wood, plastic, or metal fastened around a rebate and used to hold glass in a sash. (2) At a glazed opening, removable trim that holds the glass in place.

glazing brad See *glazier's point*.

glazing clip A metal clip used to hold glass in place in a metal frame during putty application.

glazing color A transparent wash applied over a ground coat of paint.

glazing compound Any putty or caulking compound used in glazing to seal the joint at the edges of the glass.

glazing fillet A narrow wood, metal, or plastic strip fastened to the rebate of a glazing bar and used instead of face putty to hold glass in place.

glazing gasket A narrow, sometimes grooved, prefabricated strip of material such as neoprene, which offers a dry, alternative to glazing compound in glazing operations. The gasket is impervious to moisture and temperature and is often used with large panes or sheets of heavy glass.

glazing molding (1) Molding that serves the same function as a glazing fillet. (2) A glass stop.

glazing point See *glazier's point*.

glazing size See *glass size*.

glazing spacer block One of several blocks which support glass in its frame.

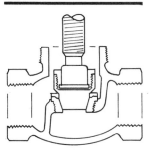

Globe Valve

**Glue Laminated,
Glu-Lam**

Gob Bucket

glazing sprig A small headless nail or brad that holds a pane of glass in its wooden frame while putty hardens.

glazing stop See *glass stop*.

glazing tape A resilient tape used to seal glass into a frame, sash, or opening.

globe, light globe (1) A protective enclosure or covering over a light source. The globe is usually made of glass and can also serve to diffuse, redirect, or change the color of the light. (2) An incandescent lamp.

globe valve A valve that is operated by turning a horizontal wheel handle on the top of a vertical shaft. The valve controls flow by a washer-fitted spindle that can be lowered into a fixed seat.

glory hole (1) In mining, a slang expression for a vertical pit whose mined material drops into a shaft below the bottom of the pit. (2) A small shaft excavated to assess the quality of an ore.

gloss Surface luster, usually expressed in terms ranging from matte to high gloss.

glow lamp A type of low-consumption electric discharge lamp whose light is produced close to the electrodes within an ionized gas. A glow lamp is often used as an indicating light.

glue A general term for any natural or synthetic viscous or gelatinous substance used as an adhesive to bind or join materials.

glue bleedthrough The seepage of glue through the face of open-pore wood.

glue block A joint-strengthening block of wood glued into the interior angle formed by two boards.

glue bond A measure of how well articles are fastened together after being glued.

glued floor system A method of floor construction in which a plywood underlayment or other structural panel is both glued and nailed to the floor joists, thus resulting in a stronger, stiffer floor less prone to squeaking and nail-popping than floors fastened only with nails.

glue laminated, glu-lam The result of a process in which individual pieces of lumber or veneer are bonded together with adhesives to make a single piece in which the grain of all the constituent pieces is parallel.

glue line The layer of glue or adhesive between two pieces of lumber or veneer.

glue nailed A term applying to plywood joints and connections that have been both glued and nailed to produce the stiffest possible construction.

glue-up The process of spreading glue on the surfaces of veneers of similar sizes and pressing them together to form a sheet of plywood.

glyph (1) An ornamental, usually vertical channel or groove, as in a column or freeze. (2) A sculptured pictograph.

gneiss A coarse-grained metamorphic rock similar to granite and containing quartz, mica, feldspar, or other minerals. In construction terms, gneiss is classed as "trade granite."

gob bucket On a crane, the bucket that is used to carry concrete.

go devil (1) A pipe-cleaning device inserted at the pump end of a pipeline and forced through the pipe by water pressure in order to clean it. (2) In lumbering, a sled or skid-pan designed to keep the leading end of a lag off the ground when skidding.

going The horizontal distance between two consecutive risers or stair. The horizontal distance between the first and last riser of

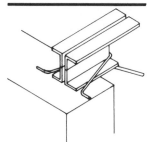

Government Anchor

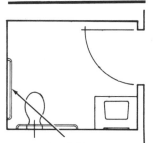

Grab Bar

an entire flight is called the "going of the flight" or the "run."

going rod A rod used in planning the going of a flight of steps.

gold bronze Powdered copper or copper alloy used for bronzing or in the manufacture of gold or bronze paint.

gold leaf An extra-fine gold sheet used for gilding and inscribing on glass or other materials.

gold size An oleo-resinous varnish that becomes sticky upon application, but hardens slowly, thus making it an appropriate adhesive for attaching gold leaf to a surface.

goliath crane A heavy but portable crane, often of about 50 ton capacity, used for jobs requiring heavy lifting and for shop fabrication involving heavy steel.

good one side, G1S A grade of sanded plywood with a higher grade of veneer on the face than on the back. G1S is used in applications where the appearance of only one side is important.

good two sides, G2S The highest grade of sanded plywood. G2S is allowed to have inlays or neat wood patches and is used in applications where the appearance of both sides is important.

gooseneck (1) In plumbing, a curved, sometimes flexible fitting connection or section of pipe. (2) In HVAC, a screened, U-shaped intake or exhaust duct. (3) The curved end of a handrail at the top of a stair. (4) The curved connector from a tractor to a trailer.

Gothic In architecture, the prevalent style in Western Europe from the 12th through the 15th century, whose characteristic features included flying butresses, ribbed vaulting, pointed arches, and lavishly fenestrated walls.

Gothic arch A typically high, narrow arch with a pointed top and a jointed apex (as opposed to keystone).

gouge A cutting chisel having a long, curved blade and used for hollowing out wood, or for making holes, channels, or grooves in wood or stone.

government anchor A V-shaped anchoring device usually fabricated from 1/2- inch round bar, and used in steelwork to secure a wall-bearing (or other) beam to masonry.

Gow caisson, Boston caisson, caisson pile A series of steel cylinders from 8-16 feet high used to protect workers and equipment during deep excavation in soft earth. Each successive cylinder is 2 feet in diameter smaller than its predecessor, through which it is dropped or driven deeper into the surrounding soft clay or silt to prevent excessive loss of ground and to facilitate deep excavation. When excavation and construction are complete, the cylinders are withdrawn.

grab bar A short length of metal, glass, or plastic bar attached to a wall in a bathroom, near a toilet, in a shower, or above a bathtub.

grab bucket A bucket-like device with hinged lower halves or "jaws" hydraulically or cable-operated from a crane or used for re-handling granular materials. See **clamshell**.

grab crane A crane outfitted with a grab bucket.

gradation An assessment, as determined through sieve analysis, of the amounts of particles of different sizes in a given sample of soil or aggregate.

gradall A trade name for a wheel-mounted, articulated hydraulic backhoe often used with a wide bucket for dressing earth slopes.

grade (1) The surface or level of the ground. (2) A classification of quality as, for instance, in lumber. (3) The existing or proposed

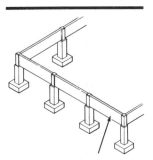

Grade Beam

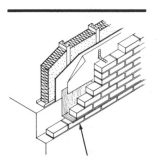

Grade Course

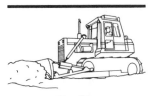

Grading

ground level or elevation on a building site or around a building. (4) The slope or rate of incline or decline of a road, expressed as a percent. (5) A designation of a subfloor, either above grade, on grade, or below grade. (6) In plumbing, the slope of installed pipe, expressed in the fall in inches per foot length of pipe. (7) The classification of the durability of brick. (8) Any surface prepared to accept paving, conduit, or rails.

grade beam A horizontal end-supported (as opposed to ground-supported) loadbearing foundation member that supports an exterior wall of a superstructure.

grade block A type of concrete masonry unit from which the top course of a foundation wall is constructed and above which a thicker or thinner masonry wall is constructed.

grade correction A distance measured on a slope and corrected to a horizontal distance between vertical lines through its end points.

grade course The first course of brick, block, or stone, at grade level, usually waterproofed.

graded aggregate Aggregate having a uniformly graded particle size ranging from course to fine.

graded sand A fine aggregate (under 1/4 inch) having a uniformly graded particle size.

graded standard sand Ottawa sand that has been subjected to accurate grading between No. 30 and No. 100 sieves and is used in testing hydraulic cements.

grade line (1) A line of stakes with markings, each at an elevation relative to a common datum and from whose elevations a grade between their terminal points can be established. (2) A strong string used to establish the top of a concrete pour or masonry course.

grader A multipurpose earthworking machine used mostly for leveling and crowning. A grader has a single blade, centrally located, that can be lifted from either end and angles so as to cast to either side.

grade stake In earthwork, a stake that designates the specified level.

grade strip A thin wooden strip fastened to the inside of a concrete form to indicate the level to which concrete should be placed.

gradient (1) The change in elevation of a surface, road, or pipe usually expressed in a percentage or in degrees. (2) The rate of change of a variable such as temperature, flow, or pressure.

gradienter A micrometer, attached to a transit's telescope, that allows the angle of incline to be measured in terms of the angle's tangent instead of in degrees and minutes.

grading (1) The act of altering the ground surface to a desired grade or contour by cutting, filling, leveling, and/or smoothing. (2) Sorting aggregate by particle size. (3) Classifying items by size, quality, or resistance.

grading curve A line on a graph illustrating the percentages of a given sample of material which passing through each of a specific series of sieves.

grading instrument A surveyor's level having a telescope that can be adjusted upward or downward to lay out a required gradient.

grading plan A plan showing contours and grade elevations for existing and proposed ground surface elevations at a given site.

grading rules Quality criteria that determine the classification of lumber, plywood, or other wood products.

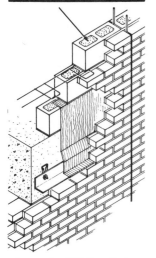

Granular Fill Insulation

graduated courses Courses of roofing slates having diminishing gauges from the eaves to the roof.

graffito A plaster surface applied over a colored undercoat for decorative effect. While still soft, the surface is scored with a pattern so as to expose the contrasting undercoat.

graft (1) The joining of a bud or scion of one plant or tree onto a part of a similar plant or tree. (2) Extortion by the unscrupulous leverage of one's position or knowledge.

grain (1) The directional arrangement of fibers in a piece of wood or woven fabric, or of the particulate constituents in stone or slate. (2) The texture of a substance or pattern as determined by the size of the constituent particles. (3) Any small, hard particle (like sand). (4) A metric unit of weight; 7,000 grains equal 1 pound.

graining The process of simulating grain on painted or man-made surfaces using special tools such as graining combs to manipulate an applied coat of translucent stain or "graining color."

grain size (1) A size classification of mineral particles in soil or rock. (2) One of the physical characteristics of a particle of soil, which relates to its mechanical properties.

grain slope The angle of the grain in a piece of wood, as determined from a hypothetical line parallel to its length.

grandmaster key (1) A key that not only operates all the locks within a given group having its own masterkey, but several such groups. (2) A master key of master keys.

granitic finish A granite-resembling face mix on precast concrete.

granny bar A slang term for a large crow bar.

granolithic concrete A type of concrete made from a hard aggregate whose particle shape and surface texture render it conducive for wearing-surface finishes on floors.

granolithic finish A concrete wearing surface, placed over a concrete slab, containing aggregate chips to improve its wearing properties.

grantee (1) The party to whom a deed or similar document transfers property or property rights. (2) The buyer.

grantor (1) The party from whom a deed or similar document transfers property or property rights. (2) The seller.

granular (1) A technical term relating to the uniform size of grains or crystals in rock. (2) Composed of grains.

granular fill insulation An insulation material such as perlite or vermiculite that can be easily placed or poured because it comes in the form of chunks, pellets, or modules.

graphics Engineering or architectural drawings created with attention to mathematical rules, such as perspective or projection.

graphite paint A type of paint made from boiled linseed oil, powdered graphite, and a drier. Graphite paint is used to inhibit corrosion on metal surfaces.

grapple An excavation or re-handling bucket on the end of a crane. A grapple has more than two opening parts or jaws and is good for handling rubble and demolition debris. Also called an "orangepeel bucket."

grappler A pointed or wedge-shaped spike whose eye is left exposed after the spike is driven into a masonry joint, thus providing support for scaffold brackets.

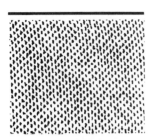

Grass Cloth

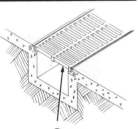

Grate

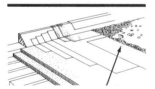

Gravel Roofing

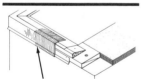

Gravel Stop

**Grease Interceptor,
Grease Trap**

grass cloth A wall covering made from woven vegetable fibers, especially arrowroot bark. The cloth is laminated onto a paper backing.

grass hopper A slang term for any type of small crane.

grate (1) A type of screen made from sets of parallel bars placed across each other at right angles and in approximately the same plane. A grate is used to allow water to flow to drainage, but to cover an area for pedestrian or vehicular traffic. (2) A surface with openings to allow air to flow through while supporting a fuel bed, as in a coal furnace.

gravel Coarse particles of rock that result from naturally occurring disintegration or that are produced by crushing weakly bound conglomerate. Gravel is retained on a No. 4 sieve.

gravel, bank run Naturally occurring gravel deposits that are often located along rivers or lakes.

gravel board A easily replaceable board that is secured horizontally near the bottom edge of a wooden fence so as to prevent contact between the ground and the vertical boards, thus preventing rotting.

gravel roofing Roofing composed of several ("built up") layers of saturated or coated roofing felt, sealed and bonded with asphalt or coal-tar pitch which, for solar protection and insulation purposes, is then covered with a layer of gravel or slag. Usually used on flat or nearly flat roofs.

gravel stop A metal strip or flange around the edge of a built-up roof. The stop prevents loose gravel or other surfacing material from falling off or being blown off a roof.

gravity dam A pyramid-shaped dam whose own weight resists the force of the water behind it.

gravity wall A massive concrete retaining wall whose own weight prevents it from overturning.

gravity water supply, gravity water, gravity system A water distribution system in which the supply source is located at an elevation higher than the use.

gray scale A series of achromatic samples that vary discretely from white to black.

grease extractor A device installed in conjunction with a cooking exhaust system and employing grease collecting baffles positioned so as to create a path of sharp turns through which the cooking exhaust is passed at high velocity. The grease, which is particulate and heavier than air, is collected on the baffles by centrifugal force, while the carrying air continues around the sharp turns on its way to being exhausted, thus becoming cleaner at each baffle.

grease interceptor, grease trap A device installed between the kitchen drain and the building sewer to trap and retain fats and grease from kitchen-waste lines.

grease monkey (1) A slang expression for a person who lubricates, refuels, and/or repairs equipment. (2) A grease dauber who greases skids so that logs can be moved over them easily.

green A term referring to unseasoned timber or to fresh, unhardened concrete, plaster, or paint.

greenbelt (1) An elongated section of trees or other plantings which serves as a boundary of or division within a community. (2) Any large area of undeveloped land, including parks and farmland that surrounds a community.

green board (1) Gypsum board whose type is distinguishable by the green color of its face paper. Green board is designed to be used in areas that are often damp,

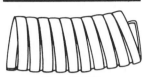

Greenfield

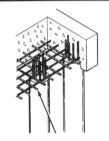

Grid (4)

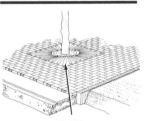

Grille

such as in bathrooms for tile backing.

green brick Molded clay block or brick before it has been fired in a kiln.

green concrete Descriptive of concrete that has set but not hardened.

greenfield Flexible metal conduit for electrical wiring.

green glass A type of glass whose green color results from impurities in its raw materials.

greenhouse A glass-enclosed space with a controlled environment for growing plants, vegetables, and fruits out of season.

greenhouse effect (1) The conversion of the sun's rays into heat that is retained by the glass roof of a greenhouse. (2) The steady, gradual rise in temperature of the earth's atmosphere due to heat that is retained by layers of ozone, carbon dioxide, and water vapor.

green lumber Undried, unseasoned lumber.

green manure (1) Crops (such as winter rye or alfalfa) which are plowed under to return nutrients into the soil. (2) The plowed-under remains of a crop from which a harvest has been reaped.

green mortar Descriptive of mortar that has set but not dried.

grid (1) In surveying, a system of evenly spaced perpendicular reference lines at whose intersections elevations are measured. (2) The structural layout of a given building. (3) A system of crossed reinforcing bars used in concrete footings.

grid ceiling (1) A ceiling with apertures into which are built luminaries for lighting purposes. (2) Any ceiling hung on a grid framework.

grid foundation A foundation consisting of several intersecting continuous footings loaded at the intersections. A grid foundation covers less than 75% of the area within its outer limits.

gridiron (1) The plotting of city streets in rectangles. (2) The framework above a stage from which lights and scenery can be hung.

grid line Any line that is part of a reference pattern for surveying or layout.

grillage Steel or wooden beams used horizontally under a structure to distribute its load over its footing or underpinning.

grille (1) Any grating or openwork barrier used to cover an opening in a wall, floor, paving, etc., for decoration, protection, or concealment. (2) A louvered or perforated panel used to cover an air duct opening in a wall, ceiling, or floor. (3) Any screen or grating that allows air into a ventilating duct.

grillwork In construction, any heavy framework of timbers or beams used to support a load on soil instead of on a concrete foundation.

grind (1) To reduce in size by removing material by friction or crushing. (2) To sharpen (as a tool) by abrasion.

grinder (1) A device that sharpens or removes particles of material by abrasion. (2) A machine or tool for finishing concrete surfaces by abrasion.

grindstone A flat sandstone wheel which is rotated to sharpen implements or to reduce the size of a material by abrasion or grinding.

grinning through (1) In plastering, a term to describe lathing that is discernable under plaster. (2) In painting, a term to describe an undercoat that is showing through a topcoat.

grip length, bond length The minimum length, expressed in bar diameters, of rebar necessary for

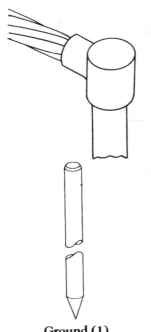

Ground (1)

anchorage in concrete.

grisaille (1) A system of surface painting in various shades of gray so as to decorate or to suggest bas-relief. (2) Uncolored, non-transparent glass in a stained glass window.

grit (1) A granular abrasive used in making sandpaper or on grinding wheels to give a surface non-slip finish. (2) Particles of sand or gravel contained in sewage.

grizzly (1) A screen or grate used to remove oversize particles during aggregate processing or loading. (2) A gate or similar device on a chute.

groin (1) In architecture, a ridge or curved line formed at the junction of two intersecting vaults. (2) A structure built outward from a shore into water to direct erosion or to protect against it.

groin centering (1) A method of supporting ribless groining during vaulting. (2) In ribbed vaulting, the support of the stone ribs by timbers until construction is complete.

grommet A metal or plastic eyelet that provides a reinforced hole in a material, such as cloth or leather, that might otherwise tear from the stress on the hole when a fastener or other device is passed through it or attached to it.

groove In carpentry, a narrow, longitudinal channel cut in the edge or face of a wood member. The groove is called a "dado" when cut across the grain, and a "plow" when cut parallel to it.

groove joint A joint formed by the intentional creation of a groove in the surface of a wall, pavement, or floor slab for the purpose of controlling the direction of random cracking.

groover A tool for creating grooved joints in unhardened concrete slabs.

groove weld A weld between the existing preformed grooves in members being joined.

gross area (1) The total area without deducting for holes or cut-outs. (2) The whole or entire area of a roof. (3) In shingles, the entire area of a shingle, including any parts which might have had to be cut out.

gross cross-sectional area The total area of that portion of a concrete masonry unit which is perpendicular to the load, inclusive of any areas within its cells and re-entrant spaces.

gross floor area The total area comprising the areas of all the floors of a building, including intermediately floored tiers, mezzanine, basements, etc., as measured from the exterior surfaces of the outside walls of the building.

gross load In heating, the net load to which allowances are added for pickup and piping losses.

gross output The number of BTU's available at the outlet nozzle of a heating unit for continuously satisfying the gross load requirements of a boiler operating within code limitations.

gross section The total area of the cross section of a structural member as calculated without subtracting for any voids in that cross section.

gross volume (1) The total volume within the revolving part of a drum concrete mixer. (2) The total volume of the trough on an open-top concrete mixer.

grotto A natural or man-made cavern or cave.

ground (1) The conducting connection between electrical equipment or an electrical circuit and the earth. (2) A strip of wood that is fixed in a wall of concrete or masonry to provide a place for attaching wood trim or burring strips. (3) A screed, strip of wood, or bead of metal fastened around an opening in a wall and

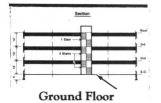

Ground Floor

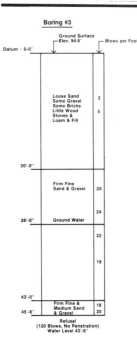

Ground Water

acting as a thickness guide for plastering or as a fastener for trim. (4) Any surface that is or will be plastered or painted.

ground beam (1) A reinforced concrete beam or heavy timber positioned horizontally at ground level to support a superstructure. (2) A groundsill.

ground brush A round or oval paint brush used to carry large quantities of paint.

ground bus A electrical bus to which individual equipment grounds are connected and which itself is grounded at one or more points.

ground casing In carpentry, a term for the blind casing of a window.

ground coat The base or undercoat of paint or enamel, often designed to be seen through the topcoat or glazecoat.

ground course A first horizontal course of masonry at ground level.

ground cover A planting of low plants that in time will spread to form a dense, often decorative mass. Ground cover is also used to prevent erosion.

ground floor In a building, that floor closest to the level of the surrounding ground.

ground glass Glass having a light-diffusing surface produced by grinding with an abrasive.

ground joint (1) In masonry, a close fitting joint usually without mortar. (2) A closely fitted joint between machined metal surfaces.

ground joist Floor joists laid on sleepers, dwarf walls, or stones.

ground key valve A fluid-flow control valve containing a tapered plug with a hole bored into it. The keg-type handle requires only a one-quarter turn between its open and shut positions.

ground line The natural grade line or ground level from which excavation measurements are taken price to determine excavation quantities.

ground plate See **ground sill**.

ground pressure The weight of a machine or a piece of equipment divided by the area, in square inches, of the ground that supports it.

ground sill The bottom horizontal structural member of a framed building on or near the ground level.

ground water Water contained in the soil below the level of standing water.

ground water recharge The reintroduction of water into the ground, as from trenches beyond the construction area or through injection.

ground water table The top elevation of ground water at a given location and at a given time.

ground wire (1) A electrical conductor leading directly or indirectly to the earth. (2) Strong, small-gauge wire used in establishing line and grade, as in shotcrete work.

grounded Descriptive of an object that is electrically connected to the earth or to another conducting body that is connected to the earth.

grounded conductor In an electrical system, a circuit conductor that is intentionally grounded, either solidly or through a device that limits current.

grounded system A system of electric conductors, at least one of which is intentionally grounded, either solidly or by means of a device that limits current.

**Group House,
Row House**

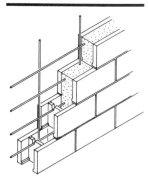

Grouted Masonry

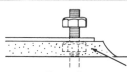

Grouting

grounded work Any joinery, such as a crown molding, attached to grounds. See **ground**, def. 2.

grounding electrode A conductor that is firmly embedded in the earth, and can thus function to maintain ground potential on the conductors connected to it.

grounding outlet An electrical outlet whose polarity type receptacle includes both the current-carrying contacts as well as a grounded contact that accepts an equipment grounding conductor.

grounding plug A receptacle plug comprising a male member which, when plugged into a live grounding outlet, provides a ground connection for an electric device.

groundwork Batters used in roofing and siding as a base over which slate, tiles, and shingles are applied.

group house, row house A single dwelling unit contained in a long, unbroken line of vertically identical houses connected by common walls.

group vent In plumbing, a branch vent connected to two or more traps.

grouser The cleat on a track shoe, such as the shoe on a bulldozer track, that is used to improve traction.

grout (1) An hydrous mortar whose consistency allows it to be placed or pumped into small joints or cavities, as between pieces of ceramic clay, slate, and floor tile. (2) Various mortar mixes used in foundation work to fill voids in soils, usually through successive injections through drilled holes.

grout box A cone-shaped sleeve in a concrete slab into which an anchor bolt for machinery can be grouted.

grouted-aggregate concrete Concrete which is produced by injecting grout into pre-positioned coarse aggregate.

grouted frame An originally hollow-metal door frame whose vacant interior has been filled with some type of cement or mortar mixture.

grouted masonry (1) Masonry hollow units with some or all of the cells filled with grout. (2) Masonry comprising two or more withes, the spaces between which are filled solidly with grout.

grouting (1) The placing of grout so as to fill voids, as between tiles and under structural columns and machine bases. (2) The injection of grout to stabilize dams or mass fills, or to reinforce and strengthen decaying walls and foundations. (3) The injection of grout to fill faults and crevices in rock formations.

grout lift The height of grout placed in masonry wall voids during a single placement.

grout slope The naturally occurring slope of hydrous grout after its injection into preplaced aggregate concrete.

growth ring A ring that designates the amount of a tree's growth in a single year.

grozing iron (1) In plumbing, a tool for finishing soldered joints. (2) A glass-cutting device.

grub In site work, the clearing of stumps, roots, trees, bushes, and undergrowth.

grub axe, grub hoe, mattock A hand tool for shallow digging or root pulling. One end of the head of the tool has a transverse cutting blade, while the other end has a cutting blade like an axe.

grub saw A hand-operated saw used for cutting stone (such as marble) into useable slabs.

grunt A slang term for a common laborer or an apprentice lineman.

guarantee A legally enforceable assurance of quality or performance of a product or work, or of the duration of

231

Guard Board

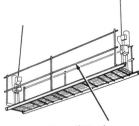

Guard Rail

satisfactory performance. Also called "guaranty" and/or "warranty."

guaranteed maximum cost The maximum amount above which an owner and contractor agree that cost for work performed (as calculated on the basis of labor, materials, overhead, and profit) will not escalate.

guaranty bond Each of the four following bonds are types of guaranty bonds: (1) bid bond, (2) labor and material payment bond, (3) performance bond, and (4) surety bond.

guard (1) Any bars, railing, fence, or enclosure that serve as protection around moving parts of machinery or around an excavation, equipment, or materials. (2) A watchman hired to maintain safety and security at a construction site.

guard board A raised board around the edge of a scaffold or ganty crane to keep men and tools from falling off.

guarded front and strike A lock designed and constructed with the front and strike protected, thus preventing the latchbolt from being "jimmied" through the crevice between the door and the jamb.

guard rail A horizontal rail of metal, wood, or cable fastened to intermittent uprights of metal, wood, or concrete around the edges of platforms or along the lane of a highway. (2) The rail that separates traffic entering or exiting through side-by-side automatic doors.

gudgeon (1) A metal pin or dowel used for joining two adjacent stones, blocks, or slabs. (2) A reinforced bushing or a block designated to absorb shock. (3) On a hinge, the leaf that is fastened to an immovable member. (4) The part of a gate hook fastened to the gate post.

guide bead See **inside stop**.

guide coat A very thin coat of paint that serves to identify through emphasis various imperfections on the surface under it.

guide pile A heavy square timber driven vertically near sheet piles that support an excavation, so as to carry the full earth pressure from the walers, or sometimes simply to guide the sheet piles.

guide rail The protective railing placed along highways, particularly at high embankments, culverts, and bridges.

guide wire (1) A steel wire or cable used to guide the vertical movement of a stage curtain in a theater. (2) A line or wire that guides the movement of a counterweight arbor. (3) A wire placed along the edge of a roadway to be paved. A sensor on the pavement spreader adjusts the elevation of the pavement from the wire.

guillotine In woodworking or cabinetry, a trimming machine that can cut the two matching ends of a joint to any given angle.

guinea In surveying, a wooden marker that is first driven to grade, then topped with blue paint for identification during finish grading.

guinea chaser In surveying, a site worker who uncovers the blue-topped stakes and informs the blade operator whether fill should be added or removed.

gullet (1) The concave gap between the teeth of a file or saw. (2) The height of a saw tooth as measured from the base to the point.

gum bloom A hazy or lusterless area on a painted surface that results from the use of the wrong reducer.

gumbo Soil composed of fine-grained clays. When wet, the soil is highly plastic, very sticky, and has a soapy appearance. When dried, it develops large shrinkage cracks.

Gusset, Gusset Plate

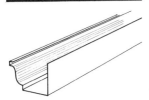

Gutter

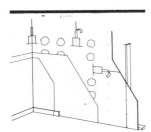

Gypsum Board

Gypsum Concrete

gumwood Wood from a gum tree, especially eucalyptus, used mostly for interior trim.

gun (1) Equipment designed to deliver shotcrete. (2) A pressure cylinder for pneumatic delivery of freshly mixed concrete. (3) A spray gun. (4) A slang expression for a "transit," as it is used to "shoot" grades.

gun consistency The degree of viscosity of caulking or glazing compound which renders it suitable for application by a caulking gun.

gun finish The finish on a layer of shotcrete left undisturbed after application.

gun grade See **gun consistency**.

gunite, gunnite (1) Concrete mixed with water at the nozzle end of a hose through which it has been pumped under pressure. Gunite is applied or placed pneumatically, as "shot," onto a backing surface.

gunstock stile In joinery, a diminishing stile whose width is tapered.

Gunter's chain In surveying, a measuring device with 100 metal links and a total length of 66 feet.

gusset, gusset plate A plate fastened across a joint, as in wood or steel framework members.

gutter A shallow channel of wood, metal, or PVC positioned just below and following along the eaves of a building for the purpose of collecting and diverting water from a roof. (2) In electrical wiring, the rectangular space allowed around the interior of an electrical panel for the installation of feeder and branch wiring conductors.

gutter bed (1) A strip of flexible metal over the wall side of a gutter which prohibits any gutter overflow from penetrating the wall.

gutter hook A bent metal strip used for securing or supporting a metal gutter.

gutter plate (1) A single side of a box or valley gutter, lined with flexible metal and carrying the feet of the rafters. (2) A beam that supports a lead gutter.

guy A cable or rope anchored in the ground at one end and supporting or stabilizing an object at the other end.

guy derrick A derrick with a guyed mast and hinged boom at its base. This type of derrick is used for erecting and hoisting materials.

gypsite Gypsum whose degree of purity is from 60-90 percent. Gypsite also contains sand, clay, and loam.

gypsum A naturally occurring, soft whitish mineral (hydrous calcium sulfate) which, after processing, is used as a retarding agent in portland cement and as the primary ingredient in plaster, gypsum board, and related products.

gypsum backerboard A type of gypsum board, not as smooth as wall board, surfaced with gray paper, and manufactured specifically as a base onto which tile or gypsum wallboard is adhered.

gypsum block A lightweight, hollow or solid masonry unit made from gypsum and used to construct nonbearing partitions.

gypsum board A panel whose gypsum core is paper-faced on each side, and which is used to cover walls and ceilings while providing a smooth surface that is easy to finish.

gypsum cement See **Keene's cement**.

gypsum concrete A mixture of calcined gypsum binder, wood chips or other aggregate, and water. The mixture is poured to form gypsum roof decks.

gypsum fiber concrete A gypsum concrete whose aggregate is composed of wood shavings, fiber, or chips.

gypsum-lath nail A low-carbon steel nail with a large flat head and long point. The characteristics make it especially suitable for fixing gypsum lath and plasterboard.

gypsum panel See **gypsum board**.

gypsum perlite plaster A base-coat plaster manufactured from gypsum and an aggregate of perlite.

gypsum plaster A plaster made from ground calcined gypsum. The set and workability of gypsum plaster are controlled by various additives. When mixed with aggregate and water, the resulting mixture is used for base-coat plaster.

gypsum sheathing A type of wallboard whose core is made from gypsum with which additives have been mixed to make it water-repellant. The sheathing is surfaced with a water-repellant paper to make it an appropriate base for exterior wall coverings.

gypsum trowel finish Factory prepared plasters consisting primarily of calcined gypsum and used in finishing applications.

gypsum wallboard See **gypsum board**.

gyratory crusher A rock-crushing mechanism whose central conical member moves eccentrically within a circular chamber.

ABBREVIATIONS

The abbreviations listed below are those most commonly used in the construction industry. Alternative forms (usually nonstandard) are shown in parentheses.

h harbor, hard, height, hours, house, hundred

H "head" on drawings, high, high strength bar joist, Henry, hydrogen

HA hour angle

HC, H.C. high capacity

HD, H.D. heavy duty, high density

H.D.O. High Density Overlaid

Hdr header

Hdwe. Hardware

hdwr hardware

He. helium

HE. high explosive

Help. Helper average

hem. hemlock

HEPA high efficiency particulate air

hex hexagon

hf half, high-frequency

H.F. hot finished

hg hectogram

Hg mercury

hgt height

HI height of instrument

hip. hipped (roof)

hl hectoliter

hm hectometer

HM hollow metal

HO high output

hor, horiz horizontal

hp horsepower

HP high pressure, steel pile section

H.P. Horsepower

H.P.F. High Power Factor

hr hour

Hrs./Day Hours Per Day

HSC High Short Circuit

ht, Ht. height

HT high-tension

htg, Htg. heating

Htrs. Heaters

hv, HV high voltage

HVAC heating, ventilating, and air conditioning

hvy, Hvy. heavy

HW high water, Hot Water

HWM high-water mark

hwy highway

hyd hydraulics, hydrostatics

Hyd, Hydr. Hydraulic

hydraul hydraulics, hydrostatics

hyp, hypoth hypothesis, hypothetical

hz hertz (cycles)

H DEFINITIONS

Half-landing

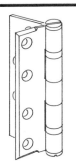

Half Mortise Hinge

hachure One of the short parallel lines used on an architectural drawing for shading or for indicating a section of a drawn object, or on a topographic map for indicating the degree and direction of slopes and depressions.

hack (1) To cut or strike at something irregularly or carelessly, or to deal heavy blows. (2) A person who lacks, or does not apply, knowledge or skill in performing his job.

hacking (1) Striking a surface with a special tool so as to roughen it. (2) A style of brick-laying in which the bottom edge is set in from the plane surface of the wall. (3) In a stone wall, the breaking of a single course into two or more courses, sometimes for effect but usually because of the scarcity of larger stones.

hacking knife A glazier's tool used for removing old putty prior to reglazing.

hack saw A lightweight, metal-cutting hand saw having a narrow, fine-toothed blade retained in an adjustable metal frame.

haft The handle of a cutting tool.

ha-ha, haw-haw A trench or similar depression serving as a sunken fence or barrier for livestock.

hair interceptor In plumbing, a trap-like device installed in the waste drain side of a fixture's plumbing system to capture and collect hair on screens or in perforated steel baskets which are removable from the bottom of the device.

hairline cracks Very fine, barely visible random cracks appearing on, but not penetrating, the finish surface of materials such as paint and concrete.

hairpin (1) A type of wedge used in tightening some kinds of form ties. (2) Hairpin-shaped rebar sometimes used in beams, columns, and prefabricated column shear heads.

half baluster An engaged baluster having an outward protrusion equal to approximately half its diameter.

half bat, half brick A half-brick produced by cutting a brick in two, across its length.

half-brick wall A brick wall having the thickness of a brick laid as a stretcher.

half column An engaged column protruding only slightly more than half its diameter.

half hatchet A carpenter's hatchet similar to a plasterer's lath hammer but having a broader blade with a notched underpart for pulling nails.

half header Half a brick or concrete block made by cutting the unit longitudinally through its faces. Half headers are used to close the work at the end of a course.

half-landing, halfpace landing, halfspace landing A platform in a stairway, usually where it changes or reverses direction, halfway between the floors of a building.

half-lapped joint, halved joint, halved splice A transverse joint formed at the intersection of two equally thick pieces of wood, both having been notched to half their original depth, so as to form a joint with flush faces.

half mortise hinge A door hinge with one plate surface-mounted on the jamb and the other plate mortised into the door stile.

half principal A roof rafter or similar member with the upper

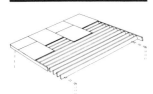

Half-span Roof

Halon Fire Extinguisher

end not extending all the way to the ridgeboard, but instead supported by a purlin.

half rabbeted lock A type of mortise lock having a front turned into two perpendicular planes, used on a door with a rabbeted edge.

half-ripsaw A handsaw with teeth more closely spaced than those on a ripsaw.

half round (1) Molding having one flat side and one rounded side, making it semicircular in profile. (2) Veneer produced by slicing a flitch or log that has been sawn into two equal halves.

half-round file A file having one side in the shape of a segment of a circle and the other flat.

half slating See **open slating**.

halfspace landing See **half-landing**.

half-span roof A roof that slopes in only one plane and abuts a higher exterior wall. Also called a "lean-to roof".

half story An attic or story immediately below a sloping roof and usually having some partitions and a finished ceiling and floor.

half-surface hinge A door hinge having one plate surface-mounted onto the door leaf, and the other plate mortised into the jamb. A half-surface hinge is the opposite of a half-mortise hinge.

half timbered Descriptive of a building style common in the 16th or 17th centuries, with foundations, supports, knees, and studs all made of timbers. The wall spaces between the timbers are filled with masonry, brick, or lathed plaster.

half truss One side of a jack truss spanning from a main roof truss to a wall, usually at an angle to the main truss.

half turn Describes a stair making a 180-degree turn or two 90-degree turns at each landing.

halide lamp See **metal halide lamp**.

halide torch A device used to detect leaks of halocarbon refrigerant. The color of the sampling torch's normal, alcohol-produced, blue flame becomes a bright green when the refrigerant is detected.

hall (1) A very large room in which people assemble for meetings. (2) A small entrance room or corridor. (3) A term often used in the proper names of public or university buildings.

halon fire extinguisher A suppressing system for use on all classes of fires. Its extinguishing agent is bromotrifluoromethane, a colorless, odorless, and electrically non-conductive gas of exceptionally low toxicity and considered to be the safest of the compressed gas fire suppressing agents. Halon systems are often used in modern computer equipment rooms.

hammer A hand tool with a handle perpendicular to its head, for driving nails or other applications involving pounding or striking.

hammer ax A hammer having a head with a flattened end for driving nails and a notched, narrow hatchet end for pulling nails. Also called a "lath hammer".

hammer beam Either one of a pair of short horizontal members used in place of a tie beam in roof framing. A hammer beam is attached to the foot of a principle rafter and supported from below by a brace to the supporting column. Also called "hammer-beam trusses".

hammer-beam roof Timbered roof construction in which hammer beams carry the principle rafters and support the feet of arched ribs.

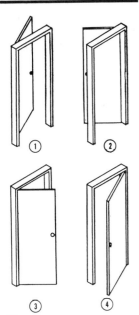

① **Left Hand**
② **Right Hand**
③ **Left Hand Reverse**
④ **Right Hand Reverse**

Handrail

hammer brace The brace, often curved, between a hammer beam and pendant post.

hammer dressed Descriptive of stone masonry having a finish created only by a hammer, sometimes at the quarry.

hammer drill A pneumatically powered mechanism using percussion to penetrate rock.

hammer finish A finish produced by the application of an enamel containing powdered metal and rendering an appearance similar to that of hammered metal.

hammerhead crane A heavy-duty crane with a swinging boom and counterbalance, giving it a "T" shape.

hammerhead key A hardwood key, dovetailed on both ends, and driven into similarly shaped recesses in the two timbers it serves to join.

hammer man The workman on a pile hammer who operates the hoist or controls the steam jet that, in turn, operates the hammer.

hammer post A pendant post at the foot of the truss in a hammer-beam roof.

Hamm tip A nozzle, used to deliver shotcrete, having a diameter midway between inlet and outlet larger than at either end.

hance A small arch or half arch connecting a larger arch or lintel to its jamb.

hand (1) Prefaced by "left" or "right" to designate how a door is hinged and the direction it opens. (2) Preceded by "left" or "right" to designate the direction of turn one encounters when descending a spiral stair, with "right-hand" being clockwise.

hand brace A wood-boring hand tool made of a single frame of small diameter bar or rod bent to form a stationary bracing handle at one end and a bit-holding

chuck at the other. A short distance from, but parallel to, the central axis, a handle repeatedly turns in wide circles, causing the bit to turn.

hand drill A hand-operated boring device made up of a central steel tube containing a shaft. At one end is a handle and at the other a bit-holding chuck.

hand float A wooden tool used to lay on and to smooth or texture a finish coat of plaster or concrete.

handicap door opening system A door equipped with a knob or latch and handle located approximately 36 inches from the floor, and an auxiliary handle on the other side at the hinge edge, for convenience to persons in wheel chairs.

handicap water cooler A water cooler set low and operated by push-bars or levers for convenience to persons in wheel chairs.

hand level In surveying, a hand-held sighting level having limited capability.

hand line A line manipulated to control stage rigging in a theater.

handling tight A degree of tightness to which couplings are screwed onto a pipe causing their removal to necessitate the use of a wrench.

handrail A bar of wood, metal, or PVC, or a length of wire, rope, or cable, supported at intervals by upright posts, balusters, or similar members or, as on a stairway, by brackets from a wall or partition, so as to provide persons with a handhold.

handrail scroll, handrail wreath The spiraled end of a handrail.

handsaw Any woodcutting saw having a handle at one end by which it is gripped and manipulated. A handsaw is never a power saw.

hand screw A woodworker's clamp made up of two parallel

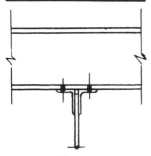

Hanger

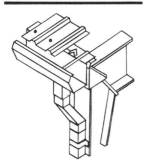

Hanging Gutter

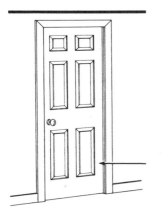

Hanging Stile

wooden jaws connected by two parallel screws tightened from opposite sides.

hand split and resawn (HS and RS) A type of cedar shake. Handsplits are split from cedar bolts by a mallet and froe (a type of steel blade). The pieces are then ripped on a resaw to produce two shakes, each with a rough, split face and a smooth, sawn back.

hand tight Descriptive of couplings tightened by hand by the application of force roughly equal to the force an average man can exert.

hang To install a door or window within its respective frame and/or by its respective hardware.

hangar An enclosure, usually for housing and/or repairing aircraft.

hanger (1) A strip, strap, rod, or similar hardware for connecting pipe, metal gutter, or framework, such as for a hung ceiling, to its overhead support. (2) Any of a class of hardware used in supporting or connecting members of similar or different material as, for instance, a stirrup strap or beam hanger for supporting the end of a beam or joist at a masonry wall. (3) A person whose trade it is to install gypsum board products.

hanging gutter A metal gutter attached to the roof eaves with metal straps and sometimes is further supported from the fascia.

hanging post The post from which a gate or door is hung.

hanging rail The horizontal section at the top and bottom of a door, to which the hinges are secured.

hanging shingling Shingling fixed to very steep or vertical surfaces.

hanging steps Cantilevered steps.

hanging stile (1) The vertical structural member on the side opposite the handle, to which the hinges are fastened. (2) That vertical section of a window frame

to which the casements are hinged.

hard asphalt A solid form of asphalt having a normal penetration of less than 10.

hardback Molding and BTR lumber which is D-select graded from the good face only, which must be clear. The back may contain knots that do not extend through the piece.

hardboard Dense sheets of building material made from heated and compressed wood fibers.

hard-burnt (1) Descriptive of clay products, such as bricks or tiles, having been burnt or fired at high temperatures, resulting in their durability, high compressive strength, and low absorption. (2) A hard plaster, such as Keene's.

hard compact soil Defined by OSHA as all earth materials not classified as running or unstable.

hard conversion The conversion from one system of measurement to another, with an inherent consequence being the necessity of changing the physical sizes of the products involved.

hard-dry The stage at which a paint film is sufficiently dry to resist thumb-inflicted mutilation, hence to accept a topcoat or other method of finishing.

hard edge A special preparation used in the core of gypsum board under the papered edges to provide extra resistance.

hardener (1) Any of several chemicals serving to reduce wear and dusting when applied to concrete sustaining heavy traffic, such as a floor. (2) The curing agent of a two-part synthetic resin, adhesive, or similar coating. (3) A substance used to harden plaster casts or gelatin molds.

hard facing Creating a hard, abrasion-resistant cutting edge on tools such as drill bits and saw blades, by welding tungsten

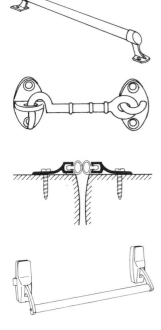

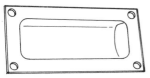

Hardware

carbide onto the steel or other metal.

hard finish A mixture of gypsum, plaster, and lime applied as a finish coat, usually over rough plastering, then troweled to provide a dense, hard, smooth finish.

hard lead See **antimonal lead**.

hard light Light creating well-defined shadows.

hard maple The sugar maple tree, *Asersaccharum*, or its wood.

hardness (1) The resistance of a substance, material, or surface, to cutting, scratching, denting, pressure, wear, or other deformation. (2) The degree, expressed as parts per million or grains per gallon of calcium carbonate in water, to which calcium and magnesium salts are dissolved in water.

hardpan Highly compacted soil, bolder clay, or other usually glacially deposited mixture, sometimes including sand, gravel, or boulders. The extreme density of hardpan makes its excavation difficult.

hard pine Any of the resinous pines, such as Loblolly or yellow pine.

hard plaster Quick-setting calcined gypsum, usually used in finishing, often requiring a retarding agent to be incorporated in the mix to help control the set.

hard solder Solder containing silver, copper, or aluminum, and thus requiring more intense heat for melting than does soft solder. Hard solder is usually applied with a brazing torch.

hard stopping A stiff paste having a calcined gypsum content, causing it to harden quickly. Hard stopping is used in painting operations to fill deep holes and wide cracks.

hard top A road which has been hard-surfaced.

hardwall A basecoat plaster made from gypsum, often without aggregate.

hardware (1) A general term encompassing a vast array of metal and plastic fasteners and connectors used in or on a building and its inherent or extraneous parts. The term includes rough hardware, such as nuts, bolts, and nails, and finish hardware, such as latches and hinges. (2) Computers and other machines and physical equipment that directly perform industrial or technological functions.

hardware cloth Usually galvanized, thin screen made from wire welded or woven to produce a mesh size of between 1/8 to 3/4 inches.

hard water Water containing a concentration higher than 85.5 ppm of dissolved calcium carbonate and other mineral salts.

hardwood A general term referring to any of a variety of broad-leaved, deciduous trees, and the wood from those trees. The term does not designate the physical hardness of wood, as some hardwoods are actually softer than some softwood (coniferous) species.

harsh mixture A concrete mixture lacking mortar or aggregate fines, resulting in an undesirable consistency and workability.

Hartford loop The configuration of a steam boiler's return piping connections serving to equalize the pressure between the supply and return sides of the system, thus preventing water from backing out of the boiler and into the return line.

hasp A metal fastening device made up of a staple secured to and protruding from one member, and a hinge with a slotted plate fastened to another member. The slotted plate can be slipped over the staple and then locked with a tapered pin or a padlock.

Hatch

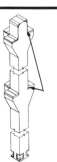

Haunch

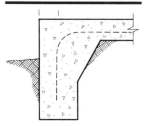

Haunched Floor

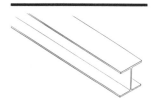

H-beam

hasp lock A device permanently secured to the hasp on a door which causes the hasp to be locked as soon as the door is closed. Hasp locks are often used in prisons.

hatch An opening in a floor or roof of a building, as in a deck of a vessel, having a hinged or completely removable cover. When open, a hatch permits ventilation or the passage of persons or products.

hatchet A wood-handled tool having a steel head flattened at one end and suitable for striking or driving, and formed at the other end into a wide sharp blade suitable for chopping. The underpart of the blade may or may not be notched for pulling nails.

hatchet iron The hatchet-shaped tip of a type of plumber's soldering iron.

haul See **haul distance**.

haul distance The shortest practical distance that excavated or filled material is transported.

haul road (1) A crude temporary road built to facilitate the movement of men and equipment along the route of a job. (2) A makeshift road between the cut and the fill.

haunch (1) A bracket built into a wall or column to support a load falling outside the wall or column, such as a hammer brace in a hammer-beam roof. (2) Either side of an arch between the crown, or centerstone, and the springing, or impost. (3) A thickening of a concrete slab to support an additional load, as under a wall.

haunched beam A beam or similar member broadened or thickened near the supports.

haunched floor A floor slab thickened around its perimeter.

haunched tenon A tenon narrower, at least in part, than the wood member from which it is fashioned.

haw-haw See **ha ha**.

hawk A flat, thin piece of wood or metal approximately one foot square and having a short, perpendicular handle centered on its underside. A hawk is used by plasterers for holding plaster from the time it is taken from the mixer to the time it is troweled.

hawk snips, hawkbill snips, duck bill snips Tin snips having curved jaws for cutting along curves.

haydite Heated shale having an expanded cellular structure, making it a suitable lightweight aggregate for concrete.

hazardous area (1) The part of a building where highly toxic chemicals, poisons, explosives, or highly flammable substances are housed. (2) Any area containing fine dust particles subject to explosion or spontaneous combustion.

hazardous substance Any substance that, by virtue of its composition or capabilities, is likely to be harmful, injurious, or lethal.

H-bar A steel or aluminum bar used in structural systems, such as suspended ceilings, and having a cross section in the shape of an "H".

H-beam A misused designation for an HP pile section.

H-block A hollow masonry unit having no ends and with opposite pairs of unconnected faces. The result is a block shaped like an "H".

H-brick Brick with horizontal perforations.

head (1) The top of almost anything, such as the head of a nail or a window head. (2) In roofing, a tile of normal width but only half the normal length, and used in constructing the eaves

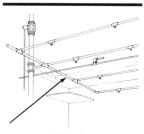

Header

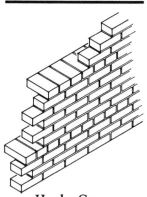

Header Course

course. (3) The horizontal member across the top of a window or door between the jambs, sometimes offering structural support for construction above it.

headache ball The rounded, heavy, metal or concrete demolition device swung on a cable from the boom of a crane to break through concrete or masonry construction. Also called a "breaker ball".

head casing The horizontally placed board at the head or top of a door or window opening between the two vertical casings.

header (1) A rectangular masonry unit laid across the thickness of a wall, so as to expose its end(s). (2) A lintel. (3) A member extending horizontally between two joists to support tailpieces. (4) In piping, a chamber, pipe, or conduit having several openings through which it collects or distributes material from other pipes or conduits. (5) The wood surrounding an area of asphaltic concrete paving.

header block A concrete masonry unit from which part of one face shell has been removed to facilitate bonding with adjacent masonry, such as brick facing.

header bond A bond whose face shows only headers, the center of which is placed directly above the joint of the two adjacent headers below.

header course In masonry, a course comprising only headers.

header joist In a roof or floor in, or framing around, an opening, such as for stairs, a beam or timber positioned horizontally between two longer beams so as to support the ends of tailpieces or to accept common joists. See **header, def. 3.**

header pipe A pipe functioning as a central connection for two or more smaller pipes. See **header, def. 4.**

header tile In a masonry-faced wall, a tile having recesses to accept headers.

head flashing In a masonry wall, the flashing over a projection, protrusion, or window opening.

heading (1) In mining, the digging face and its immediate work area in a tunnel, drift, or gallery. (2) The increase of expansion of a localized cross-sectional area of metal bar due to hot-forging. (3) A general classification of a category of data, under which follow more specific classifications. (4) Pieces of lumber from which a keg, or barrel head, is cut. (5) Stock after it has been cut and assembled to form a barrel head.

heading bond See **header bond.**

heading course See **header course.**

heading joint (1) The joint formed between two pieces of timber connected end-to-end, in a straight line. (2) The joint between two adjacent masonry units in the same course.

head jamb The horizontal member which constitutes the doorhead or top of a door opening.

headlap That portion of a shingle not exposed to the weather, because it is covered by the shingle(s) in the course above it.

head mast The tower portion of a cable excavator, which carries the working lines.

head mold The molding over an opening such as a door or window.

head nailing Nailing shingles near the top instead of at the middle.

head plate See **wall plate.**

head race A channel through which water is fed to a waterwheel, mill, or turbine.

head room (1) The vertical distance, or space, allowable for passage, as in a room or under a doorway. (2) The space between

243

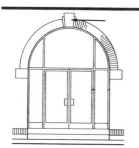

Headstone

Headwall

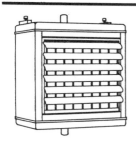

Heater

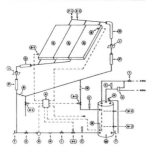

Heat Exchanger

the top of one's head and the nearest obstacle above it, as inside a vehicle. (3) The unobstructed vertical space between a stair tread and the ceiling or stairs above.

headstone Any principal stone in masonry construction, such as the keystone in an arch or the cornerstone of a building.

headwall The wall, usually of concrete or masonry, at the outlet side of a drain or culvert, serving as a retaining wall, as protection against the scouring or undermining of fill, or as a flow-diverting device.

heart The center portion of the cross-section of a log. The term usually refers to heartwood.

heart bond A masonry bond, used in walls too thick for through stones, in which a third header covers the joint of two headers meeting within a wall.

heart check Seasoning checks in the central core of a timber.

heart face The face side of a piece of lumber that is free of sapwood.

hearth The floor of a fireplace or a stove, and the immediately surrounding noncombustible area.

hearthstone (1) A large stone used as the floor of a fireplace (2) Other naturally occurring or synthetic materials used to construct a hearth. (3) Figuratively, the fireside.

hearting The interior of a masonry wall to whose face or faces finishing is subsequently applied.

heartshake A radial crack or split emanating from the center of a log or timber, usually as a result of uncontrolled or improper drying.

heart side The side of a piece of lumber that, if it were still part of the log, would be closest to the heart of the log. In flat grain lumber, the side on which the grain is more likely to rise or separate from the piece than it is on the back side.

heartwood The portion of a tree from the center to the sapwood, which is no longer vital to the life and/or growth of the tree and which is often darker and of a different consistency than the sapwood.

heat The form of energy inherent in the motion of atoms or molecules, measured in British thermal units, and transferred automatically (wherever temperature differences exist) from warmer to cooler bodies, areas, or elements by conduction, convection, or radiation.

heat absorbing glass Slightly blue-green tinted plate glass or float glass designed with the capacity to absorb 40% of the infrared solar rays and about 25% of the visible rays that pass through it. Cracking from uneven heating can occur if the glass is not exposed uniformly to sunlight.

heat balancing (1) An efficient procedure for determining a numerical degree of combustion by totaling all the heat losses, in percentages, and subtracting the result from 100%. (2) A condition of thermal equilibrium where heat gains equal heat losses.

heat capacity The amount of heat required to increase the temperature of a given mass by one degree. The capacity is arrived at numerically by multiplying the mass by the specific heat.

heated doorway See **air curtain**.

heater (1) A general term including stoves, appliances, and other heat-producing units. (2) A person who heats something, such as a steelworker who heats rivets on a small forge before passing them to the sticker.

heat exchanger A device in which heat from a hot fluid is transferred to another fluid. The fluids are usually separated by the thin walls of tubing.

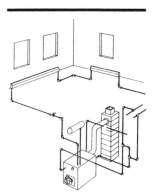

Heating Plant

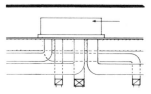

Heat Source

heat gain The net increase in BTUs, caused by heat transmission, within a given space.

heating element (1) A piece of resistance material connected between terminals to produce heat electrically. (2) That portion of a heating device, such as a stove or soldering gun, consisting of a wire or other metal piece heated by an electric current.

heating load The number of BTUs per hour required to maintain a specified temperature within a given enclosed space.

heating medium The fluid or gas conveying the heat from a source, such as a stove or boiler, to an area or substance being heated. The heating medium may or may not be confined within carriers such as pipes.

heating plant The entire heating system of a building or complex, including either a boiler, piping, and radiators, or a furnace, ducts, and air outlets.

heating rate The rate of temperature increase in degrees per hour, as in a kiln or autoclave.

heating system The method and its related necessary equipment used in a given heating application, such as a forced hot air system.

heat loss (1) The net decrease in BTUs within a given space, caused by heat transmission through spaces around windows, doors, etc. (2) The loss by conduction, convection, or radiation from a solar collector after its initial absorption.

heat of fusion The amount of heat needed to melt a unit mass of a solid at a specified temperature.

heat of hydration (1) Heat resulting from chemical reactions with water, as in the curing of portland cement. (2) The thermal difference between dry cement and partially hydrated cement.

heat pump A mechanical refrigeration system designed to transfer heat to water or the ground to provide cooling in the summer, and to absorb heat from the water or ground for heating in the winter.

heat recovery The extraction of heat from any source not primarily designed to produce heat, such as a chimney or lightbulb.

heat-reflective glass Window glass in which the exterior surface has been treated with a transparent metallic coating to reflect substantial portions of the light and radiant heat striking it.

heat-resistant concrete Concrete immune to disintegration when subjected to constant or cyclic heating to below ceramic-bonding temperature.

heat-resistant paint A paint, usually containing silicon resins, which is used on items such as stoves and radiators because of its stability at high temperatures.

heat sealing The use of heat and pressure to bond plastic sheets or films.

heat sink (1) The substance or environment into which heat is discharged after its removal from a heat source, as by a heat pump. (2) Any medium capable of accepting discharged heat.

heat source (1) Any area, environment, or device which supplies heat. (2) The area from which a refrigeration system removes heat.

heat transfer fluid The liquid substance used to carry heat away from its source to be cooled, usually by another fluid, as in a heat exchanger.

heat transmission The rate at which heat passes through a material by the combination of conduction, convection, and radiation.

heat transmission coefficient Any of several coefficients used to

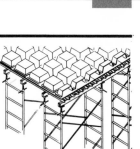

Heavy-duty Scaffold

calculate heat transmission by conduction, convection, and radiation through a variety of materials and structures.

heat treatment Subjecting any solid metal or alloy to heating and cooling to produce specific, desired changes in its physical conditions or properties.

heave The localized upward bulging of the ground due to expansion or displacement caused by phenomena such as frost or moisture absorption.

heavy concrete Concrete having a high unit weight, up to 300 pounds per cubic foot, primarily due to the types of aggregate employed and the density of their ultimate incorporation. Such diverse materials as trap rock, barite, magnetite, steel chips, nuts, and bolts can be used as aggregate. The density makes heavy concrete especially suitable for protection from radiation. Also called "high-density concrete".

heavy construction Construction requiring the use of large machinery, such as cranes or excavators.

heavy-duty scaffold As defined by OSHA, a scaffold constructed to carry a working load not to exceed 75 lb. per square foot.

heavy-edge reinforcement In highway pavement slabs, reinforcement made of wire fabric with up to four edge wires that are heavier than any of the other longitudinal wires.

heavy joist A timber at least four inches thick and eight inches wide.

heavy soil A fine-grained soil consisting primarily of clay and silt, which are damper, hence heavier, than sand.

heavy timber (1) A type of construction requiring non-combustible exterior walls with a minimal fire-resistance rating of two hours, solid or laminated interior members, and heavy

plank or laminated wood floors and roofs. Also called "mill construction". (2) Rough or surface pieces with a least dimension of five inches.

heavyweight aggregate Aggregate possessing a high specific gravity, such as barite, magnetite, limonite, ilmenite, iron, and steel, making it suitable for use in high-density concrete.

heck (1) A type of door having an upper section that swings independently of its lower section. (2) A gate of latticework.

heel (1) The lower end of a door's hanging stile or of a vertically placed timber, especially if it rests on a support. (2) A socket, floor brace, or similar device for wall-bracing timbers. (3) The bottom inside edge of the footing or a retaining wall. (4) The back end of a carpenter's plane.

heel bead A glazing compound used at the base of the channel after setting a pane but prior to the installation of the removable stop, so as to prohibit leakage past the stop.

heeling The temporary, severely angular planting of trees and shrubs, often in trenches, to facilitate their removal prior to permanent transplanting.

heel post (1) A post or stanchion at the open end of a stall partition. (2) The post, either of a gate or stairway, to which the gate hinges are secured.

heel strap A steel fastening device for connecting a rafter to its tie beam.

height (1) The distance between two points in vertical alignment or from the top to the bottom of any object, space, or enclosure. (2) The vertical distance between the average grade around a building, or the average street curb elevation, and the average level of its roof. (3) The rise of an arch.

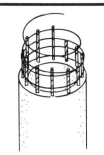

Helical Reinforcement

Helical Starcase

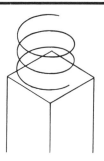

Helix

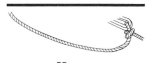

Hemp

height board A measuring device for setting the heights of stair risers.

height of instrument The height of a leveling instrument above the datum being used in the survey.

helical reinforcement Column reinforcement bent in the form of a helix. More commonly called "spiral reinforcement".

helical staircase A staircase built in the form of a helix. Also called a "spiral staircase".

helicline A helical ramp, as in a parking garage.

heliograph In surveying, a device to make a distant surveying station easily identifiable by reflecting sunlight in flashes.

heliostat An instrument having an automatically adjusting mirror that follows the apparent movement of the sun and continuously reflects its rays onto a collector.

heliport An airport specifically for helicopters.

helix (1) A reinforcing rod bent to form a spiral used for reinforcing the circumference of concrete columns. (2) A volute found on a Corinthian or Ionic capital. (3) Any spiral structure, ornament, or form.

helm roof A roof with four steeply-pitched faces rising diagonally from four gables to form a spire where they converge.

helve (1) A tool handle, such as that of an ax, hatchet, or hammer. (2) The handle of a wagon.

hem-bal A combination of Western Hemlock and Balsam Fir produced in British Columbia for overseas markets.

hem-fir A species combination used by grading agencies to designate any of various species, such as White Fir and Western Hemlock, having common characteristics. The designation is used for identification and

standardization of recommended design values and because some species, in lumber form, cannot be visually distinguished.

hemihydrate Any hydrate having only half a molecule of water for every molecule of compound. The most common hemihydrate is partially hydrated gypsum, or plaster of paris.

hemlock A coniferous North American tree, the wood of which is used in general construction and for pulp.

hemlock spruce A coniferous tree of eastern North America having soft, coarse wood of uneven texture that is unusable in construction but widely used for pulp. Also called "eastern hemlock" and "eastern spruce".

hemp A natural fiber once widely used in cordage, but now almost totally replaced by synthetic fibers, such as nylon and dacron. Hemp is still laminated to a paper backing to produce a type of wallcovering.

hem-tam A combination of Eastern Hemlock and Tamarack produced in the northeastern United States and eastern Canada.

Herculite Trade name for a type of thick, tempered plate glass, commonly used for doors without framing.

herringbone bond In masonry, a type of raking bond in which a zigzag effect is created by laying rows of headers perpendicular to each other.

herringbone drain A V-shaped drain. Also called a "chevron drain".

herringbone work In masonry, the zigzag pattern created by laying consecutive courses of masonry units at alternating 45-degree angles to the general run of the course.

hertz A unit of frequency equal to one cycle per second.

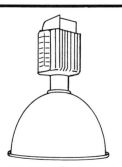

High-bay Lighting

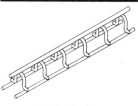

High Chair

hesitation set See **early stiffening**.

hession Same as burlap.

hewn Roughly cut, fabricated, shaped, or dressed.

hex roofing Hexagonally shaped asphalt roofing shingles.

hickey, hicky (1) A threaded electrical fitting for connecting a light fixture to an outlet box. (2) An apparatus used to bend small pipe, conduit, or reinforcing bar.

hick joint In masonry, a mortar joint cut in any direction to be flush with the face of the wall, resulting in a hairline crack that renders the joint no longer watertight.

hickory A North American tree having hard, strong wood that is inherently shock resistant and of high bending strength.

hiding power The capacity of a paint film to completely obscure the surface, including flaws, to which it is applied.

high-alumina cement See **calcium aluminate cement**.

high-bay lighting Usually an industrial lighting system having direct or semidirect luminaires located high above floor or work level.

high-bond bar See **deformed bar**.

high-build coating A coating composed of a series of films that are thicker than those normally associated with paints (minimum 5 mls) and thinner than troweled material.

high-calcium lime A type of lime composed primarily of calcium oxide or calcium hydroxide and containing a maximum of 5% magnesium oxide or hydroxide.

high-carbon steel Steel having a carbon content between .6% and 1.5%.

high chair Slang for a heavy, wire, vaguely chair-shaped device used to hold steel reinforcement off the framework during the placement of concrete.

high-density foam Usually a type of synthetic rubber applied as a liquid foam to the back side of carpeting. When cured, the foam is an integral part of the whole.

high-density overlay A cellulose fiber sheet impregnated with a thermosetting resin and bonded to plywood, rendering a hard, smooth, waterproof, wear-resistant surface for use in concrete formwork and decking.

high-density plywood Plywood manufactured from resin-impregnated veneer and formed with heat at high pressures to render a product having at least twice the density of conventional plywood.

high-discharge mixer See **inclined-axis mixer**.

high-early-strength cement Fast-hardening cement used in mortar or concrete to produce earlier strength than that of regular cement.

high-early-strength concrete Concrete containing high-early-strength cement or admixtures causing it to attain a specified strength earlier than regular concrete.

high-frequency gluing An extremely rapid gluing process in which high-frequency electronic waves are passed through the wood and glue to cause bonding, as in the construction of laminated beams.

high gloss Descriptive of a substantial degree of luster or of a paint which dries with a lustrous, enamel-like finish.

high hat (1) A recessed lighting fixture that sheds its light vertically downward. (2) A black circular tube attached to the front of a spotlight to contain the stray light around the perimeter of the beam.

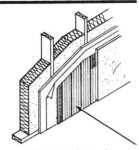

High Rib Lath

high-hazard contents Those building contents that, in a fire, might explode, burn with such vigor as to approximate explosion, produce toxic fumes, or produce other dangerous effects.

high-intensity discharge lamp A mercury, high pressure sodium, or other electric discharge lamp requiring a ballast for starting and for controlling the arc, and in which light is produced by passing an electric current through a contained gas or vapor.

high-joint pointing Pointing in which a high joint is grooved on each side by scraping along the edges of the brick while the mortar is still soft.

high-lift grouting In masonry, a method of grouting in which each lift is raised at least twelve feet.

high line A high-tension electric power supply line.

high-magnesium lime The product resulting from calcining dolomitic limestone or dolomite and containing 37%-41% magnesium oxide or hydroxide, as compared to the 5% contained in high-calcium lime.

high-output fluorescent lamp A rapid-start fluorescent lamp with a greater-than-normal flux as a result of its operation on higher-than-normal current.

high polymer (1) A substance consisting of a large molecule often made of repeat units of low molecular weight. (2) A polymer with a molecular weight greater than 10,000.

high-pressure laminate Laminate manufactured at pressures between 1,200 and 2,000 pounds per square inch during its molding and curing processes.

high-pressure mercury lamp A mercury vapor lamp designed to function at a partial mercury vapor pressure of about 1 atmosphere or more (usually 2-4).

high-pressure overlay A plastic laminate consisting of layers of melamine sheet or phenolic impregnated kraft paper into which a melamine-impregnated printed pattern sheet and/or a translucent melamine overlay may have have been impressed. The laminate is produced at a temperature above 300 degrees F and a pressure of about 400 psi, resulting in a hard, smooth, wear-resistant surface that is often bonded to wood and used in doors and on tabletops.

high-pressure sodium lamp A sodium vapor lamp, operating at a partial vapor pressure of 0.1 atmosphere, that produces a wide-spectrum yellow light.

high-pressure steam curing The steam curing of products made from cement, sand-lime, concrete, asbestos cement, or hydrous calcium silicate in an autoclave at temperatures of 340 to 420 degrees F. Also called "autoclave curing".

high-pressure steam heating system A steam heating system in which heat is transported from a boiler to a radiator by steam at pressures above 100 psi.

high relief Sculptured relief in which the modeled figures protrude from the background by at least half their thickness.

high rib lath An expanded metal lath used as a backup for wet wall plaster and as formwork for thin concrete slabs.

high rise (1) A high building having many stories usually serviced by elevators. (2) A building with upper floors higher than fire department arial ladders, usually 10 or more stories. (3) Slang for a traffic-control device consisting of a barricade with stationary flagged arms positioned at 10, 12, and 2 o'clock and located at each end of a construction zone.

high side (1) Those parts in a refrigeration system which are

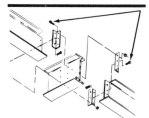

High-strength Bolts

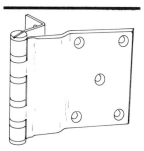

Hinge

exposed to pressure at least as great as condenser pressure. (2) Slang for the "outside" of a pipe bend or conduit bend. Also called "high pressure side".

high-silicon bronze A copper alloy made of 96% copper, 3% silicon, and 1% zinc, manganese, aluminum, iron, and/or nickel.

high-strength bolts Bolts made from high-strength carbon steel or from alloy steel that has been quenched and tempered.

high-strength steel Steel with an inherent high yield point, 60,000 psi in reinforcing bars.

high-tensile bolt, high-tension bolt A bolt made from high-strength steel and tightened to a specified high tension. High-tensil bolts have replaced steel rivets in steel-frame construction.

high-velocity duct system A duct system carrying air at more than 2,400 feet per minute.

hinge A flexible piece or a pair of plates or leaves joined by a pin so as to allow swinging motion in a single plane of one of the members to which it is attached, such as a door or gate.

brass A hinge made from or plated with brass, either for ornamental effect or because its imperviousness to corrosion is desired or required, as in marine applications.

cabinet Any decorative hinge used in cabinet work.

electric A hinge designed to pass electric wires from the frame to the door for use in an electric-controlled lock.

hospital See *hospital hinge*.

paumelle Usually a contemporary-designed pivot-type door hinge with a single joint.

residential A lighter hinge than those designed for commercial and industrial use.

security (1) A hinge with a pin

that can not be removed. (2) A hinge with a stud in one leaf projecting into the other leaf when the door is closed so the door can not be moved with the pin removed. (3) See *hinge, electric*.

hinge backset The horizontal distance from the edge of a hinge to the face of the door that closes against a rabbet or stop.

hinge jamb The doorjamb to which the hinges are fastened.

hinge joint Any joint allowing action similar to that permitted by a hinge and only a very slight separation between the adjacent members.

hinge plate, hinge strap See **hinge strap**.

hinge reinforcement A metal plate secured to a door or its frame to supply a base to which a hinge is attached.

hinge strap A usually ornamental metal strap fastened to the surface of a door to render the appearance of a strap hinge.

hip (1) The exterior inclining angle created by the junction of the sides of adjacent sloping roofs, excluding the ridge angle. (2) The rafter at this angle. (3) In a truss, the joint at which the upper chord meets an inclined end post.

hip-and-valley roof A roof incorporating both hips and valleys.

hip bevel (1) The angle between two adjacent sloping roofs separated only by a hip. (2) The angle at the end of a rafter which allows its conformation to the oblique construction at a hip.

hip capping The top layer of a hip's protective covering.

hip iron A galvanized steel or wrought iron bar or strip secured to the foot of a hip rafter to hold the hip tiles in place. Also called a "hip hook".

hip jack In a hip roof, a rafter shorter in length than most of the

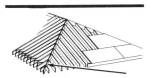

Hip Rafter

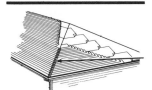

Hip Roll
Hip Roof

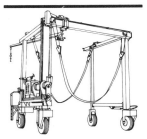

Hoist

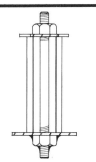

Hold-down Bolt

other rafters used in the same construction, whose upper end is secured to a hip rafter.

hip knob An ornament, such as a finial, at the top of hip of a roof or at the apex of a gable.

hip molding The molding on a hip rafter.

hipped end Either of the triangular ends of a hipped roof.

hipped gable A modified gabled end that is gabled only about halfway to the ridge and then inclines backwards and forms hips where it meets the two principal slopes. Also called a "jerkinhead".

hipped-plate construction See **folded-plate construction**.

hip rafter The rafter which, in essence, is the hip of a roof, by virtue of its location at the junction of adjacent inclined planes of a roof.

hip roll The finish covering over the hip of a roof, fashioned from wood, metal, tile, or thin material.

hip roof A roof formed by several adjacent inclining planes, each rising from a different wall of a building, and forming hips at their adjacent sloping sides.

hip tile Shaped tile, of material such as clay or concrete, covering the other roof tiles meeting at the hips. The lowest tile is held by a hip hook. See **hip iron**.

hip vertical The upright tension member in a truss, the lower end of which carries a floor beam, and the upper end of which joins an inclined end post and an upper cord at a hip.

hoarding A crude, temporary wall or fence on the site of a construction project.

hod In masonry construction, a V-shaped, trough-like container with a pole handle projecting vertically downward from the bottom to allow steadying with one hand while being carried on the shoulder of a laborer (hod carrier) at a construction site. A hod is

used to transport bricks or mortar.

hoe (1) An implement similar to a garden hoe but having a larger blade to facilitate the mixing of cement, lime, and sand, in a mortar tub at a construction site. (2) A backhoe.

hog (1) In masonry, a course which is not level, usually because the mason's line was incorrectly set and/or pulled. (2) A closer in the middle of a course. (3) A machine to grind waste wood into chips for fuel or other purposes.

hog-backed Cambered. The term is often used in reference to a sagging roof.

hogging The sagging of the end extremities of a beam or timber supported only in the middle.

hogsback tile A slightly less than half-round ridge tile.

hoist (1) Any mechanical device for lifting loads. (2) An elevator. (3) The apparatus providing the power drive to a drum, around which cable or rope is wound in lifting or pulling a load. Also called a "winch". See **chain hoist**.

hoistway See **elevator shaft**.

holdback A safety device on a conveyor to prevent reverse motion of the belt automatically.

hold-down bolt See **anchor bolt**.

hold-down clip A fastener used in an exposed suspension acoustical ceiling system or in roofing to join and anchor adjacent sections of capping.

holder-up A dolly bar used by an iron worker to back up a rivet while the driver forms a head on it.

hold harmless A clause of indemnification by which an insurance carrier agrees to assume his client's contractual obligation and to assume responsibility in certain situations which otherwise might be the obligation of the other party to the contract.

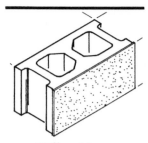

Hollow Masonry

Hollow Metal Door

Hollow Metal Frame

holding tank A tank used for temporary storage of chemicals or materials being processed.

hole saw See **crown saw**.

holiday A small area inadvertently missed during painting or other surfacing applications.

holing A process of punching holes in roofing slates to facilitate nailing during installation.

hollow-backed Descriptive of the unexposed surface of a piece of wood, stone, or other material, intentionally hollowed to render a snug fit against an irregular surface.

hollow bed In masonry, a bed joint in which mortar is placed so as to provide contact only along the edges.

hollow block See **hollow masonry unit**.

hollow brick A hollow clay masonry unit in which the net cross-sectional area is at least 60% of the gross cross-sectional area when compared in any given plane parallel to the bearing surface.

hollow chamfer Any concave chamfer.

hollow-core door A flush door with plywood or hardwood faces secured over a skeletal framework, the interior remaining void or honeycombed.

hollow masonry unit A masonry unit in which the net cross-sectional area is less than 75% of the gross cross-sectional area when compared in any given plane parallel to the bearing surface.

hollow-metal (1) Light-gauge metal fabricated into a door, window frame, or similar assembly. (2) Descriptive of an assembly thus produced.

hollow metal door A hollow-core door constructed of channel-reinforced sheet metal. The core may be filled with some type of lightweight material.

hollow metal frame A doorframe constructed of sheet metal with reinforcing at hinges and strikes.

hollow partition A partition constructed of hollow blocks or in two separate sections between which a void is left for accepting a sliding door and/or acoustic or thermal insulation.

hollow plane A woodworking plane with a convex blade for fashioning hollow or concave molding.

hollow roll A process of joining two flexible-metal roofing sheets in the direction of the roof's maximum slope by lifting them at the joint and bending them there to create a cylindrical roll. The fastening of the roll sometimes requires a fastener or metal clip.

hollow tile See **structural clay tile**.

hone A smooth, fine-grained stone against which a tool's cutting edge is worked to achieve a finish edge much sharper than that yielded by the coarser stone used in preliminary sharpening procedures. Usually an oil is used in the process to carry off minute particles of loose stone and metal to prevent them from clogging the pores on the stones surface.

honed finish The very smooth surface of stone effected by manual or mechanical rubbing.

honeycomb (1) In concrete, a rough, pitted surface resulting from incomplete filling of the concrete against the formwork, often caused by using concrete that is too stiff or by not vibrating it after it has been poured. (2) Voids in concrete resulting from the incomplete filling of the voids among the particles of coarse aggregate, often caused by using concrete that is too stiff. (3) In sandwich panel construction or in some hollow-cored doors, resin-impregnated paper fabricated into a network of small, interconnected, open-ended, tubular hexagons laminated

Hook Bolt

Hooked Bar

between two face panels to provide internal support.

honeycomb wall A brick wall whose face contains a pattern of openings created by missing units or gaps between stretchers, sometimes used under floors to provide ventilation and/or joist support.

hood (1) A protective cover over an object or opening. (2) A cover, sometimes including a fan, a light fixture, fire extinguishing system, and/or grease filtration/extraction system, and supported, hung, or secured to a wall such as above a cooking stove chimney, or to draw smoke, fumes, and ordors away from the area and into a flue. (3) A curved baffle used to minimize scattering and separation of material discharged by a conveyor belt.

hoodmold The interior or exterior drop molding projecting over a door.

hook (1) Any bent or curved device for holding, pulling, catching, or attaching. (2) A terminal bend in a reinforcing bar. (3) Slang term for a "crane".

hook-and-butt joint A scarf joint between timbers with their ends fashioned to lock together positively and resist tension.

hook bolt A bolt with an unthreaded end bent into a L shape.

hooked bar In reinforced concrete, a reinforcing bar which has a hooked end to facilitate its anchorage. See **hook, def. 2**.

hook knife A knife with a blade in which the cutting edge is bent back toward itself in the same plane, forming a hook shape. Also called a "linoleum knife".

hook strip A narrow board fastened horizontally to a closet wall to provide a surface to which clothes hooks are secured.

hoop iron Thin iron strips used to bond masonry, as in a chain bond.

hoop reinforcement Closely-spaced steel rings providing circumferential or lateral reinforcement to prevent buckling of vertical reinforcing bars in concrete columns.

hopper (1) A top-loading, bottom-discharging funnel or storage bin, as for crushed stone or sand. (2) One of a pair of draft barriers at the sides of a hopper light. (3) A toilet bowl, usually funnel-shaped.

hopper frame The bottom-hinged, inward-opening upper sash of a window frame.

hopper head A funnel-shaped enlargement at the top of a downspout where the gutter rainwater is received.

hopper lite, hopper light (1) A bottom-hinged, inward-opening window sash which allows air to pass above it when open. (2) A side-hinged, inward opening window sash which, when open, allows most of the passing air over its top, but also allows the passage of some air through a narrow opening along its bottom.

hopper window A hopper light with hoppers along the sides to minimize draft. Also called a "hospital window". See **hopper, def. 2**.

horizon (1) The apparent intersection of the earth and sky, as perceived from any given position. (2) The same illusion as it might be portrayed in a perspective drawing.

horizontal Parallel to the plane of the horizon and perpendicular to the direction of gravity.

application A method of installing gypsum board with its length perpendicular to the framing members.

auger A drilling machine with a horizontally mounted auger,

253

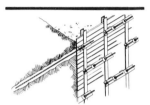

Horizontal Bridging

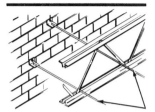

Horizontal Sheeting

Horizontal Shoring

Hose-bib

used to drill blast holes in strip mining.

boring Soil-boring on the horizontal as opposed to the vertical.

bracing Any bracing lying in a horizontal plane.

branch A branch drain accepting waste products delivered to it vertically, by gravity, from one or more similar but usually smaller fixtures, and conducting them horizontally to the primary disposal drain.

bridging Perpendicular braces between joists or beams placed horizontally to stiffen the system and distribute the load.

cell tile Structural masonry ceramic tile with the cells installed horizontally in a wall.

circle In surveying, a device for measuring horizontal angles and consisting of a graduated circle on the lower plate of a transit or telescope.

control In surveying, a control system in which the relative positions of points has been precisely established by traverse, triangulation, or another system.

diaphram A metal plate serving to disperse forces in a horizontal plane.

distance The distance between points anywhere on a horizontal plane.

lock A lock in which the primary dimension is horizontal.

panel A wall panel in which the major dimension is horizontal.

pipe Any pipe placed or laid horizontally or at an angle to the horizontal of less than 45 degrees.

sheeting In excavation, any type of earth-restraining sheeting placed horizontally between and supported by solder piles.

shoring (1) Extendible beams or trusses capable of providing concrete form support over fairly long spans, thus reducing the number of vertical supports required. (2) The collective support provided by several horizontal shores in an application.

horn (1) The extension beyond a right-angled joint that is part of a stile, jamb, or sill. (2) The stub of a broken branch left on a log.

hornblende A mineral composed of iron, silicate of magnesium, calcium, and aluminum.

horse (1) Framework functioning as a temporary support, such as a sawhorse. (2) In a stair, one of the slanting supports or strings carrying the treads and risers.

horsehead (1) A frame-like device for supporting a pulley back over a pit so that men and materials may be lowered into it and raised out from it. (2) Forepole support when tunneling through soft material.

horse mold A template for a cornice mounted on a wooden frame, and used in plastering to shape a cornice.

horsepower A unit measurement of power or energy in the United States Customary System. Mechanically, a single horsepower represents 550 foot pound per second. Electrically, a single horsepower represents 746 watts.

horsepower hour A unit representing the amount of work performed by one horsepower in one hour.

horse scaffold As defined by OSHA, a scaffold for light or medium duty, composed of horses supporting a work platform.

hose bib An outdoor water faucet protruding from a building at about sill height, which is usually threaded to accept a hose connection.

hose cock See **hose bib**.

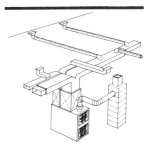

Hot-air Furnace

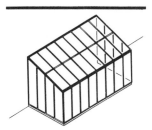

Hothouse

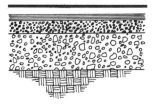

Hot Mix

hose coupling A connection between hoses or between a hose and a pipe.

hose thread A standard screw thread (12 threads per inch on 3/4" pipe) used in garden hose connections.

hospital A building or institution in which housing and medical care and services are constantly available and provided.

arm pull A door handle having an extension to allow opening by hooking one's arm around it, thus leaving the hands free.

door A flush door through an opening large enough to allow passage of beds and/or other large equipment.

door hardware The special hardware with which hospital doors are often equipped, such as arm pulls, hinges, terminated stops, latches, and strategically placed protective metal strips or plates.

frame A doorframe incorporating terminated stops.

hinge A fast pin hinge furnished with a special tip to eliminate the possibility of injuries caused by the projection of conventional hinge tips.

partition A system of tracks and curtains used to provide a degree of privacy around beds in a hospital. More commonly called "hospital cubicles".

window See *hopper window*.

hot Slang for a "live" or electrically charged wire or other electrical component.

hot-air furnace A heating unit in which air is warmed and from which the warmed air is drawn into ducts to be carried throughout a building or selected portion thereof.

hotbed A usually small, shallow (rarely larger than 5' x 10' x 18") enclosure, usually of wood and often covered by glass, containing a plant-growing medium heated by electric cables buried in it or by fermenting manure. A hotbed is used to provide a controlled environment favorable to seed germination and maximum seedling and plant growth.

hot-cathode lamp A type of fluorescent, electric discharge lamp in which the electrodes operate at incandescent temperatures and in which the arc and/or circuit elements provide the energy required to maintain the cathodes at incandescence.

hot cement Cement having a high physical temperature resulting from improper cooling after manufacture.

hot-dip galvanized Descriptive of iron or steel immersed in molten zinc to provide it with a protective coating.

hot driven rivet Any rivet heated just prior to placement.

hot glue A glue requiring heating before being used.

hothouse A greenhouse in which the interior atmosphere is kept very warm.

hotmelt A thermoplastic substance almost always heated before being applied as a coating, sealer, or adhesive.

hot mix Paving made of a combination of aggregate uniformly mixed and coated with asphalt cement. To dry the aggregate and obtain sufficient fluidity of asphalt cement for proper mixing and workability, both the aggregate and asphalt must be heated prior to mixing.

hot press The method of producing plywood, laminates, particleboard, or fiberboard, in which adhesion of layers in the panel is accomplished by the use of thermosetting resins and a heat process, under pressure, to cure the guidelines.

hot rolled Descriptive of structural steel members or sections shaped

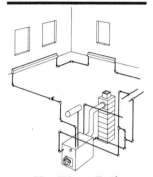

Hot Water Boiler

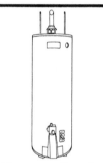

Hot-water Supply

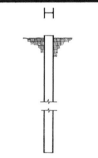

H-pile

from steel billets or plates, heated to a plastic state, by passing them through successive pairs of massive steel rollers, each of which serves to bring the product closer to its final, intended shape, such as an angle, channel, or plate.

hot-setting adhesive An adhesive whose proper setting necessitates a minimum temperature of 212 degrees F.

hot spraying The spraying of paints or lacquers in which the viscosity has been reduced by heat rather than by thinners, allowing formation of a thicker coat, requiring less spraying pressure, hence less overspray.

hot surface (1) A highly alkaline or highly absorbent surface. (2) A surface having a high temperature.

hot water boiler Any heating unit in a hot water heating system in which or by which water is heated before being circulated through pipes to radiators or baseboards throughout a building or portion thereof.

hot-water supply The combination of equipment and its related plumbing supplying domestic hot water.

hot-wire anemometer An anemometer with a velocity sensing device consisting of a temperature-sensitive wire resistor connected to an electrical circuit and placed directly in the path of the airflow.

hot wires An electrically-charged or power-carrying wire.

house connection See **house sewer**.

housed (1) Descriptive of a piece or member fitted into another piece or member that has been modified, as by hollowing, gouging, or chiseling. (2) Enclosed.

housed joint A usually perpendicular joint formed where the full thickness of one member's edge or end is accepted into a

corresponding housing, groove, or dado in another member.

house drain In any given plumbing system, as of a house or building, the major lowest horizontal pipe(s) connecting directly to the building sewer just outside the building wall.

housed stair An entire stair between two walls. See **box stair**.

housed string See **housed stair, close string**.

house sewer The exterior horizontal extension of a house drain outside the building wall leading to the main sewer, either public or private, and connecting directly to the sewer pipe.

H-pile A steel H-beam driven into the earth by a pile driver.

HP-shape A typical pile section made from hot-rolled steel and used for a specific type of pile in which the size is prefaced by "HP".

H-runner A lightweight, H-shaped, metal member used on its side in a suspended ceiling system, so that its flat top fastens to a channel and the flat bottom fits into the kerfs in the ceiling tiles.

hub (1) The central core of a building, usually the area into which stairs and/or elevators are incorporated, and from which hallways or corridors emanate. (2) The usually strengthened central part of a wheel, gear, propeller, etc. (3) That end of a pipe enlarged into a bill or socket. (4) A rotating piece within a lock, through whose central aperture the knob spindle passes to actuate the mechanism. (5) In surveying, a stake designating a theodolite position.

Hudee rim A metal frame used to secure a sink in a countertop.

humidifier A mechanical apparatus to add moisture to the air or other material.

humidistat The automatic regulating device of a humidifier

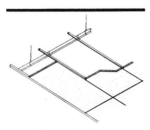

Hung Ceiling

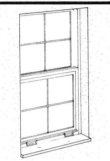

Hung Window

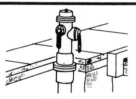

Hydrant

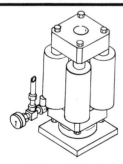

Hydraulic Jack

or dehumidifier which is sensitive to and actuated by changes in humidity.

humidity The water vapor contained in a given space, area, or environment.

humus Soil-like organic substance composed of decayed or decaying organic matter.

hungry Descriptive of a surface so absorbent that paint applied to it leaves a thin film revealing every detail of the background.

hung ceiling A non-structural ceiling having no bearing on walls, being entirely supported from above by the overhead structural element(s) from which it is suspended.

hung sash A sash hung from its sides by cords or chains, whose other ends are secured to counterweights to allow movement in the vertical plane.

hung slating (1) Slates installed to cover a vertical surface, rather than an inclining or horizontal surface. (2) Slates secured by wire clips rather than nails.

hung window A window containing one or more hung sashes.

hybrid The result of breeding plants or animals of two varieties or species.

hybrid beam A fabricated beam having flanges made from a steel with a specified minimum yield strength different from that of the steel used in the web plate.

hydrant A discharge connection to a water main, usually consisting of an upright pipe having one or more nozzles and controlled by a gate valve.

hydrated lime A dry, relatively stable product derived from slaking quicklime.

hydration (1) Any chemical action occurring as the result of combining a material with water. (2) The chemical reaction that

occurs when cement is mixed with water.

hydraulic Characterized or operated by fluid, especially under pressure.

cement Cement whose constituents react with water in ways that allow it to set and harden under water.

dredge A floating dredge or pump by which water and soil, sediment, or seabed are pumped, either on board for sifting, as for clams or oysters, before being discharged overboard, or through a series of floating pipes for discharge on shore.

ejector A pipe through which the working chamber of a pneumatic caisson is cleansed of sand, mud, or small gravel. See also *elephant trunk*.

excavator A powered piece of excavating equipment having a hydraulically operated bucket.

fill Fill composed of solids and liquid, usually water, and usually delivered by a dredge. After placement, the water eventually drains to leave only the solid fill.

friction Resistance to the flow, effected by roughness or obstructions in the pipe, channel, or similar conveying device.

glue Glue unaffected by water.

hydrated lime The dry, hydrated, cementitious product resulting from the process of calcining a limestone containing silica and alumina to a temperature just below incipient. The resultant lime will harden under water.

jack A mechanical lifting device incorporating an external lever to which force is applied to cause a small internal piston to pressurize the fluid, usually oil, in a chamber. The pressure exerts force on a larger piston, causing it to move vertically

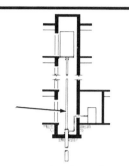

Hydraulic Lift

upward and raise the bearing plate above it.

jump An abrupt increase in the depth of a fluid flowing in a channel as its velocity is slowed and kinetic energy is converted to potential energy.

lift An elevator car or platform moved by a piston, or plunger, powered by a pressurized fluid, usually oil, in a cylinder.

lime Lime composed of at least 10% silicates and which will set and harden under water.

mortar A mortar capable of hardening under water, hence used for foundations or under water masonry construction.

pile driving The employment of hydraulic force to drive sheet piles.

pump The device causing the fluid to be forced through a hydraulic system.

radius The ratio of the cross-sectional area of a stream of fluid within a conduit to the wetted perimeter of that conduit.

splitter A concrete or rock-cracking mechanism incorporating a wedge inserted into a predrilled hole and then expanded by hydraulic power to cause the cracking.

spraying Paint spraying accomplished by high fluid pressure rather than by compressed air.

test Employing pressurized water to test a plumbing line for pressure integrity.

hydrostatic head The pressure in a fluid, expressed as the height of a column of fluid, which will provide an equal pressure at the base of the column.

hydrostatic pressure Pressure exerted by water, or equivalent to that exerted on a surface by water in a column of specific height.

hygrometer An instrument used for measuring the moisture content of air.

hygroscopic Having the tendency to absorb and retain moisture from the air.

hygrostat See **humidistat**.

hypalon caulking See **hypalon roofing**.

hypalon roofing An elastomeric roof covering available commercially in liquid, sheet, or puttylike (caulking) consistency in several different colors. Hypalon roofing is more resistant to thermal movement and weathering than Neoprene.

I

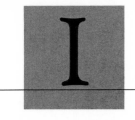

I ABBREVIATIONS

The abbreviations listed below are those most commonly used in the construction industry. Alternative forms (usually nonstandard) are shown in parentheses.

I moment of inertia

IC interrupting capacity, ironclad, incense cedar

ID inside dimension, inside diameter, identification

IF inside frosted

ihp indicated horsepower

IMC intermediate metal conduit

imp imperfect

in. inch

inc included, including, incorporated, increase, incoming

incan incandescent

incl included, including

Ins insulate, insurance

inst installation

insul insulation, insulate

int intake, interior, internal

IP iron pipe

IPS iron pipe size

IPT iron pipe threaded

IR inside radius

I

DEFINITIONS

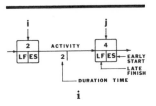

i

I-beam

i In the Critical Path Method, the symbol which represents the event at the start, or initiation, of an activity.

I-beam A structural member of rolled steel whose cross section resembles the capital letter I.

ice dam An accumulation of ice and snow at the eaves of a sloping roof.

ichnography The graphic, geometrically scaled representation of a horizontal section.

idler (1) A gear or wheel used to impart a reversal of direction or rotation of a shaft. (2) The duration in which a belt or chain turns.

igneous rock A rock formed by the solidification of molten materials.

illuminated sign A sign which is illuminated by an internal or external source and used to mark a path of emergency exit.

illumination The luminous flux density on a surface exposed to incident light.

immersion heater A thermostatically controlled electric resistance water heater submerged in a water tank.

impact factor A number by which a static load is multiplied to approximate that load applied dynamically.

impact isolation (1) The use, in buildings, of insulating materials and structures that reduce the transmission of impact noise. (2) The degree of effective reduction of impact noise transmission, accomplished by the structures and materials designed and used specifically for that purpose.

impact load The dynamic effect upon a stationary or mobile body as imparted by the short, forcible contact of another moving body.

impact noise The sound created when a body strikes a surface.

impact noise rating The single-number rating used to evaluate and compare the effectiveness of assemblies and floor/ceiling constructions in isolating impact noise. The higher the number is, the greater is the effectiveness.

impact resistance The resistance of a member or assembly to dynamic loading.

impact test Any of a number of dynamic tests, usually a load striking a specimen in a specified manner, used to estimate the resistance of a material to shock.

impact transmission The transfer of sound waves through walls, floors, and other structures.

impact wrench An electric or pneumatic wrench whose adjustable torque is supplied to a nut or bolt in short, rapid impulses.

impedance Measured in ohms, the total opposition to the flow of current when voltage is applied to an alternating-current electric circuit.

impeller (1) The vaned member of a rotary pump which employs centrifugal force to convey fluids from intake to discharge. (2) A related device used to force pressurized gas in a given direction. (3) In ventilation, a device which rotates to move air.

impending slough The consistency of shotcrete to which no more water can be added without causing it to sag or flow after placement.

impervious Highly resistant to penetration by water.

impervious soil A very fine-grained soil, such as clay, which is so resistant to water penetration that slow capillary creep is the only means by which water can enter.

imposed load Any load which a structure must bear, exclusive of dead load.

impost The often distinctively decorated uppermost member of a column, pillar, etc., which supports an end of an arch.

impregnation (1) The penetration of a (timber) product under pressure with an oil, mineral, or chemical solution, usually for preservation. (2) Treating soil with a liquid waterproofing agent to reduce leakage.

improved land Land where water, sewers, sidewalks, and other basic facilities have been installed prior to subsequent residential or industrial development.

improvement A physical change or addition made on a property so as to increase its value or enhance its appearance.

inactive leaf, inactive door In a pair of doors, the stationary leaf to which the strike plate is secured. It is usually bolted at the head and sill.

in-and-out bond A bond in which masonry units are arranged so that headers and stretchers alternate in successive courses, an arrangement most often used when forming a corner.

inband A header stone used in a reveal.

inbark, bark pocket Ingrown bark enclosed in the wood of a tree by growth and subsequently exposed by manufacture.

inbond A masonry bond across the entire thickness of a wall, and usually consisting of headers or bondstones.

incandescence The emission of visible light as a consequence of being heated.

incandescent lamp A lamp in which electricity heats a

(tungsten) filament to incandescence.

incandescent lighting fixture A complete luminaire, comprising an incandescent lamp, socket, reflector, and often a diffusing apparatus.

incense cedar A tree, indigenous to the area from northern Oregon to southern California, whose aromatic, durable wood is used in pencil manufacture and for many of the same applications as Western Red Cedar. Incense cedar is highly resistant to moisture.

inch of water A unit of pressure which is equal to the pressure exerted by a one inch high column of liquid water at a temperature of 39.2 degrees F (4 degrees C).

inch-pounds (1) A unit of work derived by multiplying the force in pounds by the distance in inches through which it acts. (2) A unit of energy which will perform an equivalent amount of work.

inch stuff Descriptive of building materials having a nominal thickness of one inch, but in reality somewhat less (usually 7/8 inch).

incident radiation Solar energy, both direct and diffuse, upon its arrival at the surface of a solar collector or other surface.

incinerator A type of furnace in which combustible solid, semisolid, or gaseous wastes are burned.

incipient decay The early stage of decay in timber in which the disintegration has not proceeded far enough to affect the strength or hardness.

incising (1) Cutting in, carving, or engraving, usually for decorative purposes. (2) Cutting slits into the surface of a piece of wood prior to preservative treatment to improve absorption.

inclination (1) The deviation of a surface or line from the vertical or

Increaser

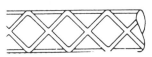

Indented Bar

horizontal. (2) The angle produced by the deviation of a surface or line from the vertical or horizontal.

incline A slope, slant, or gradient.

inclined-axis mixer A truck-mounted concrete mixer. A revolving drum rotates around an axis which is inclined from the horizontal axis of the truck's chassis.

incombustible, non-combustible Incapable of burning.

increaser In plumbing, a coupling with one end larger than the other, generally the small end has outside threads and the large end has inside threads.

incrustation Mineral, chemical, or other deposits left in a pipe, vessel, or other equipment by the liquids which they convey.

indemnification See **insurance**.

indent In masonry, a gap left in a course by the omission of a masonry unit. An indent is used for bonding future masonry.

indented bar A deformed concrete reinforcing bar having indentations for improving the bond between the steel and the concrete.

indented bolt A type of anchor bolt comprising a plain bar into which indentations have been forged so as to increase its grip in concrete or grout.

indented joint A type of butt joint in which a notched fish plate is fitted to notches in the timbers and the entire assembly is fastened with bolts.

indented wire Wire whose surface has been provided with indentations to increase its bond when used as concrete reinforcement or for pretensioning tendons.

indenture (1) An official agreement between a bond issuer and his bond holders. (2) Any

deed or contract between two or more parties. (3) A document in duplicate, triplicate, etc., whose edges have been irregularly indented so that the copies can later be matched to corroborate authenticity.

index of plasticity The numerically expressed difference between the liquid and plastic limits (of a cohesive material).

Indiana limestone A durable, easily sawn, planed, carved, and lathed limestone quarried in and exported from the state of Indiana.

indicated horsepower That horsepower, determined by an indicator gauge, which is developed in the cylinders exclusive of losses sustained due to engine friction.

indicator bolt A type of door bolt used primarily on doors of bathrooms and toilets, which indicates occupancy when locked and vacancy when unlocked.

indicator button An occupancy-indicating mechanism used mostly on the locks of hotel room doors.

indicator valve A valve which includes some device indicating its open or closed condition.

indigenous Descriptive of any product, substance, growth, outcropping, characteristic, etc., which is geographically native to the area where it occurs, as opposed to having been introduced there.

indirect expense Overhead or other indirect costs incurred in achieving project completion, but not applicable to any specific task.

indirect heating (1) A method of heating areas which are removed from the source of heat by steam, hot air, etc. (2) Central heating.

indirect lighting Lighting achieved by directing the light emitted from a luminaire toward a ceiling, wall, or other reflecting surface, rather

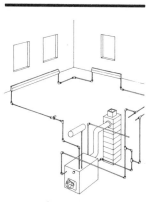

Indirect System

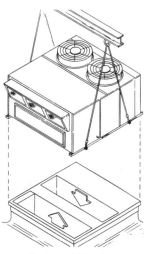

**Induced Draft
Fan**

than directly at the area to be illuminated.

indirect luminaire A luminaire which distributes 90%-100% of its emitted light upward.

indirect system A system of heating, air-conditioning, or refrigeration whereby the heating or cooling of an area is not accomplished directly. Rather, a fluid is heated or cooled, then circulated to the area requiring the conditioning, or used to heat or cool air which then is circulated to achieve the same end.

indirect waste pipe A waste pipe which discharges through an air break or air gap into a trapped receptacle or fixture, rather than directly into the building drainage system.

individual vent A pipe which vents a fixture drain and which is connected to the main vent system at some point higher than the fixture.

indoor/outdoor carpet A carpet whose components all have been designed or treated so as to remain essentially unaffected by water, sun, temperature, etc.

induced draft boiler A boiler that uses a fan at its discharge end to pull air through the burner and boiler and transfer the exhaust products into the atmosphere through a chimney.

induced draft fan See **induced draft boiler.**

induced draft water cooling tower A water-cooling tower incorporating one or more fans in the path of the saturated air stream leaving the tower.

induction The entrainment of air in a room by the strong flow of primary air from an air outlet.

induction brazing A brazing process whose required heat is derived from the resistance of the work to an induced electric current.

induction heating A technique used to heat-treat completed welds in piping. The heat is generated by the use of induction coils around the piping.

induction motor A motor which operates an alternating current and whose primary winding, usually on the starter, is connected to the electric power source, and whose secondary winding, usually on the rotor, conveys the induced current.

induction soldering A soldering process whose required heat is derived from the resistance of the work to an induced electric current.

induction welding A type of welding in which coalescence is achieved by heat derived from the work's resistance to an induced electric current, either with or without applied pressure.

industrial area Any area which is primarily devoted to manufacturing.

industrial waste Liquid waste from manufacturing, processing, or other industrial operations, which might include chemicals, but not rainwater or human waste.

inelastic behavior Deformation of a material which remains even after the force which caused it has been relieved or removed.

inert (1) Chemically inactive. (2) Resistant to motion or action.

inertia The tendency of a body at rest to remain at rest, or of a body in motion to remain in motion in a straight line unless directly influenced by an external force.

inertia block Usually a concrete block supported on some sort of resilient material and used as a base for heavy, vibrating mechanical equipment, such as pumps and forms, to reduce the transmission of vibration to the building structure.

inertial guidance system A guidance system in which a predetermined course is

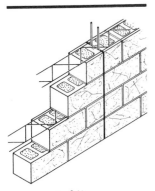

Infilling

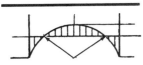

Inflecture Point

maintained by a computer utilizing gyroscopic and accelerometer data.

inert pigment A pigment or extender which does not react chemically with the paint. Also called a "paint extender".

infilling Fill material providing insulation, stiffness, and/or fire resistance, used in buildings to fill the void areas, within a frame, between structural members.

infiltration The leakage of air into a building through the small spaces around windows, doors, etc; caused by pressure differences between indoor and outdoor air.

inflammable Easily ignitable and highly combustible. The preferred term is "flammable".

inflatable gasket A type of gasket whose effective seal results from inflation by compressed air.

inflatable structure An airtight structure of impervious fabric, supported from within by slightly greater than atmospheric pressure generated by fans.

inflecture point That point in a flexural structural member where reversal of curvature occurs and the bending moment is zero.

influence line A technique, useful in solving problems which involve moving loads, which indicates the effect at a given section of a unit load imposed at any point on the structure.

infrared Descriptive of invisible electromagnetic radiation which, when produced by a light source, is usually undesirable, except in certain industrial applications, such as drying and baking finished surfaces.

infrared drying Drying which is accomplished or accelerated by infrared lamps so as to decrease drying time.

infrared heater A source of heat-producing wavelengths, longer than visible light, which do not heat the air through which they

pass, but only those objects in the "line of sight".

infrared lamp A type of incandescent lamp which often has a red glass bulb to reduce its radiated visible light. Such a lamp emits more radiant power in the infrared region than does a standard incandescent lamp, and it has a lower filament temperature, which contributes to its longer life.

infrared photography Photography in which the film used is more sensitive to infrared rays than to visible light rays.

ingle (1) A fireplace. (2) A hearth.

inglenook A nook or part of a corner near a chimney or fireplace, often having built-in seats.

ingot A large mass of molten metal cast in a vertical mold but requiring further processing before becoming a finished product.

ingot iron An iron, primarily used in sheets, containing very small amounts of carbon, manganese, and other impurities.

inhibiting pigment Rust-, corrosion-, or mildew-resistant pigments, such as lead and zinc chromate or red lead, which are added to paint to color it and to provide it with their protective qualities.

inhibitor Oxidants added to paint to retard drying, skinning, and other undesirable effects or conditions.

initial drying shrinkage The difference between the length of a concrete specimen when first poured and the final, permanent length of the same specimen after it has dried, usually expressed as a percentage of the initial moist length.

initial prestress The prestressing stress or force applied to prestressed concrete at the time of its stressing.

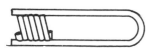

Insert

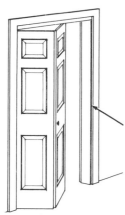

Inside Casing

initial set That point in the setting of a concrete and water mixture when it has attained a certain degree of stiffness but is not yet finally set. Initial set is usually expressed in terms of the time required for a cement paste to stiffen enough to resist a pre-established degree of penetration by a weighted test needle.

initial setting time The time it takes for a fresh mixture of cement paste, mortar, or concrete to attain initial set.

initial stress The stress existing in a prestressed concrete member prior to the occurrence of any loss of stress.

injection burner A gas burner in which air for combustion is forced into the burner and mixed with the gas by a gas jet.

injection molding A process in which the desired shape is obtained by forcing heat-softened plastic into a comparatively cool cavity.

injector The mechanism in a diesel engine that sprays the fuel into the combustion chamber.

inlaid work Small pieces of material, different from that of the background, which are laid or set into a surface so as to produce a decorative pattern.

inlay (1) To decorate with inlaid work. (2) An ornamental design cut into the surface of linoleum, wood, or metal, and filled with a material of different color, often by glueing.

inlet (1) The surface connection to a closed drain or pipe. (2) The upstream end of any structure through which there is a flow.

inlet well An opening at the surface of the ground through which run-off water enters the drainage system.

inner court An outdoor area which is open above but surrounded on four sides by the exterior walls of a building or structure.

inorganic material Substances whose origin and composition are mineral, not animal or vegetable.

input system A ventilating system which comprises a fan to draw air from a roof into rooms through ducts. An air cleaner or filter and an automatic air heater are usually included.

insert (1) A patch, plug, or shim used to replace a defect in a plywood veneer. (2) A unit of hardware embedded in concrete or masonry to provide a means for attaching something. (3) A non-structural patch in laminated timber, made for the sake of appearance.

insert grille A grille which is fabricated separately from the door and installed in the field.

inside-angle tool In masonry and plastering, a float designed especially for shaping inside (internal) angles.

inside caliper A caliper whose pointed legs turn outward and which is used for measuring inside diameters.

inside casing The interior trim around the frame of a door or window, which might consist of dressed boards, molding and trim.

inside corner molding Concave or canted molding used to cover the joint at the internal angle of two intersecting surfaces.

inside glazing External glazing which is placed from within the frame or from inside the building.

inside lining (1) Inside casing. (2) Any part of a cased frame which has its face toward the building or structure.

inside micrometer An instrument used to accurately measure the inside diameters of relatively small cylinders or pipes.

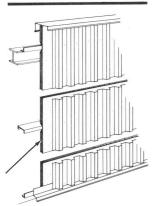

Insulating Board

inside stop A usually beaded or molded strip of wood secured to the casing along the inside edge of the inner sash to hold it in place and restrict its movement to the vertical plane.

inside thread The threaded inner surface of a pipe, or fitting that accepts the outside threads of another pipe or fitting.

inside trim (1) Any trim used inside a building. (2) Inside casing.

in situ (1) In place, as natural, undisturbed soil. (2) Descriptive of work accomplished on the site rather than in prefabrication elsewhere, as in cast-in-place concrete.

in situ concrete Concrete placed where it will harden to become an integral part of the structure, as opposed to precast concrete.

in situ soil test Soil testing performed in a borehole, tunnel, or trial pit, as opposed to being performed elsewhere on a sample which has been removed.

insoluble residue The part of an aggregate or cement that does not dissolve in diluted hydrochloric acid.

inspection (1) Examination of work completed or in progress to determine its conformance with the requirements of the contract documents. (2) Examination of the work by a public official, owner's representative, or others.

inspection eye A pipe fitting equipped with a plug which can be removed to allow examination or cleaning of the pipe run.

inspection fitting See **inspection eye**.

inspection junction See **cleanout**.

inspection list A list of items of work to be completed or corrected by the contractor.

inspection manhole A covered shaft leading from the outside down to a sewer or duct, so constructed as to allow a person to enter it from the surface.

inspector A person authorized and/or assigned to perform a detailed examination of any or all portions of the work and/or materials. See **building inspector, owner's inspector,** and **resident engineer**.

instant lock, instant locker (1) An automatic lock which is actuated by the closing of the door. (2) A time lock or chromatic lock working on the same principle.

instant-start fluorescent lamp An electric-discharge lamp which is started without the preheating of electrodes, but rather by application of a high enough voltage to eject electrons from the electrodes by field emission, initiate electron flow through the lamp, ionize the gases, and initiate a discharge through the lamp.

instructions to bidders Instructions contained in the bidding documents for preparing and submitting bids for a construction project or designated portion thereof.

insulate To provide with special features and/or materials which afford protection against sound, moisture, heat, or loss of heat.

insulated metal roofing A type of roofing panel made from mineral fiber, cellular glass, foamed plastic, etc., and faced with light-gauge flexible metal.

insulating board A thin, lightweight, rigid or semi-rigid board, usually of processed plant fibers, which offers little structural strength but does provide thermal insolation and is usually applied under a finish material.

insulating cement A putty-like mixture of hydraulic-setting cement or other bonding material

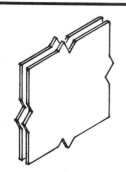

**Insulating
Glass**

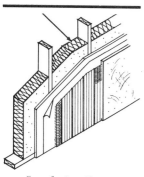

Insulation Batt

and a loose-fill insulation, used to fill voids, joints, cracks, etc.

insulating concrete Concrete possessing low thermal conductivity and used as thermal insulation.

insulating fiberboard Building board manufactured from processed plant fiber in a variety of sizes, thicknesses, densities, and strengths. See **insulating board**.

insulating formboard Insulation board which serves as a permanent form for poured-in-place gypsum or lightweight concrete roof decks.

insulating glass Glazing comprising two or more lights, between which there exist(s) an hermetically sealed airspace(s), joined around the edges.

insulating oil The oil contained and used within the enclosure of a transformer, switch, or other electric device for cooling and insulation.

insulating plasterboard Gypsum board with aluminum foil backing, which provides resistance to heat flow and moisture.

insulating strip The soft, resilient material used in an expansion joint.

insulating varnish A varnish which is applied to wires or electric circuits as an insulator.

insulation Any material, device, or technique which provides protection against fire or the transfer of electricity, heat, cold, moisture, or sound.

insulation batt Flexible insulation of loosely matted plant or glass fibers faced on one or both sides with kraft paper or aluminum foil and usually available in specifically sized sections.

insulation blanket Usually composed of the same materials and in the same widths and thicknesses as batts, but is available in rolls.

insulation lath Gypsum lath with an aluminum foil backing which provides resistance to heat flow and moisture.

insulation resistance The resistance of an insulation to direct current.

insulator An insulating device designed and used to physically support a conductor and electrically separate it from other conductors or objects.

insurance A contractual obligation by which one person or entity agrees to secure another against loss or damage from specified liabilities for premiums paid.

bodily injury (insurance terminology): Physical injury, sickness, or disease sustained by a person. See also *personal injury*.

builder's risk A specialized form of property insurance which provides coverage for loss or damage to the work during the course of construction. See also *property insurance*.

care, custody and control (insurance terminology): The term used to describe a standard exclusion in liability insurance policies. Under this exclusion, the liability insurance does not apply to damage to property over which the insured is for any purpose exercising physical control.

certificate of A document, issued by an authorized representative of an insurance company, stating the types, amounts, and effective dates of insurance in force for a designated insured.

completed operations Liability insurance coverage for injuries to persons or damage to property occurring after an operation is completed (1) when all operations under the contract have been completed or abandoned; or (2) when all operations at one project site are completed; or (3) when the portion of the work out of which the injury or damage arises has been put to its intended use by the person or organization for whom that portion of the work was done. Completed operations insurance does not apply to damage to the completed work itself.

comprehensive general liability A broad form of liability insurance covering claims for bodily injury and property damage which combines, under one policy, coverage for all liability exposures (except those specially excluded) on a blanket basis and automatically covers new and unknown hazards that may develop. Comprehensive General Liability Insurance automatically includes contractual liability coverage for certain types of contracts. Products Liability, Completed Operations Liability, and broader Contractual Liability coverages are available on an optional basis. This policy may also be written to include Automobile Liability.

contractor's liability Insurance purchased and maintained by the contractor to protect the contractor from specified claims which may arise out of or result from the contractor's operations under the contract, whether such operations are by the contractor or by any subcontractor or by anyone directly or indirectly employed by any of them, or by anyone for whose acts any of them may be liable.

employer's liability Insurance protection for the employer against claims by employees or employees' dependents for damages which arise out of injuries or diseases sustained in the course of their work, and which are based on common law negligence rather than on liability under workers' compensation acts.

errors and omissions See *professional liability insurance*.

extended coverage An endorsement to a property insurance policy which extends the perils covered to include windstorm, hail, riot, civil commotion, explosion (except steam boiler), air craft, vehicles and smoke. See **property insurance.**

liability Insurance which protects the insured against liability on account of injury to the person or property of another. See also (1) *completed operations insurance*; (2) *comprehensive general liability insurance*; (3) *contractor's liability insurance*; (4) *employer's liability insurance*; (5) *owner's liability insurance*; (6) *professional liability insurance*; (7) *property damage insurance*; (8) *public liability insurance*; and (9) *special hazards insurance*.

loss of use Insurance protecting against financial loss during the time required to repair or replace property damaged or destroyed by an insured peril.

owner's liability Insurance to protect the owner against claims arising out of the operations performed for the owner by the contractor and arising out of the owner's general supervision.

personal injury (insurance terminology): Bodily injury, and also injury or damage to the character or reputation of a person. Personal injury insurance includes coverage for injuries or damage to others

269

caused by specified actions of the insured, such as false arrest, malicious prosecution, willful detention or imprisonment, libel, slander, defamation of character, wrongful eviction, invasion of privacy, or wrongful entry. See also *bodily injury*.

personal liability See *personal injury*.

professional liability Insurance coverage for the insured professional's legal liability for claims for damages sustained by others allegedly as a result of negligent acts, errors, or omissions in the performance of professional services.

property Coverage for loss or damage to the work at the site caused by the perils of fire, lightning, extended coverage perils, vandalism and malicious mischief, and additional perils (as otherwise provided or requested). See Also **(1) builder's risk insurance; (2) extended coverage insurance; (3) special hazards insurance.**

property damage Insurance covering liability of the insured for claims for injury to or destruction of tangible property, including loss of use resulting therefrom, but usually not including coverage for injury to or destruction of property which is in the care, custody, and control of the insured. See also *care, custody, and control*.

public liability Insurance covering liability of the insured for negligent acts resulting in bodily injury, disease, or death of persons other than employees of the insured, and/or property damage. See *(1) comprehensive general liability insurance (2) contractor's liability insurance*.

special hazards Insurance coverage for damage caused by additional perils or risks to be included in the property insurance (at the request of the contractor or at the option of the owner). Examples often included are sprinkler leakage, collapse, water damage, and coverage for materials in transit to the site or stored off the site. See *property insurance*.

work in progress See *insurance, builder's risk*.

workers' compensation Insurance covering the liability of an employer to employees for compensation and other benefits required by worker's compensation laws with respect to injury, sickness, disease, or death arising from their employment. Also still known in some jurisdictions as "workmen's compensation insurance".

intake The opening or device through which a gas or fluid enters a system.

intake belt course In building, a belt course whose molded face is cut so as to function as an intake between the varying thicknesses of two walls.

intarsia Mosaic accomplished with small pieces of wood inlaid in contrasting colors.

integral frame A metal doorframe whose trim, backbands, rabbets, and stops are all fabricated from one piece of metal for each jamb and for each head.

integral lock A type of mortise lock in which the cylinder is contained in the knob.

integral waterproofing The waterproofing of concrete achieved by the addition of a suitable admixture.

integrated ceiling A suspended-ceiling system in which the grid and the individual acoustical, illumination, and air-handling elements are combined to form a single, integrated system.

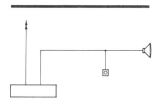

Intercom

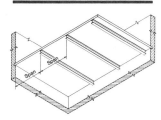

Interior span

Interior Trim

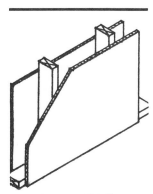

Interior Wall

intercept In surveying, the length of the staff visible between the two stadia hairs of a transit's telescope.

intercepting drain A ditch, trench, or similar depression surrounding a subdrainage pipe and filled with a pervious filter material.

intercepting sewer A sewer into which empty the dry-weather flows from several branch sewers or outlets and which also may receive a certain amount of storm water.

interceptor An apparatus which functions to trap, remove, and/or separate harmful, hazardous, or otherwise undesirable material from the normal waste which passes through it, allowing acceptable waste and sewage to discharge by gravity to the disposal terminal.

intercom A general term for an intercommunication system, such as one employing telephones or microphones and loudspeakers and functioning without a central switchboard.

interconnection Any arrangement of pipes whereby water may flow from one system to another as determined by the pressure differential between the two systems.

interface A common boundary shared by two adjacent parts of a system of construction.

interface strength See **bond**.

interfenestration The area between windows in a facade consisting primarily of the windows and their ornamentation.

interior casing See **inside casing**.

interior door A door installed inside a building, as in a partition or wall, having two interior sides.

interior finish The interior exposed surfaces of a building, such as wood, plaster, and brick, or applied materials, such as paint and wallpaper.

interior glazed Descriptive of glass which has been placed from inside a building.

interior hung scaffold A scaffold which, rather than being supported from below, is suspended from a ceiling or roof structure.

interior lot A lot bounded by a street either at its front or back, but not on either of its sides.

interior plywood Plywood in which the laminating glue is adversely affected by moisture, hence should be restricted to indoor or interior applications.

interior span A continuous beam or slab which is continuous with neighboring spans.

interior stop In glazing, a removable bead or molding strip which serves to hold a light or panel in position when the stop is on the interior of the building, as opposed to an exterior stop.

interior trim (1) Any trim, but especially that around door and window casings, baseboards, stairs, and on the inside of a building. (2) Inside finish.

interior wall A wall having two interior faces and existing entirely within the exterior walls.

interlaced fencing, interwoven fencing Fencing which is constructed from very thin, flat boards which are woven together.

interlocked (1) Firmly joined. (2) Closely united. (3) Placed in close relative proximity or in a specific relationship with another or others.

interlocking joint (1) In ashlar or other stonework, a joint accomplished by joggles in the joining units. (2) In sheet metal, a joint between two parts whose preformed edges engage to form a continuous locked splice.

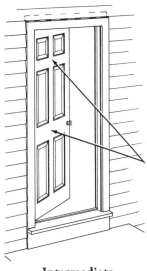

**Intermediate
Rail**

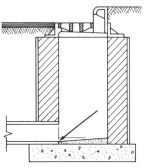

Invert

interlocking tile Single-lap tiles designed so that a groove along the edge of one tile accommodates an edge of an adjacent tile in the same course.

intermediate floor beam In floor framing, any other floor beams which are positioned between the end floor beams.

intermediate rafter See **common rafter**.

intermediate rail In a door, any rail, one of which might be a lock rail, between the top and bottom rails.

intermediate sight In leveling, a staff reading which is neither a back sight nor a foresight.

intermediate stiffener On a beam or girder, any stiffener between the end stiffeners.

intermittent weld A weld whose continuity is interrupted by recurrent unwelded spaces.

internal dormer A vertical window in a sloped roof which is set in a depression in the main roof. An internal dormer is the inverse of a standard dormer.

internal glazing Glazing placed in interior partitions or walls (different from inside glazing). See **borrowed light**.

internal quality block A masonry block which is structurally sufficient but whose inferior surfaces make it suitable only for concealed work.

internal quality brick Structurally sound brick whose inferior surfaces make it suitable only for concealed work.

internal thread See **inside thread**.

internal treatment The treatment of water by feeding chemicals into the boiler rather than into the preheated water itself.

internal vibration Rapid agitation of freshly mixed or placed concrete performed by mechanical vibrations inserted at strategic locations. See **concrete vibration**.

intern architect An apprentice architect who works under the direction of registered architects.

interpier sheeting, interpile sheeting Usually wooden sheeting placed horizontally between underpinning pits or piles, used in applications not requiring continuous underpinning.

intertie An intermediate member used horizontally between studs to strengthen them, especially at door heads or other places between floor heads.

invert The lowest inside surface or floor of a pipe, drain, sewer, culvert, or manhole.

invert block A wedge-shaped hollow masonry tile incorporated into the invert of a masonry sewer.

inverted asphalt emulsion A type of emulsified asphalt, anionic or cationic, in which asphalt (usually in liquid form) is the continuous phase, and whose discontinuous phase comprises minute globules of water in relatively small amounts.

inverted ballast A lamp ballast which operates on direct current.

inverted crown The fall or pitch from the sides to the center of a road, driveway, etc.

invert elevation The elevation of an invert (lowest inside point) of pipe or sewer at a given location in reference to a bench mark.

invisible hinge A door hinge designed, fabricated, and installed with no visible or exposed parts when the door is in the closed position.

invitation to bid A portion of the bidding documents soliciting bids for a construction project.

invited bidders The bidders selected by the owner, after consultation with the architect, as

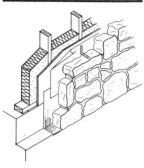

**Irregular
Coursed Rubble**

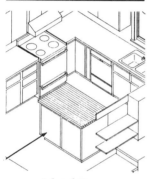

**Island Base
Kitchen Cabinet**

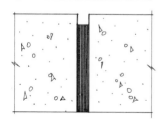

Isolation Joint

the only ones from whom bids will be received.

involute (1) The locus of a fixed point on a string as the string is unwound from a fixed plane curve, such as a circle (the Spiral of Archimedes), generally used to generate cams. (2) Spirally curved, intricate, complex.

iron A lustrous, malleable, magnetic, magnetizable metallic element mined from the earth's crust as ore in hematite, magnetite, and lemonite. These minerals are heated together to 3,000 degrees F in a blast furnace to produce pig iron, which emerges from the furnace as 95% iron, 4% carbon, and 1% other elements.

  cast An iron alloy usually containing 2-1/2% to 4% carbon and silicon, possessing high compressive but low tensile strength, and which, in its molten state, is poured into sand molds to produce castings.

  cement A type of cement that contains cast iron boring or filings, sal-ammoniac sulfur, and other additives, and used for joining or repairing cast iron parts.

  core The steel bar under the wooden handrail connecting the tops of balusters in a stair.

  malleable Cast iron which has undergone an annealing process to reduce its brittleness.

  oxide A primary ingredient in a whole range of inorganic pigments.

  work A comprehensive term for iron fashioned to be used decoratively or ornamentally, as opposed to structurally.

  wrought The purest form of iron metal, which is fibrous, corrosion-resistant, easily forged or welded, and used in a wide variety of applications, including water pipes, rivets,

stay bolts, and water tank plates.

irregular coursed rubble Rubble walls constructed in courses of various depths.

irregular pitch A type of roof whose slope does not have a constant rise per foot throughout.

irrigation (1) The process or system, and its related equipment, by which water is transported and supplied to otherwise dry land. (2) The use of water thus supplied for its intended purpose.

I-section The cross section of a beam comprising a vertical web connecting top and bottom flanges.

island base kitchen cabinet A free-standing kitchen cabinet having exposed ends and a counter or work surface.

isodomum An ancient, very regular masonry pattern with units of uniform length and height placed to form continuous horizontal joints, with vertical joints centered above the blocks beneath forming discontinuous straight lines.

isolated solar gain Passive solar heating in which heat to be used on one area is collected in another area.

isolating membrane An underlay.

isolation Sound privacy effected by reducing direct sound paths.

isolation joint A joint positioned so as to separate concrete from adjacent surfaces or into individual structural elements which are not in direct physical contact, such as an expansion joint.

isolator That device on a circuit which can be removed to break the circuit in the absence of flowing current.

isometric drawing A form of three-dimensional projection in which all of the principal planes are drawn parallel to

273

corresponding established axes and at true dimensions. Horizontals are usually drawn at 30 degrees from the normal horizontal axes; verticals remain parallel to the normal vertical axis.

isothermal Descriptive of a process or procedure which occurs at a constant temperature.

Italian tiling (pan-and-roll roofing tile) A roof covering with two different kinds of single-lop tiles, one being the curved and tapered overtile, and the other being the flanged, tapered, tray-shaped undertile.

item A particular kind of work performed or material(s) supplied, as specified in the contract plans and specifications.

ivorywood A South American wood used mostly for cabinet and lathe work. White to pale yellow-brown in color, sometimes even with a tinge of green, it is heavy, hard, and strong, but lacks durability.

Izod impact testing A type of impact test, used to estimate the impact resistance of a material, in which a falling or swinging pendulum delivers energy in a single impact.

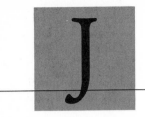

ABBREVIATIONS

The abbreviations listed below are those most commonly used in the construction industry. Alternative forms (usually nonstandard) are shown in parentheses.

J jack, joule
jct junction
J.I.C Joint Industrial Council
jour journeyman
JP jet propulsion
jt/jnt joint
junc junction
jsts joists

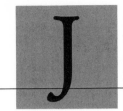 Definitions

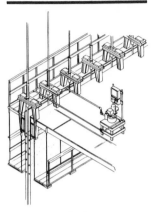

Jack

jack (1) A portable mechanism for moving loads short distances by means of force applied with a lever, screw, hydraulic press, or air pressure, as in applying the prestressing force to the tendons, or making small adjustments in the elevations of forms or form support as in lift slab or slipform operations. (2) In electricity, a female connecting device or socket to which circuit wires are attached and into which a plug may be inserted (as on a telephone switchboard).

jack arch An arch whose intrados is flat or almost flat, instead of being curved or rounded.

jack beam A beam which is used to support another beam or truss, thereby eliminating the need for a column.

jack boom A boom which supports sheaves between the hoist and the main boom in a dredge or pull shovel.

jacked pile A pile which is forced into the ground by jacking against the building above it.

jacket (1) A covering, either of cloth or metal, applied to exposed heating pipes or ducts, or to exposed or unexposed cashing pipes, or over the insulation of such pipes. (2) A watertight outer housing around a pipe or vessel, the space between being occupied by a fluid for heating, cooling, or maintaining a specific temperature.

jacketing The surrounding of a pipe or vessel by a confined bath or stream of fluid for temperature control or heat absorption. See **jacket, def. 2.**

jackhammer A hand-held pneumatically powered drill which hammers and rotates a bit or chisel. A jackhammer is used for drilling rock or breaking up concrete, asphalt paving, etc.

jacking device A mechanism used to stress the tendons for prestressed concrete, or to raise a vertical slipform. See **jack, def. 1.**

jacking dice Blocks, usually of concrete-filled steel cylinders or pipes, used as temporary fillers during the jacking operations, as in foundation work.

jacking force The temporary force applied to tendons by a jacking device to produce the tension in prestressing tendons.

jacking plate A steel bearing plate used during jacking operations to transmit the load of the jack to the pile.

jacking stress The maximum stress that occurs during the stressing of a precast concrete tendon.

jackknife (1) The unintentional raising of a boom on a derrick, as caused by the load. (2) A tractor and trailer accidently arriving at such an angle with each other that the tractor can not move forward.

jack lagging The timber-bearing members employed in building the forms for arches or for other unusual shapes.

jackleg An outrigger post.

jack plane A medium-sized carpenter's plane used to do the coarse work on a piece of timber, such as trying up the edges.

jack rafter A rafter, shorter in length than the normal rafters used in the same building, and used to support the roof in a hip or between a valley and ridge.

jack rib Any curved jack rafter used in a framed arch or small dome roof.

jack screw A screw-operated mechanical device equipped with a

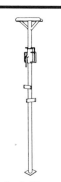

Jack Shore

Jacob's Ladder

Jamb Anchor

load-bearing plate and used for lifting or leveling heavy loads.

jack shaft A short drive shaft, as between a clutch and a transmission.

jack shore An adjustable, usually telescopic, single-post metal shore.

jack timber In framework, a timber which is shorter than the rest, because it has been intercepted by another member.

jack truss (1) A roof truss which is smaller than the main trusses, such as a truss in the end slopes of a hip roof. (2) A truss which, by supporting a beam or another truss, provides for the elimination of a column support.

Jacob's ladder According to OSHA, a marine ladder of rope or chain with wooden or metal rungs.

jagger A toothed stone-dressing chisel.

jalousie A window shutter or blind having stationary or adjustable slats angled so as to permit ventilation and provide shade and even some privacy, while simultaneously preventing the entrance of rain.

jamb An exposed upright member on each side of a window frame, doorframe, or door lining. In a window, these jambs outside the frame are called "reveals".

jamb anchor A metal anchoring device which, when inserted in the rear of the jamb of a doorframe or window frame, supports the frame to the wall.

jamb block A concrete masonry unit used at an opening, where its slotted end receives a jamb.

jamb depth The face-to-face depth of a doorframe.

jamb horn In a window frame, that portion of the jamb which extends beyond the sill or head jamb.

jamb lining A wood facing at the inside edge of a window jamb for the purpose of increasing its width.

jamb post The vertical timber that serves as a jamb at the side of an opening.

jamb shaft Seen often in medieval architecture, a small shaft having a capital and a base, positioned against or incorporated into the jamb of a door or window (sometimes called "esconsons" when employed in the inside axis of a window jamb).

jambstone A stone that forms the jamb of a door.

japan (1) A dark-colored, short-oil varnish used to provide a hard, glossy surface. (2) A type of resin varnish often used in paints as a drying agent.

japanned Painted with a black japan and then baked.

jaw One of a pair of opposing members of a device used for holding, crushing, or squeezing an object, as the jaws of a vise or a pair of pliers.

jaw crusher A rock-crushing device comprising one fixed inclined jaw and one movable inclined jaw, and used to reduce rock to specific sizes.

jemmy A short (less than 18") pinchbar or crowbar, both ends of which are curved.

jenny (1) A machine that cleans surfaces by emitting a steady or pulsating jet of steam. (2) A British term for a gin block.

jerkinhead See **hipped gable**.

jerrybuilt Constructed in a shoddy or flimsy manner.

jesting beam Any beam employed strictly for decorative or ornamental, as opposed to structural, purposes.

jet (1) A high-velocity, pressurized stream of fluid or mixture of fluid and air, as emitted from a nozzle or other small orifice. (2) The nozzle or orifice which shapes the stream.

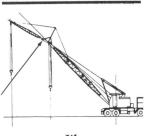

Jib

jetted pile A pile whose sinking has been accomplished by jetting.

jetting The process by which piles or well points are sunk with the aid of jets, or occupy holes which have been created by jets. Jetting is the option usually employed in locations where nearby buildings might be adversely affected during pile-driving operations.

jetty (1) Any portion of a building which protrudes beyond the part immediately below it, such as a bay window or the second story of a garrison house. (2) A dike-like structure, usually of rock, which extends from a shore into water, usually to provide some kind of protection, but sometimes to induce scouring or bank-building.

jewel An ornamental glass protrusion in a leaded window.

jib (1) The hoisting arm of a crane or derrick, whose outer end is equipped with a pulley, over which the hoisting cable passes. (2) The arm that holds a drifter on a rock drill.

jib boom The hinged extension attached to the upper end of a crane boom.

jib crane A crane having a swinging jib, as opposed to an overhead traveling crane, which does not.

jib door A door constructed and installed so as to be flush with its surrounding wall and whose unobtrusiveness is furthered by the lack of hardware on its interior face.

jig A device which facilitates the fabrication or final assembly of parts by holding or guiding them in such a way as to insure their proper mechanical and relative alignment. A jig is especially useful in assuring that duplicate pieces are identical.

jig saw An electrically powered table-mounted saw having a small, thin, narrow, vertically reciprocating blade which facilitates the cutting of curves or intricate patterns.

jimmer See **gemel**.

jimmy (1) Slang for a General Motors Corporation motor, engine, or truck. (2) See **jemmy**.

jitterbug (1) A tamping mechanism, usually powered, for concrete. (2) A hand tool similarly used to cause sand and cement grout to rise to the surface of wet concrete during the placement of slabs.

job An entire construction project or a component thereof.

jobber (1) A person reasonably knowledgeable and somewhat skilled in most of the more common construction operations, such as carpentry, masonry, or plumbing. (2) In construction, a jack-of-all-trades.

job condition Those portions of the contract documents that define the rights and responsibilities of the contracting parties and of others involved in the work. The conditions of the contract include general conditions, supplementary conditions, and other conditions.

job made Made or constructed on the construction site.

job site The area within the defined boundaries of a project.

jog An offset, such as an intentional change in direction or other unintentional irregularity in a line or surface.

joggle (1) A notch or protrusion in one piece or member which is fitted to a protrusion or notch in another piece to prevent slipping between the members. (2) A protrusion or shoulder which receives the thrust of a brace. (3) A horn or stub tenon at the end of a mortised piece to strengthen it and prevent its lateral movement. (4) The enlarged portion of a post by which a strut is supported.

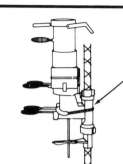

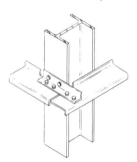

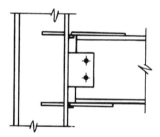

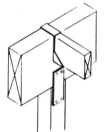

Joints

joggle beam A built-up beam in which joggles are used to secure the components in their respective positions.

joggle joint In masonry or stonework, a joint which employs joggles in the adjacent members so as to prevent their lateral movement. See **joggle, def. 1**.

joggle post (1) A post constructed of two or more joggled timbers. (2) A king post whose lower end has been joggled to support the feet of struts. See **joggle, def. 4**.

joggle truss A roof truss comprising a single, centrally positioned post whose upper end is joggled to connect with the overhead chord, and whose lower end is supported by two braces that angle upward to join the ends of the cord.

joggle work Masonry or stonework in which the units of vertically adjacent courses are joggled on at least one side, resulting in a joggled horizontal joint.

joiner (1) A primarily British term for a craftsman who constructs joints in woodwork. (2) A carpenter who deals primarily with joining fitted parts, such as of a door, window, or cabinet.

joiner's chisel A long-handled woodworking chisel whose cutting is accomplished without the aid of a striking tool, but by hand force only. Also called a "paring chisel".

joiner's gauge Same as **marking gauge**.

joiner's hammer A hammer whose head has a flat end for striking and a clawed end for pulling nails.

joinery (1) Woodworking that deals more with joining and finishing, such as that required by doors, cabinets, or trim. (2) A European designation for quality grades of lumber suitable for cabinetry, millwork, or interior trim.

joint (1) The point, area, position, or condition at which two or more things are jointed. (2) The space, however small, where two surfaces meet. (3) The mortar-filled space between adjacent masonry units. (4) The place where separate but adjacent timbers are connected, as by nails or screws, or by mortises and tenons, glue, etc.

joint bolt A threaded metal rod or bolt having a nut at each end, used to bring two ends of a handrail together, and also used in joinery to bolt two mating surfaces together, as in a butt joint. Also called a "handrail bolt".

joint box A cast iron box constructed around a joint between the ends of two electric cables. The cables' protective lead (or other) sheathing is secured by bolted clamps on the exterior of the box, which itself may be filled with insulation.

joint efficiency The ratio of the strength of a welded joint to the strength of the base metal, expressed as a percentage.

jointer (1) A power-driven woodworking tool or long, hand operated, bench plane used to square the edges of lumber or panels. (2) An offset metal tool used to smooth or indent mortar joints in masonry. (3) A metal tool about six inches long and two to four inches wide, having interchangeable depth-regulating bits and used for cutting joints in fresh concrete. (4) In masonry, a bent strip of iron used in a wall to strengthen a joint.

jointer plane See **jointer**.

joint fastener A small strip of corrugated steel, having one sharpened edge and used to fasten (usually corner) pieces in rough carpentry. A joint fastener is positioned vertically, sharp edge down over the joining edges of the two pieces, and then hammered down into them.

joint filler (1) A powder that is mixed with water and used to treat joints, as in plasterboard construction. (2) Any putty-like material similarly used. (3) A compressible strip of resilient material used between precast concrete units to provide for expansion and/or contraction.

jointing (1) Finishing the surface of mortar joints, as between units or courses, by tooling before the mortar has hardened. (2) The finishing, as by machining, of a squared, flat surface on one face or edge of a piece of wood. (3) The initial operation in sharpening a cutting tool, consisting of filing or grinding the teeth or knives to the desired cutting circle.

jointing compound In plumbing, any material, such as paste, paint, or iron cement, used to ensure a tight seal at the joints of iron or steel pipes.

jointing material A sheet of rubber, asbestos, or synthetic compressible material from which gaskets or washers may be cut for use in joints of flanged pipes, pumps, etc.

jointing rule In masonry, a long straightedge used for drawing lines and pointing.

jointless flooring Although often laid or poured without joints, magnesite or expoxy-based flooring that is often best laid with joints which allow for shrinkage.

joint reinforcement Any steel reinforcement used in or on mortar joints, such as reinforcing bars or steel wire.

joint rule A metal rule having one end formed at a 45-degree angle and used in plastering to form and shape miters at the joints of cornice moldings.

joint runner In plumbing, an incombustible packing material used around the outside of a pipe joint to contain the molten lead which is poured in the bell of a joint, such as pouring rope.

joint sealant An impervious substance used to fill joints or cracks in concrete or mortar, or to exclude water and solid matter from any joints.

joint tape Paper, paper-faced cotton tape, or plastic mesh fabric, used with mastic or plaster to cover the joints between adjacent sheets of wallboard.

joint venture A collaborative undertaking by two or more persons or organizations for a specific project (or projects), having many of the legal characteristics of a partnership.

joist A piece of lumber two or four inches thick and six or more inches wide, used horizontally as a support for a ceiling or floor. Also, such a support made from steel, aluminum, or other material. See **random lengths**.

joist anchor A beam or wall anchor or metal tie used to anchor beams or joists to a wall. An example of a joist anchor is a metal strip with one end embedded in a concrete or masonry wall and the other end secured to a joist or rafter, so as to provide a lateral tie between the wall and a floor or roof.

joist chair A wire device used to support reinforcing steel in the formwork of a concrete beam, ensuring that the steel bars touch neither the bottom nor sides of the forms.

joist hanger A metal angle or strap used to support and fix the ends of wood joists or rafters to beams or girders.

journal That section of a shaft or axle which rotates within a load-supporting bearing.

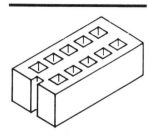

Jumbo Brick

judas A small hole or opening in a door, as in a prison door, used for inspection or surveillance.

jumbo (1) A mobile support for concrete forms. (2) An assortment of tunnel drilling devices mounted on a mobile carriage.

jumbo brick A brick of larger than standard size, whether intentionally or accidentally.

jumper (1) The short length of wire or cable used to make a usually temporary electrical connection within, between, or around circuits and/or their related equipment. (2) A steel bar used manually as a drilling or boring tool. (3) A stretcher covering two or more cross joints in snecked or square rubble. (4) The inverted mushroom-shaped component of a domestic water tap, on which the washer fits.

junction box A metal box in which splices in conductors or joints in runs of raceways or cable are protectively enclosed, and which is equipped with an easy access cover.

junction chamber That section in a sewer system where the flow from one or more sewers joins or converges into a main sewer.

junction manhole A manhole located over the convergence of two or more sewers.

junior channel An obselete term for a lightweight structural channel.

jurisdictional dispute An argument between or among labor unions over which should perform certain work.

jut Any protruding part of a building or structure, such as a jut window. See **jetty, def. 1.**

jute A plant fiber, from which a strong, durable yarn is made and used mostly for carpet backing, burlap, and rope.

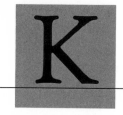

K ABBREVIATIONS

The abbreviations listed below are those most commonly used in the construction industry. Alternative forms (usually nonstandard) are shown in parentheses.

k kilo, knot
K Kalium
Ka cathode
kc kilocycle
kcal kilocalorie
kc/s kilocycles per second
KD kiln dried
KDN knocked down
kg keg, kilogram
KIT. kitchen
kl kiloliter
KLF kips per lineal foot
km kilometer
kmps kilometers per second
kn knot
Kr krypton
kv kilovolt
kvar kilovar
kw kilowatt
kwhr/kwh kilowatt hour

K DEFINITIONS

Kalamein Door

Key

kalamein door A fire door whose wood core is usually covered by galvanized sheet metal.

kalsomine See **calcimine**.

kaolin A usually white mineral found in rock formations, composed primarily of low-iron hydrous aluminum silicate, and used as a basic ingredient in the manufacture of white cement and as a filler or coating for paper and textiles.

keel molding A molding having two ogee curves that meet at a point or fillet, forming a shape that resembles that of a ship's keel.

Keene's cement (1) A white cementitious material manufactured from gypsum which has been burned at a high temperature and ground to a fine powder. Alum is added to accelerate the set. The resulting plaster is hard and strong and accepts and maintains a high polish, hence it is used as finishing plaster. (2) Anhydrous calcined gypsum.

keeper See **strike plate**.

keeping the gauge In masonry, maintaining the proper spacing of courses of brick.

keeping the perpends In masonry, the accurate laying of the units so that the perpends (cross joints) in alternating courses line up vertically.

kelly ball A round-bottomed metal plunger that is dropped into fresh concrete, the degree of penetration indicating the consistency of the concrete.

kerf (1) A saw-cut in wood, stone, etc., which is usually performed crosswise and usually not completely through the member. (2) A groove cut into the edges of acoustical tiles to accommodate the splines or supporting elements in a suspended acoustical ceiling system.

kerfed beam A beam in which several kerfs have been cut so as to permit bending.

kettle (1) The storage container for asphalt to be used for "hot mopped" roof construction. (2) Any open vessel used to contain paint or in which glue is melted.

kevel An axe whose head has a flat face at one end and a pointed peen at the other. A kevel is used by stone masons for removing angular projections or diminishing surfaces.

key (1) The removable actuating device of a lock. (2) A wedge of wood or metal inserted in a joint to limit movement. (3) A keystone. (4) A wedge or pin through the protruding part of a projecting tenon to secure its hold. (5) A back piece on a board to prevent warping. (6) The tapered last board in a sequence of floorboards, which, when driven into place, serves to hold the others in place. (7) The roughened underside of veneer or other similar material intended to aid in bonding. (8) In plastering, that portion of cementitious material which is forced into the openings of the backing lath. (9) A joggle. (10) A keyway. (11) A cotter as in def. 4. (12) A small, usually squared piece which simultaneously fits into the keyways or grooves of a rotating shaft and the pulley.

key brick A brick whose proper fit in an arch is attained by tapering it toward one end.

key course A course of adjacent keystones as might be used in an archway too deep for a single

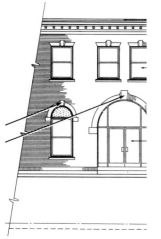

Keystone

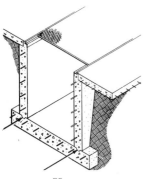

Keyway

Kicker

keystone, or in the crown of a barrel vault.

key drop A pivoting cover of a keyhole, usually attached to the escutcheon above the hole and covering it when at rest.

keyed (1) Fastened or fixed in position in a notch or other recess, as forms become keyed into the concrete they support. (2) Equipped with a notch, groove, indentation, or other receptacle designed to accept a correspondingly shaped member.

keyed beam (1) A lap-jointed beam with joggles or slots cut into both components. Keys are driven into the joggles or slots to increase the bending strength of the joint. (2) A compound beam whose adjacent layers possess mating grooves to help resist horizontal shearing stress at the interfaces.

keyed brick A brick, one of whose faces has been supplied with a usually dovetail-shaped recess which serves as a mechanical key for plastering or rendering.

keyed joint (1) A joint between two timbers which employs a key to ensure its security. (2) The concave pointing of a mortar joint.

keyhole saw A thin, narrow-bladed saw used to cut holes in panels or other surfaces. Also called a "hole saw".

keying (1) A process used to add strength to mitered joints. (2) Fastening or fixing in position in a notch or other recess.

keying in The tying in or bonding of a brick or block wall to an existing one.

key plan A small plan which depicts the units in a layout.

key plate An escutcheon.

keystone The usually wedge-shaped uppermost, hence last, set stone or similar member of an arch, whose placement not only completes the arch but also binds

or locks its other members together.

key switch An on-off switch in an electric circuit, which is actuated by a removable key rather than a toggle or button.

key valve A valve operated by a key rather than a handle or lever.

keyway (1) A recess or groove in one lift or placement of concrete which is filled with concrete of the next lift, giving shear strength to the joint. Also called a "key". (2) In a cylindrical lock, the aperture that receives and closely engages the key for its entire length, unlike a keyhole of a common lock. (3) A key-accepting groove in a shaft, pulley, sprocket, wheel, etc.

kibble A bucket-like device in which material, water, tools and/or men are raised from a shaft.

kick (1) In brick, a shallow depression, fray, or panel. (2) The raised fillet of a brick mold which forms the frog. (3) The pitch variation between patent glazing and the surrounding roof.

kicker (1) A wood block or board attached to a formwork member in a building frame or formwork to make the structure more stable. In formwork, a kicker acts as a haunch to take the thrust of another member. Sometimes called a "cleat". (2) A catalyst. (3) An activator, as the hardener for a polyester resin.

kicker plate A timber used to anchor a stair to concrete.

kicking piece A short timber attached to a wale for absorbing the thrust of a raking shore.

kickout (1) In excavation work, the accidental release or failure of a shore or brace. (2) in a downspout, the (usually lowest) section which directs the flow away from a wall.

kickpipe A short length of pipe that provides protection for an

Kitchen Cabinet

electric cable where it protrudes from a floor or deck.

kickplate (1) A metal strip or plate attached to the bottom rail of a door for protection against marring, as by shoes. (2) A plate, usually metal, used to create a ridge or lip at the open edge of a stair platform or floor, or at the back edge or open ends of a stair tread.

kick rail Used primarily in institutions, a usually short rail affixed near the bottom of a door to facilitate its opening by kicking.

kick strip See **kicker**.

kill (1) To terminate electrical current from a circuit. (2) To shut off an engine. (3) To prevent resin from bleeding through paint on wood by the preliminary application to knots of a shellac or other resin-resistant coating.

killesse A grooved or channeled piece of wood, such as one in which a frame slides.

kiln A furnace, oven, or heated enclosure for drying (wood), charring, hardening, baking, calcining, sintering, or burning various materials.

kiln dried (1) Control-dried or seasoned artificially in a kiln. (2) Lumber that has been seasoned in a kiln to a predetermined moisture content.

kiln run Descriptive of bricks or tiles from one kiln which have not been sorted or graded for size or color variation.

kiloampere A unit of electric current equal to 1,000 amperes.

kilovolt A unit of potential difference equal to 1,000 volts.

kilovolt-ampere A unit of apparent power equal to 1,000 volt-amperes.

kilowatt A measurement or unit of power equal to 1,000 watts or approximately 1.34 horsepower.

kilowatt-hour A unit of electrical energy consumption equal to 1,000 watts operating for one hour.

king bolt A vertical tie rod which takes the place of the king post of a truss.

king closer A rectangular brick, one corner of which has been removed diagonally to leave a 2" end, and which functions as a closer in brickwork.

king pile In strutted sheetpile excavation, a long guide pile driven at the strut spacing in the center of the trench before it is excavated.

kingpin A vertically mounted swivel, pivot, or hinge pin usually supported both above and below.

king post In a roof truss, a member placed vertically between the center of the horizontal tie beam at the lower end of the rafters and the ridge, or apex of the inclined rafters.

kiosk (1) A small gazebo. (2) A small, free-standing structure either open or partially enclosed, where merchandise is displayed, advertised, or sold.

kip (1) A kilopound, or 1,000 pounds. (2) Slang term for a bunkhouse on a construction site.

kiss mark Marks on the faces of bricks where they were in contact with one another during their firing in a kiln.

kitchen cabinet In a kitchen, a case or box-type assembly, or similar cupboard-like repository, having shelves, drawers, doors, and/or compartments, and used primarily for storing utensils, cutlery, food, linen, etc.

kitchenette A small kitchen.

kite A sheet of kraft paper applied to a sheet of coated roofing during manufacturing to measure the weight of the granules applied to the surface of the roofing.

kite winder The center tread on a staircase winder with three steps. A kite winder is so named because its shape resembles that of a kite.

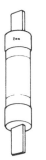

Knife-blade Fuse

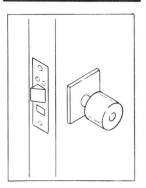

Knob

Knockout

knapping hammer A steel hammer whose head design may vary, but which is used for breaking and shaping stone, splitting cobbles, etc.

knee (1) A naturally or artificially bent piece of wood, as used for a brace or haunch. (2) A sharp, right-angled bend in a pipe. Also called an "elbow". (3) A convex handrail.

knee brace A brace between vertical and horizontal members in a building frame or formwork to make the structure more stable. In formwork, it acts as a haunch.

kneeler (1) The pattern-breaking stone or brick at the point where a normal masonry wall changes and begins to form the curve or angle of an arch or vault. (2) In a gable, the sloped-top, level-bedded stone which supports the inclined coping.

knee rafter A rafter whose lower end is bent downward to rest more firmly against a wall. Sometimes called a "knee piece".

knee wall A wall that shortens the span of the roof rafters by acting as a knee brace, in that it supports the rafters at some intermediate point along their length.

knife-blade fuse A type of cartridge fuse in which the metal blades at each end of the cylinder make contact with the fuse within.

knife consistency A compound whose degree of firmness makes it suitable for application with a putty knife.

knife switch A type of electric switch designed with a hinged or removable blade which enters or embraces the contact clips.

knob (1) A usually round or somewhat spherical handle by which a latch, lock, or other device is operated. (2) Any similarly shaped ornament.

knob bolt A door lock whose bolt is operated not by a key, but by a knob or thumb piece on either or both sides of the door.

knob latch A door latch whose spring bolt is operated not by a key, but by a knob on one or both sides of the door.

knob lock A door lock whose spring bolt is operated by one or more knobs, but whose dead bolt is actuated by a key.

knob rose The usually raised round plate which is attached to a door face so as to surround a hole in the door and form a knob socket.

knob shank The stem of a doorknob, into whose hole or socket the spindle is received and fastened.

knob top The usually round or somewhat spherical terminal end of a knob which is grasped by the hand and turned.

knocked down Descriptive of precut, prefitted, and premeasured, but unassembled consruction components, such as might be delivered to a job for on-site assembly.

knocked down frame A door frame which comes from the manufacturer in three or more parts.

knocker A hinged, usually metal fixture on the exterior face of a door, used for striking or knocking.

knockout A prestamped, usually circular section in an electrical junction box, panel box, etc., which can be easily removed to provide access for a fitting or raceway cable.

knot (1) The hard, cross-grained portion of a tree where a branch meets the trunk. (2) An architectural ornament of clusters of leaves or flowers at the base of intersecting vaulting ribs. (3) Intentional or accidental compact intersection(s) of rope(s) or similar material.

knot brush A rather thick brush whose bristles or fibers are bunched into one, two, or three round or oval knots.

knot cluster A compact grouping of two or more knots surrounded by deflected wood fibers or contorted grain.

knot sealer Any sealer, such as shellac, used to cover knots in new wood to prevent sap or resin bleed-through.

knotting See **knot sealer**.

knotty pine Pine wood sawn so as to expose firm knots as an appearance feature. Knotty pine is used for interior paneling and cabinets.

knuckle One of the enlarged, protruding, cylindrical parts of a hinge through which the pin is inserted.

knuckle joint (1) A hinged joint by which two rods are connected.

knurling Very small ridges or beads as machined on a surface to facilitate gripping.

kraft paper A strong brown wrapping paper made from sulfate wood pulp and sometimes asphalt or resin-impregnated for moisture resistance when used in construction.

K truss A truss in which the arrangement of panels, chords, and web members resembles the letter K.

k value The thermal conductivity of a substance or material.

kyanize Soaking or impregnating wood with a solution of mercuric chloride to preserve it against decay.

ABBREVIATIONS

The abbreviations listed below are those most commonly used in the construction industry. Alternative forms (usually nonstandard) are shown in parentheses.

l labor only, left, length, liter, long, lumen

L Lambert, large

L&CN lime and cement mortar

L&H light and heat

L&L latch and lock

L&O lead and oil (paint)

L&P lath and plaster

Lab. labor

LAG lagging

LAM laminated

LAT latitude, lattice

Lath lather

LAV lavatory

lb, lbs pound, pounds

lb/hr pounds per hour

lbf/sq in pound-force per square inch

lb/LF pounds per linear foot

Lbr lumber

LCL less-than-carload lot

LCM least common multiple, loose cubic meter

LCY loose cubic yard

LDG landing

ld load

LE leading edge

LECA light expanded clay aggregate

LEMA Lighting Equipment Manufacturers' Association

lf lightface, low frequency, lineal foot, linear foot

LG liquid gas

lg large, length, long

lgr longer

lgt lighting

lgth length

LH left hand, long-span, high strength bar joist

LIC license

lin linear, lineal

lin ft linear foot, linear feet

lino linoleum

LJ obsolete designation for long span standard strength bar joist

LL live load

LL&B latch, lock, and bolt

LLD lamp lumen depreciation

lm lumen

LM lime mortar

lm/sf lumen per square foot

lm/W lumen per watt

lng, Lng lining

LOA length over all

log logarithm

LP liquid petroleum, low pressure

LPF low power factor

LPG liquid petroleum gas

LR living room

LS (1) left side, (2) loudspeaker

LT long ton, light

Lt Ga light gauge

LTL less than truckload lot

Lt Wt lightweight

LV low voltage

LW low water

LWC lightweight concrete

LWM low water mark

DEFINITIONS

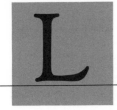

Lag Bolt/Lag Screw

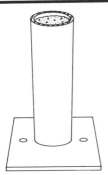

Lally Column

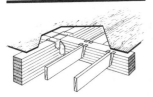

Laminated Beam

labeled Descriptive of doors, windows, frames, and other building components that carry certification of approval from a recognized testing laboratory based on fire tests conducted on identical materials and articles.

labeled door A door that carries a certified fire-rating issued by Underwriters' Laboratories, Inc.

labeled frame A door frame that conforms to standards and tests required by Underwriters' Laboratories, Inc., and has received its label of certification.

labeled window A fire-resistant window that conforms to the testing standards of Underwriters' Laboratories, Inc., and bears a label designating its fire rating.

labor and material payment bond A contractor's bond in which a surety guarantees to the owner that the contractor will pay for labor and materials used in the performance of the contract. The claimants under the bond are defined as those having direct contracts with the contractor or any subcontractor. A labor and material payment bond is sometimes referred to as a **payment bond**.

labyrinth A maze of passageways or paths.

laced corner A method of laying shingles on sidewalls. The corner shingles of each course are laid alternately on the faces of the two walls.

lacing A system of members used to connect the different elements of a composite column or girder in such a way that they structurally act in unison.

lacquer A glossy enamel, composed of volatile solvents and diluents, that evaporates and dries quickly upon application to a surface.

ladder core A hollow structure of wood or insulation board used as the core of interior doors and built with strips running vertically or horizontally through the core area.

ladder jack scaffold A simple scaffolding system that uses ladders for support. This form of scaffold is not strong enough for most construction uses.

lag bolt A threaded screw or bolt with a square head. Also called a "coach screw."

lagging (1) Heavy wood boards used to line the sides of excavations and prevent cave-ins. (2) Preformed insulation for pipes and tanks.

lag screw See **lag bolt**.

laitance In concrete, a weak, crumbly, and dusty surface layer caused by excessive water that has bled to the surface and subsequently weakened it. Overworking the surface during finishing can aggravate the problem. If laitance forms between pours, it must be brushed and washed away.

Lally column A trade name for a pipe column from 3 to 6 inches in diameter, sometimes filled with concrete.

laminate (1) To form a product or material by bonding together several layers or sheets with adhesive under pressure and sometimes with nails or bolts. (2) Any material formed by such a method.

laminated beam A straight or arched beam formed by built-up layers of wood. The method of lamination may be by gluing under pressure, by mechanical

293

nailing or bolting, or a combination.

laminated glass A shatter-resistant safety glass made up of two or more layers of sheet glass, plate glass, or float glass bonded to a transparent plastic sheet. Also called "safety glass" or "shatterproof glass."

laminated plastic Layers of synthetic resin-coated or resin-impregnated filler materials bonded together into a single piece by application of heat and pressure.

laminated wood Any of several products formed by built-up layers (plies) of wood. Thin wood veneers may be laminated to a wood subsurface, several plies may be laminated together to form plywood, or thicker pieces may be used to form structural members such as beams or arches.

laminboard A compound board consisting of a core of small strips of wood glued together and covered by veneer faces.

lamp ballast See **ballast**.

lampblack A black pigment composed of carbon from the soot of burning oil.

Lamp Post

lamp post A supporting device, for an external light or luminaire, with wiring attachments concealed inside and with outside attachments for the bracket.

land-clearing rake A device outfitted with blades and attached to the front of a tractor to cut, collect, and remove brush from the site of proposed construction.

land drain See **agricultural pipe drain**.

landing newel A newel positioned on a stair landing or at any point where stairs change direction. Also called an "angle newel."

landing tread That board or

Landing Tread

portion of a stair landing that is closest to the next step down. The tread has an appearance and dimensions similar to the other treads, but it is actually part of the landing, and not a true tread.

land reclamation Gaining land from a submerged or partially submerged area by draining, filling, or a combination of these procedures.

landscape architect A person whose professional specialty is designing and developing gardens and landscapes, especially one who is duly licensed and qualified to perform in the landscape architectural trade.

land survey See **boundary survey** and/or **survey**.

land surveyor A person whose occupation is to establish the lengths and directions of existing boundary lines on landed property, or to establish any new boundaries resulting from division of a land parcel.

land tie See **deadman**.

land tile Clay tile laid with open joints and usually surrounded by porous materials. See also **agricultural pipe drain**.

land-use analysis A systematic study of an area or region that documents existing conditions and patterns of use, identifies problem areas, and discusses future options and choices. Part of the general planning process, such analysis, might cover topics such as traffic flow, residential and commercial zoning, sewer services, water supply, solid-waste management, air and water pollution, or conservation areas. In short, any factors that could affect how particular areas of land should or should not be used.

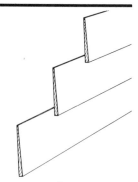

Lap

lap In construction, a type of joint in which two building elements are not butted up against each other, but are overlapped, part of one covering part of the other.

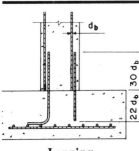

Lapping

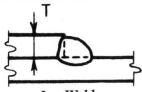

Lap Weld

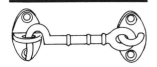

Latch

Typical examples include roof and wall shingles, clapboard siding, welded metal sheets or plates, and concrete reinforcing bars lapped together at their ends.

lap cement Asphalt used in roll roofing as an adhesive between the laps.

lapped dovetail An angle drawer dovetail in carpentry commonly used in constructing the front of a drawer, on which the pegged end of one member does not pass completely through the thickness of the adjacent perpendicular members.

lapped tenons Overlapping tenons that enter a mortise from opposite sides.

lapping The overlapping of reinforcing bars or welded wire fabric for continuity of stress in the reinforcing when a load is applied.

lap-riveted The riveting together of two metal members or plates where they have been deliberately overlapped, thus forming a lap joint.

lap seam The same as a lap joint, but typically used to refer to sheet metal, and sometimes plates, that are welded, soldered, or riveted at the overlapping joint.

lap splice In concrete construction, the simplest method for providing continuity of steel reinforcement. Ends are overlapped a specified number of diameters, usually no fewer than 30, and tied with wire.

lap weld A weld used to join two pieces of metal at their common lap joint.

larch, tamarack A heavier than average softwood from a coniferous tree, characterized by fine texture and hard, strong straight-grained consistency.

larmier, lorymer A specific drip strip or molding that is part of a cornice. Also called a corona. By projecting from the surrounding cornice, it catches rain and forces it to drip off away from the wall.

last in, first out, (LIFO) A method of accounting for inventory in which it is assumed that goods bought last are sold first. This allows automatic updating of inventory values.

latch A fastening device for a door or window, usually operable from both sides and built without a dead bolt or provisions for locking with a key.

latch bolt In a door or window, a bolt that is springloaded and beveled. When the door or window is closed, the bevel forces the bolt into the member, only to be released when in the fully closed position as the spring forces the bolt into a notch in the frame.

latchkey A key used to open or raise the latch on a door.

latchstring A string that is attached to the inside latch of a door and passed through a hole above the latch to the outside in order to operate the latch from outdoors.

latent defect A defect in materials or equipment that could not have been discovered under reasonably careful observation. A latent defect is distinguished from a patent defect, which may be discovered by reasonable observation.

lateral buckling The failure of any structural column, wall, or beam which has undergone excessive side-to-side (lateral) deflection, movement, or twist.

lateral load See **wind load** and/or **earthquake load**.

lateral reinforcement See **reinforcement, lateral**.

lateral sewer A sewer that discharges into another sewer or branch, but is engineered without any other common tributary to it.

KEY

Latest Start Date

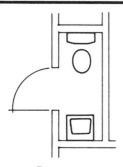

Lavatory

lateral support Any bracing, temporary or permanent, that provides greater support in resisting side-to-side (lateral) forces and deflections. Floor and roof members typically provide lateral support for walls, columns, and beams. Vertical pilasters or secondary walls may also provide support.

latest event occurrence time In the CPM (Critical Path Method) of construction scheduling, the final deadline by which a particular item of work must be completed in order to avoid delaying the entire project.

latest finish date In the CPM (Critical Path Method) of scheduling, a completion deadline for a particular activity. Work performed after this deadline will result in project delay.

latest start date In the CPM (Critical Path Method) of scheduling, the deadline for starting a particular activity. A late start will throw off the schedule and delay the project.

latex (1) The sap of a rubber tree. (2) An emulsion in water of very fine particles of rubber or plastic.

latex foam Sponge rubber manufactured with a latex base.

latex paint A paint with a latex binder, usually a polymeric compound, characterized by its ability to be thinned or washed from applicators with water.

lathe A machine used to shape circular pieces of wood, metal, or other material. The stock is rotated on a horizontal axis while a stationary tool cuts away the unwanted material or creates ornamental turned work.

lath hammer A hammer used chiefly for cutting and nailing wood lath, designed with a nail-driving hammer head, as well as a hatchet blade with a lateral nick that is used for pulling nails out.

lathing The process of assembling or placing laths.

lathing hammer See **lath hammer**.

lath laid-and-set A two-coat method of plastering walls and ceilings. The first coat is called "laying" and is scratched so as to provide a rough bonding surface.

latrine A public toilet or privy.

latrobe A type of heater located inside a fireplace. Rooms above may be heated by hot air, but the room containing the stove is heated by direction radiation.

lattice Typically, a diagonal network or grid of strips of material. Sallice is often used as ornamental screening, or to provide privacy.

lattice truss A structural truss in which the web is a lattice work of diagonal members. Also known as a "lattice girder" or "lattice beam."

latticework Any item or member formed by the repetitive crossing of diagonal, thin strips, often of wood or metal.

lauan Philippine mahogany.

laundry chute A shaft running between the upper and lower floors of a building and used to convey soiled laundry to the lower level by gravity.

lavatory (1) A basin, with running water and drainage facilities, used for washing the face and hands. (2) A room with a wash basin and a toilet, but no bathtub. (3) A room containing a toilet or water closet.

lay A term used to define the direction of twist of the strands or wires in wire rope. The strands or wires have either a right-hand or left-hand lay.

laying line, laying guide Lines printed on felt or roll roofing as a guide for the amount of lap required.

laying out Marking of materials in preparation for work, showing where they are to be cut.

layout A design scheme or plan showing the proposed arrangement of objects and spaces spaces within and outside a structure.

lay-up (1) The reinforcing material that is placed into position in the mold during the manufacturing of reinforced plastics. Also, the resin-impregnated reinforcement. (2) The method or process of assembling veneers to manufacture plywood.

lazy susan A revolving circular shelf or tray, often used in cabinets to utilize space more efficiently at blind corners.

L-beam A beam whose section has the form of an inverted L, usually occurring in the edge of a floor, of which a part forms the top flange of the beam.

L-column The portion of a precast concrete frame composed of the column, the haunch, and part of the girder.

leaching The process of separating liquids from solid materials by allowing them to percolate into surrounding soil.

leaching cesspool An underground storage tank or chamber for domestic wastes. The sides are made porous with small holes so that solids are retained but liquids may leach out into the surrounding earth.

leaching field Same as **absorption field**.

leaching pit See **leaching well**.

leaching well, leaching pit Like a cesspool, a pit with porous walls that retain solids but permit liquids to pass through.

lead (1) A soft, dense heavy metal easily formed and cut. Historically, lead was used for flashing and for the joints in stained glass windows. (2) End sections of a masonry wall, usually at the corners, which are built up, in steps, before the main part of the wall is begun. Also, a string stretched between these end sections that serves as a guide for the rest of the wall.

lead chromate One of a number of opaque pigments that range from orange to yellow in color and have strong tinting properties.

lead-covered cable An electric cable protected from damage and excess moisture by a lead covering.

leaded light A window whose diamond-shaped or rectangular glass panes are set in lead cames.

leader (1) In a hot-air heating system, a duct that conveys hot air to an outlet. (2) A downspout.

leader head That part of a drainpipe assembly that is placed at the top of a leader and serves as a catch basin to receive water from the gutter.

lead glazing A **leaded light**.

leading edge The vertical edge of a hinged, swinging door or window which is opposite the hinged edge and in proximity to the knob or latch.

lead joint A joint in a water pipe, such as a bell-and-spigot joint, into which molten lead is poured.

lead-lag ballast A ballast that serves to cut down on the stroboscopic effect of two fluorescent lamps, one of which is on a leading current and the other on a lagging current.

lead-lined door A door with lead sheets lining its internal core to prevent the penetration of x-ray radiation.

lead-lined frame A frame that is used with lead-lined doors and is itself lined internally with lead sheets to prevent penetration of x-ray radiation.

lead paint A paint having white lead as one of its constituents.

Lead Shield

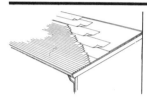

Lean-To Roof

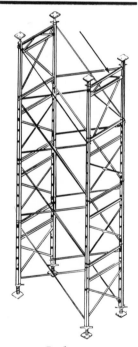

Ledger

lead plug (1) A cylinder of lead placed inside a hole in a masonry, plaster, or concrete wall. A screw or nail driven into it will then be firmly held in place. (2) In stone masonry, a piece of lead that holds adjacent stones together. Grooves are cut in both rock surfaces to be joined, and molten lead is poured into it.

lead roof A flat roof with a surface of sheet lead.

leads A collective term for short sections of electric conductors, generally insulated.

lead-sheathed cable See **lead-covered cable**.

lead shield A lead sleeve used to provide anchorage for expansion bolts or screws. Similar to a **lead plug**.

lead slate, copper slate, lead sleeve A cylinder of sheet lead or sheet copper that surrounds a pipe at the point where it passes through a roof, ensuring a watertight intersection.

lead sleeve Same as **lead slate**.

lead soaker A piece of lead sheeting that forms a weathertight joint at the intersection of a roof and of any vertical wall that passes through the roof at a hip or valley.

lead spitter The tapered part of a drainpipe assembly that connects a lead gutter with a downpipe.

lead tack (1) A lead strip used to attach a lead pipe to some means of support. (2) A lead strip placed along the edge of metal flashing. One side of the strip is attached to the structure, while the other side is folded over the free edge of the flashing.

lead wool Fine strands of lead formed into a wool-like consistency, often used as caulking at the joints of pipes.

leaf See **wythe**.

lean concrete Concrete of lower than usual cement content.

lean mix, lean mixture (1) A mixture of concrete or mortar with a relatively low cement content. (2) A plaster with too much aggregate and not enough cement, which thus renders it unworkable.

lean mortar A mortar with a low cement content, which makes it sticky, overly adherent to the trowel, and difficult to apply.

lean-to roof A roof with one pitch, supported at one end by a wall extending higher than the roof.

lease A contract that transfers the right of possession and use of buildings, property, vehicles, and items of equipment for a time agreed upon in the contract, in return for rent as monetary compensation.

leasehold A tenure established by lease. Also, real estate that is held under a lease.

ledge (1) A molding that projects from the exterior wall at a building. (2) A piece of wood nailed across a number of boards to fasten them together. (3) An unframed structural member used to stiffen a board or a number of boards or battens. (4) Bedrock.

ledged-and-braced door A batten door outfitted with diagonal bracing for extra strength.

ledger (1) A horizontal framework member that carries joists and is supported by upright posts or by hangers. (2) A slab of stone laid flat, such as that over a grave. (3) A horizontal scaffold member, positioned between upright posts, on which the scaffold planks rest.

ledger board (1) One of multiple boards attached horizontally across a series of vertical supports, as in the construction of a fence. (2) A ribbon strip.

ledger plate (1) Same as **ledger strip**. (2) Same as **ledger, 1**.

ledger strip On a beam that carries joists flush with its upper edge, the strip of wood attached along

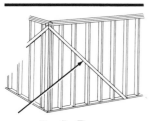

Let-In Brace

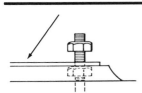

Leveling Plate

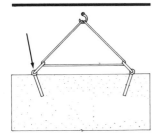

Lewis Bolt

the bottom edge of the beam that serves to seat the joists and to support them.

left-hand door See **hand**.

left-hand lock A lock designated for use on a left-hand door.

left-hand reverse door See **hand**.

left-hand stairway A stairway on which the handrail is positioned on the left-hand side in the direction of ascent.

length The longest dimension of an object.

lessee A person who receives use and possession of property by lease.

lessor The party who grants the use and possession of property by lease.

let in In joinery, to fasten a timber securely in place by inserting or embedding it in another.

let-in brace A diagonal brace inserted or let in to a stud.

letter chute Same as **mail chute**.

letter form of agreement, letter agreement A letter stating the terms of an agreement between addressor and addressee, usually prepared to be signed by the addressee to indicate acceptance of those terms as legally binding.

letter of intent A letter signifying an intention to enter into a formal agreement, usually setting forth the general terms of the agreement.

letting of bid Same as **bid opening**.

level (1) A term used to describe any horizontal surface that has all points at the same elevation and thus does not tilt or slope. (2) In surveying, an instrument that measures heights from an established reference. (3) A spirit level, consisting of small tubes of liquid with bubbles in each. The small tubes are positioned in a length of wood or metal which is hand held and, by observing the position of the bubbles, used to

find and check level surfaces.

level control Bench marks or other devices used to identify points of known elevation on a project site.

leveling The procedure used in surveying to determine differences in elevation.

leveling course Same as **asphalt leveling course**.

leveling plate A plate set to an elevation used for bearing.

leveling rod, leveling staff A graduated straight rod used in construction with a leveling instrument to determine differences in elevation. The rod is marked in feet and fractions of feet, and may be fitted with a moveable target or sighting disc. See also **New York leveling rod** and **Philadelphia leveling rod**.

leveling rule A long level used by plasterers to detect irregularities in the height of horizontal surfaces measured at various points.

leveling staff Same as **leveling rod**.

lever arm In a structural member, the distance from the center of the tensile reinforcement to the center of action of the compression.

lever tumbler In a lock, a type of pivoted tumbler.

lewis A metal device used to hoist heavy units of masonry. A lewis is equipped with a dovetailed tenon that is made in sections and fitted into a corresponding recess cut into the piece of masonry to be moved.

lewis bolt (1) A bolt shaped into a wedge at its end, which is inserted into a prepared hole in a heavy unit or stone and secured in place with poured concrete or melted lead. (2) An eyebolt inserted into heavy stones and used in the manner of a lewis to lift and move the stones.

lewis hole A dovetailed hollow cut into a stone block, column, or heavy piece of masonry to receive a lewis.

299

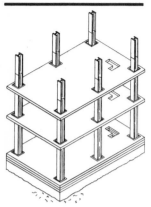

Lift Slab

Light

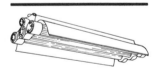

Lighting Fixture

L-head The top of a shore formed with a braced horizontal member projecting from one side and forming an inverted L-shape assembly.

liability insurance Insurance that protects the insured against liability on account of injury to the person or property of another. See also **insurance**.

licensed architect See **architect**.

licensed contractor A person or entity certified by a governmental authority, where required by law, to engage in construction contracting.

licensed engineer See **professional engineer**.

lien See **mechanic's lien**.

lien waver See **waiver of lien**.

life-cycle costing The determination of the value of a system, such as roof covering, amortized over the projected life of the system, as opposed to the value determined by the initial cost only. Life-cycle costs include such costs as service and maintenance.

lift (1) The concrete placed between two consecutive horizontal construction joints, usually consisting of several layers or courses, such as in slip forming. (2) A metal handle or projection from the lower sash in a hung window, used as an aid in lifting the sash. (3) The amount of grout, mortar, or concrete placed in a single pour. (4) A British term for elevator.

lifting block A combination of pulleys or sheaves that provide a mechanical advantage for lifting a heavy object.

lifting pin Same as a **lewis**.

lift joint A surface at which two successive concrete lifts meet.

lift slab A method of concrete construction in which floor and roof slabs are cast on or at ground level and hoisted into position by jacking. Also, a slab that is a component of such construction.

lifts, tiers The number of frames of scaffolding erected one above each other in a vertical direction.

ligger (1) A horizontal timber that supports floor boards or scaffolding. (2) In thatched roofs, a strip of wood placed along the ridge.

light (1) A man-made source of illumination, such as an electric light. (2) A pane of glass.

light bulb See **incandescent lamp**

light dimmer See **dimmer**.

light framing Lumber that is 2 to 4 inches thick, 2 to 4 inches wide, and is graded Construction, Standard, or Utility No.3.

lighting A system for providing illumination to an area.

lighting fixture See **luminaire**.

lighting outlet An electrical outlet that serves to accommodate the direct connection of a lighting fixture or of a lamp holder and its pendant cord.

lighting panel An electric panel housing fuses and circuit breakers, that serve to protect the branch circuits of lighting fixtures.

lighting unit See **luminaire**.

light loss factor In illumination calculations, an adjustment factor that estimates losses in light levels over time due to aging of the lamp, dirt on the room surfaces, and other causes. Losses corrected by lamp replacement or cleaning are termed recoverable. Non-recoverable losses would be due to deterioration of the fixture or voltage drops.

lightning arrester A device that connects to and protects an electrical system from lightning and other voltage surges.

lightning conductor, lightning rod A cable or rod built of metal that protects a building from

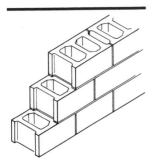

Lightweight Block

Limestone

lightning by providing a direct link to ground.

light-reflective glass Same as **reflective glass**.

light source A See **standard source A**.

light source B See **standard source B**.

light source C See **standard source C**.

lightweight aggregate See **aggregate, lightweight**.

lightweight block A cement masonry unit manufactured using lightweight aggregate and often used to reduce the weight of partitions.

lightweight concrete Concrete of substantially lower unit weight than that made using gravel or crushed stone aggregate.

light well An inside shaft with an open top through which light and air are conveyed from the outdoors to windows opening on the shaft.

lignin (1) A substance that occurs naturally in wood and joins with celluloses to make up the chief constituents of wood tissue. (2) A crystalline byproduct of paper pulp, used in manufacturing plastics, wood chipboard, and protective chemical coatings that prevent corrosion.

lime Specifically, calcium oxide (CaO). Also, a general term for the various chemical and physical forms of quicklime, hydrated lime, and hydraulic hydrated lime.

lime-and-cement mortar A lime, cement, and sand mortar used in masonry and cement plaster. In addition to imparting a favorable consistency to the mix, the lime also increases the flexibility of the dried mix, thus limiting cracks and minimizing water penetration.

lime concrete A lime, sand, gravel, and concrete mix made without portland cement. Lime concrete is found in older structures, but is no longer in general use.

lime mortar An uncommon mix of lime putty and sand that is not often used because it hardens at a very slow rate.

lime putty A thick lime paste used in plastering, particularly for filling voids and repairing defects.

limestone A sedimentary rock composed mostly of calcium carbonate, calcium, or dolomite. Limestone can be used as a building stone, crushed into aggregate, crushed for agricultural lime, or burned to produce lime.

limewash A milk-like mixture of water and lime used to coat the exterior or interior surfaces of a structure. Also called **whitewash** or **whiting**.

limit control A safety device for a variety of mechanical systems that detects unsafe conditions, sounds an alarm, and shuts off the system.

limit design Any of a number of structural design methods based on limits related to stability, elasticity, fatigue, deformation, and other structural criteria.

limit of liability The maximum amount that an insurance company agrees to pay in case of loss, damage, or injury.

limit switch (1) An electrical switch that controls a particular function in a machine, often independently of other machine functions. (2) A safety device, such as a switch that automatically slows down and stops an elevator at or near the top or bottom terminal landing.

lineal foot A straight-line measurement of 1 foot, as distinguished from a cubic foot volume or a square foot area.

linear diffuser An air outlet, usually with a width no greater than 4 inches, wherein the length of the outlet to its width exceeds a ratio of 10:1.

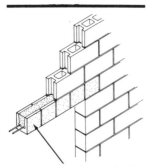

Lintel

Lintel Block, U-Block

linear measurement A unit or system of units for measuring length: 12 inches = 1 foot, 3 feet = 1 yard, 1 yard = 0.9144 meters, 1 mile = 5,280 feet. Also called a "long measure."

linear prestressing Prestressing as applied to linear structural members, such as beams, columns, etc.

line drawing A graphic representation made with lines and solids, as opposed to one made with tone gradations, such as a photograph or rendering.

line drilling In the blasting of rock, the boring of a series of holes along the desired line of breakage. Holes are spaced several inches apart to create a plane of weakness.

line drop A decrease in voltage caused by the resistance of conductors in an electric circuit.

line level A spirit level used in excavation and pipe laying. Each end of the level has a hook used to hang it from a horizontal line.

line of levels A series of differences in elevation as measured and recorded by surveyors.

line of sight The line extending from a telescope or other long-distance sighting device, along which distant objects can be viewed. Same as a sight line.

line pin A metal pin used in masonry work to support a horizontal string or line. The mason positions the line and then uses it as a guide in maintaining proper alignment of the work.

line pipe A welded or seamless pipe typically used to convey gas, oil, or water.

liner (1) Extra stone bonded or otherwise attached to the back face of thin stone veneers. The purpose is to add strength and to create a deeper joint.

lining Any sheet, plate, or layer of material attached directly to the inside face of formwork to improve or to alter the surface texture and quality of the finished concrete.

lining paper A water-proof or water-resistant building paper used in frame buildings and placed under siding and roofing shingles.

link dormer A dormer that is formed around a chimney. Also, a dormer that connects one roof area to another.

linked switch A series of mechanically connected electrical switches designed to act simultaneously or sequentially.

link fuse A type of exposed fuse attached to electrically insulated supports.

linoleum An inexpensive form of resilient floor covering that is manufactured of ground cork and oxidized linseed oil. Linoleum is applied to a coarse fabric backing and possesses a low resistance to staining, dents, and abrasion.

linseed oil A drying oil processed from flaxseed and used in many paints and varnishes.

lintel A horizontal supporting member, installed above an opening such as a window or a door, that serves to carry the weight of the wall above it.

lintel block, U-block A special U-shaped concrete block used with other blocks to form a continuous bond beam or lintel. The opening faces up and is filled with grout with steel reinforcing.

lintel course A course of stone masonry set level to a lintel, but different from the rest of the wall by virtue of size or finish.

lip A rounded, overhanging edge or member.

lip union A pipe connection with a lip on the inside to keep the gasket from being forced into the pipe.

FIRST FLOOR PLAN

SECOND FLOOR PLAN

Living Unit

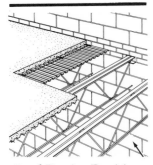

Load-Bearing Partition

liquefaction The sudden failure of a loose soil mass due to total loss of shearing resistance. Typical causes are shocks or strains that abruptly increase the water pressure between soil particles, causing the entire mass to behave similarly to a liquid.

liquid asphaltic material An asphaltic product too soft to be measured by a penetration test at normal temperatures. The material is principally used for pavement surface treatments. See also **liquid roofing**.

liquidated damages A sum established in a construction contract, usually as a fixed sum per day, as the measure of damages suffered by the owner due to failure of the contractor to complete the work within a stipulated time.

liquid indicator A device located in the liquid line of a refrigerating system and having a sight port through which liquid flow may be observed for the presence of bubbles.

liquid limit The water content at which soil passes from the plastic to the liquid state under standard test conditions. The limit is expressed as a percentage of the dry weight of the soil. See also **Atterberg limits**.

liquid line In a refrigeration system, the pipe transporting refrigerant away from the condenser.

liquid petroleum gas (LPG) A general term referring to propane, butane, and other similar hydrocarbons used as fuel.

liquid roofing Any of a number of different liquid or semi-liquid roofing materials used to create a seamless waterproof membrane.

liquid-volume measurement A measurement of grout on the basis of the total volume of solid and liquid constituents.

liquid waste The discharge from any plumbing fixture, area, appliance, or component that does not contain fecal matter.

litmus A piece of test paper containing a chemical indicator that changes color when exposed to liquids. Litmus paper is used to determine acidity or alkalinity values expressed as pH variations.

live (1) Descriptive of a wire or cable connected to a voltage source. (2) A descriptive term for a room with a very low amount of sound absorption.

live-front Descriptive of an item of electrical equipment constructed so that one or more of its "live parts" can be touched from the front of the device.

live load The load superimposed on structural components by the use and occupancy of the building, not including the wind load, earthquake load, or dead load.

live part Any part or component of an electrical device or system that is engineered to function at a voltage level different from that of the earth.

live steam Steam that has not condensed and still retains its energy, such as steam issuing from a boiler or radiator.

living unit A dwelling place or any self-contained area or part thereof that comprises complete living facilities for a family, including space and fixtures for sleeping, cooking, eating, living, bathing, and sanitation.

load (1) The force, or combination or forces, that act upon a structural system or individual member. (2) The electrical power delivered to any device or piece of electrical equipment. (3) The placing of explosives in a hole.

load-bearing partition A partition that can support a load in addition to its own weight. This type of partition is often

Load-Bearing Wall

Loader

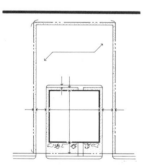

Location Plan

constructed perpendicular to the joists in a structure.

load-bearing tile A form of tile used in masonry walls that is capable of supporting loads superimposed on the wall structure.

load-bearing wall A wall specifically designed and built to support an imposed load in addition to its own weight.

load binder A device used to tighten chains that are holding loads in place on a truck bed.

loader A construction machine used to push or transport earth, crushed stone, or other construction materials. The bucket, or scoop, is located on the front of the vehicle and can be raised, lowered, or tilted.

load factor (1) In structural design, the factor applied to the working load to determine the design ultimate load. (2) In a drainage system, the percentage of the total flow that occurs at a particular location in the system. (3) A ratio of the average air conditioning load on a system to the maximum capacity.

load-indicating bolt A bolt that permits measurement of its tension. Upon tightening, a small projection on the bolt compresses and is measured by a feeler gauge.

loading cycles A calculation of the number of repetitions of load that a structure is expected to support in its lifetime. The calculation is used as a determining criterion in measuring the structure's fatigue strength.

loading dock leveler Typically, an adjustable mechanized platform built into the edge of a loading dock. The platform can be raised, lowered, or tilted to accommodate the handling of goods or material to or from trucks.

loading dock seal A flexible pad installed around the door of a loading dock to form a tight seal

between the receiving doors and the opening of a truck that is backed into the dock.

loading hopper A hopper in which concrete or other free flowing material is placed for loading by gravity into buggies or other conveyances.

loading platform, loading dock A platform adjoining the shipping and receiving door of a building, usually built to the same height as the floor of the trucks or railway cars on which shipments are delivered to and from the dock.

loading ramp A fixed or adjustable inclined surface that adjoins a loading platform and is installed to ease the conveyance of goods between the platform and the trucks or railway cars that transport goods.

loam Soil consisting primarily of sand, clay, silt, and organic matter.

local buckling In structures, the failure of a single compression member. The local failure may cause the failure of the whole structural member.

local lighting Lighting used to illuminate a limited area without significantly altering the illumination of its wider surroundings.

location plan Same as **site plan**.

location survey The establishment of the position of points and lines on an area of ground, based on information taken from deeds, maps, and documents of record, as well as from computation and graphic processes.

lock bevel In a door lock, the angled surface of the latch bolt.

lock corner A corner held together by interlocking construction of adjacent members, such as the dovetail joint on the front panel of a drawer.

Locke level A **hand level**.

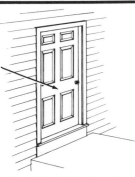

Lock Rail, Lock Stile

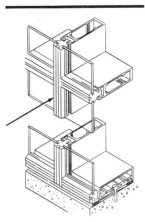

Lock-Strip Gasket, Structural Gasket

lock face The surface of a mortise lock, which remains visible in the edge of a door when the lock is installed.

lock front On a door lock or latch, the plate through which the lock or latch bolt projects.

lock front bevel A lock front that is installed flush with the beveled edge of a door.

locking device In scaffolding, a device used to secure a cross brace to the frame or panel.

lock jamb See **strike jamb**.

lock joint See **lock seam**.

lock keeper The box on a door jamb that accommodates the extending bolt on a lock.

locknut (1) A special nut that locks when tightened so that it will not come loose. (2) A second nut used to prevent a primary nut from loosening.

lock plate See **strike plate** and/or **box strike plate**.

lock rail On a door, the horizontal structural member situated between the vertical stiles at the same height as the lock.

lock reinforcement A metal plate installed inside the lock stile or lock edge of a door and designed to receive a lock.

lock reinforcing unit In a metal door, a metal device that houses and supports a lock.

locksaw A saw used for cutting the seats for locks in doors. The saw is designed with a tapering blade that can be flexibly maneuvered. See also **compass saw**.

lock seam, lock joint In sheet metal roofing, a joint or seam formed by bending the two adjoining edges over in the form of hooks, which are interlocked. The hooks are then pressed down tightly to form a seam.

lockset A complete system including all the mechanical parts and accessories of a lock, such as knobs, reinforcing plates, and protective escutcheons.

lock stile, closing stile, locking stile, striking stile On a door or a casement sash, the vertical member that closes against the jamb of the frame that surrounds it. The stile is located on the side away from the hinges.

lock strike Same as **strike plate**.

lock-strip gasket, structural gasket Typically, a thick and stiff black neoprene glazing gasket that holds and attaches panes of glass to each other or to the surrounding structure. During installation the gasket is tightened by the insertion of a wedge-like strip (the lock-strip) along the entire length of the gasket.

locust The coarse-grained wood of the locust tree, which is used in construction because of its strength, hardness, durability, and resistance to decay.

loess Silty material that is deposited by the wind, but maintains significant cohesion due to the presence of clay or other cementitious materials.

loft (1) Space beneath a roof of a building, most commonly used for storage of goods. (2) In a barn, the upper space at or near the ceiling with an elevated platform on which hay and grains are stored. (3) The upper space in a church or auditorium, sometimes enclosed and cantilevered, which accomodates a pipe organ or area for a choir. (4) The space between the grid and the upper part of the proscenium in a theatre stagehouse. (5) Within a loft building, the unpartitioned upper spaces visible from the floor immediately below. See also **attic** and/or **garret**.

loft ladder See **disappearing stair**.

long-and-short work In rubble masonry, quoins that are

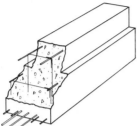

**Longitudinal
Reinforcement**

Long-Radius Elbow

Long Span

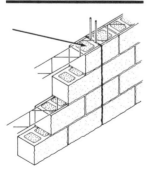

Loose-Fill Insulation

alternately placed horizontally and vertically.

long column In structural design, a column of sufficient slenderness to necessitate a reduction of its load bearing capacity.

long float A concrete finishing float designed to be handled by two men.

longitudinal bar See **longitudinal reinforcement**.

longitudinal bond In masonry, a bond in which a number of courses are laid only with stretchers and used principally for thick walls.

longitudinal bracing Bracing that extends lengthwise or runs parallel to the center line of a structure.

longitudinal joint Any joint parallel to the long dimension of a structure or pavement.

longitudinal reinforcement Steel reinforcement placed parallel to the long axis of a concrete member.

longitudinal shear In structural design, shear forces that act parallel to the longest axis of a member.

long-life lamp An incandescent lamp that has a lower luminous output than standard lamps of equal wattage, but a longer design life than the value set for lamps of its general class.

long nipple, space nipple A nipple with a long unthreaded middle section.

long-radius elbow In plumbing, a pipe elbow with a larger radius than is standard. The elbow is designed to mitigate losses from friction and to facilitate the flow of liquids through the pipe.

long screw A pipe nipple, usually measuring 6 inches long, with one thread that is longer than average.

long span (1) The distance between supports in a structure, usually spanned by a truss or heavy timber. (2) A logging

operation where logs are yarded over a long distance.

long ton A unit of weight equal to 2,240 pounds (1,016 kilograms).

loop See **loophole** and/or **circuit vent**.

loophole An aperture in a wall or parapet to provide air, light, and a view of the outside. See **arrow loop**.

looping in In interior electrical wiring, the connection of an outlet by two conductor cables, one to and one from the outlet. Splices (junction boxes) are thus avoided, but more wire is used.

loop vent In plumbing, a venting configuration for multiple fixtures, as in a public restroom. The vent pipe is connected to the waste branch in only two places, before the first and last fixtures. The fixtures are not individually vented. The two vents are connected together, in a loop, and the loop is then connected to the vent stack.

loose cubic yard (meter) A unit of measure with which to express the volume of loose soil, rock, or blasted earth material.

loose-fill insulation Any of several thermal insulation materials in the form of granules, fibers, or other types of small pieces that can be poured, pumped, or placed by hand.

loose-joint hinge A door hinge that can be separated by lifting. The door can thus be removed without unscrewing the hinges.

loose knot A knot on a piece of lumber that is not fixed firmly and is likely to fall out.

loose lintel A lintel that is placed across a wall opening during construction to support the weight of the wall above, but which is not attached to another structural member.

loose material Soil, rock, or earth materials in loose form, whether

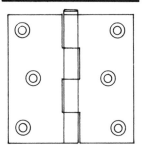

Loose-Pin Hinge

Loose Yards

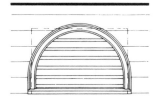

Louver

Louver Window

blasted or broken by artificial or natural means.

loose-pin hinge A hinge, usually for a door, that can be separated by the removal of a vertical pin.

loose-tongue In a timber joint, the piece of wood that extends into the opposite member, thus strengthening a tenoned frame. Same as a cross tongue.

loose yards A term defining the cubic measurement of earth or blasted rock after excavation, as when loaded on a truck.

Los Angeles Abrasion Test A test for abrasion resistance on concrete aggregates.

loss of prestress In prestressed concrete, the reduction in prestressing force which results from the combined effects of strain in the concrete and steel, including slip at anchorage, relaxation of steel stress, frictional loss due to curvature in the tendons, and the effects of elastic shortening, creep, and shrinkage of the concrete.

loss-of-use insurance Insurance against financial loss during the time required to repair or replace property damaged or destroyed by an insured peril.

loss on ignition The percentage loss in weight of an ignited sample to constant weight at a specified temperature, usually 900-1,000 degree C.

lot A parcel of land that is established by a survey or delineated on a recorded plot.

lot line The limit or boundary of a land parcel.

louver A framed opening in a wall, fitted with fixed or moveable slanted slats. Though commonly used in doors and windows, louvers are especially useful in ventilating systems at air intake and exhaust locations.

louver board One of multiple narrow boards or slats on a louver door, window, or ventilator. The boards are installed at an angle. Also called "luffer board."

louver door A door or louver, usually assembled with its blades in a horizontal position, which allows air to pass through the door when it is closed.

louver shielding angle The angle, measured from the horizontal, above which objects are concealed by a louver.

louver window (1) A window, all or part of which is outfitted with louvers in place of glass panes. (2) An open window in a church tower.

louvre Same as **louver**.

low-alkali cement A cement containing smaller than usual amounts of sodium and/or potassium. Its use is necessary with certain types of aggregate which would otherwise react with high levels of alkali.

low-alloy steel Steel composed of less than 8 percent alloy.

low bid In bidding for construction work, the lowest price submitted for performance of the work in accordance with the plans and specifications.

low-carbon steel Steel with less than 0.20% carbon. This type of steel is not used for structural members, due to its ductility. It is good for boilers, tanks, and objects that must be formed. Also called "mild steel."

lowest qualified bidder See **lowest responsible bidder**.

lowest responsible bidder The bidder who submits the lowest bona fide bid and is considered by the owner and architect to be fully responsible and qualified to perform the work for which the bid was submitted.

lowest responsive bid The lowest bid that is responsive to and complies with the requirements of the bidding documents.

low-hazard contents Building contents with such an

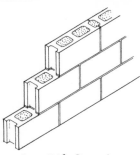

Low-Lift Grouting

exceptionally low level of combustibility that they are unable to propagate or sustain a fire in and of themselves.

low-heat cement, type IV cement A special cement that minimizes the amount and rate of heat generation during hydration (setting). Strength is also achieved at a slower rate. Use is limited to structures involving large masses of concrete, such as dams, where the heat generated would be excessive if normal cement were used.

low-lift grouting The common and simple method of unifying concrete masonry, in which the wall sections are built to a height of not more than 4 feet (1.2 meters) before the cells of the masonry units are filled with grout.

low-pressure mercury lamp A mercury-vapor lamp, including germicidal and fluorescent lamps, whose partial pressure during operation is no more than 0.001 atmosphere.

low-pressure steam curing See **atmospheric-pressure steam curing**.

low steel A characteristically soft steel that contains less than 0.25 percent carbon.

L-shore A shore with an L-head. See also **L-head**.

Lucite The trade name for a strong, clear plastic material manufactured in sheets and other forms.

lug (1) Any of several types of projections on a piece of material or equipment. Such projections are used during handling and installation. (2) A connector for fastening the end of a wire to a terminal.

lug bolt A bolt with a flat iron bar welded to it.

lug sill A windowsill or doorsill with ends that extend beyond the window or doors, converging with

and built into the masonry of the jambs.

lumber Timbers that have been split or processed into boards, beams, planks, or other stock that is to be used in construction and is generally smaller than heavy timber.

lumber core, stave core Wood core made up of narrow strips of lumber glued together at the edges and commonly held together by a veneer, which is glued to both faces with its grain at 90 degrees to that of the core wood.

lumen A unit of luminous flux that defines the quantity of light.

lumen-hour A measurement of light equal to 1 lumen for 1 hour.

luminaire A lighting fixture, with or without the lamps in it.

luminaire efficiency In lighting calculations, a special ratio of the light emitted by a light fixture to the light emitted by the lamps inside the fixture.

luminous ceiling An area lighting system, mounted on a ceiling, that has a surface of light-transmitting materials with light sources installed above it.

lump sum agreement See **stipulated sum agreement**.

lute (1) A straight-edged scraper used to level wet concrete. (2) A straightedge used to strike off clay from a brick mold.

M ABBREVIATIONS

The abbreviations listed below are those most commonly used in the construction industry. Alternative forms (usually nonstandard) are shown in parentheses.

m meter
M thousand, bending moment (on drawings)
ma milliampere
MA mechanical advantage
mach machine, machinist
mag magazine, magneto
MAN manual
manuf manufacture
mas masonry
mat, matl material
max maximum
mb millibar
MBH 1,000 BTU's per hour
MBM, M.b.m. thousand feet board measure
MC moisture content, metal-clad, mail chute
me marbled edges
ME mechanical engineer
meas measure
mech mechanic, mechanical
med medium
memb member
mep mean effective pressure
MER mechanical equipment room
met metallurgy
mezz mezzanine
mf mill finish
mfg manufactured
Mg magnesium
MG motor generator
mgt management
MH manhole
MHW mean high water
mi mile
mid middle
min minimum, minor, minute
misc miscellaneous
mix, mixt mixture
mks meter-kilogram-second
ML, ml material list
Mldg, mldg molding

MLW mean low water
MMF magnetomotive force
Mn manganese
MN magnetic North, main
MO month
Mo molybdenum
MOD model
mod, modif modification
MOL maximum overall length
MOT motor
mp melting point
mpg miles per gallon
mph miles per hour
MRT mean radiant temperature
MSF per 1,000 square feet
msl mean sea level
mtg, mtge mortgage
mult multiple, multiplier
mun, munic municipal
mxd mixed

M DEFINITIONS

Machine Bolt

Machine Excavation

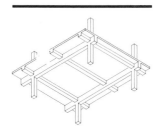

Main Beam

macadam A method of paving in which layers of uniformly graded, coarse aggregate are spread and compacted to a desired grade. Next, the voids are completely filled by a finer aggregate, sometimes assisted by water (waterbound), and sometimes assisted by liquid asphalt (asphalt bound). The top layers are usually bound and sealed by some specified asphaltic treatment.

macadam aggregate A uniformly graded crushed stone, gravel, or slag used for constructing macadam pavement.

machine bolt A threaded straight bolt usually specified by gauge, thread, and head type.

machined A term used to describe a smooth finish on a metal surface.

machine excavation Digging or scooping performed by a machine, as opposed to that performed by hand.

machine finish A finish on a stone surface produced by a smooth-edged planer.

machine rating A standard describing the power that an electrical motor is designed to produce.

made ground Land or ground created by filling in a low area with rubbish or other fill material. Often, such created land is not suitable for building without the use of a pile foundation.

magazine A building for storage of explosives.

magnesia Finely processed magnesium oxide.

magnesite flooring A finished surface material consisting of magnesium oxide, saw dust, and sand, combined in various proportions, and subsequently applied to integral concrete floors.

magnesium A lightweight silver-colored metal which is highly flammable and immune to alkalies and is usually used in an alloy.

magnetic bearing The horizontal angle from magnetic North for a given survey line.

magnetic catch A door catch that uses a magnet to hold it in the closed position.

magnetic driver A tool, employing a magnet, used to hold and drive nails.

magnetic switch An electric switch using an electromagnet for operation.

magnetite A naturally occurring, black iron oxide used as an iron ore and as a high-density aggregate in concrete.

mahogany A straight-grained, medium-density wood originating in the West Indies and Central and South America and used principally in interior plywoods and cabinetry.

mail chute A shaft for dropping mail from upper floors of a building to a central collection box.

mail slot A slot in a wall or door for receiving incomming mail. The slot usually has a hinged cover to prevent draft.

main (1) In electricity, the circuit that feeds all sub-circuits. (2) In plumbing, the principal supply pipe that feeds all branches. (3) In HVAC, the main duct that feeds or collects air from the branches.

main bar A reinforcing bar in a concrete member designed to resist stresses from loads and moments, as opposed to those designed to resist secondary stresses.

main beam A structural beam that transmits its load directly to

Main Tie

columns, rather than to another beam.

main contractor See **prime contractor**.

main couple The main truss in a timber roof.

main office expense A contractor's main office expense consists of the expense of doing business that is not charged directly to the job. Depending on the accounting system used, and the total volume, this can vary from 2 to 20%, with the median about 7.2% of the total volume.

main rafter A structural roofing member that extends from the plate to the ridge pole at right angles.

main reinforcement See **main bar**.

main runner In a suspended ceiling system, one of the main supporting members.

main sewer In a public sanitary sewer system, the trunk sewer into which branch sewers are connected.

main stack In plumbing, a vent that runs from the building drains up through the roof.

maintainer A small motor grader used for driveways and for preparing the fine grade inside buildings.

maintenance bond A contractor's bond in which a surety guarantees to the owner that defects of workmanship and materials will be rectified for a given period of time. A one-year bond is commonly included in the performance bond.

maintenance curve For a light source, a plot of lumens vs. time.

maintenance factor In lighting calculations, the ratio of illumination of a light source or lighted surface at a given time to that of the initial illumination. This factor is used to determine the depreciation of a lamp or a

reflective surface over a period of time.

maintenance period The period after completion of a contract during which a contractor is obligated to repair any defects in workmanship and materials that may become evident. See also **maintenance bond**.

main tie In a roof truss, the bottom straight member that connects the two feet.

main vent See **vent stack**.

makeup air unit A unit to supply conditioned air to a building to replace air that has been removed by an exhaust system or by combustion.

makeup water Water that is added to a system to replace water that has been lost through evaporation or leaking.

malachite A copper-bearing ore also used as ornamental stone.

male nipple A short length of pipe with threads on the outside of both ends.

male plug An electrical plug that inserts into a receptacle.

male thread A thread on the outside of a pipe or fitting.

mall (1) A shaded or covered walk for pedestrians that is often lined with shops. (2) The median strip dividing a highway.

malleability The property of a metal that enables it to be hammered, bent, and extruded without cracking.

malleable iron Cast iron that has been annealed to give it some malleability.

mallet A small wooden hammer used to drive another tool, such as a chisel or a gouge.

mallet headed chisel A steel mason's chisel with a rounded head.

mall front A glazed store front facing an enclosed mall.

mandrel A retractable insert for driving a shell pile.

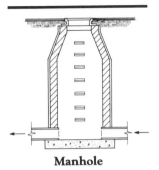

Manhole

Mansard Roof

manganese An alloy that is added to most steels as a hardener and deoxidizer.

manhole A vertical access shaft from the ground surface to a sewer or underground utilities, usually at a junction, to allow cleaning, inspection, connections, and repairs.

manhole block Concrete block cast with curved faces and used to form a cylindrical manhole.

manhole cover A removable cast iron cover for a manhole. As many manholes are in paved areas, the cover must be strong enough to bear the weight of traffic.

manhole frame The cast iron frame into which a manhole cover fits.

manhole invert In a sewer manhole, the elevation or grade of the inlet or outlet pipes.

manhole removal An item of work in site preparation that includes demolition and/or filling in an existing manhole.

manhole step A preformed metal or fiberglass step that is permanently fixed to the inside of a manhole or catch basin.

manhour A unit describing the work performed by one person in one hour.

manifold A distribution or collection pipe or chamber having one inlet and several outlets, or one outlet and several inlets.

manipulative joint A joint in copper tubing where the ends of the tubing are belled outwards.

man lock A chamber in which personnel pass from one environmental pressure to another, such as when entering or leaving a caisson.

manometer A U-shaped tube filled with a liquid used to measure the differential pressure of gasses.

mansard roof A roof having a change of slope on each of four sides, the lower slope being steeper.

mantel, mantelpiece The shelf above and the finished trim or facing around a fireplace.

manual A system of controls that can be operated by hand.

manual batcher A batcher with gates and scales that can be operated by hand.

manual fire pump A pump for water supply to a sprinkler or standpipe system that must be activated by hand.

manufactured sand A fine aggregate that is produced by crushing stone, gravel, or slag.

map cracking See **checking**.

maple A hardwood that grows in North America and Europe and has a dense uniform texture. Maple is used primarily in flooring and furniture.

marble A metamorphic rock, chiefly calcium carbonate, with various impurities that give it distinctive colors. Marble is used in the architectural facing of both interior and exterior walls.

marbling The application of paints to a surface to give it the appearance of marble.

marezzo A cast imitation of marble used extensively for commode tops and wall facing.

margin (1) The amount added to the cost of materials as a mark-up. (2) An edge projecting over the gable of a roof. See also **verge**.

marginal bar A glazing bar that separates a large glazed area in the middle of a window from smaller panes around outside.

margin draft In stone masonry, a dressed border on the edge of the face of a hewn stone.

margin light See **side light**.

margin strip In wood flooring, a narrow strip that forms a border.

margin trowel A plasterer's hand trowel on which the edges are

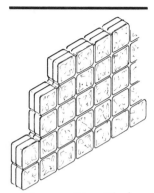

Masonry-Glass Block

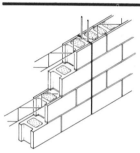

**Masonry Block
and Reinforcing**

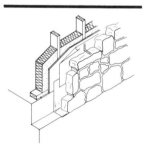

Masonry-Stone

turned up to finish plaster in corners.

marine glue A waterproof glue used on exterior plywoods and other wood-gluing applications where water may be encountered.

marine paint A paint containing elements to withstand exposure to sunlight, salt, and fresh water.

marine plywood A high-grade plywood especially adaptable to boat hull construction. All inner plys must be B grade or better.

marking gauge A carpenter's hand tool for scribing a line parallel to an edge. The gauge has a scribe on a rod whose distance is adjustable from the head rides along the edge of the material.

mark out To lay out the locations where cuts are to be made on lumber.

marl A silty clay, found in the bottom of lake beds or swamps, with a high percentage of calcium carbonate.

marl brick A high-grade brick made from marl.

marquee A canopy extending out from an entrance for protection from the weather.

marquetry Mosaics of inlaid wood and sometimes ivory and mother of pearl.

Martin's cement Similar to Keene's cement and used in plaster. This type of cement contains potassium carbonate as an additive in place of the alum used in Keene's cement.

mash hammer A heavy short-handled mason's hammer with two striking faces.

masking The temporary covering of areas adjacent to those to which paint is to be applied. Masking is applied either by sticking something on, as with masking tape, or by covering with a firm mask.

masking tape An adhesive-backed tape used for masking that comes

in rolls and various widths. The tape is applied to the surface that is to be left unpainted and removed after the painting has been done, leaving a clean straight line.

mason A workman skilled in the trade of masonry and/or the finishing of concrete floors.

Masonite A trade name for a non-structural building board about 1/4" thick, usually with one surface hard and smooth. Masonite can be either tempered or untempered, the tempered form being harder and more water resistant.

masonry Construction composed of shaped or molded units, usually small enough to be handled by one man and composed of stone, ceramic brick, or tile, concrete, glass, adobe, or the like. The term masonry is sometimes used to designate cast-in-place concrete.

anchor A metal device attached to a door or window frame that is used to secure it to masonry construction.

block See *masonry unit*.

bonded hollow wall A hollow masonry wall in which the inner and outer wythes (thicknesses) are tied together with masonry units rather than metal ties.

cement A mill-mixed mortar to which sand and water must be added.

drill See *star drill*.

fill Insulation material used to fill the voids in masonry units.

filler unit Masonry units that are placed between joists or beams prior to placing the concrete for a concrete slab. The filler unit is used to reduce the amount of concrete required and the weight of the slab.

guard See *plaster guard*.

insulation Sound and thermal insulation used in masonry walls. The material can be either

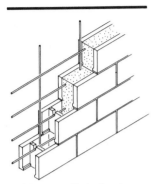

Masonry Reinforcing

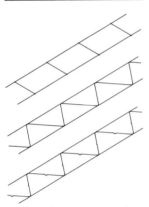

Reinforcing Lateral

Mason's Scaffold

rigid insulation, or an expanded aggregate such as perlite.

nail A hard steel nail with a fluted shank that can be driven into masonry or concrete.

panel A prefabricated masonry wall section that is constructed on the ground or in a shop and erected by crane.

pointing Troweling mortar into a masonry joint after the masonry units have been laid.

reinforcing Refers to both the lateral steel rods or mesh laid between the courses of masonry units and the vertical rods that are grouted into the voids.

tie See *wall tie*.

toothing Cutting or leaving out of alternate masonry units in a wall to provide a bond for new work.

unit Natural or manufactured building units of burned clay, stone, glass, gypsum, concrete, etc.

veneer A single wythe of masonry for facing purposes.

mason's adjustable multiple-point suspension scaffold A scaffold with a platform supported by bearers that are suspended from wire rope and overhead supports. This type of scaffolding is designed and operated in such a way as to permit the platform to be raised or lowered to any desired working position.

mason's ax See **axhammer**.

mason's hammer A steel hammer having one square face for striking and one curved chisel face for trimming masonry units.

mason's joint A projecting V-shaped masonry joint.

mason's level Three separate levels set in a straight bar of wood or metal for determining level or plumb lines. The level is usually about 4-feet long.

mason's lime Lime used in preparing plaster or mortar.

mason's line A heavy string or cord used by masons to align courses of masonry.

mason's measure A method of making a quantity survey, of masonry units required for a job, that counts corners twice and does not deduct for small openings.

mason's miter A corner formed out of a solid masonry unit, the inside of which looks like a miter joint. The joints are actually butt joints away from the corner.

mason's putty A lime-based putty mixed with portland cement and stone dust and used in ashlar masonry construction.

mason's scaffold A self-supporting scaffold for the erection of a masonry wall. The scaffold must be strong enough to support the weight of the masons, the masonry units, and the mortar tubs during construction.

mass concrete Any volume of concrete with dimensions large enough to require that measures be taken to cope with generation of heat from hydration of the cement and attendant volume change, to minimize cracking.

mass curing Adiabatic curing in sealed containers.

mass diagram A plotted diagram of the cumulative cuts and fills at any station in a highway job. The diagram is used in highway design and to determine haul distances and quantities.

mass foundation A foundation that is larger than that required for support of the structure and one that is designed to reduce the effects of impact or vibration.

mass haul curve A curve developed from the mass diagram to display haul distances and quantities in a highway job.

mass profile A road profile graphically showing volumes of cut and fill between stations.

Mast Arm

Matched Boards

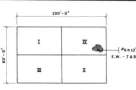

Mat Foundation

mass shooting The simultaneous detonation of explosives in blast holes, as opposed to detonation in sequence with delay caps.

mast The vertical member of a tower crane that carries the load lines.

mast arm The bracket attached to an exterior lamp post that supports a light.

master clock system An electrical system that synchronizes all the clocks in a building.

master key A key that operates all the locks in a master-keyed series.

master-keyed lock A locking system intended for use in a series, each lock of which may be actuated by two different keys, one capable of operating every lock of the series, and the other capable of operating only one or a few of the locks.

master plan A zoning plan of a community classifying areas by use, or zoning code used as a guide for future development.

master switch An electrical switch that controls two or more circuits.

mastic A thick, bituminous-based adhesive used for applying floor and wall tiles. Also, a waterproof caulking compound used in roofing that retains some elasticity after setting.

mat A heavy flexible cover for retaining blasted rock fragments that is usually made of wire, chain, or cordage.

matched boards Boards having been worked with a tongue on one edge and/or end and a groove at the opposite edge and/or end to provide a tight joint when two pieces are fitted together.

matched siding See **drop siding**.

material hose See **delivery hose**.

materials cage The platform on a hoist used for transporting materials to upper floors.

materials lock The chamber through which materials are passed from one environmental pressure to another.

materials tower A materials hoist.

mat foundation A continuous, thick-slab foundation supporting an entire structure. This type of foundation may be thickened or have holes in some areas and is typically used to distribute a building's weight over as wide an area as possible, especially if soil conditions are poor.

Matheson joint A bell and spigot joint in wrought iron pipe.

matrix In concrete, the mortar in which the coarse aggregate is embedded. In mortar, the cement paste in which the fine aggregate is embedded.

mat sink The depression at an entrance door into which a floor mat is placed.

matte A dull surface finish with low reflectancy.

matte-surfaced glass Glass that has been etched, sand blasted, or ground to create a surface that will diffuse light.

mattock A heavy digging tool with a hoe blade on one side of the head and a pick or ax on the other.

mattress A grade-level concrete slab used to support equipment outside a building, such as transformers and air conditioning units.

maturing The curing and hardening of construction materials such as concrete, plaster, and mortar.

maul A long-handled heavy wooden mallet.

maximum demand (1) The greatest anticipated load on a sanitary waste system during a given period of time. (2) The greatest anticipated load on an electrical system during a given period of time.

maximum rated load The greatest live load, plus dead load, which a

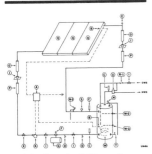

Mechanical Drawing

Mechanical Joint

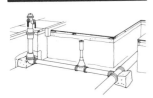

Medium Duty Scaffold

scaffold is designed to carry, including a safety factor.

maximum size of aggregates The maximum size of aggregate permitted in a concrete mix design determined by the thickness of slab, distance from the reinforcing steel to the face of the concrete, and the method of placement.

measurement A quantity survey from plans or a field survey of work completed.

measuring chain See **chain**.

mechanic (1) A person skilled in the repair and maintenance of equipment. (2) Any person skilled in a particular trade or craft.

mechanical advantage The ratio of the weight lifted by a machine divided by the force applied.

mechanical analysis See **sieve analysis**.

mechanical application The placing of plaster or mortar by pumping or spraying, as opposed to placement by hand with a trowel.

mechanical bond A bond formed by keying or interlocking, as opposed to a chemical bond, by adhesion as plaster bonding to lath or concrete bonding to deformed reinforcing rods.

mechanical drawing (1) A graphic representation made with drafting instruments. (2) Plans showing the HVAC and plumbing layout of a building.

mechanical joint A plumbing joint that uses a positive clamping device to secure the sections, such as a flanged joint using nuts and bolts.

mechanical properties The properties of a material defining its elasticity and its stress-strain relationships.

mechanical trowel A machine with interchangeable metal or rubber blades used to compact and smooth plaster.

mechanic's lien A lien on real property, created by statute in all states, in favor of persons supplying labor or materials for a building or structure for the value of labor or materials supplied by them. In some jurisdictions a mechanic's lien also exists for the value of professional services. Clear title to the property cannot be obtained until the claim for the labor, materials, or professional services is settled.

median The untraveled portion in the center of a divided highway which separates the traffic traveling in opposite directions.

medium In paint, the liquid in which the other ingredients are suspended or dissolved.

medium carbon steel Steel with a carbon content of from 0.3 to 0.6%.

medium curing asphalt Liquid asphalt composed of asphalt cement and a kerosene-type diluent (thinner) of medium volatility.

medium curing cutback An asphalt that has been liquified using a kerosene-based solvent.

medium duty scaffold A scaffold designed and constructed to carry a working load not to exceed 50 psf.

meeting posts With a double gate, the stiles which meet in the middle.

meeting rail With a double-hung window, the horizontal rails which meet in the middle.

meeting stile Any abutting stiles in a pair of doors or windows.

megalith A very large hewn or unhewn stone used in architecture or as a monument.

member A general term for a structural component of a building, such as a beam or column.

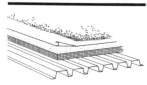

Membrane Roofing

**Membrane
Waterproofing**

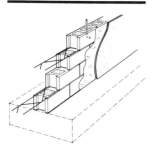

Mercury Vapor Lamp

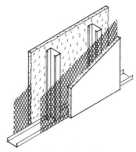

Metal Lath

membrane The impervious layer or layers of material used in constructing a flat roof.

membrane curing A process of controlling the curing of concrete by sealing in the moisture that would be lost to evaporation. The process is accomplished either by spraying a sealer on the surface or by covering the surface with a sheet film.

membrane fireproofing A lath and plaster layer applied as a fireproofing barrier.

membrane roofing A term that most commonly refers to a roof covering, employing flexible elastomeric plastic materials from 35 to 60 mils thick, that is applied from rolls and has vulcanized joints. The initial cost of an elastomeric-membrane roof-covering system is higher than a built-up roof, but the life-cycle cost is lower.

membrane theory In thin-shell design, the assumption that a shell has no strength in bending because of deflection and that the only stresses in any section are in tension, compression, and shear.

membrane waterproofing The application of a layer of impervious material, such as felt and asphaltic cement, to a foundation wall.

mending plate A steel strap with predrilled screw holes used to span and strengthen wood joints.

mensuration The determination of length, area, and volume.

mercury switch An electrical switch that contains mercury in a vial to make a silent contact.

mercury vapor lamp An electric discharge lamp that produces a blue-white light by creating an arc in mercury vapor enclosed in a globe or tube. These lamps are classified as low-pressure or high-pressure.

mesh (1) A network of wire screening or welded wire fabric used in construction. (2) The number of openings per lineal inch in wire cloth.

mesh reinforcing See **welded wire fabric**.

metal clad cable See **armored cable**.

metal clad fire door A flush door with a wood core or stiles and rails and heat insulating material covered with sheet metal.

metal curtain wall A metal exterior building wall which is attached to the structural frame but does not support any roof or floor loads.

metal deck Formed sheet-metal sections used in flat-roof systems.

metal grating Open metal flooring for pedestrian or vehicular traffic used to span openings in floors, walkways, and roadways.

metal gutter Typically, a preformed aluminum or galvanized steel trough attached at the eaves of a sloped roof.

metal halide lamp An electric-discharge lamp that produces light from a metal vapor such as mercury or sodium.

metal lath Any of a variety of metal screening or deformed and expanded plate used as a base for plaster. The metal lath is attached to wall studs or ceiling joists.

metallic paint A paint containing metal flakes that reflect light.

metallize To coat with a metal, usually by spraying with molten metal.

metal pan A form used for placing concrete in floors and roofs. A metal pan may also made of molded fiberglass. See **perforated metal pan**.

metal primer The first coat of paint on a metal surface. Primer usually contains rust inhibitors and/or agents to improve bonding.

metal sash block A concrete masonry unit with a groove in the

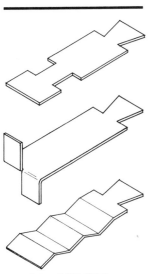

Metal Wall Ties

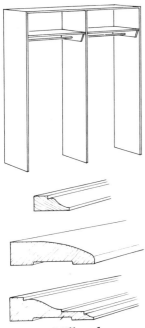

Millwork

end into which a metal sash can fit.

metal trim Grounds, angle beads, picture rails, and other metal accessories that are attached prior to plastering.

metal valley A roof valley gutter lined with sheet-metal flashing.

metal wall ties The prefabricated metal strips that secure a masonry veneer to a structural wall.

metal window A solid metal-framed window such as those used in factories.

metamorphic rock A rock mass whose crystalline structure, composition, or texture has been altered by great heat or pressure during its formation.

meter A device for measuring the flow of liquid, gas, or electrical current. See appendix Table of Equivalents.

meter stop A valve, in a water service line, that cuts off the flow of water before it reaches the meter.

metes and bounds The boundaries, property lines, or limits of a parcel of land, defined by distances and compass bearings.

metric ton A weight equal to 1,000 kilograms or 2,205 pounds.

mezzanine A suspended floor, usually between the first floor and the ceiling, that covers less area than the floor below.

mica A naturally occurring, clear silicate used in thermal and electrical insulation, paint suspensions, and composite roofing materials.

mica pellets See **exfoliated vermiculite**.

middle rail The intermediate horizontal rail between two door stiles that can be either exposed as in a panel door or concealed as in a flush door.

middle strip In flat slab framing, the slab portion that occupies the middle half of the span between columns. See also **column strip**.

mil A lineal measurement equal to .001 inch.

mildew A fungus that grows on damp fabric and other materials, particularly when there is a lack of air circulation.

mild steel A steel having a low-carbon content, and therefore being relatively soft and ductile. This type of steel is sometimes used for the manufacture of boilers and tanks, but *not* for structural beams, columns, or lintels.

milkiness A whitish haze caused by moisture and often occurring in a varnish finish.

milk of lime A hydrated lime slaked in water to form a lime putty.

mill To shape metal to a desired dimension by a machine that removes excess material.

mill construction Historically, a type of construction used for factories and mills and consisting of masonry walls, heavy timbers, and plank floors.

mill finish The type of finish produced on metal by the extrusion or cold rolling of sections.

milling (1) In metal, the process of shaping an item by rotary cutting machines. (2) In stonework, the shaping of a stone to the desired dimensions.

mill run Products from a mill, such as a saw mill, which have not been graded or sized.

mill scale A thin loose coat of iron oxide which forms on iron or steel when heated.

millwork All the building products made of wood that are produced in a planing mill such as moldings, door and window frames, doors, windows, blinds, and stairs. Millwork does not include flooring, ceilings, and siding.

Mission Tile

millwright A carpenter skilled in the layout, installation, and alignment of heavy equipment such as that used in manufacturing.

mineral aggregate See **aggregate**.

mineral dust Aggregate passing the No. 200 screen, usually a by-product of crushed limestone or traprock.

mineral fiber tile A preformed ceiling tile composed of mineral fiber and a binder with good acoustical and thermal properties.

mineral filled asphalt Asphalt with mineral dust in suspension to improve its body and plasticity.

mineral filler See **mineral dust**.

mineral insulated cable Seamless copper tubing, carrying one or more conductors, that is embedded in refractory mineral and is used in areas that may be subjected to high heat.

mineral surfaced felt A roofing felt used on flat or sloped roofs which has a mineral aggregate surface that improves its wearing and heat-reflecting properties.

mineral wool Fibers formed from mineral slag, the most common being glass wool, which is used in loose or batt form for thermal and sound insulation and for fireproofing.

miner's dip needle An instrument with a magnetic needle that indicates the presence of magnetic material in the ground.

minor change A job change requiring field approval only. No change order is necessary.

mirror glazing quality A definable high standard of quality used in the glass industry.

misfire An explosive charge which has failed to detonate.

mission tile A clay roofing tile shaped like a longitudinal segment of a cylinder. The tile is used on sloped roofs with the concave side alternately up then down.

mist coat A very thin sprayed coat of paint or lacquer.

miter See **miter joint**.

miter box A device used by a carpenter or cabinetmaker to cut the bevels for a mitered joint.

miter brad A corrugated fastener that spans a mitered joint.

miter clamp A clamping device that holds a mitered joint during fastening and gluing.

miter cut The beveled cut, usually 45 degrees, made at the end of a piece of molding or board that is used to form a mitered joint.

miter dovetail A dovetail joint in which the pins do not project all the way through, so that it appears like a mitered joint.

mitered hip A roofing hip that has been close cut.

mitered valley A roofing valley that has been close cut.

miter gauge A gauge that measures the angle of a miter.

miter joint A joint, usually 90 degrees, formed by joining two surfaces beveled at an angle, usually 45 degrees.

miter knee The miter joint formed when the horizontal handrail at a landing is joined to the sloping handrail of the stairs.

miter plane A carpenter's planing tool used for preparing the surfaces for miter or butt joints.

miter square A carpenter's square with one edge having an angle of 45 degrees for laying out miter joints.

mix A general term referring to the combined ingredients of concrete or mortar. Examples might be a fire-bag mix, a lean mix, or a 3,000-psi mix.

mix design The selection of specific materials and their proportions for a concrete or mortar batch, with the goal of achieving the required properties with the most economical use of materials.

Mixer

Mixer Truck

mixed glue A premixed synthetic resin glue including the hardener.

mixer A machine for blending the ingredients of concrete, mortar, or grout. Mixers are divided into two categories: batch mixers and continuous mixers. Batch mixers blend and discharge one or more batches at a time, whereas continuous mixers are fed the ingredients and discharge the mix continuously.

mixer efficiency The ability of a mixer to produce a homogeneous mix within a given number of revolutions or a given period of time.

mixer truck See **transit mix concrete**.

mixing box See **dual duct system**.

mixing cycle The elapsed time between discharges of a batch mixer.

mixing plant See **batch plant**.

mixing speed In mixing a batch of concrete, the rate of rotation of a mixer drum or of the mixing paddles expressed in revolutions per minute (RPM). The rate can also be expressed as the distance traveled, in feet per minute (FPM), of a point on the circumference of a mixer drum at its maximum diameter.

mixing time The elapsed time for mixing a batch of concrete or mortar.

mixing valve A valve that mixes two liquids or a liquid and a gas, such as steam with water or hot water with cold water.

mixing water The water used in mixing a batch of concrete, mortar, or grout exclusive of water previously absorbed by the aggregates. As a general rule, mixing water should be clean enough to drink.

mix proportions The quantities of cement, coarse aggregate, fine aggregate, water, and other additives in a batch of concrete by weight or volume.

mobile crane Any crane mount on wheels or tracks. They are classified by lifting capacity.

mobile hoist A personnel or material platform hoist that can be towed to and around the site on its own wheels.

mobile home A term commonly applied to any prefabricated dwelling, whether or not wheels are attached.

mobile scaffold A scaffold that can be moved on wheels or casters.

mock-up A model, either full size or to scale, of a construction system or assembly used to analyze construction details, strength, and appearance. Mock-ups are commonly used for masonry and exposed concrete construction projects.

model (1) A scale representation of an object, system, or building used for structural, mechanical, or aesthetic analysis. (2) A compilation of parameters used in developing a system.

modification (to construction contract documents) (1) A written amendment to the contract signed by both parties. (2) A change order. (3) A written interpretation issued by the architect. (4) A written order for a minor change in the work issued by the architect.

modular construction (1) Construction in which similar units or subcomponents are combined repeatedly to create a total system. (2) A construction system in which large prefabricated units are combined to create a finished structure. (3) A structural design which uses dimensions consistent with those of the uncut materials supplied. Common modular measurements are 4" to 4'.

Moisture Proofing

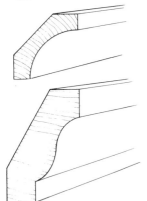

Moldings

modular masonry unit A brick or block manufactured to a modular dimension of 4".

modular ratio The ratio of the modulus of elasticity of steel to that of elasticity of concrete, denoted by "n" in the formula: $n = E_s/E_c$.

module (1) A repetitive dimensional or functional unit used in planning, recording, or constructing buildings or other structures. (2) A distinct component forming part of an ordered system.

modulus of elasticity The unit stress divided by the unit strain of a material that has been subjected to a strain below its elastic limit.

mogul base A screw-in type base for a large incandescent lamp usually of 300 watts or more.

Mohs' scale An arbitrary scale devised to determine the relative hardness of a mineral by its resistance to scratching by another mineral. Talc is rated No. 1, and diamond is rated No. 15.

moist room An enclosure maintained at a given temperature and relative humidity and used for curing test cylinders of concrete or mortar.

moisture barrier A damp-proof course or vapor barrier, but not the same as a water-proof barrier or membrane.

moisture content The weight of water in materials such as wood, soil, masonry units, or roofing materials, expressed as a percentage of the total dry weight.

moisture expansion The increase in dimensions (or "bulking") of a material as a result of the absorption of water.

moisture gradient The difference in moisture content between the inside and the outside of an object, such as a wall or masonry unit.

moisture migration The movement of moisture through the components of a building system such as a floor or wall. The direction or the movement is always from high-humidity areas to low-humidity areas.

moisture movement See **moisture migration**.

moisture proofing The application of a vapor barrier.

mold The hollow form in which a casting or pressing is made.

moldboard The curved shape or blade of a plow, bull dozer, grader, or other earth-moving equipment.

molded brick Brick that has been cast rather than pressed or cut, often with a distinct design or shape.

molded insulation Thermal insulation premolded to fit plumbing pipes and fittings. Common materials are fiberglass, calcium silicate, and urethane foams, with or without protective coverings.

molded plywood Plywood that has been permanently shaped to a desired curve during curing.

molding An ornamental strip of material used at joints, cornices, bases, door and window trim, and the like, and most commonly made of wood, plaster, plastic, or metal.

mole ball An egg-shaped ball pulled behind a special subsoil plow to provide a water course for drainage.

molecular sieve A device for selective collection, by adsorption, of a substance in a gas or a liquid.

mole drain A subsurface drain created by a mole ball.

moler brick Brick, made from moler earth or diatomite, that has better insulation properties than common brick.

molly A threaded insert for plaster, sheet rock, or concrete walls for receiving a bolt, screw, or nail.

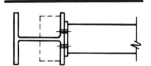
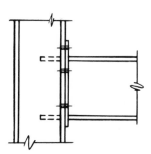
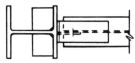
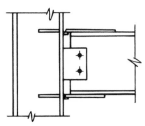

Moment Connection

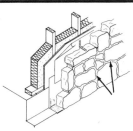

Mortar

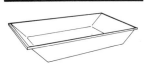

Mortar Box

moment An applied load or force which creates bending in a structural member. It is numerically expressed as the product of the force times the lever arm, and given in units such as foot-pounds.

moment connection A rigid connection between structural members which transfers moment from one member to the other, and thus resists the moment force. A pinned connection cannot resist moment forces, only shear forces.

moment of inertia In a structural member, the product of each element of mass times the square of the distance from an axis.

momentum The mass times the velocity of a moving object.

monitor A raised section of a roof, often along the ridge of a gable roof, with louvers or windows in the side for ventilation or light.

monk bond A modified Flemish bond with two stretchers and a header.

monkey tail A vertical scroll at the bottom of a handrail.

monolith A large architectural member or monument cut from one stone or cast as one unit from concrete.

monolithic concrete Concrete that has been cast with no joints other than construction joints.

monolithic surface treatment A concrete finish obtained by shaking a dry mixture of cement and sand on a concrete slab after strike-off, and then troweling it into the surface.

monolithic terrazzo Terrazzo applied directly over a concrete surface instead of over a mortar underbed.

mopboard See **baseboard**.

mop plate A protective plate at the bottom of a door, such as a kickplate.

mop sink A low, deep sink used by janitors.

mopstick A wood handrail with a circular cross section, except for a small flat section on the bottom for attaching the supports.

mortar (1) A plastic mixture used in masonry construction that can be troweled and hardens in place. The most common materials that mortar may contain are portland, hydraulic, or mortar cement; lime; fine aggregate; and water. (2) The mixture of cement paste and fine aggregate which fills the voids between the coarse aggregate in fresh concrete.

   aggregate Natural or manufactured fine aggregate; usually washed screened sand.

   bed A layer of fresh mortar into which a structural member or flooring is set.

   board A board about 3-feet square on which mortar is placed for use by a mason on a scaffold.

   box A shallow box in which mortar or plaster is mixed by hand.

   cube A standard-sized cube made of mortar for testing the compressive strength of a mix.

   mill A machine with paddles in a rotating drum for mixing and stirring mortar.

   mixer See *mortar mill*.

   sand See *mortar aggregate*.

mortgage A loan in which property is pledged as collateral.

mortgagee The lender in a mortgage loan.

mortgage lien A filed charge using property as security. Often mortgage liens are obtained by contractors or material suppliers for a particular project.

mortgagor The borrower in a mortgage loan.

mortise (1) A recess cut in one member, usually wood, to receive a tenon from another member. (2) A recess such as one cut into a

Motor Controller

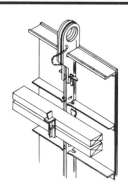

Moveable Form

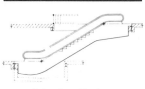

Moving Stairs

door stile to receive a lock or hinge.

mortise and tenon joint A joint between two members, usually wood, which incorporates one or more tenons on one member fitting into mortises in the other member. Used on joints such as door stiles, door rails, window sash, and cabinetry.

mortise chisel A carpenter's tool for cutting mortises.

mortised astragal A two-piece astragal used on a door with two leaves. One piece of the astragal is mortised into the edge of each door.

mortise gauge A carpenter's tool for scribing the location on mortises. It is similar to a marking gauge, but scribes two parallel lines.

mortise lock A lock designed to be mortised into a door stile rather than mounted on the surface.

mortise machine A power-driven machine for cutting rectangular or round mortises in a wood member.

mortise pin A pin that secures a mortise and tenon joint by being driven through either the extension of the tenon or through the whole joint.

mosaic (1) An aerial photographic map pasted up using the center portion of overlapping vertical photographs. (2) A design created by inlaying pieces of stone, glass, or tile in a mortar bed. (3) A design of inlaid pieces of wood.

motor controller A device that controls the power delivered to a motor or motors.

motor grader See **grader**.

mottle A clouding, spotting, or irregular grain appearing in stone such as marble or in wood and wood veneers.

mottler A thick paint brush used for creating a mottled or grained look.

mottling A defect in spray painted surfaces appearing as round marks.

mouse A device with a piece of curved lead and string for pulling a sash cord over a pulley.

moveable form A large prefabricated concrete form of a standard size which can be moved and reused on the same project. The form is moved either by crane or on rollers to the next location.

moveable partition See **demountable partition**.

moving ramp A continuously moving belt or other system designed for carrying passengers on a horizontal plane or up an incline.

moving stairs See **escalator**.

moving walk See **moving ramp**.

mucilage A gum adhesive with low bonding strength.

muck (1) A soil high in organic material, often very moist. (2) Any soil to be excavated.

mud Soil containing enough water to make it soft and plastic.

mud-capping The process of blasting boulders or rock surfaces by placing explosives on the surface and covering them with mud rather than placing the explosives in a blast hole.

mud-jacking An effective method of raising a depressed concrete highway slab or slab-on-grade by boring holes at selected locations and pumping in grout or liquid asphalt.

mud room An entrance, particularly to a rural residence, where muddy footwear can be removed and stored.

mudsill A plank or beam laid directly on the ground, especially for posts or shores for formwork or scaffolding.

mud slab A base slab of low-strength concrete from 2 to 6 inches thick placed over a wet

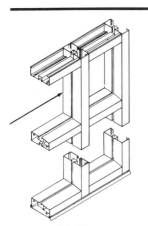

Mullion

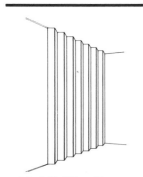

Multifolding Door

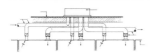

Multizone System

Muntin

sub-base before placing a concrete footing or grade slab.

muffle A layer of grout over a plaster mold used to rough in the plaster. The muffle is chipped off when the final coat of plaster is applied.

mulch Organic material such as straw, leaves, or wood chips spread on the ground to prevent erosion, control weeds, minimize evaporation, as well as temperature extremes, and to improve the soil.

mule A template used to form concrete curbs and gutters.

mullion The vertical member separating the panels or glass lights of a window or door system.

multifolding door A large door or room divider composed of hinged, rigid panels supported on an overhead track. When the door is open, the panels fold against each other.

multiple dwelling A building that houses two or more independent residential dwelling units.

multiple of direct personnel expense A method of compensation for professional services based on direct personnel expense multiplied by an agreed factor to cover indirect expenses, other direct expense and profit.

multiple of direct salary expense A method of compensation for professional services based on direct salary expense multiplied by an agreed factor to cover the cost of contributions and benefits related to direct salary expense, indirect expenses, other direct expense and profit.

multiple surface treatments A term applied to successive pavement surface treatments of asphaltic materials and aggregate.

multiplier The factor by which an architect's direct personnel expense or direct salary expense is multiplied to determine compensation for professional services or designated portions thereof.

multistage stressing The prestressing of concrete members in stages as the construction progresses.

multistory A term commonly applied to buildings with five or more stories.

multiunit wall A masonry wall with two or more wythes.

multizone system A heating or HVAC system having individual controls in two or more zones in a building.

muntin A short vertical or horizontal bar used to separate panes of glass in a window or panels in a door. The muntin extends from a stile, rail, or bar to another bar.

mushroom construction A system of flat-slab concrete construction, with no beams, in which columns are flared at the top to resist shear stresses near the column head.

music wire Steel wire used for alignment. Also called "piano wire."

mute A mortised rubber silencing device for a door.

ABBREVIATIONS

The abbreviations listed below are those most commonly used in the construction industry. Alternative forms (usually nonstandard) are shown in parentheses.

n noon, number

N North, nail, nitrogen, normal

Na sodium

NAAMM National Association of Architectural Metal Manufacturers

NAT natural

NBC National Building Code

NBS National Bureau of Standards

NC noise criterion

NCM noncorrosive metal

NEC National Electrical Code

NEMA National Electrical Manufacturers Association

NESC National Electrical Safety

NFC National Fire Code

Ni nickel

NIC not in contract

NOM nominal

NOP not otherwise provided for

norm normal

NPS nominal pipe size

nr near

NRC noise reduction coefficient

NS not specified

ntp normal temperature and pressure

NTS not to scale

nt. wt.; n.wt. net weight

num numeral

N1E nosed one edge (used in lumber industry)

N2E nosed two edges (used in lumber industry)

N

DEFINITIONS

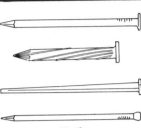

Nails

Nailer

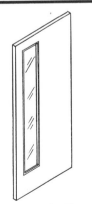

Narrow Light Door

nail A slender piece of metal with a point on one end which is driven into construction materials by impact. Nails are classified by size, shape, and usage.

nailable concrete Concrete, usually made with a suitable lightweight aggregate, with or without the addition of sawdust, into which nails can be driven.

nailer A strip of wood or other fitting attached to or set in concrete, or attached to steel to facilitate making nailed connections.

nail float See **devil float**.

nailing block A wood block set into masonry or steel and used to facilitate fastening other structural members by nailing.

nailing ground A nailing strip to which trim is attached.

nail puller One of a variety of hand tools for pulling nails. The shape and size depends on the nails to be pulled.

nail punch See **nail set**.

nail set A hand-held, tapered steel rod specifically designed to drive nail heads below the surface of wood. The rod is used specifically in finish-carpentry work.

naked flooring The system of floor joists and beams before flooring has been placed.

naked wall A wall or partition, with lath in place, ready for plastering.

narrow light door A door with a narrow vertical light near the lock stile.

narrow ringed timber Lumber with fine-grained, closely spaced growth rings.

natural asphalt Asphalt occurring in nature through natural evaporation of petroleum. This type of asphalt can be refined and used in paving materials.

natural cement A hydraulic cement produced by heating a naturally occurring limestone at a temperature below the melting point, and then grinding the material into a fine powder.

natural convection The movement of air resulting from differences in density usually caused by differences in temperature.

natural finish A finish on wood that allows the grain to show; a clear finish.

natural gas A naturally occurring combustible gas used for industrial and domestic heating and power.

natural seasoning In the lumber industry, a curing process using natural air convection.

natural stone Stone shaped and sized by nature as opposed to stone which has been quarried and cut.

neat A term referring to a process by which a material is prepared for use without addition of any other materials except water. Examples include neat cement or neat plaster.

neat cement A cement mortar or grout made without addition of sand or lime.

neat line The line or plane defining the limits of work, particularly in excavation of earth or rock. Excavation beyond the neat line is usually not a pay item in a unit price contract.

neat plaster Plaster mixed with no aggregate.

neat size The final size after trimming, planing, or finishing.

needle (1) In underpinning, the horizontal beam that temporarily

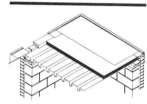

Negative Reinforcing

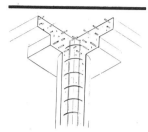

Neoprene Roof

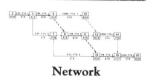

Network

holds up the wall or column while a new foundation is being placed. (2) In forming or shoring, a short beam passing through a wall to support shores or forms during construction. (3) In repair or alteration work, a beam that temporarily supports the structure above the area being worked on.

needle bath A shower bath in which many water jets are sprayed horizontally onto the bather.

needle beam See **needle**.

needle scaffold A scaffold that is supported by needles passing through a wall.

needle valve A type of globe valve in which a long pin or needle, tapered at the end, moves in and out of a conical seat to regulate the flow of liquid.

needling The placing of needles, or a support system using needles.

negative friction The additional load placed on a pile by the settling of fill placed around it. The effect of negative friction is to pull the pile down.

negative reinforcing Steel reinforcing for negative moment in a reinforced concrete structural member.

negligence Failure to exercise due care under the circumstances. Legal liability for the consequences of an act or omission frequently depends upon whether or not there has been negligence. See also **due care**.

negotiation phase See **bidding**.

neon lamp A lamp that gains its illumination by electric current passing through neon gas.

neoprene A synthetic rubber with high resistance to petroleum products and sunlight. Neoprene is used in many construction applications, such as roofing and flashing, vibration absorption, and sound absorption.

neoprene roof A roof covering, made of neoprene sheet material with heat-welded joints, that can be either ballasted or non-ballasted. This type of roof covering has good elastic and durability properties over a long time span.

neoprene vibration pad A vibration absorbing device placed under permanently installed machinery.

neoprene waterproofing Sheet waterproofing material placed on the outside of a foundation wall with a mastic.

net cross-sectional area In defining a masonry unit, the gross cross-sectional area minus the area of the ungrouted cores or cellular voids.

net cut In excavation, the total cut less the compacted fill required between particular stations.

net fill In excavation, the compacted fill required less the cut material available between particular stations.

net floor area The occupied area of a building not including hallways, elevator shafts, stairways, toilets, and wall thicknesses. The net floor area is used for determining rental space and fire-code requirements.

net load In heating calculations, the heating requirement, not considering heat losses, between the source and the terminal unit.

net mixing water See **mixing water**.

net site area The area of a building site less streets and roadways.

network In CPM (Critical Path Method) terminology, a graphic representation of activities showing their interrelationships.

neutral axis In structural design, an imaginary line in a structural member where no tension, compression, or deformation exist. If holes are to be drilled through a structural member for

Newel

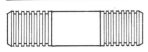

Nipple

conduits or pipes, they should be drilled at the neutral axis.

neutralizing The treatment of concrete, plaster, or masonry surfaces with an acid solution in order to neutralize the lime before application of paint.

newel The post supporting a handrail at the top and bottom of a flight of stairs. Also, the center post of a spiral staircase.

newel cap The often decorative top or cap of a newel.

newel drop A decorative downward projection of a newel through a soffit.

newel joint The joint between a newel and the handrail.

newel post See **newel**.

New York rod A two-piece surveyor's rod with narrow markings and a movable target.

N grade wood (1) In molding, stock intended for natural or clear finishes. The exposed face must be of one single piece. (2) In plywood, cabinet quality panels for natural finishes.

nib Any particle or piece projecting from a surface, particularly used to describe a defect on a painted or varnished surface.

nibbed tile A small lug at the upper end of a roofing tile that hooks over a batten.

nib grade, nib guide, nib rule A wooden straight edge nailed on the ceiling base plaster coat as a guide for a cornice mold.

niche A recess in a wall, usually intended for statuary. Often a base and a canopy project out from the wall as well.

nickel A silver-colored, hard ductile metal used in alloys, batteries, and electroplating. Nickel is used extensively on plumbing fixtures, where its anti-corrosive properties are important.

nickel steel A steel alloy with 3 to 5% nickel content, which gives it greater strength, ductility, and anti-corrosive properties.

nidge, nig To dress or shape the edge of a masonry stone with the sharp point of a hammer, as opposed to doing so with a chisel and mallet.

nidged ashlar Ashlar stone that has been shaped with the sharp point of a hammer.

night latch A door lock with a spring bolt that cannot be operated from the outside except by a key.

night vent A small light with horizontal hinges that is mounted in an operable sash to allow ventilation without opening the entire sash.

nippers A hand tool for cutting wire and small rods.

nipple A short length of pipe with threads on both ends used in pipe connections.

Nissen hut A semicylindrically shaped building made of corrugated metal. See also **Quonset hut**.

nobble The process of shaping building stones to the roughly desired dimensions while the stone is still at the quarry.

node In CPM (Critical Path Method) scheduling, a junction of arrows containing the i-j number, early start, and late finish. In electrical wiring, a junction of several conductors.

no-fines concrete A concrete mixture with little or no fine aggregate.

nog See **nailing block**.

nogging The process of filling the space between timber framing members with bricks.

nogging piece A horizontal timber fitted between vertical studs or beams to give lateral support, especially when nogging is used.

no-hub pipe Cast iron pipe that is fabricated without hubs for coupling.

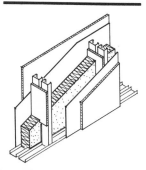

Noise Insulation

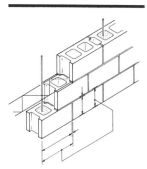

Nominal Dimensions

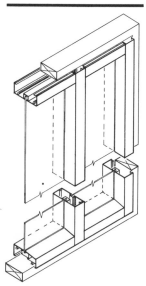

Nonbearing Wall

noise absorption The reduction of noise in an enclosure by introducing sound absorbing materials or methods of construction which restrict the transmission of sound.

noise energy The total sound from all sources within a room at a given time, including reverberations and echos.

noise insulation Sound-absorbing materials installed in partitions, doors, windows, ceilings, and floors.

noise reduction (1) The difference, expressed in decibels, between the noise energy in two rooms when a noise is produced in one of the rooms. (2) The difference in noise energy from one side of a partition to another when a noise is produced on one side.

noise reduction coefficient The average sound absorption coefficient of a material to the nearest 0.05 at four frequencies: 250, 500, 1000, and 2000 cycles.

nominal dimension (1) The size designation for most lumber, plywood, and other panel products. In lumber, the nominal size is usually greater than the actual dimension; thus, a kiln dried 2 x 4 ordinarily is surfaced to 1-1/2 x 3-1/2. In panel products, the size is generally stated in feet for surface dimensions and increments of 1/16-inch for thickness. Product standards permit various tolerances for the latter, varying according to the type and nominal thickness of the panel. (2) In masonry, a dimension larger than the one specified for the masonry unit by the thickness of a joint.

nominal mix The proportions of the constituents of a proposed concrete mix.

nominal size The dimensions of sawn lumber before it is surfaced and dried. See **nominal dimension**.

nonagitating unit A truck-mounted container, for transporting central-mixed concrete, not equipped to provide agitation (slow mixing) during delivery.

non-air entrained concrete Concrete in which neither an air-entraining admixture nor air-entraining cement has been used. See **air entraining agent**.

nonbearing partition A partition which is not designed to support the weight of a floor, wall, or roof.

nonbearing wall A wall designed to carry no load other than its own weight.

noncohesive soil A soil in which the particles do not stick together, such as sand or gravel.

noncollusion affidavit A notarized statement by a bidder that the bid was prepared without any kind of secret agreement intended for a deceitful or fraudulent purpose.

noncombustible Any material which will neither ignite nor actively support combustion in air at a temperature of 1200 degrees Fahrenheit when exposed to fire.

nonconcordant tendons In a statistically indeterminate structure, tendons that are not coincident with the pressure line caused by the tendons.

nonconductor A material that does not easily conduct electric current. Such materials are used as insulators.

nonconforming A building, or the use of a building, that does not comply with existing laws, rules, regulations, or codes.

nonconforming work Work that does not fulfill the requirements of the contract documents.

nondrying A term applied to a material containing oils that do not oxidize or evaporate, and therefore do not form a surface

Nonmetallic Sheathed
Cable

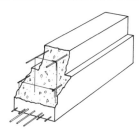

Nonprestressed
Reinforcement

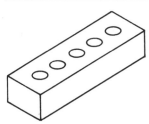

Norman Brick

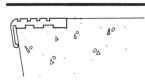

Nosing

skin. The term is often applied to glazing compounds.

nonelectric delay blasting cap A detonating cap with a delay device built in so that it detonates at a designated time after receiving an impulse or signal from a detonating cord.

nonevaporable water The water that is chemically combined during cement hydration and which is not removable by specified drying.

nonferrous A term referring to any metal or alloy that does not contain iron, such as brass or copper.

nonflammable A material which will not burn with a flame.

non-load-bearing partition See **nonbearing partition**.

non-load-bearing tile Tile designated for use in masonry walls carrying no superimposed loads.

non-load-bearing wall See **nonbearing wall**.

nonmetallic sheathed cable Two or more electrical conductors enclosed in a nonmetalic, moisture resistant, flame-retardant sheath.

nonpressure process A method of treating wood by allowing it to soak in a preservative and to absorb the preservative naturally.

nonprestressed reinforcement Reinforcing steel in a prestressed, concrete structural member which is not subjected to prestressing or post-tensioning.

nonreturn valve A check valve that allows flow in only one direction.

nonsimultaneous prestressing The posttensioning of tendons individually rather than simultaneously.

nonskid floor A concrete floor surface treated with carborundum powder, iron filings, or other

material to improve its traction qualities, especially when wet.

nonstaining mortar A mortar, with low free-alkali content, that avoids efflorescence or staining of adjacent masonry units by migration of soluble materials.

nontilting mixer A rotating drum concrete mixer on a horizontal axis. Concrete is discharged by inserting a chute that catches the concrete from the rotating fins.

non-union See **open shop**.

normal consistency (1) The degree of wetness exhibited by a freshly mixed concrete, mortar, or neat cement grout when the workability of the mixture is considered acceptable for the purpose at hand. (2) The physical condition of neat cement paste determined with the Vicat apparatus in accordance with a standard method of testing.

normalizing The heating of steel or other ferrous alloys to a specified temperature above the transformation range, followed by cooling in ambient air. The process reduces the brittleness and strength of the metal, but increases its ductility.

normal weight concrete Concrete having a unit weight of approximately 150 pounds per cubic foot and made with aggregates of normal weight.

Norman brick A brick with nominal dimensions of 2-3/4 x 4 x 12 inches. Three courses of Norman brick lay up to 8 inches.

Normandy joint A patented plumbing joint made between two unthreaded pipes using a sleeve and gaskets.

nosing The horizontal projection of an edge from a vertical surface, such as the nosing on a stair tread.

nosing line The slope of a stair as defined by a line connecting the nosing of each stair tread.

no-slump concrete Fresh concrete with a slump of 1 inch or less.

333

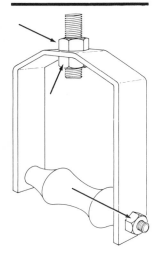

Nut

notch board A stair stringer with the notches cut in for the treads.

notched bar test An impact test for metals.

notching A timber joint in which one or both of the members have a section or notch cut out.

notch joist A joist with a section cut out to fit a ledger of girder.

notice to bidders A notice contained in the bidding documents informing prospective bidders of the opportunity to submit bids on a project and setting forth the procedures for doing so.

notice to proceed Written communication issued by the owner to the contractor authorizing the work to proceed and establishing the date of commencement of the work.

novelty siding See **drop siding**.

nozzle An attachment to the outlet of a pipe or hose that controls or regulates the flow.

nozzle liner A replaceable rubber insert for the inside of a nozzle. The liner is used to prevent wear to the metal parts, particularly to shotcrete nozzles.

N-truss A Pratt truss.

nut A short metal block with a threaded hole in the center for receiving a bolt or threaded rod.

nylon fiber A synthetic fiber used extensively in floor coverings, wall coverings, drapery, and other furnishings.

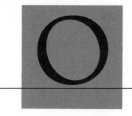 **ABBREVIATIONS**

The abbreviations listed below are those most commonly used in the construction industry. Alternative forms (usually nonstandard) are shown in parentheses.

O oxygen

OA overall

O/A on approval

OAI outside air intake

O.B.M. ordinance bench mark

OBS (1) obsolete (2) open back strike

OC, o.c. on center

OCT octagon

OD outside diameter

OFF. office

OG, o.g. ogee

O/H overhead

OHS oval-headed screw

O.J.T. on-the-job-training

opp opposite

opt optional

OR. (1) outside radius (2) owner's risk

ord (1) order (2) ordnance

ORIG original

OSHA (1) Occupational Safety and Health Administration, Department of Labor (2) Occupational Safety and Health Act

OVHD overhead

OZ, oz ounce

O DEFINITIONS

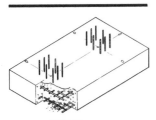

Obscure Glass

Off-Center

oak A hard dense wood used for heavy framing, flooring, interior trim, plywood, and furniture. The two types available from the mill are white oak and red oak.

oakum A caulking material, made from hemp fibers sometimes saturated with tar. Oakum is most commonly used with a bell and spigot joint in cast iron pipe. The oakum is packed into the joint with a hammer and chisel before molten lead is poured into the joint to seal it.

oblique photograph An aerial photograph taken with the camera axis inclined away from the vertical. A high oblique photograph is one that shows the horizon; a low oblique does not include the horizon.

oblique section An architectural or mechanical drawing of a section through an object at an angle other than 90 degrees to its long axis.

obscure glass Glass that transmits light but does not allow vision through it, such as ground glass or frosted glass; translucent glass.

obscuring window A window glazed with obscure glass, installed for privacy.

observation of the work A function of the architect in the construction phase of a project. During visits to the site, the architect becomes familiar with the progress and quality of the work and determines whether or not work is proceeding in accordance with the contract documents.

occupancy The designed or intended use of a building. Also, the ratio of present space being used or rented to the designed full use, expressed as a percent.

occupancy permit See **certificate of occupancy**.

occupancy rate The number of persons per building, unit, room, or floor.

occupant load The maximum number of persons in an area at a given peak period.

occurrence In insurance terminology, an accident or a continuous or repeated exposure to conditions that result in injury or damage, provided the injury or damage is neither expected nor intended.

occurrence form An insuring agreement which covers claims arising from both accidents and occurrences, as opposed to the more limited accident-only form.

odorless mineral spirit A thinner used in interior paints because of its odorless qualities.

odorless paint A water-base paint, or a base using odorless mineral spirit, which produces very little odor.

off-center A load that is applied off the geometric center of a structural member, or a structural member that is placed off the geometric center of an applied load.

off-count mesh Wire mesh in which the spacing or count is greater in one direction than in the other.

offer A proposal, as in a wage and benefits package, to be accepted, negotiated, or rejected.

offering list A list of items for sale, published by a mill or wholesaler and distributed to potential customers. The list usually includes a description of the item, shipping information, price, and terms of sale.

office occupancy The use of a building or space for business, as

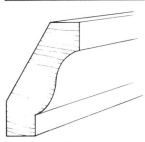

Ogee

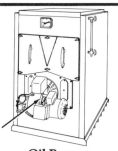

Oil Burner

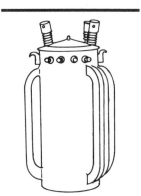

Oil-Filled Transformer

opposed to manufacturing, warehousing, or other uses.

offset bend An intentional distortion from the normal straightness of a steel reinforcing bar in order to move the center line of a segment of the bar to a position parallel to the original position of the center line. The offset bend is commonly applied to vertical bars that are used to reinforce concrete columns.

offset scale A scale used on a map or plan for plotting points which have been determined by field measurements.

offset screwdriver A screwdriver whose head is set at 90 degrees to its shaft.

ogee (1) A double curve much like the letter S. (2) The union of concave and convex lines.

ohm A unit of electrical resistance in a conductor that produces a decrease in voltage of 1 volt with a constant current of 1 ampere.

Ohm's law A law stating that the current in an electrical circuit is directly proportional to the voltage and inversely proportional to the resistance. Stated as an equation: I (current) E (voltage) / R (resistance).

ohmmeter An instrument for measuring in ohms the electrical resistance in a conductor or appliance.

oilborne preservative One of the two general classifications of wood preservatives (the other being waterborne salts). Examples of oilborne preservatives include creosote and various chlorinated phenols, such as pentachloraphenol.

oil-bound distemper A distemper, or water base paint, that contains some drying oils to enhance its spreading and drying characteristics.

oil burner A fuel-oil-burning unit installed in a furnace or boiler.

oil-canning, tin canning A term defining the waving or buckling of formed sheet metal such as roofing or siding.

oiled and edge sealed A process used to resist moisture and preserve plywood concrete form panels. Oiling and sealing increases the number of times a panel can be used and makes the panel easier to release from the concrete.

oiler The second person assigned to a piece of equipment such as a crane. An oiler's principal job is to lubricate the equipment.

oil-filled transformer A transformer having its core and coils immersed in an insulating oil such as mineral oil.

oil-immersed transformer See **oil filled transformer**.

oil paint A paint with drying oil as a base, as opposed to a water base or other base.

oil preservative See **oilborne preservatives**.

oil separator In a refrigeration system, a device for purging the refrigerant of oil and oil vapor.

oil soak treatment A method of treating wood with preservative, as is done with plywood concrete form panels. Oiling and sealing increases the number of times a panel can be used, and makes the panel easier to release from the concrete.

oil stain A stain, with an oil-base containing dye or pigment, used to penetrate and permanently color wood or other porous materials.

oilstone A stone with a fine-grained, oil-lubricated surface used to sharpen the cutting edge of tools.

oilstone slip An oilstone used to sharpen gauges and sharpen chisels.

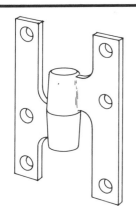

Olive Knuckle Hinge

One-Way Joist Construction

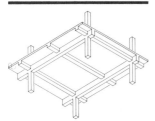

One-Way Slab

oil switch A switch immersed in oil or in another insulating fluid. Also called "oil-immersed switch."

oil varnish A high-gloss varnish, manufactured with a blend of drying oil and gum or resin, used principally for interior finishes.

old bundling A method of packaging siding in which each piece is of equal length. Also called "full-length bundling." Most shippers use "new bundling" in which pieces of various lengths are nested.

old wood Reused wood that has previously been worked.

olive butt See **olive knuckle hinge**.

olive hinge See **olive knuckle hinge**.

olive knuckle hinge A single-pivot Paumelle hinge with knuckles that join to form an oval shape.

on center (1) A measurement of the distance between the centers of two repeating members in a structure. (2) A term used for defining the spacing of studs, joists, and rafters.

one-hour rating A measure of fire resistance, indicating that an object can be exposed to flame for an hour without losing structural integrity or transmitting excessive heat.

one-on-two (one-to-two) A slope in which the elevation rises 1 foot in 2 horizontal feet.

one-pipe system (1) In drainage systems, two vertical pipes, with waste and soil water flowing down the same pipe, and all the branches connected to the same anti-siphon pipe. (2) A heating circuit in which all the flow and return connections to the radiators come from the same pipe. The radiator at the far end is therefore much cooler than the radiator which is nearest the heat source.

one-way joist construction A concrete floor or roof framing system with monolithic parallel joists cast with the slab. The joists are supported on girders, which in turn are supported by columns.

one-way slab A slab panel, bounded on its two long sides by beams and on its two short sides by girders. In such a condition the dead and live loads acting on the slab area may be considered as being entirely supported in the short or transverse direction by the beams; hence, the term "one-way."

one-way system The arrangement of steel reinforcement within a slab that presumably bends in only one direction.

on-grade A concrete floor slab resting directly on the ground.

opacity In painting, the ability of a paint to cover or hide the original background.

opalescent glaze A smooth surface with a milky appearance.

opaque ceramic-glazed tile A fire-bonded facing tile with an opaque, colored ceramic glaze and a glassy or satin finish.

open assembly time The time required between the application of glue to veneer or to joints and the assembly of the pieces.

open bidding A bidding procedure wherein bids or tenders are submitted by and received from all interested contractors, rather than from a select list of bidders privately invited to compete.

open boarding A system of laying boards on a roof, leaving a gap between adjacent boards.

open cell A cell that interconnects with other cells in foam rubber, cellular plastics, and similar materials.

open-cell foam Cellular plastic with a great many open and interconnected cells.

Open Cut

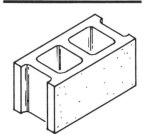

Open-End Block

Open Floor

open cell process A process for fixing preservative in wood under pressure. The chemical used as a preservative is retained in the cell walls only, with the cells left empty.

open-circuit grouting A grouting system with no provision for recirculation of grout to the pump.

open cornice, open eaves Overhanging eaves with exposed rafters that are visible from below.

open cut An excavation in the ground that is open to the sky at its surface, as opposed to a tunnel or horizontal mine shaft.

open defect Any hole or gap in lumber, veneer, or plywood that has not been filled and repaired.

open eaves See **open cornice**.

open-end block (1) An A-block or H-block of concrete masonry. (2) A block of standard material and sizes built with recessed end webs.

open-end mortgage A mortgage arrangement wherein the mortgagor may borrow additional sums for the repair and upkeep of property after the original loan is granted. The mortgages are paid over an extended period.

open floor A floor with joists that are visible from the floor beneath.

open-frame girder An open-web girder or truss built with verticals connected rigidly to top and bottom chords, but no diagonals.

open-graded aggregate (1) An aggregate that contains almost no mineral filler. (2) A compacted aggregate with relatively large void spaces.

open grain (1) Lumber which is not restricted as to the number of rings per inch or the rate of growth. (2) Timber with a coarse texture and open pores.

opening light An operable pane or sash in a window which may be open or shut, as opposed to a fixed light.

opening of bids Same as **bid opening**.

opening protective A device installed over an opening to guard it against the passage of smoke, flame, and heated gases.

opening size The size of a door opening as measured from jamb to jamb and from the threshold or floorline to the head of the frame, allowing for the size of the door and for the necessary clearance to open and close it freely. See **door opening**.

open joint (1) A joint in which two pieces of material that are joined together are not entirely flush. (2) A joint that is not tight.

open mortise A mortise, or notch, that is open on three sides of the piece of timber into which it is cut. See **slot mortise**.

open-newel stair, hollow newel stair, open-well stair (1) A stairway that is built around a wall and has open spaces between its strings. (2) A stair without newels.

open plan A building plant that has relatively few interior walls or partitions to subdivide areas slated for different uses.

open plumbing Plumbing that is exposed to view beneath the fixtures it serves with ventilated drains and traps readily accessible for inspection and repairs.

open riser The space between two successive treads of a staircase built without solid risers.

open-riser stair A stair built without solid risers.

open roof, open timbered roof A style of roof with exposed rafters, sheathing, and supporting timbers visible from beneath, with no ceiling.

open shaft A vertical duct or passage in a building that is used

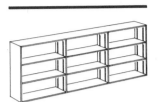

Open Shelving

Open Stairway

Open-Web Steel Joist

to ventilate interior spaces, drawing in outside air from its open top.

open sheathing See **open sheeting**.

open sheeting, open sheathing, open timbering Vertical or horizontal supporting planks and timbers that are set along an excavated surface at intervals. Open sheeting is used whenever the ground is firm and dry enough to be effectively shored up without close sheeting.

open shelving Shelving that is exposed to view, not concealed by doors or cabinets.

open shop, merit shop, non-union shop A term describing a firm whose employees are not covered by collective bargaining agreements.

open slating, spaced slating A pattern of laying roof slates, leaving spaces between adjacent slates in a course.

open space A term used in urban planning for parks, woods, lawns, recreation spaces, and other areas on which no building stands.

open stair, open-string stair A stairway with treads that are visible from either or both sides.

open stairway A stairway with either or both sides open to the room in which it is located.

open string, open stringer, cut string A stairway string with its upper edge cut or notched to conform with the treads and risers of the stairs.

open-string stair See **open stair**.

open system A piping system for water or fuels in which the conduits that circulate the liquid are attached to an elevated tank or tower, with open vents to facilitate storage, access, and inspection.

open tank A tank for treating wood with preservative at atmospheric pressure.

open-timbered Description of structures built with their timber work or wooden frameworks exposed to view, without plaster or another covering; exposed timberwork.

open-timbered floor A style of floor construction in which the joists and other supporting timbers are exposed and visible from the underside.

open-timbered roof Same as **open roof**.

open time The time that is required for the completion of the bond after an adhesive has been spread.

Open-top mixer A mixer consisting of a trough or a segment of a cylindrical mixing compartment within which paddles or blades rotate about the horizontal axis of the trough. See also **mixer** and/or **horizontal shaft**.

open valley A type of roof valley on which shingles and slates are not applied to the intersection of two roof surfaces, leaving the underlying metal or mineral-surfaced roofing material exposed along the length of the valley.

open web A form of construction on a truss or girder in which multiple members, arranged in zigzag or crisscross patterns, are used in place of solid plates to connect the chords or flanges.

open-web steel joist A steel truss with an open web constructed of hot-rolled structural shapes or shapes of cold-formed light gauge steel.

open wiring A network of electrical wiring that is not concealed by the structure of a building, but is protected by cleats, flexible tubing, knots, and tubes, which also support its insulated conductors.

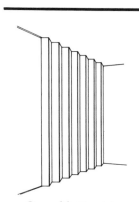

Operable Partition

Operable Window

openwork (1) Ornamental work characterized by perforations. (2) A section of work on a fortress wall that is unprotected by a parapet or another defensive structure at the site of a gorge or bastion opening.

operable partition A partition made of two or more large panels suspended on a ceiling track, and sometimes also supported on a floor track, which may be opened up by sliding the panels so that they overlap. The panels form a solid partition when closed.

operable transom A glass light or panel that is installed above a door and may be opened or shut for ventilation.

operable wall Same as **operable partition**.

operable window A window that may be opened and shut to accommodate ventilation needs, as opposed to a fixed light or fixed sash.

operating engineer The workman or technician who operates heavy machinery and construction equipment.

operating pressure The pressure registered on a gauge when a system is in normal operations.

opposed-blade damper A damper installed over an air passage to regulate the volume of air entering an enclosure. The damper is operated by means of two sets of blades linked so that adjacent blades open and turn in opposite directions.

optical coatings Very thin coatings applied to glass or other transparent materials to increase the transmission or to reduce the reflection of sunlight. Coatings are also used to reflect infrared radiation back to the heat exchanger from which it was emitted.

optical losses Losses resulting from solar radiation reflecting from the surface of the cover plate.

optical plummet In surveying, a device used in place of a plumb bob to center transits and theodolites over a given point, preferred for its steadiness in strong winds.

optimum moisture content The percentage or degree of moisture in soil at which the soil can be compacted to its greatest density. Optimum moisture content is used in specifications for compacting embankment.

option A financial agreement between a property owner and user that grants the latter party, upon payment of a stated sum, the right to buy or to rent the property within a time limit specified on the contract.

opus quadratum Stone masonry in which the stones are cut into squares and laid out in regular ashlar courses.

Orangeburg A standard paneling pattern used in decorative plywood. Orangeburg panels have random width grooving, with the width between each panel following a pattern of 4-8-4-7-9-6-4-6 inches.

orange peel, orange peeling (1) A surface flaw on paint, resulting from poor flow or application, that leaves the finish pocked with tiny holes like citrus skins. Also, a wavy surface defect on porcelain enamel. (2) A segmented hemispherical bucket that resembles a half-orange peel, equipped with self-opening and closing capabilities and used to excavate earth. (3) A distinctive texture applied to drywall.

orbital sander A hand-held electrical sander whose base, covered with sandpaper or abrasive material, moves rapidly in an elliptical pattern and is used chiefly for coarse work.

ordinary construction Construction in which some interior materials and members

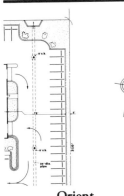

Orient

are of smaller dimensions composed wholly or in part of wood or other combustible materials, but in which the exterior bearing walls and non-bearing exterior walls are of stable, noncombustible construction with a minimum fire endurance of 2 hours.

ordinary-hazard contents Building contents that burn at moderate speed and give off smoke, but release no poison fumes or gases that would cause an explosion under fire conditions.

ordinary portland cement, Type 1 portland cement A basic formula for portland cement used in general construction, containing none of the distinctive properties or refinements of other cements.

Oregon Cedar Same as **Port Orford Cedar.**

Oregon Larch, Pacific Silver Fir, and Noble Fir Lumber from the latter two species shared the Oregon Larch designation for many years in sales to the Orient.

Oregon Maple *Acer macrophyllum.* A wood used in furniture, especially the burls, and as core veneer in some plywood. Also called "Big Leaf Maple."

Oregon Pine Another name for **Douglas Fir.**

Oregon Spruce Same as **Sitka Spruce.**

Oregon White Cedar Same as **Port Orford Cedar.**

Oregon White Pine See **Ponderosa Pine.**

organic Descriptive of materials or compounds produced from vegetable or animal sources.

organic clay Clay containing a high volume of composted animal or vegetable materials.

organic silt Silt composed in a large part of organic substances.

organic soil A highly compressible soil with heavy organic content, considered generally undesirable for construction because of its inability to bear sizeable loads.

oriel In architecture, a projecting bay, frequently outfitted with one or more windows, which is corbeled out from the wall of a structure or supported by brackets and which serves both to expand interior space and to enhance the appearance of the building.

orient (1) To locate a building by points of the compass. (2) To locate a church so the altar end is toward the east.

oriented-core barrel A surveying instrument that takes and marks a core to show its orientation, and at the same time records the bearing and slope of the test hole.

Oriented Strand Board (OSB) Panels made of narrow strands of fiber oriented lengthwise and crosswise in layers, with a resin binder. Depending on the resin used, OSB can be suitable for interior or exterior applications.

orthographic projection A method of representing the exact shape of an object by dropping perpendiculars from two or more sides of the object to planes, generally at right angles to each other. Collectively, the views on these planes describe the object completely. The term "orthogonal" is sometimes used for this system of drawing.

orthography The drafting procedure in which elevations or sections of buildings are geometrically represented.

orthotropic A contraction of the terms "orthogonal anistropic" as in the phrase "orthogonal anistropic plate"; a hypothetical plate consisting of beams and a slab acting together with different flexural rigidities in the longitudinal and transverse directions, as in a composite beam bridge.

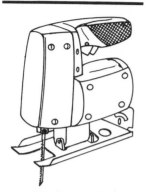

Oscillating Saw

Outbuilding

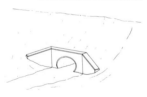

Outfall

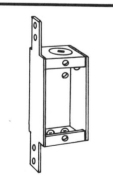

Outlet Box

oscillating saw A power saw with a straight blade that oscillates in short strokes.

Ottawa sand Silica sand produced by processing of material obtained by hydraulic mining of massive orthoquartzite situated in deposits near Ottawa, Illinois. The sand is composed almost entirely of naturally rounded grains of nearly pure quartz and is used in mortars for testing of hydraulic cement. See also **standard sand** and **graded standard sand**.

outband A masonry jambstone that serves as a stretcher and is cut to accommodate a frame.

outbuilding A building that is separate from a main building but attendant on it or collateral to it in its function, such as a stable, a garage, or an outside lavatory.

outcrop A segment of an underground rock stratum or formation that breaks through the surface of the earth and forms a visible protuberance.

outdoor-air intake Same as **outside-air intake**.

outer court An outdoor space that is bounded on three sides by property lines or building walls, but maintains a view of the sky and is open on one side to an adjacent street or public area.

outer string On a stairway, the string that stands away from the wall on the exposed outer edge of the stair.

outfall The final receptacle or depositing area for sewage and drainage water.

outfall sewer A sewer that receives the sewage from the collecting system and conducts it to a point of final discharge or to a disposal plant.

outlet (1) The point in an electrical wiring circuit at which the current is supplied to an appliance or device. (2) A vent or opening, principally in a parapet wall, through which rainwater is

released. (3) In a piping system, the point at which a circulated liquid is discharged.

outlet box The metal box, located at the outlet of an electrical wiring system, that serves to house one or more receptacles.

outlet ventilator An opening covered with a louvered frame, serving as an outlet from an enclosed attic space to the outside.

outline lighting Gaseous tubes or incandescent lamps that are arranged around the outline of a building, a door, a window, or another part of a structure in order to emphasize its configuration.

outline specifications An abbreviated listing of specification requirements normally included with schematic or design development documents.

outlooker In roof construction, a projecting member that supports the portion of the roof beyond the face of a gable.

out-of-center Same as **off-center**.

out-of-plumb Deviating from a true vertical line of descent, as determined by a plumb line.

out-of-sequence services Services performed in an order or sequence other than that in which they are normally carried out.

out-of-true Descriptive of a structural member that is twisted or otherwise out of alignment.

output (1) The net volume of work produced by a system. (2) The maximum capacity or performance of which a system is capable under normal conditions of operation.

outrigger (1) A beam that projects beyond a wall in order to support an overhanging roof or extended floor, built in a direction perpendicular to the joists of the structural member it serves. (2) An extended beam that supports scaffolds or hoisting tackle as

Outside Casing

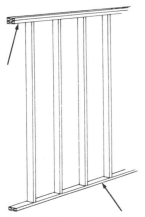

Outside Studding Plate

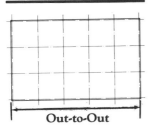

Out-to-Out

work is being performed on or near a building's wall. (3) A beam that gives stability to a crane by widening its base.

outrigger scaffold A scaffold suspended from outrigger beams or brackets fixed to the outer wall of a building.

outrigger shore, horsing A bracket installed temporarily to support an outrigger beam or another projecting member.

outside-air intake An opening or inlet to the outside of a building, through which fresh air is introduced to the boiler room or to an air-conditioning system. Also called a fresh-air intake.

outside caliper A measuring instrument set on adjustable legs and used especially to measure the outside circumference and diameter of round or cylindrical objects and structures. See also **caliper, inside caliper.**

outside casing, outside architrave, outside facing, outside lining, outside trim The supporting members of the jamb or head on a cased window that face the outside and have the appearance of trim.

outside corner molding A molding that covers and protects the projecting outside angle of two intersecting surfaces, as in wood veneer. Same as **corner head.**

outside facing See **outside casing.**

outside finish, exterior finish Ornamental trim or surface treatment on the outside of a building.

outside foundation line A line that indicates where the outer side of a foundation wall is located.

outside glazing External windows or glass doors installed in a building from the outside.

outside studding plate The soleplate or double top plate in the construction of a wood-frame

wall or partition, usually built with stock of a size equal to the studding.

outstanding leg A leg of a structural angle member, generally unconnected to any other member.

outsulation (1) The placing of insulation to the exterior of a wall. (2) The elimination of all thermal bridges between the inner and outer surfaces of a wall.

out-to-out In measurements, a term meaning that the dimensions are overall.

ovals Marble chips that have been tumbled until a smooth oval shape has been obtained.

ovendry The condition resulting from having been dried to essentially constant weight in an oven at a temperature which has been fixed, usually 221 and 230 degrees F (105 and 115 degrees C).

ovendry wood, bone dry wood Wood that gives off no moisture when subjected to a temperature of 212 degrees F (100 degrees C).

overall, overall dimension The total external dimension of any building material, including all projections.

overbreak Excavation performed beyond the work limits established by the neat line.

overburden (1) A mantle of soil, rock, gravel, or other earth material covering a given rock layer or bearing stratum. (2) An unwanted top layer of soil that must be stripped away to gain access to useful construction materials buried beneath it.

overcloak The portion of a metal roofing sheet that overlaps the edge of an adjacent sheet set underneath it.

overconsolidated soil deposit A soil deposit that has been forced to withstand an effective pressure

Overhang

Overhead Door

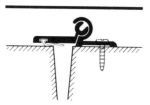

Overlapping Astragal,
Wraparound Astragal

exceeding the pressure imposed by its present overburden.

overcurrent protection Safety provisions within an electrical system, such as would be furnished by ground-fault circuit interrupters, that guard against damage and injury resulting from excessive current by shutting off the flow of current when it reaches a certain level.

overdesign A term used to describe adherence to structural design requirements higher than service demands, as a means of compensating for statistical variation, anticipated deficiencies, or both.

overflow, overflow pipe (1) A pipe installed to prevent flooding in storage tanks, fixtures, and plumbing fittings, or to remove excess water from buildings and systems. (2) An outlet fitted to a storage tank to set the proper level of liquid and to prevent flooding.

overgrainer A special brush with flat, thin elongated bristles that is used to create imitations of natural wood grains.

overhand work The process by which bricklayers install brick in an external wall while standing on a scaffold or on the floor inside a building or structure.

overhang (1) A jetty. (2) The extension of a roof or an upper story of a building beyond the story situated directly beneath.

overhead balance A tense steel coil or spring installed in the head jamb of a window frame to serve as a balance for the sash.

overhead concealed closer A door closer, installed out of view in the head of the doorframe, designed with a hinged arm that connects the door with the top rail of the frame.

overhead door A door, constructed of a single leaf or of multiple leaves, that is swung up

or rolled open from the ground level and assumes a horizontal position above the entrance way it serves when closed. Commonly used as a garage door.

overhead expense See **indirect expense**.

overhead shovel A tractor loader that digs at one end, swings the bucket overhead, and dumps at the other end.

overhead traveling crane A lifting machine generally power-operated at least in its hoisting operation. The crane is carried on a horizontal girder, reaching between rails above window level at each side of a workshop, and consists of a hoisting cab that can travel from end to end on the girder. The whole area between the rails can thus be traversed by the cab.

overhead-type garage door See **overhead door**.

overhung door A sliding door or multiple folding door that is suspended from an overhead track.

overlaid plywood Plywood with a surfacing material added to one or both sides. The material usually provides a protective or decorative characteristic to the side, or a base for finishing. Materials used for overlays include resin-treated fiber, resin film, impregnated paper, plastics, and metal.

overlap In plywood manufacture, a defect in panel construction caused by one of two adjacent veneers overriding the other.

overlapping astragal, wraparound astragal A molding that is attached lengthwise along the masting edge on one of a pair of doors to close the gap between them, providing a weather-resistant seal and redressing the transmission of light or smoke from one side of the doors to the other.

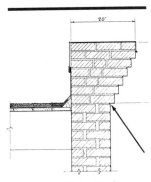

Oversailing Course

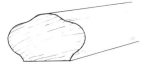

Ovolo

overlay (1) A layer of concrete or mortar, seldom thinner than 1 inch (25 millimeters), placed on and usually bonded to the worn or cracked surface of a concrete slab either to restore or to improve the function of the previous surface. (2) The surfacing of a plywood face with a solid material other than wood. See **overlaid plywood.**

overlay flooring Finish flooring of maple, mahogany, oak, or other hardwood cut into narrow tongue-and-groove strips.

overlay glass Glass that is made up of two or more layers of different colors fused together, and which is often cut to permit a lower layer of glass to show on top for decorative effect.

overlay technique A method of producing composition roofing in which a layer of asphalt is spread over an existing asphalt layer.

overload (1) A load exceeding that for which the bearing structure was designed. (2) Excess power, current, or voltage in an electrical device or circuit that is not designed to accommodate it.

overload relay A relay in a motor circuit that disconnects the motor from its power source if the current that feeds the motor surpasses a certain predetermined level.

overrun See **quantity overrun**.

overrun brake, overriding brake A brake fitted to a towed vehicle such as a concrete mixer to a trailer. It operates as soon as the towing truck slows down and the towed vehicle tends to push into it. Movement of the towed vehicle applies the overrun brake, making safe high-speed towing possible.

oversailing course (1) A course of masonry that extends beyond the face of the wall in which it is set. (2) A string course.

oversanded Descriptive of mortar or concrete containing more sand than would be necessary to produce adequate workability and a satisfactory condition for finishing.

oversite concrete A layer of concrete that is laid below a slab or flooring to prevent the ground beneath from being disturbed, to block out air and moisture, and to provide a hard, level surface for subsequent flooring layers.

oversize brick A brick measuring greater than 2-1/2” x 3-1/2” x 7-1/2”.

overstretching Stressing of tendons to a value higher than designed for the initial stress to: (a) overcome friction losses, (b) overstress the steel temorarily to reduce creep that occurs after anchorage, and (c) counteract loss of prestressing force that is caused by subsequent prestressing of other tendons.

overtime (1) Payment for time worked over the normal number of hours. (2) Paid for at a premium, e.g., time-and-one half or double the normal rate.

overturning The failure of a retaining wall as the result of hydraulic or earth pressure on one side. Overturning occurs when walls are built on a narrow base or with materials too light to withstand surrounding pressure.

overvibration Excessive use of vibrators during placement of freshly mixed concrete, causing segregation and excessive bleeding.

ovolo A convex molding approximately the shape of a quarter circle.

owner (1) The architect's client and party to the owner-architect agreement. (2) The owner of the project and party to the owner-contractor agreement.

owner-architect agreement Contract between owner and architect for professional services.

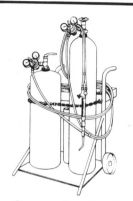

Oxyacetylene Torch

owner-contractor agreement
Contract between owner and contractor for performance of the work for construction of a project or a portion thereof.

owner's inspector A person employed by the owner to inspect construction on the owner's behalf. See also **clerk of the works**.

owner's liability insurance Insurance to protect the owner against claims that arise from the operations performed for the owner by the contractor and from the owner's general supervision of such operations.

owner's representative The person designated as the official representative of the owner in connection with a project.

oxidation (1) The reaction of a chemical compound mixed with or exposed to oxygen. (2) Part of the asphalt refining process, wherein oxygen is incorporated in hot bituminous liquids by blowing it through the melted substance. (3) The hardening of asphalt coating on a roof under exposure to sun and air.

oxidation stain A stain that occurs when a mineral in wood combines with oxygen.

oxidized asphalt Asphalt that has been specially treated by having air blown through it at high temperatures, making it suitable for use in roofing, hydraulics, pipe coating, membrane envelopes, and undersealing.

oxter piece A vertical length of timber used in ashlaring.

oxyacetylene torch A welding torch that produces a super-heated flame through combustion of oxygen and acetylene.

oxyachloride cement, sorel cement A hard, durable cement constituted of calcined magnesia and magnesium chloride occasionally blended with fillers.

oxygen cutting A process of metal cutting in which the separation of the metal is effected by its chemical reaction with oxygen at high temperatures.

oxygen starvation Areas of corrosion on metals exposed to an electrolyte, caused by a smothering effect or by the formation of a crevice between two areas of the metal or between the metal and any other adjacent material.

ozone Triatomic oxygen, O_3, an unstable form of oxygen that is produced by ultraviolet activity and electrical discharges and is used as an oxidizing agent, a deodorizer in air conditioning and cold storage systems, and as an agent for stemming the growth of bacteria, fungus, and mildew. Excessive concentrations of ozone are poisonous to humans.

P ABBREVIATIONS

The abbreviations listed below are those most commonly used in the construction industry. Alternative forms (usually nonstandard) are shown in parentheses.

P&G post and girder
P&T post and timbers
P1E planed one edge
P1S planed one side
1S2E planed one side and two edges
P4S planed four sides
p part, per, pint, pipe, pitch, pole, post, port, power
P phosphorus, pressure, pole, page
PA particular average, power amplifier, purchasing agent, public address system
pan. panel
par. parapet
p.a.r planed all round
PAR. paragraph
part. partition
partn partition
PASS. passenger
pat. patent
Pat. pattern
PAX private automatic (telephone) exchange
Pb lead
pc piece
pc/pct percent
PC portland cement
PCE pyrometric cone equivalent
pcf pounds per cubic foot
pcs pieces
pd paid
Pd palladium
PD per diem, potential difference
p.e. plain edged
PE professional engineer, probable error, plain end, polyethylene
P.E. professional engineer
pecky cyp pecky cypress
PEP Public Employment Program
per. perimeter, by the, period
PERF perforate
PERM permanent

PERP perpendicular
PERT project evaluation and review technique
PF power factor
PFA pulverized fuel ash
PFD preferred
ph phase, phot
Ph phenyl
PH phase, Phillips head
1PH single phase
3PH three phase
pil pilaster
piv pivoted
pk park, peak, plank
pk.fr. plank frame
pkwy parkway
pl place, plate
PL pile, plate, plug, power line, pipe line, private line
P/L plastic laminate
platf platform
PLG piling
plmb, plb, PLMB plumbing
PLYWD plywood
pmh production man-hour
PNEU pneumatic
PNL panel
PO purchase order
POL polish
PORC porcelain
PORT. CEM portland cement
pos, POS positive
pot. potential
PP-AC air-conditioning power panel
ppd prepaid
PPGL polished plate glass
ppm parts per million
ppt, pptn precipitate, precipitation
PR payroll, pair
PRCST precast
preb prebend

prec preceding
PREFAB. prefabricated
prelim preliminary
prin principal
prod. production
proj project, projection
PROJ project
prop. property
prov provisional
prs pairs
PRV Pressure Regulating Valve
ps pieces
p.s.e. planed and square-edged
psf pounds per square foot
psi pounds per square inch
p.s.j. planed and square-jointed
pt paint, pint, payment, port, point
PT part, point
P.T. pipe thread
ptfe polytetrafluorethylene
p.t.g. planed, tongued, and grooved
PTN partition
PU pickup
PUD pickup and delivery
pur purlins
PVA polyvinyl acetate
PVC polyvinyl chloride
PWA Public Works
   Administration
pwr power
pwt pennyweight

P DEFINITIONS

Packaged Air Conditioner

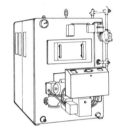

Packaged boiler

pace A landing in a staircase.

pache Color coding used on drawings to aid in quantity takeoffs for estimating.

Pacific Coast Cypress Alaska Yellow Cedar.

Pacific Coast Spruce Sitka Spruce.

Pacific Coast Yellow Cedar Alaska Yellow Cedar.

Pacific Red Cedar Western Red Cedar.

Pacific Silver Fir *Abies amabilis.* This species is found in British Columbia, Washington, and Oregon. The name comes from the silvery appearance of the undersides of the tree's needles. Its wood is classed in the Hem-Fir group.

Pacific Yew *Taxus brevifolia.* This species, generally small in size, is not a commercially important tree. Its wood is heavy and strong and is used for such purposes as archery bows. Yews are usually found growing in the shade of larger trees.

pack The bundling in which shakes and shingles are shipped. In shakes, the most prevalent pack is a 9/9. This describes a bundle packed on an 18-inch wide frame with nine courses, or layers, at each end. The most common pack for shingles is 20/20. Because of their smoother edges, shingles can be packed tighter than shakes; a bundle of shakes usually contains a net of about 16 inches of wood across the 18-inch width of the frame.

packaged air conditioner A factory-assembled air conditioning unit ready for installation. The unit may be mounted in a window, an opening through a wall, or on the building roof. These units may serve an individual room, a zone, or multiple zones.

packaged boiler A factory-assembled water or steam heating unit ready for installation. All components, including the boiler, burner, controls, and auxiliary equipment, are shipped as a unit.

packaged concrete, mortar, grout Mixtures of dry ingredients in packages, requiring only the addition of water to produce concrete, mortar, or grout.

package dealer A contractor responsible for the design and construction of a project to specific end requirements, usually for a fixed sum.

packaged lumber Lumber strapped in standard units, usually pulled to length and wrapped in paper or plastic.

packer A device inserted into a hole in which grout is to be injected, which acts to prevent return of the grout around the injection pipe. A packer is usually an expandable device actuated mechanically, hydraulically, or pneumatically.

packer-head process A method of casting concrete pipe in a vertical position in which concrete of low water content is compacted with a revolving compaction tool.

packing (1) Stuffing or shaped elastic material to prevent fluid leakage at a shaft, valve stem, or joint. (2) Small stones, usually embedded in mortar, used to fill cracks between larger stones.

pack set The condition where stored cement will not flow from a container, caused by interlaced particles or electrostatic charges on particles. See **sticky cement**.

pad (1) A plate or block used to spread a concentrated load over

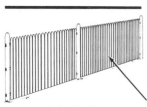

Pale, Paling

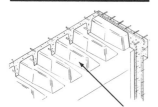

Pan

an area, such as a concrete block placed between a girder and a loadbearing wall. (2) A shoe of a crawler type truck.

paddle A tool or implement with a wide blade at one end.

paddock An enclosed area for animals, usually horses.

pad foundation A thick slab-type foundation used to support a structure or a piece of equipment.

padlock A unit lock with a U-shaped bar that is passed through a staple of a hasp or link in a chain, and the bar pressed into the body to lock.

pad saw A keyhole saw.

padstone A concrete or stone block used to spread a concentrated load over an area of wall.

pad support A wire grid to keep a sound-absorbing insert from contact with the perforated pan in a metal acoustical ceiling.

paint (1) A mixture of a solid pigment in a liquid vehicle which dries to a protective and decorative coating. (2) The resultant dry coating.

paint base The liquid vehicle into which a pigment is mixed to produce a paint.

painter A tradesman experienced in painting.

paint grade A description of a wood product that is more suitable for painting than for a clear finish.

paint kettle An open container with a wire handle used while painting.

paint remover A liquid solvent applied to dry paint to soften it for removal by scraping or brushing.

paint roller A tube, with a fiber surface, that is mounted on a roller and handle and used to apply paint.

paint system A specific combination of paints applied in sequence. A paint system consists of a combination of some of the following coats: sealer or primer, stain, filler, undercoat, and one or more top coats.

paint thinner A liquid compatible with the vehicle of a paint, used to make a paint flow easier. Paint thinner lowers the viscosity of paints, adhesives, etc.

pairing veneers Matching full sheets of veneers (faces and backs) together to reduce handling when laying up panels at the glue spreader.

pale, paling (1) One of the stakes in a palisade. (2) A picket in a fence.

palisade A fence of poles driven into the ground and pointed at the top.

pallet (1) A platform used for stacking material and arranged to be handled by a fork-lift truck. (2) A wood insert in a brick wall used for support of a surface system.

pallet brick A brick made with a groove, to hold a pallet.

palletized A term used frequently in the shingle and shake industry. Both items are often shipped on pallets from the mill for ease in handling while in transit. These shipments are referred to as palletized loadings.

pallet stock Lumber used to make pallets.

pan (1) A prefabricated form unit used in concrete joist floor construction. (2) A container that receives particles passing the finest sieve during mechanical analysis of granular materials.

panache The triangular-like surface of a vault between a supported dome and two supporting arches.

pan and roll roofing tile (1) A roofing tile system consisting of two types of tile. (2) A flat or

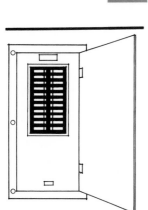

Panel Board

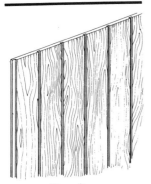

Paneling

slightly curved tile with a flange on each side and curved tile that fits the flanges and closes the joints.

pan construction A type of concrete floor or roof in which pan forms are used to create intersecting ribs and resulting in a waffle-like under-surface.

pane (1) A flat sheet of glass installed in a window or door. The installed sheet is also referred to as a light. (2) A British term for the peen of a hammer. (3) One face or side of a building.

panel (1) A section of form sheathing, constructed from boards, plywood, metal sheets, etc., that can be erected and stripped as a unit. (2) A concrete member, usually precast, rectangular in shape, and relatively thin with respect to other dimensions. (3) A sheet of plywood, particleboard, or other similar product, usually of a standard size, such as 4 x 8 feet.

panel board A board on which electric components and/or controls are mounted.

panel box A box in which electric switches and fuses are mounted.

panel clip A specially shaped metal device used in joining panels in roof construction. The clip substitutes for lumber blocking and helps to spread the load from one panel to the next one.

panel divider Molding or trim used to fill or cover the joint between two surface sheets.

panel door A door constructed with panels, usually shaped to a pattern, installed between the stiles and rails which form the outside frame of the door.

panel, drop See **drop panel**.

paneled door A door which consists of raised or indented panels. Also referred to as a "Colonial door".

panel heating A method of heating a space using floor, wall,

or roof panels in which are embedded electric elements or pipes for hot water, steam, or hot air.

paneling The material used to cover an interior wall. Paneling may be made from a 4/4 select milled to a pattern and may be either hardwood or softwood plywood, often prefinished or overlaid with a decorative finish, or hardboard, also usually prefinished.

panel insert A metal unit used instead of glass in a panel door.

panel mold A mold in which plaster panels are cast.

panel molding A decorative molding, originally used to trim raised panel wall construction.

panel patch See **patch**.

panel pin A very thin nail used to fasten wood paneling to supports.

panel point Point of intersection of the members of a truss.

panel product Any of a variety of wood products such as plywood, particleboard, hardboard, and waferboard, sold in sheets, or panels. Although sizes vary, a standard size for most panel products is 4 x 8 feet.

panel saw A power saw held in a framework and used in cutting panels to size.

panel strip (1) A strip extending across the length or width of a flat slab for design purposes. (2) A narrow piece of wood or metal used to hide a joint between two sheathing boards forming a panel.

panel wall An exterior, non-loadbearing wall with individual panels hung from or supported by the framing of the building.

pan form stair A metal stair assembly with metal sheet pans at the treads to hold precast or cast-in-place masonry or stone treads.

pan formwork The supportwork for the forms while a concrete pan construction floor or roof is being built.

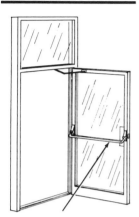

Panic Hardware

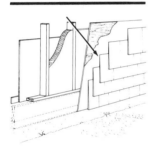

Paper Sheathing

pan fraction (1) The reported results of mechanical analysis of granular materials. (2) The weight of the material retained on any one sieve divided by the initial weight of the sample.

pan head A head of a screw or rivet shaped like a truncated cone.

panic bolt The bolt, in panic hardware, that is released by pressure on a horizontal bar.

panic hardware A door locking assembly that can be released quickly by pressure on a horizontal bar. Panic hardware is required by building codes on certain exits.

panier A corbel form for smoothing the angle between a beam and pilaster.

pan mixer See **mixer, pan**.

panopticon A building plan with corridors that radiate from a central point. All corridors can be observed from that location.

pan steps Prefabricated treads for a pan form stair.

pantile A roofing tile shaped like an elongated S. Joints are protected by the overlapping edges.

pantograph A device for tracing drawings at different scales.

pants Steel plates attached to the hammer of a pile-driver to aid in driving sheetpiling.

pan-type humidifier A pan with water placed in a flow of air to increase humidity. A heating element may be placed in the pan for greater evaporation.

pan-type stairs See **pan form stair**.

pap The vertical outlet from a roof gutter.

paper backed lath Any lath with building paper attached. A paper backed lath serves as formwork and reinforcing for a concrete floor over open web joists.

Paper Birch *Betula papyrifera.* A North American birch with a tough bark.

paperboard A stiff cardboard composed of layers of paper, or paper pulp, compressed into a sheet. Also called "cardboard" or "pasteboard".

paper form A heavy paper mold used for casting concrete columns and other structural shapes.

paper hanger A tradesman experienced in preparing surfaces for and hanging fabric wall coverings.

paper overlay Paper prepared for application to the face of a panel after first being printed in four colors with the grain and color of a more valuable wood, or in a decorative design.

paper sheathing Felt or heavy paper sheets used as an air and/or vapor barrier in walls.

paper wrap A method of packaging wood products for shipment on a truck or railroad flat car, with the paper designed to protect the product from dirt and the elements.

papreg A paper product produced by impregnating sheets of high-strength paper with synthetic resin and then laminating the sheets to form a dense, moisture-resistant product.

parabolic reflector A light reflector shaped so as to project light from a small source in an approximately parallel beam, as in a spotlight.

paraform Paraformaldehyde, an additive used with wood flour as a hardener in adhesives. See **resorcinal resin adhesive**.

paragraph In the AIA Documents, the first subdivision of an article, identified by two numerals (e.g., 2.2). A paragraph may be further subdivided into subparagraphs (e.g., 2.2.1) and clauses (e.g., 2.2.1.1.).

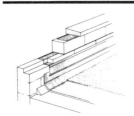

Parapet

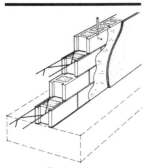

Pargetting

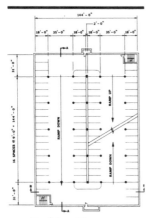

Parking Garage

parallel (1) The condition in which two lines or planes are equal distance apart at all points. (2) Electric blasting caps arranged such that the firing current passes through all of them at the same time.

parallel application An installation of gypsum board with the long dimension of the board in the same direction as the framing members.

parallel connection A connection at which a flow is diverted to two or more parallel conduits.

parallel flow An arrangement of a heat exchanger where the hot and cold materials enter at the same end and flow to the exit.

parallel gutter See **box gutter**.

parallel-laminated veneer A product in which the veneers have been laminated with their grains parallel to one another. Parallel-laminated veneer is used in furniture and cabinetry to provide flexibility over curved surfaces, and in the production of laminated-veneer structural products.

parallel series Two or more series of blasting caps arranged in parallel.

parallel siding (1) Siding that is not beveled. (2) Siding whose edges are of the same thickness. Also called "square-edged siding".

parallel-wire unit A post-tensioning tendon composed of a number of wires or strands which are approximately parallel.

parapet (1) That part of a wall that extends above the roof level. (2) A low wall along the top of a dam.

parapet gutter A gutter built or placed behind a parapet.

parapet skirting Roofing felt turned up against a parapet.

parapet wall See **parapet**

parcel A contiguous land area, subject to single ownership and legally recorded as a single unit.

paretta Cast masonry with a surface of protruding pebbles.

parge To coat with plaster, particularly foundation walls and rough masonry.

parge coat A coat of masonry cement applied to masonry for resistance to penetration of moisture.

pargetting (1) Lining of a flue to aid in smooth flow and increase fire resistance. (2) Application of a dampproofing masonry cement. (3) Ornamental, often elaborate, facing for plaster walls.

parging A thin coat of plaster or masonry cement.

Parian cement A hard finish plaster, similar to Keene's Cement Plaster, except borax is used as an additive in place of alum.

paring Trimming wood by shaving small portions from the surface with a chisel.

paring chisel A long-handled chisel used to shape wood by hand without the use of a mallet.

paring gouge A long thin woodworking gouge, the cutting edge of which is beveled on the concave side.

Parkerized Steel products, such as fasteners, that have been given a zinc phosphate coating for corrosion protection.

parking garage A garage for short term storage of automobiles only.

parking lot A ground-level space for short term storage of automobiles.

parking space A marked off area for short term storage of a single automobile.

parliament hinge An H-shaped hinge.

parquet flooring A floor covering composed of small pieces of wood, usually forming a geometric design.

Partition

Partition Block

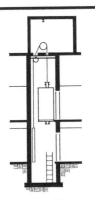

Passenger Elevator

parquet strip A wood flooring composed of tongued and grooved hardwood boards.

partial cover plate A cover plate attached to the flange of a girder but not extending the full length of the girder.

partially air dried (PAD) Seasoned to some extent by exposure to the atmosphere without artificial heat, but still considered green or unseasoned.

partial occupancy Occupancy by the owner of a portion of a project prior to final completion. See **final completion**.

partial payment See **progress payment**.

partial prestressing Prestressing to a stress level such that, under design loads, tensile stresses exist in the precompressed tensile zone of the prestressed member.

partial release Release in a prestressed concrete member of a portion of the total prestress initially held wholly in the prestressed reinforcement.

particle board A generic term used to describe panel products made from discrete particles of wood or other ligno-cellulosic material rather than from fibers. The wood particles are mixed with resins and formed into a solid board under heat and pressure.

particle shape The shape of a particle. See also **cubical piece, elongated piece,** and **flat piece.**

particle size (1) Minimum particle diameter that will be removed by an air filter. (2) Diameter of a pigment particle in paint. (3) Diameter of a grain of sand in the mechanical analysis test.

particle size distribution A tabulation of the result of mechanical analysis expressed as the percentage by weight passing each of a series of sieves.

parting bead A narrow strip between the upper and lower sashes in a double-hung window frame.

parting slip A thin piece of wood in the cased frame of a sash window separating the sash weights.

parting stop See **parting bead.**

parting tool A turning tool with a narrow blade and V-shaped gauged used for cutting recesses or grooves in wood.

partition A dividing wall within a building, usually non-loadbearing.

partition block Light concrete masonry unit with a nominal thickness of 4 to 6 inches.

partition plate The top horizontal member of a partition, which may support joists or rafters.

partition stud A steel or wood upright in a partition.

partition tile A hollow clay unit for use in inferior partitions. The surface of a partition tile is often grooved for plastering.

party wall A common wall between two living units.

pass (1) Layer of shotcrete placed in one movement over the field of operation. (2) A single progression of a welding operation along a joint, resulting in a weld bead.

passage, passageway A horizontal space for moving from one area of a building to another.

pass door A door through the wall separating a stage from the auditorium.

passenger elevator An elevator mainly used for people.

passings The dimension of the overlap between sheets of flashing.

pass-through An opening in a partition for passing objects between adjacent areas.

paste content (of concrete) Proportional volume of cement paste in concrete, mortar, or the like, expressed as volume percent

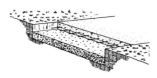

Paumelle

Pavement

of the entire mixture. See also **neat cement paste.**

paste filler A filler in paste form used in preparing wood surfaces for painting.

paste paint A paste-like mixture of pigment and solvent, usually requiring additional solvent for use.

paste volume (of concrete) See **paste content.**

pat A specimen of neat cement paste about 3 in. (76 mm) in diameter and 1/2 in. (13 mm) in thickness at the center, and tapering to a thin edge on a flat glass plate for indicating setting time.

patch (1) A piece of wood or synthetic material used to fill defects in the plies of plywood. Also called a "plug". (2) A compound used in stone masonry to replace chips and broken corners or edges in fabricated pieces of cut stone or to fill natural voids. The patch is applied in plastic form.

patch board, patch panel A board with jacks and plugs for terminals of electric circuits. The circuits may be temporarily interconnected by "patch cords".

patching machine A machine that cuts out the defect in a piece of veneer and replaces the defect with a solid piece of veneer used as a patch. The machine is often referred to by the brand name of the machine, such as Raimann or Skoog.

patch panel See **patch board.**

patent defect A defect in materials, equipment, or completed work which reasonably careful observation could have discovered. A patent defect is distinguished from a latent defect, which could not be discovered by reasonable observation. See **latent defect.**

patent glazing Any of a number of devices, usually formed neoprene

gaskets, for securing glass in frames without putty.

patent hammer A hammer with chisel-like faces, used for dressing stone.

patent knotting A solution of shellac and benzine or similar solvent used to seal knots in wood.

patent plaster A packaged hard plaster.

patent stone Stone chips embedded in a binder of mortar, cement, or plaster. The surface may be ground and/or polished. Also called "artificial stone".

patina Color and texture added to a surface as a result of oxidation or use, such as the green coating on copper or its alloys.

patio An outdoor area, usually paved and sometimes shaded, adjoining a building.

patten The base of a column.

pattern (1) A plan or model to be a guide in making objects. (2) A form used to shape the interior of a mold.

pattern cracking Fine openings on concrete surfaces in the form of a pattern, resulting from a decrease in volume of the material near the surface and/or an increase in volume of the material below the surface.

pattern staining Dark areas on finished plaster, particularly on the interior of external walls, which are caused by different thermal conductances of backings.

paumelle A door hinge with a single joint, usually of modern design.

pavement (concrete) A layer of concrete over such areas as roads, sidewalks, canals, playgrounds, and those used for storage or parking. See also **rigid pavement.**

pavement base The layer of a pavement immediately below the surfacing material and above the subbase.

Pavement Saw

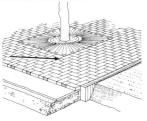

Paver

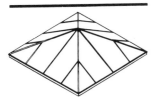

Pavilion Roof

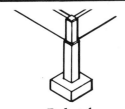

Pedestal

pavement light Transparent or translucent inserts in a pavement to light a space below.

pavement saw A self-propelled machine with a circular saw blade for cutting control joints.

pavement sealer A bituminous coating used to seal and renew the surface of asphalt paving.

pavement structure The collection of courses of specified materials placed on a graded surface.

paver (1) A block or tile used as a wearing surface. (2) A machine that places concrete pavements.

pavilion roof (1) A roof composed of equally hipped areas. (2) Pyramid-shaped roof.

paving The hard surface covering of areas such as walks, roadways, ramps, waterways, parking areas, and airport runways.

aggregate The various solid materials, such as sand, gravel, or slag, used in construction of a pavement.

asphalt A sticky residue from the refining of crude oil. Paving asphalt is used in built-up roofing systems, as the binder in asphaltic concrete, or as a waterproofing agent.

breaker, chipper A hand-held, pneumatic tool for cutting pavements.

brick A vitrified clay brick with good resistance to abrasion.

stone A block of natural stone shaped or selected for use in a pavement surface. Also called "rock".

train An assemblage of equipment designed to place and finish a concrete pavement.

unit A fabricated or shaped unit used in a pavement surface.

payment bond See **labor and material payment bond.**

payment request See **application for payment.**

peacock's eye A circular marking in wood, such as bird's eye in maple, found particularly in Sugar Maple but also in other species.

pea gravel Screened gravel, most of the particles of which will pass a 3/8 in. (9.5 mm) sieve and be retained on a No. 4 (4.75 mm) sieve.

pea gravel grout A grout with pea gravel added.

peaked roof A roof with two or more slopes rising to a ridge or point.

peak joint The joint of a roof truss that is at the ridge.

peak load The maximum demand or design load of a device, system, or structure over a designated time period.

peak-load controller An electrical controller used to limit the maximum power demands to a device or system.

peat Fibrous organic matter in various stages of decomposition, found in swamps and bogs, and used to enrich soil for plantings.

peat moss (1) A type of moss growing in wet areas. (2) The partially decomposed moss used as mulch.

Pecan *Carya illinoensis.* One of the largest native hickories, its wood is used in furniture and flooring, while it is grown commercially for its nuts in the Mississippi River valley.

peck Channeled or pitted areas or pockets sometimes found in cedar or cypress, the decay resulting from fungus in isolated spots.

pecky Characterized by peck, channeled, or pitted areas or pockets found in cedar and cypress.

pecky timber Decay-spotted timber. In cedar and cypress the decay stops when the wood is dried.

pedestal An upright compression member whose height does not

Pedestal Pile

Pencil Rod

exceed three times its average least lateral dimension, such as a short pier or plinth used as the base for a column.

pedestal pile A cast-in-place concrete pile constructed so that concrete is forced out into a widened bulb or pedestal shape at the foot of the pipe which forms the pile.

pedestal urinal A urinal that is supported by a pedestal rather than wall-hung.

pedestal washbasin A wash basin that is supported by a pedestal rather than wall-hung.

pedestrian control device Any device, especially turnstiles, but including gates, railings, or posts, used to control or monitor the movement of pedestrians.

peeler A log from which veneer is peeled on a lathe, for the production of plywood. A peeler-grade log most frequently is from an old growth tree, with a high proportion of clear wood.

peeler core That portion of a peeler block that remains after the veneer has been taken. Peeler cores are often used as raw material for the production of studs.

peeling A process in which thin flakes of mortar are broken away from a concrete surface, such as by deterioration or by adherence of surface mortar to forms as they are removed.

peen The end of a hammer, other than a claw hammer, opposite the hammering face. A peen may be pointed or ball or cone-shaped. A peen is used for chipping, indenting, and metalworking.

peen-coated nail A mechanically galvanized nail coated by tumbling in a container with zinc dust and glass balls.

peening To hammer, bend, or shape with a peen.

peg (1) A pointed pin of wood used to fasten wood members

together. (2) A short, pointed wood stick used as a marker by surveyors.

pegboard A hard fiberboard sheet, usually 1/4 in. thick with regular rows of holes for attaching pegs or hooks.

pelmet A valence or cornice, sometimes decorative, at the head of a window to conceal a drapery track or other fittings.

pelmet board A board at the head of a window, acting as a pelmet.

pelmet lighting Lighting furnished by sources that are concealed by a pelmet.

penal sum The amount named on a contract or bond as the penalty to be paid by a signatory thereto in the event that the contractual obligations are not performed.

penalty and bonus clause See **bonus and penalty clause**.

penalty clause A provision in a contract for a charge against the contractor for failure to complete the work by a stipulated date. See also **liquidated damages**.

Pencil Cedar Eastern Red Cedar.

penciling Painting mortar joints, usually white.

pencil rod Plain metal rod of about 1/4 in. (6 mm) diameter.

pencil rot A type of decay found in cedar.

pencil stock Pieces of Eastern Red Cedar or Incense Cedar from which pencils are manufactured. Pencil stock consists of squares eight inches in length, or in multiples of eight inches, and equal to the thickness of the piece in width.

pendant (1) An electric device suspended from overhead. (2) A suspended ornament in Gothic architecture, used in vaults and timber roofs.

pendant luminaire A suspended lighting unit.

pendant switch An electric switch suspended by an electric cord and

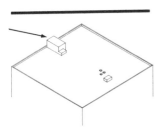

Penthouse

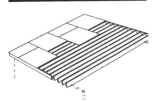

Pent Roof

used to control lamps or other devices that are out of reach.

pendent Variation of **pendant, def. 2.**

pendent post A short post placed against the wall and supporting an arch or tie beam.

penetrating finish A low-viscosity oil or varnish that penetrates into wood with only a film of material at the surface.

penetration (1) A test of the hardness of an asphalt utilizing a weighted needle at standard conditions. (2) The cut-off depth of piles or sheet piling. (3) The depth of a caisson below ground level. (4) The intersection of two surfaces of vaulting.

penetration probe A device for obtaining a measure of the resistance of concrete to penetration, customarily determined by the distance that a steel pin is driven into the concrete from a special gun by a precisely measured explosive charge.

penetration resistance The resistance, usually expressed in pounds per square inch (psi) or megapascals (MPa), of mortar or cement paste to penetration by a plunger or needle under standard conditions.

penetration test A test to estimate the bearing capacity of soil by recording the number of blows required to drive a standard tool into soil.

peninsula-base kitchen cabinet A kitchen cabinet or series of cabinets at right angles to a wall and having access from two sides.

penny A measure of the length of a nail. The larger the number is, the longer the nail is.

penta Short for pentachlorophenol, a wood preservative.

pentachlorophenol A chemical used in wood preserving, usually

applied under pressure so that it will penetrate the wood.

penthouse A structure on the roof of a building, usually less than one-half the projected area of the roof and housing equipment or residents.

pent roof A roof with a single plane surface sloping on one side only.

peppermint test A test for leaks in a drainpipe using oil of peppermint as a trace odor source.

percentage agreement An agreement for professional services in which the compensation is based upon a percentage of the construction cost.

percentage fee Compensation, based upon a percentage of construction cost, that is applicable to either construction contracts or professional service agreements. See also **fee** and **compensation.**

percentage humidity The ratio, expressed as a percentage, of the weight of water vapor in a pound of dry air to the weight of water vapor if the same weight of air were saturated.

percentage of reinforcement The ratio, expressed as a percentage, of cross-sectional area of reinforcing steel to the effective cross-sectional area of a member.

percentage void The ratio, expressed as a percentage, of the volume of voids to the gross volume of material.

percent fines (1) Amount, expressed as a percentage, of material in aggregate finer than a given sieve, usually the No. 200 sieve. (2) The amount of fine aggregate in a concrete mixture expressed as a percent by absolute volume of the total amount of aggregate.

percent saturation The ratio, expressed as a percentage, of the

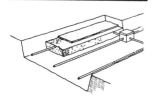

Percolation

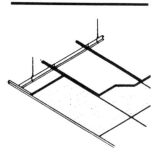

Perforated Metal Pan

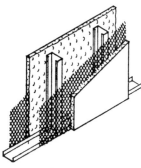

Perlite Plaster

volume of water in a soil sample to the volume of voids.

percent voids See **percentage void**

perched water table A water table, of limited area, held above the normal water table by an intervening strata of impervious confining strata.

percolation The movement of a fluid through a soil.

percolation test A test to estimate the rate at which a soil will absorb waste fluids, reported by measuring the rate at which the water level drops in a hole full of water.

percussion drill A pneumatic or electric tool that drills holes by applying a rapid series of blows.

perennial A plant or shrub with a life cycle greater than two years.

perfections Shingles 18-inches long and .45 inches thick at the butt.

perforated facing A perforated sheet or board used as a finished surface and allowing a fraction of sound to penetrate the surface to an absorbant layer.

perforated hardboard See **pegboard**.

perforated metal pan A unit that forms the exposed surface of a type of acoustical ceiling. The perforated pan contains a sound absorbant material, usually in pad form.

perforated tape A special paper tape used to reinforce the material covering joints between gypsum boards.

performance bond (1) A guarantee that a contractor will perform a job according to the terms of the contract, or the bond will be forfeited. (2) A bond of the contractor in which a surety guarantees to the owner that the work will be performed in accordance with the contract documents. Except where prohibited by statute, the

performance bond is frequently combined with the labor and material payment bond. See also **surety bond**.

periclase A crystalline mineral, magnesia (MgO), the equivalent of which may be present in portland cement clinker, portland cement, and other materials, such as open hearth slags, and certain basic refractories.

periderm The cork-producing tissue of a tree.

perimeter grouting Injection of grout, usually at relatively low pressure, around the periphery of an area which is subsequently to be grouted at greater pressure. Perimeter grouting is intended to confine subsequent grout injection within the perimeter.

perimeter heating system A system of warm-air heating where outlets for air ducts are located near the outside walls of rooms and are close to the floor. The returns are near the ceiling.

periphery wall A wall on the exterior of a building.

perlite A volcanic glass having a perlitic structure, usually having a higher water content than obsidian when expanded by heating. Perlite is used as an insulating material and as a lightweight aggregate in concretes, mortars, and plasters.

perlite plaster A plaster using perlite as an aggregate instead of sand.

perlitic structure A structure produced in a homogeneous material, generally natural glass, by contraction during cooling, and consisting of a system of irregular convoluted and spheroidal cracks.

permafrost Permanently frozen soil or subsoil found in arctic or subarctic regions.

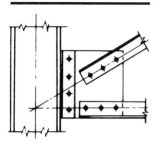

Permanent Bracing

Perspective Drawing

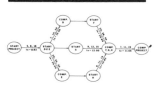

PERT Schedule

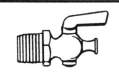

Pet Cock

permanent bracing Bracing that forms part of a structure's resistance to horizontal loads. Permanent bracing may also function as erection bracing.

permanent form Any form that remains in place after the concrete has developed its design strength. A permanent form may or may not become an integral part of the structure.

permanent load The load, including a dead load or any fixed load, that is constant through the life of a structure.

permanent set Inelastic elongation or shortening.

permanent shore An upright used to support dead loads during alterations to a structure and left in place.

permeability (1) The property of a material that permits passage of water vapor. (2) The property of soil which permits the flow of water.

permeability to water, coefficient of The rate of discharge of water under laminar flow conditions through a unit cross-sectional area of a porous medium under a unit hydraulic gradient and standard temperature conditions, usually 20 degrees C.

permeameter An instrument to measure the coefficient of permeability of a soil sample.

permeance The resistance, measured in perms, to the flow of water vapor through a given thickness of material.

permit A document issued by a governing authority such as a building inpsector approving specific construction.

building See **building permit**.

occupancy See **certificate of occupancy**.

zoning See **zoning permit**.

perpend A stone which extends completely through a wall and is exposed on each side of the wall.

personal injury (insurance terminology) Bodily injury, and also injury or damage to the character or reputation of a person. Personal injury insurance includes coverage for injuries or damage to others caused by specified actions of the insured, such as false arrest, malicious prosecution, willful detention or imprisonment, libel, slander, defamation of character, wrongful eviction, invasion of privacy, or wrongful entry. See also **bodily injury**.

personal property Temporary or moveable property, as opposed to real property.

perspective (1) The technique of preparing a perspective drawing. (2) The appearance of objects in depth.

perspective center A point of origin or termination of perspective rays used in preparing a perspective drawing.

perspective drawing A graphic representation of the project or part thereof as it would appear three-dimensionally.

PERT schedule An acronym for project evaluation and review technique. The PERT schedule charts the activities and events anticipated in a work process. See also **critical path method (CPM)**.

pervious cesspool A tank in the ground which receives domestic sewage or other organic wastes. The walls and floor of the tank are designed to permit the liquids to seep through to the soil.

pervious soil A soil that allows relatively free passage of water.

pet cock A small valve installed on equipment or piping for drainage of liquids or air.

petrifying liquid A penetrating solution used for waterproofing masonry surfaces.

petrography The branch of petrology dealing with description and systematic classification of rocks, aside from their geologic relations, mainly by laboratory methods, largely chemical and microscopical. Also loosely referred to as "petrology" or "lithology".

petroleum asphalt Asphalt refined directly from petroleum, as opposed to asphalt from natural deposits.

petroleum hydrocarbon Any of a number of solvents refined from cured petroleum and used to lower the viscosity of oils and resins.

petroleum spirit A thinner, for paints and varnishes, having a low-aromatic hydrocarbon content, obtained in petroleum distillation.

petrology The science of rocks, dealing with their origin, structure, composition, etc., from all aspects and in all relations. See also **petrography**.

pew A benchlike seat used in a church.

p-grade In molding, stock intended for opaque paint finishes or overlays. P-grade stock can be finger-jointed and/or edge-glued.

phantom line A broken line, usually fine with alternating long and short dashes, in order to show details.

phased application The installation of built-up roofing plies in two or more applications, usually at least one day apart.

phenol A product of the petroleum industry used in the production of phenolic resin, exterior plywood glue. Phenol is made from benzene. It exists naturally in coal tar and wood tar; however, phenol from these sources is rarely used in the production of glue.

phenolic resin A class of synthetic, oil-soluble resins (plastics) produced as condensation products of phenol,

substituted phenols and formaldehyde, or some similar aldehyde that may be used in paints for concrete.

phenolic resin glue An adhesive used for bonding exterior plywood. Phenolic resin is produced in a reaction between phenol and formaldehyde. An extender is usually added to the phenolic resin prior to use in the plywood-manufacturing process.

phi factor Capacity reduction factor in structural design. The factor is expressed as a number less than 1.0 (usually 0.65-0.90) by which the strength of a structural member or element, in terms of load, moment, shear, or stress, is required to be multiplied in order to determine design strength or capacity. The magnitude of the factor is stipulated in applicable codes and construction specifications for respective types of members and cross sections.

Philadelphia leveling rod A leveling rod in two sliding parts with color coded graduations. The rod can be used as a self-reading leveling rod.

Philippine mahogany The wood of several types of trees found in the Philippines. The wood resembles mahogony in grain. Density varies from very light to quite heavy. The heavier, darker woods are durable and strong and used like mahogany; the lighter-weight, colored woods are used for interior plywood.

Phillips head screw A screw with a recessed head and an X-shaped driving indentation.

phosphor mercury-vapor lamp A high-pressure mercury-vapor lamp with a phosphor-coated glass cover over the lamp proper. The phosphor in the cover adds colors not generated by the lamp.

Picture Window

phot A measure of illumination equal to 1 lumen per square centimeter.

photoelectric cell An electronic device for measuring illumination level or detecting interruption of a light beam. The electric output or resistance of the device varies according to the illumination.

photoelectric control An electric control that responds to a change in incident light.

piano hinge A continuous strip hinge used in falling doors, etc.

picea The general botanical classification of the spruces.

pick A hand tool, consisting of a steel head pointed at one or both ends and mounted on a wood handle, for loosening and breaking up compacted soil or rock.

pick and dip A method of laying brick in which the bricklayer picks up a brick in one hand and, with a trowel in the other hand, scoops enough mortar to lay the brick.

pickax See **pick**.

pick dressing The rough dressing of hard quarried stone with a heavy pick or wedge-shaped hammer.

picked finish A surface finish for stone masonry in which the surface is covered with small pits made by striking it perpendicularly with a pick or chisel.

picket A sharpened or pointed stake, post, or pale, usually used as fencing.

picket fence A fence consisting of vertical piles, often sharpened at the upper end, supported by horizontal rails.

pickled A metal surface that has been treated with strong oxidizing agents to remove scale and provide a tough oxide film.

pickling Slang for the preservative treatment of wood, metals, and piping systems.

pickup Unwanted adherence of solids to the open surface of a sealant.

pickup load The heat consumption required to bring piping and radiators to their operating temperature when a heating system is first turned on.

picture molding Molding designed to support picture hooks near the ceiling.

picture plane The plane on which rays are projected when making a perspective drawing.

picture window A large window, usually a fixed sheet of plate or insulating glass.

pieced timber (1) A timber made from two or more pieces of timber fitted together. (2) A damaged timber patched with a fitted piece of wood.

piece mark A mark given to one or more pieces in an assembly designating a location in the assembly, as shown on shop drawings.

pien check In a stair constructed of stone, a rabbet cut in the front edge of a tread that fits over the riser below it.

pier (1) A short column to support a concentrated load. (2) Isolated foundation member of plain or reinforced concrete.

pierced louver A louver set in the face sheets or panels of a door.

pier glass A mirror hung between two windows.

pigeonhole A small compartment for holding papers or small objects, usually one of an adjoining series.

piggyback (1) A method of transportation where truck trailers are carried on trains, or cars on trucks. (2) In staple application to gypsum wallboard, a second staple

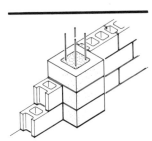

Pilaster

Pile

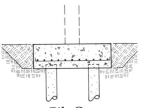

Pile Cap

Pile Friction

is driven directly on top of the first. The staple will spread to create a firm bond between the wallboard and the tile.

pigment A coloring matter, usually in the form of an insoluble fine powder, dispersed in a liquid vehicle to make paint.

pigment figure A natural pattern in woods, such as rosewood and zebra wood, consisting of variations in color rather than grain.

pigtail A flexible conductor attached to an electric component for connecting the component to a circuit.

pigtail splice A connection of two electric conductors, made by placing the ends of the conductors side by side and twisting the ends about each other.

pig tin A metal alloy which is at least 99.80% tin.

pike pole A long pole, with a spear-type point and a hook on one end, used to move logs around in a mill pond.

pike staff Similar to a pike pole, but lighter.

pilaster A column built within a wall, usually projecting beyond the wall.

pilaster block Concrete masonry units designed to form plain or reinforced concrete masonry pilasters of the projecting type.

pilaster face The form for the front surface of a pilaster parallel to the wall.

pilaster side The form for the side surface of a pilaster perpendicular to the wall.

pile A slender timber, concrete, or steel structural element, driven, jetted, or otherwise embedded on end in the ground for the purpose of supporting a load.

pile bearing capacity The load on a pile or group of piles that will theoretically produce failure if exceeded.

pile bent Two or more piles driven in a row transverse to the long dimension of the structure and fastened together by capping and (sometimes) bracing.

pile cap (1) A structural member placed on, and usually fastened to, the top of a pile or a group of piles and used to transmit loads into the pile or group of piles and, in the case of a group, to connect them into a bent. Also known as a "rider cap" or "girder". (2) A masonry, timber, or concrete footing resting on a group of piles. (3) A metal cap or helmet temporarily fitted over the head of a precast pile to protect it during driving. Some form of shock-absorbing material is often incorporated.

pile core The mandrel used to drive the shell of a cast-in-place concrete pile.

pile driver A machine for driving piles, usually by repeated blows from a free-falling or driven hammer. A pile driver consists of a framework for holding and guiding the pile, a hammer, and a mobile plant to provide power.

pile eccentricity The amount that a pile deviates from its plan location, or from plumb.

pile extractor A machine for loosening piles in the ground by exerting upward striking blows. The actual removal is by a crane.

pile foundation The system of piles, and pile caps, that transfers structural loads to bearing soils or bed rock.

pile friction The friction forces on an embedded pile limited by the adhesion between soil and pile and/or the shear strength of the adjacent soil.

pile hammer A weight which strikes a pile to drive it into the ground. The weight may fall freely or be assisted by steam or air pressure.

pile head The top of a pile.

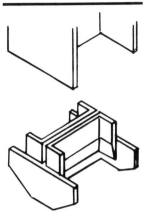

Pile Shoe

pile height The height of piles in a rug measured from the top surface of the backing to the top of the pile.

pile helmet A cap which covers and protects the top of a pile during driving and also holds any packing used.

pile load test A static load test of a pile or group of piles used to establish an allowable load. The applied load is usually 150% to 200% of the allowable load.

pile shoe A pointed or rounded device on the foot of a pile to protect the pile while driving.

pile tolerance The permitted deviation of a pile in the horizontal and vertical planes.

piling The behavior of a quick-drying paint in which viscosity increases during application, resulting in uniform coverage.

piling pipe A pipe used as the shell or a section of shell for a cast-in-place concrete pile.

pillar (1) A post or column. (2) A column of ore left in a mine to support the ground overhead.

pilot boring A preliminary boring or series of borings to determine the nature of the soil in which a foundation will be dug or a tunnel driven.

pilot hole A guiding hole for a nail or screws, or for drilling a larger hole.

pilot lamp See **pilot light**.

pilot light (1) A small, constantly burning flame used as an ignition source in a gas burner. (2) A low-wattage light used to indicate that an electric circuit, control, or device is active.

pilot nail A temporary nail used to align boards until permanent nails are driven.

pilot punch A machine punch in which the punching tool is fitted with a small control plug to be inserted in a guide hole in the material to be punched.

pin A peg or bolt of some rigid material used to connect or fasten members.

pincers A joined tool with a pair of jaws and handles used to grip an object.

pinch bar A steel bar with a chisel point at one end used as a lever for lifting or moving heavy objects.

pin-connected truss A truss in which the main members are connected by pins.

pine Any of various softwoods of the genus *Pinus*.

pine oil A high-boiling-point solvent, obtained from the resin of pine trees and used in paint to provide good flow properties and as an anti-skinning agent.

pine shingles Shingles made from pine wood.

pine tar A blackish-brown liquid distilled from pine wood. Pine tar is used as an antiseptic externally and an expectorant internally and also to make the grips of tools sticky.

pin hinge A butt hinge with a pin for the pivot.

pinhole (pin hole) A small, round hole made by a pin-hole borer.

pin joint A structural joint found in trusses and some girder seats.

pin knot A knot with a diameter no larger than 1/2 inch.

pinnacle (1) The highest point. (2) A turret or elevated portion of a building. (3) A small ornamental body or shaft terminated by a cone or pyramid.

pinner A small stone which supports a larger stone in masonry.

pinning Fastening or securing by means of a pin.

pinning in Filling in joints of masonry with chips of the stone.

pinning up The operation of driving wedges to bring an upper

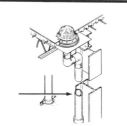

Pipe

Pipe Bend

Pipe Bracket

Pipe Cross

work to fully bear on shoring or under pinning.

pintle A vertical pin fastened at the bottom and serving as a center of rotation.

pin tumbler A lock mechanism having a series of small pins that must be properly aligned by a key to open.

Pinyon Pine *Pinus monophylla, P. cembroides, P. quadrifolia, P. edulis.* A group of small pines, 20 to 40 feet in height, occurring in scattered groves in the semiarid areas of the west. They produce an edible seed, mostly used as topping for a salad. Trees exposed to constant winds take on a sprawling form. Also called "Nut Pine"; also spelled "Pinon Pine".

pipe (1) A hollow cylinder or tube for conveyance of a fluid. (2) From ASTM B 251-557: Seamless tube conforming to the particular dimensions commonly know as "Standard Pipe Size."

pipe bend A pipe fitting used to change direction.

pipe bracket A shaped metal assembly used to support a pipe from a wall or floor.

pipe chase A vertical space in a building reserved for vertical runs of pipe.

pipe column A column made of steel pipe and often filled with concrete.

pipe coupling A fitting used to connect two lengths of pipe in a direct line.

pipe covering Any wrapping on a pipe which acts as thermal insulation and/or a vapor barrier.

pipe cross A fitting used to connect four lengths of pipe in the same plane with all lengths at right angles to each other.

pipe cutter A hand tool for cutting pipe or tubing, consisting of a frame with a cutting wheel and drive wheels. Cutting is accomplished by forcing the cutting wheel into the pipe

material and rotating the tool around the pipe a number of times.

pipe die An adjustable tool for cutting threads on or in a pipe.

pipe elbow See **pipe bend**.

pipe expansion joint An assembly, other than a fabricated U-bend, designed to compensate for pipe contraction or expansion.

pipe fitting Ells, tees, and other connectors used in assembling pipe.

pipe gasket A fabricated packing to seal flanged joints in pipe.

pipe hanger A device or an assembly to support pipes from a slab, beam or others structural element.

pipelayer (1) A tradesman skilled in laying and joining in a trench pipes of glazed clay, concrete, iron, steel, or asbestos cement. (2) An attachment for a tractor or other machine consisting of a winch and side boom for placing lengths of pipe in a trench.

pipeline heater A heater, usually a wrapping with an electric element used to prevent the liquid in the pipe from freezing or to maintain the viscosity of the liquid.

pipeline refrigeration Refrigeration provided to a group of buildings by piping refrigerant from a central plant.

pipe pile A steel cylinder, usually between 10 and 24 inches in diameter, generally driven with open ends to form bearing and then excavated and filled with concrete. This pile may consist of several sections from 5 to 40 feet long joined by special fittings, such as cast-steel sleeves. A pipe pile is sometimes used with its lower end closed by a conical steel shoe.

pipe plug A pipe fitting with outside threads and a projecting head used to close the opening in another fitting.

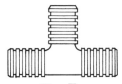

Pipe Tee

Pipe Wrench

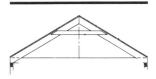

Pitched Roof

pipe reducer A pipe fitting used to connect two lengths of pipe of different diameters.

pipe ring A circular-shaped metal part used to support a pipe from a suspended rod.

pipe run Any path taken by pipe in a distribution or collection system.

pipe saddle An assembly to support a pipe from the underside.

pipe sleeve A cylindrical insert cast in a concrete wall or floor for later passage of a pipe.

pipe stock An assembly to hold a pipe die.

pipe stop A stopcock in a pipe.

pipe strap A thin metal strip used as a pipe hanger.

pipe tee A T-shaped fitting to connect three lengths of pipe in the same plane with one length at right angles to the other two.

pipe thread A V-cut screw thread cut on the inside or outside of a pipe or fitting. The diameter of the thread tapers.

pipe tongs A hand tool used to screw or unscrew lengths of pipe and/or fittings.

pipe vise A vise for holding pipe or tubing for cutting or threading. The pipe is held in curved, V-shaped serrated jaws or, for larger pipes, by chains.

pipe wrench A heavy hand tool with adjustable serrated jaws for gripping, screwing, or unscrewing metal pipe.

piping (1) An assembly of lengths of pipe and fittings, i.e., a run of pipe. (2) Movement of soil particles by percolating water that produces erosion channels.

piping loss The heat lost from piping between the heat source and the radiators.

pit (1) An excavation, quarry, or mine made or worked by the open cut method. A pit seldom goes below the ground water level.

(2) The area between the stage and the first row of seats in a theater. (3) A small hole or cavity on a surface.

pit boards Horizontal boards used as sheeting to retain the soil around a pit.

pitch (1) An accumulation of resin in the wood cells in a more or less irregular patch. Pitch is classified for grading purposes as light, medium, heavy, or massed. (2) The angle or inclination of a roof, which varies according to climate and roofing materials used. (3) The set, or projection, of teeth on alternate sides of a saw to provide clearance for its body.

pitch board A thin piece of board used as a guide in stair construction. The board is cut in the shape of a right triangle to the slope of the nosings of the treads. Usually, the two sides equal the tread length and rise of the stair.

pitch dimension The distance between the bases of the top and the bottom risers in a flight of stairs, measured parallel to the slope.

pitched roof A roof having one or more surfaces with a slope greater than 10 degrees from the horizontal.

pitched stone A rough-faced stone having each edge of the exposed face pitched at a slight bevel from the plane of the face.

pitch fiber pipe A rough-faced stone having each edge of the exposed face pitched at a slight bevel from the plane of the face.

pitching chisel A mason's chisel with a wide thick edge used for rough dressing stone.

pitch pine *Pinus rigida.* This pine is found in a wide area, from Maine to northern Georgia. Lumber from it is graded under rules established by the Northeastern Lumber Manufacturers Association.

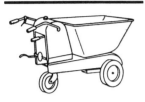

Placing

pitch pocket An opening between growth rings which usually contains or has contained resin, or bark, or both. A pitch pocket is classified for grading purposes as very small, small, medium, large, closed, open, or through.

pitch seam Shake or check filled with pitch.

pitch select D Select or better, except that the grade admits any amount of medium to heavy pitch. Massed pitch is admissible but limited to half the area of an otherwise high line piece. Dimensions are 4 inches and wider, 6 feet and longer in multiples of one foot.

pitch streak A well-defined accumulation of pitch in a more or less regular streak. A pitch streak is classified for grading purposes as small, medium, or large.

pith The small, soft core in the center of a log.

pith fleck A narrow streak resembling pith on the surface of a piece of lumber, usually brownish and up to several inches in length, resulting from the burrowing of larvae in the growing tissue of the tree.

pith knot A minor defect in lumber, a pith knot is a knot whose only blemish is a small pith hole in the center.

pitot tube A device, used with a manometer or other pressure reading device, to measure the velocity head of a flowing fluid.

pit prop A timber used as a support in an excavation or mine.

pit-run gravel Ungraded gravel used as taken from a pit.

pitting Development of relatively small cavities in a surface due to phenomena such as corrosion, cavitation, or, in concrete, localized disintegration. See also **popout**.

pivot A short shaft or pin about which a part rotates or swings.

pivoted door A swing door that swings on pivots, rather than a door hung on hinges.

pivoted window A window with a sash that rotates about fixed horizontal or vertical pivots.

placeability See **workability**.

placement (1) The process of placing and consolidating concrete. (2) A quantity of concrete placed and finished during a continuous operation. Also inappropriately referred to as "pouring".

placing The deposition, distribution, and consolidation of freshly mixed concrete in the place where it is to harden. Also inappropriately referred to as "pouring".

plain ashlar A rectangular block of stone, the face of which has been smoothed with a tool.

plain bar A reinforcing bar without surface deformations, or one having deformations that do not conform to the applicable requirements.

plain concrete (1) Concrete without reinforcement. (2) Reinforced concrete that does not conform to the definition of reinforced concrete. (3) Used loosely to designate concrete containing no admixture and prepared without special treatment.

plain masonry Masonry with no reinforcement or with reinforcement only for shrinkage and temperature changes.

plain rail A meeting rail in a double-hung window that is the same thickness as the other members of the frame.

plain-sawn Wood sawn from logs so that the annual rings intersect the wide faces at an angle less the 45 degrees.

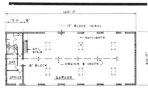

Plan

Planking

Planter

plain tile A flat rectangular tile of concrete or burnt clay.

plan A two-dimensional graphic representation of the design, location, and dimensions of the project, or parts thereof, seen in a horizontal plane viewed from above. See also **drawings**.

planar frame A structural frame with all members in the same plane.

plancier The wood or plaster soffit or underside of an overhanging eave.

plan deposit See **deposit for bidding documents**.

plane (1) A flat surface. (2) A tool used to smooth or shape wood. (3) To run sawn wood through a planer to smooth its surface.

planed all round A piece that has been surfaced on all four sides.

planed lumber Lumber which has been run through a planer to finish one or several sides.

plane of weakness The plane along which a body under stress will tend to fracture. The plane of weakness may exist by design, by accident, or because of the nature of the structure and its loading.

planer A machine used to surface rough lumber.

planer heads Sets of cutting knives mounted on cylindrical heads which revolve at high speed to dress lumber fed through them. Top and bottom heads surface or pattern the two faces, while side heads dress or pattern the two edges or sides.

planer knife One of the sharp blades used in a planer head.

plane surveying Surveying which neglects the curvature of the earth.

plane table A device, consisting of a drawing board on a tripod and a telescope attached to a ruler, for plotting lines of a survey directly from observations.

planimeter A mechanical device that measures plane areas on a map or drawing.

planing The process of smoothing a surface by shaving of small chips.

planing machine (1) Several types of a fixed machine for planing wood or steel. (2) A portable machine for planing a wood floor in place.

plank A piece of lumber two or more inches thick and six or more inches wide, designed to be laid flat as part of a loadbearing surface, such as a bridge deck.

planking Material used for flooring, decking, or scaffolding.

planking and strutting The temporary timbers supporting the soil at the side of an excavation.

plank-on-edge floor A subfloor formed by joists in contact with each other to form a continuous surface.

plank-type grating An aluminum extrusion consisting of a tread plate reinforced by integral ribs. The tread plate is perforated between the ribs.

planning The process of developing a scheme of a building or group of buildings by studying the layout of spaces within buildings, and of buildings and other installations in an open space.

planning grid A graph-like paper, with the lines at right angles or other selected angles to each other, used by architects or engineers in modular planning.

planted molding A molding which is nailed, tongued-in, or otherwise fastened to a base, as opposed to one cut into the base material.

planted stop A molding or strip nailed to a frame and used as a door or casement stop.

planter An ornamental container to hold plants.

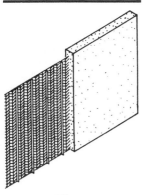

Plaster

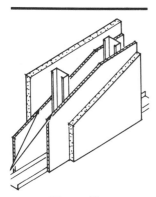

Plaster Base

planting In masonry, laying the first courses of a foundation on a prepared bed.

planting box A box, usually wood, to hold live plants. The box is placed in a planter.

plant mix (1) A mixture of aggregate and asphalt cement or liquid asphalt, prepared in a central or traveling mechanical mixer. (2) Any mixture produced at a mixing plant.

plaster (1) A cementitious material or combination of cementitious material and aggregate that, when mixed with a suitable amount of water, forms a plastic mass or paste. When applied to a surface, the paste adheres to it and subsequently hardens, preserving in a rigid state the form or texture imposed during the period of plasticity. (2) The placed and hardened mixture created in definition 1 above. See also **stucco**.

plaster aggregate Graded mineral particles and mineral, vegetable, or animal fibers to be used with gypsum or cement-base plasters to produce a plaster mix.

plaster base Any working ground to receive plaster, including wood, metal, or gypsum lath, insulating board, or masonry.

plaster-base finish tile Ceramic tile with exposed surfaces scored or roughened for an application of a plaster system.

plaster bead An edging, usually metal, to strengthen applied plaster at corners.

plasterboard Any prefabricated board of plaster with paper facings. Plasterboard may be painted or used as a base for a finish coat of applied plaster. Also called "sheetrock" or "drywall".

plasterboard nail A nail for fastening plasterboard to a supporting system. The nails are galvanized with a flat head and a deformed shank.

plaster bond The mechanical or chemical adhesion of plaster to a surface.

plaster ceiling panel A raised or sunken section of a plaster ceiling, forming a panel.

plaster cornice A molding of plaster at the intersection of a wall and a ceiling.

plaster cove A concave molding of plaster at the intersection of a wall and a ceiling.

plaster ground A wood strip or metal bead used as a guide for application of a desired thickness of plaster or for attaching trim.

plaster guard A shield attached behind the hinge and strike reinforcement on a hollow metal doorframe to prevent mortar or plaster from entering mounting holes.

plaster lath A supporting structure for plaster, such as a wood lath, metal lath, or lath board.

plaster mold A mold or form made from gypsum plaster, usually to permit concrete to be formed or cast in intricate shapes. See also **mold** and **form**.

plaster of Paris Gypsum, from which three-quarters of the chemically bound water has been driven off by heating. When wetted, it recombines with water and hardens quickly. See also **hemihydrate**.

plaster ring A metal collar attached to a base and used as a guide for thickness of applied plaster and a fastener for trim.

plaster set The initial stiffening of a plaster mix that may be reworked without the addition of water.

plaster wainscot cap A horizontal wood strip which covers the joint between the wainscoting and the plaster surface.

plasterer's putty A hydrated lime with just enough water added to make a thick paste for use as a hole or crack filler.

Plastic Skylight

plastic Possessing plasticity, or possessing adequate plasticity. See also **plasticity**.

bond fire clay (1) A fire clay of sufficient natural plasticity to bond nonplastic material. (2) A fire clay used as a plasticizing agent in mortar.

cement A synthetic cement used in the application of flashing.

centroid Centroid of the resistance to load computed for the assumptions that the concrete is stressed uniformly to 0.85 its design strength and the steel is stressed uniformly to its specified yield point.

consistency (1) Condition of freshly mixed cement paste, mortar, or concrete such that deformation will be sustained continuously in any direction without rupture. (2) In common usage, concrete with slump of 3 to 4 in. (80 to 100 mm).

cracking Cracking that occurs in the surface of fresh concrete soon after it is placed and while it is still plastic.

deformation Deformation that does not disappear when the force causing the deformation is removed.

design See *ultimate-strength design*.

flow See *creep*.

glue Resin bonding materials used in joining wood pieces. These materials include: 1. Thermosetting resins such as phenol-formaldehyde, urea-formaldehyde, and melamine resin. 2. Thermoplastics such as acryl-polymers and vinyl-polymers. 3. Casein plastics. 4. Natural resin glues.

laminate A thin board used as a finished surfacing, made from layers of resin-impregnated paper fused together under heat and pressure.

limit The water content at which a soil will just begin to crumble when rolled into a thread approximately 1/8 in. (3 mm) in diameter. See also *Atterberg limits*.

loss See *creep*.

mortar A mortar of plastic consistency.

shrinkage cracks See *plastic cracking*.

skylight A molded unit of transparent or translucent plastic that is set in a frame for use as a skylight.

wood A quick drying putty of nitrocellulose, wood flour, resins and solvents used as a filler for holes and cracks.

plastic-hinge Region where ultimate moment capacity in a member may be developed and maintained with corresponding significant inelastic rotation as main tensile steel elongates beyond yield strain.

plasticity (1) The capability of being molded, or being made to assume a desired form. (2) A property of wood that allows it to retain its form when bent. (3) A complex property of a material involving a combination of qualities of mobility and magnitude of yield value. (4) That property of freshly mixed cement paste, concrete, or mortar, which determines its resistance deformation or ease of molding.

plasticity index (1) The range in water content through which a soil remains plastic. (2) Numerical difference between the liquid limit and the plastic limit. See also **Atterberg limits**.

plasticizer A material that increases plasticity of a cement paste, mortar, or concrete mixture.

plasticizing Producing plasticity or becoming plastic.

plate (1) In formwork for concrete, a flat, horizontal member at the top and/or bottom of studs or

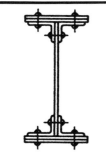

Plate Girder

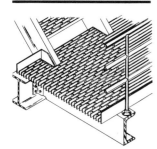

Platform

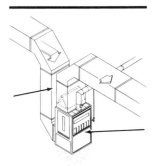

Plenum

posts. If on the ground, a plate is called "a mudsill". (2) In structural design, a member, the depth of which is substantially smaller than its length and width. (3) A flat rolled iron or steel product. See also **flat plate.**

load-transfer assembly.

anchor An anchor bolt used to fasten a plate or sill to a foundation.

beam See *plate girder.*

bolt See *plate anchor.*

girder A girder fabricated from plates, angles, or other structural shapes, welded or riveted together.

glass High-quality glass of the same composition as window glass but thicker, up to 1-1/4 inches, with ground and polished faces, usually used for large areas in a single sheet.

rail Decorative molding on the upper part of a wall and grooved to hold chinaware plates or decorations.

stock Plates are pieces of lumber used in construction, to which the studs are fastened at top and bottom to form a wall. The term plate stock usually refers to a general class of commodity-type 2 x 4 lumber used as plates.

vibrator A self-propelled, mechanical vibrator used to compact fill.

plate-type tread A stair tread fabricated from metal plate and/or floor plate; the riser may be integral.

platen A flat plate in a hot press, usually one of many in a multi-opening press used in the manufacture of panel products.

platform (1) A floor or surface raised above the adjacent level. (2) A landing in a stairway. (3) (OSHA) A working space for persons, elevated above the surrounding floor or ground level such as a balcony or platform for

the operation of machinery or equipment.

frame See *platform framing.*

framing A framing system in which the vertical members are only a single story high, with each finished floor acting as a platform upon which the succeeding floor is constructed. Platform framing is the common method of house construction in North America.

roof (1) A truncated roof. (2) A roof, the top of which is a horizontal plane.

plenishing nail A large nail used to fasten planks to joists.

plenum A closed chamber used to distribute or collect warmed or cooled air in a forced-air heating/cooling system.

plenum barrier A barrier, erected in a plenum ceiling, used to reduce sound transmission between rooms or over a large area.

plenum chamber See **Plenum.**

plen-wood system A system for distributing air for heating and cooling using the entire under-floor area of a building as a plenum chamber. The system eliminates the need for duct work in some structures. Proponents of the system cite savings in both construction costs and in the costs of heating and cooling.

plinth (1) A block or slab supporting a column or pedestal. (2) The base course of an external masonry wall when of different shape from the masonry in the wall paper.

course (1) The masonry course which forms the plinth of a stone wall. (2) The final course of a brick plinth in a brick wall.

plot (1) A measured and defined area of land. (2) A ground plan of a building and adjacent land.

plow (1) In molding, a rectangular slot of three surfaces cut with the

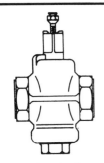

Plug Cock

grain of the wood. (2) In carpentry, a tool that cuts grooves.

plow and tongue joint See **tongue-and-groove joint**.

plucked finish A rough-textured stone surface, made by overcutting with a planer so that stone is removed by spalling rather than shaving.

plug (1) A wood peg driven into a wall for support of a fastener. (2) A stopper for a drain-opening. (3) A male-threaded fitting used to seal the end of a pipe or fitting. (4) A fixture for connection of electric wires to an outlet socket. (5) A fibrous or resinous material used to fill a hole and close a surface. (6) Material that stops or seals the discharge line of a channel or pipe.

center bit A plug-shaped bit used to enlarge a hole or counterbore around the hole.

cock A valve where full flow is through a hole in a tapered plug. Rotating the plug 90 degrees completely stops the flow.

cutter A small bit used to cut a hole to receive a plug. A plug cutter is used to conceal recessed screwheads in hardwood floors.

fuse A fuse contained in an insulated container with a metal screwbase. There is a small window on the face of the container for checking the condition of the fuse element.

tap See *plug cock*.

tenon A short tenon which projects from the material into which it is fitted, the free end fitting into a mortise. A plug tenon is used to provide lateral stability for a wood column.

valve See *plug cock*.

weld A weld made through a circular hole in one of the members to be connected.

plugged lumber Lumber in which a defect has been filled by plastic material to provide a smooth paint surface.

plugging Drilling a hole in masonry surface for a wood plug. The plug will later be used to support a fastener.

plugging chisel A steel rod, with a star-shaped point, used for drilling holes in masonry by striking with a hammer.

plum, plum stone A large random-shaped stone dropped into freshly placed mass concrete to economize on the volume of concrete used. See also **cyclopean concrete**.

plumb Vertical, or to make vertical.

bob A cone shaped metal weight, hung from a string, used to establish a vertical line or as a sighting reference to a surveyor's transit.

bond Any bond in masonry in which the vertical joints are in line.

bond pole A pole used to insure that vertical masonry joints are in line.

cut A vertical cut, as the cuts in a rafter at the top ridge where it meets the ridge plate.

joint A sheet-metal joint made by lapping the edges and soldering them together flat.

level A level that is set in a horizontal position by placing it at a right angle to a plumb line.

line The cord or line that supports a plumb bob.

rule A board or metal rule, fitted with one or more leveling bubbles, used to establish horizontal and vertical lines.

plumber's friend, plunger A tool, consisting of a large rubber suction cup on a wood handle, for clearing plumbing traps of minor obstructions.

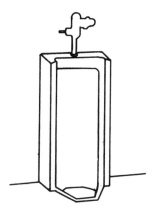

Plumbing Fixtures

plumber's furnace A portable, gas-fired furnace for melting solder, heating lead, or soldering irons.

plumber's rasp A coarse rasp used to file lead.

plumber's round iron A specially shaped soldering iron used to solder seams in tanks.

plumber's soil A mixture of lampblack and glue used to prevent solder from adhering where not wanted.

plumber's solder An alloy of lead and tin, with a melting point below that of lead, used for joining lead pipes.

plumbing (1) The work or practice of installing in buildings the pipes, fixtures and other apparatus required to bring in the water supplies and to remove water-borne wastes. (2) The process of setting a structure or object truly vertical.

fixture A receptacle in a plumbing system, other than a trap, in which water or wastes are collected or retained for use and ultimately discharged to drainage.

system The water supply and distribution pipes; plumbing fixtures and traps; soil, waste, and vent pipes; building drains and sewers; and respective devices and appurtenances within a building.

plummet A plumb bob.

plunger See *plumber's friend*.

ply (1) A single layer or sheet of veneer. (2) One complete layer of veneer in a sheet of plywood.

plyform The trademark owned by the American Plywood Association for concrete form panels produced by its association members.

plymetal Plywood covered on one or both sides with sheet metal.

plywood A flat panel made up of a number of thin sheets, or veneers, of wood, in which the grain direction of each ply, or layer, is at right angles to the one adjacent to it. The veneer sheets are united, under pressure, by a bonding agent.

plywood squares Plywood fabricated for use as floor tile.

pneumatically applied mortar See *shotcrete*.

pneumatic caisson A system in which control is effected by pressurized air.

pneumatic control system A system in which control is effected by pressurized air.

pneumatic drill A reciprocating drill actuated by compressed air.

pneumatic feed Shotcrete delivery equipment in which material is conveyed by a pressurized air stream.

pneumatic hammer An air-powered reciprocating drill fitted with a chisel or hammer.

pneumatic placement See **pneumatic feed**.

pneumatic structure A fabric envelope supported by an internal air pressure slightly above atmospheric pressure. The pressure is provided by a series of fans.

pneumatic water supply A water supply system for a building in which water is distributed from a tank containing water and compressed air.

pocket (1) A recess in a masonry wall to receive an end of a beam. (2) A recess in a wall to receive part or all of an architectural item, such as a curtain or folding door. (3) The slot on the pulley stile of a double hung window from through which the sash weight is placed in the sash weight channel.

chisel A chisel with a wide blade that is sharpened on both sides.

piece A small piece of wood that closes the *pocket, def. 3* in the pulley stile of a double hung window.

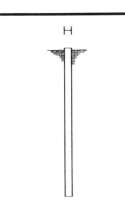

Point-bearing Pile

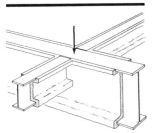

Point Load

Point of Support

rot A type of decay found in cedar.

pockmarking Undesirable depressions formed in a paint or varnish film.

podium (1) A stand for a speaker. (2) An elevated platform for a conductor. (3) The masonry platform on which a classical temple was built.

point (1) A tooth for a saw. (2) A mason's tool.

point-bearing pile A pile that transfers its load to the supporting stratum by point bearing as opposed to a friction pile.

point count Method for determination of the volumetric composition of a solid by observation of the frequency with which areas of each component coincide with a regular system of points in one or more planes intersecting a sample of the solid.

point count (modified) The point count method supplemented by a determination of the frequency with which areas of each component of a solid are intersected by regularly spaced lines in one or more planes intersecting a sample of the solid.

pointed ashlar Rectangular stone work with face markings made by a pointed tool.

pointed work The rough finish on the face of a stone that is made by a pointed tool.

pointing (1) The finishing of joints in a masonry wall. (2) The material with which joints in masonry are finished.

pointing trowel A diamond-shaped trowel used in pointing or repointing masonry joints.

point load A term used in structural analysis to define a concentrated load on a structural member.

point of contraflexure See *point of inflection.*

point of inflection (1) The point on the length of a structural member subjected to flexure where the curvature changes from concave to convex or conversely, and at which the bending moment is zero. Also called "point of contraflexure". (2) Location of an abrupt bend in a plotted locus of points in a graph.

point of support A point on a member where part of its load is transferred to a support.

point source A light source, the dimensions of which are insignificant at viewing distance. A fluorescent lamp is a point source at a large distance.

Poisson's ratio The ratio of transverse (lateral) strain to the corresponding axial (longitudinal) strain resulting from uniformly distributed axial stress below the proportional limit of the material. The ratio will average about 0.2 for concrete and 0.25 for most metals.

polarized receptacle An electric receptacle with contacts arranged so a mating plug must be inserted in only one orientation.

polarizing microscope A microscope equipped with elements permitting observations and determinations to be made using polarized light. See also *Nicol prism.*

pole (1) A long, usually round piece of wood, often a small diameter log with the bark removed, used to carry utility wires or for other purposes. A pole is often treated with preservative. (2) Either of two oppositely charged terminals, as in an electric cell or battery. (3) Either extremity of an axis of a sphere.

pole-frame construction A construction system using vertical poles or timbers.

pole plate A horizontal board or timber that rests on the tie beams of a roof and supports the lower ends of the common rafters at the

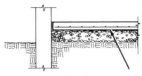

Polyethylene

wall, and also raises the rafters above the top plate of the wall.

pole shore See *post shore*.

pole trailer A specially constructed log trailer designed to carry extremely long poles, usually employing a disconnected rear section with independent steering, much like a long-ladder fire truck.

polish (1) To give a sheen or gloss to a finish coat of plaster. (2) The operation in which fine abrasives are used to hone a finished surface to a desired smoothness.

polished finish A stone work finish so smooth that it forms a reflective surface, usually produced by mechanical buffing and chemical treatment of a surface with no voids.

polish or final grind The final operation in which fine abrasives are used to hone a surface to its desired smoothness and appearance.

polishing varnish A hard varnish that can be polished by rubbing with abrasive and mineral oil without dissolving the resin.

poll The broad end or striking face of a hammer.

polychromatic finish (1) A finish obtained by blending a number of colors. (2) A finish obtained by using a paint containing metallic flakes on transparent pigments. The resulting effect is the appearance of a variety of colors when viewed from different angles.

polyester resin A synthetic resin that polymerizes during curing and has excellent adhesive properties, high strength, and good chemical resistance.

polyethylene A thermoplastic high-molecular-weight organic compound. In sheet form, polyethylene is used as a protective cover for concrete surfaces during the curing period, a temporary enclosure for construction operations, and as a vapor barrier.

polygonal masonry Masonry constructed of stones with multi-sided faces.

polymer (1) The product of polymerization. (2) A rubber or resin consisting of large molecules formed by polymerization.

polymer-cement concrete A mixture of water, hydraulic cement, aggregate, and a monomer or polymer. The cement is polymerized in place when a monomer is used.

polymer concrete (1) Concrete in which an organic polymer serves as the binder. See **concrete**. Also known as "resin concrete". (2) Sometimes erroneously employed to designate hydraulic cement mortars or concretes in which part or all of the mixing water is replaced by an aqueous dispersion of a thermoplastic copolymer.

polymerization The reaction in which two or more molecules of the same substance combine to form a compound containing the same elements, and in the same proportions, but of high molecular weight. The original substance can be generated from the compound, in some cases only with extreme difficulty.

polypropylene A tough plastic with good resistance to heat and chemicals. Polypropylene is a polymer of propylene.

polystyrene foam A low-cost, foamed plastic weighing about 1 lb. per cu ft, with good insulating properties and resistant to grease.

polystyrene resin Synthetic resins, varying in color from water-white to yellow, formed by the polymerization of styrene on heating, with or without catalysts. These resins may be used in paints for concrete, for making sculptured molds, or as insulation.

polysulfide coating A protective coating system prepared by polymerizing a chlorinated alkyl polyether with an inorganic polysulfide. This coating exhibits outstanding resistance to ozone, sunlight, oxidation, and weathering.

polyurethane Reaction product of an isocyanate with any of a wide variety of other compounds containing an active hydrogen group. Polyurethane is used to formulate tough, abrasion-resistant coatings.

polyurethane finish A synthetic varnish that is exceptionally hard and wear-resistant.

polyvinyl acetate Colorless, permanently thermoplastic resin, usually supplied as an emulsion or water-dispersible powder, which may be used in paints for concrete. Polyvinyl acetate is characterized by flexibility, stability towards light, transparency to ultraviolet rays, high dielectric strength, toughness, and hardness. The higher the degree of polymerization, the higher the softening temperature.

polyvinyl chloride, PVC A synthetic resin prepared by the polymerization of vinyl chloride, used in the manufacture of nonmetallic waterstops for concrete, floor coverings, pipe and fittings.

pommel (1) A knob at the top of a conical or dome-like roof. (2) A rounded metal block on an end of a handle, raised and dropped by hand to compact soil.

Ponderosa Pine *Pinus ponderosa*. A pine species found in a wide range that reaches from British Columbia to Mexico, and from the Pacific coast to the Dakotas. The wood is widely used in general construction, most often as boards, but is more valued for its uses in millwork and in cuttings for remanufacture. Also known as "Western White Pine" and "Western Yellow Pine".

ponding (1) The process of flooding the surface of a concrete slab by using temporary dams around the perimeter in order to satisfactorily cure the concrete. (2) The accumulation of water at low points in a roof. The low points may be produced or increased by structural deflection.

Pond Pine *Pinus serotina*. A minor species of the Southern Yellow Pine group found along the Atlantic coast from southeast Virginia to the Florida panhandle. Lumber from this species carries a "Mixed Pine Species" stamp.

Poor Pine Another name for "Spruce Pine".

pop A delaminated area in a plywood panel. Also called a "blow" or "blister".

pop-corn concrete No-fines concrete containing insufficient cement paste to fill voids among the coarse aggregate so that the particles are bound only at points of contact. See *no-fines concrete*.

poplar A member of the willow family. In North America: *Populus tremula*, Aspen; *P. balsamifera*, Cottonwood; *P. tacamahaca*, Balsam Poplar. Its wood is used such as in furniture core stock, crates, plywood.

popout The breaking away of small portions of a concrete surface due to internal pressure, leaving a shallow, typically conical, depression.

popping Shallow depressions ranging in size from pinheads to 1/4 inch in diameter, immediately below the surface of a lime-putty finish coat. Popping is caused by expansion of coarse particles of unhydrated lime or of foreign substances.

pop valve A safety valve made to open immediately when the fluid pressure is greater than the design force of a spring.

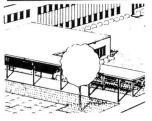

Portico

porcelain A hard glazed or unglazed ceramic used for electrical, chemical, mechanical, or thermal components.

enamel A silicate glass bonded to metal by fusion at a temperature above 800 degrees F (427 degrees C). Porcelain enamel is not a true porcelain.

tile A dense, usually impervious, fine-grained, smooth-surfaced, ceramic mosaic tile or paver.

tube A ceramic tube, with a slight shoulder at one end, used to carry an exposed, insulated wire where it passes through a wood joist, stud, etc.

porch A roofed structure, with or without walls, on the outside of a building, often at the entrance.

porch lattice An open lattice that closes the open side(s) of a porch below floor level.

porcupine boiler A vertical, cylindrical boiler with many projecting, closed stubs to provide an additional thermal surface.

pore water The free water present in soil.

pore water pressure The pressure of the water in a saturated soil.

porosity The ratio, usually expressed as a percentage, of the volume of voids in a material to the total volume of the material, including the voids.

porous fill See **pervious soil.**

porous woods Hardwoods that have pores or vessels that can be seen with the naked eye.

portico (1) A covered walk consisting of a roof supported on columns. (2) A colonnaded (continuous row of columns) porch.

portland blast-furnace slag cement See *cement, portland blast-furnace slag.*

portland cement See *cement, portland.*

portland cement concrete See *concrete.*

portlandite A mineral, calcium hydroxide, that occurs naturally in Ireland. Portlandite is a common product of hydration of portland cement.

portland-pozzolan cement See *cement, portland-pozzolan.*

Portland stone A limestone, from the Island of Portland off the coast of England, used as a building stone.

port of entry (POE) A port that provides custom house services to collect duties on imports.

Port Orford Cedar *Chamaecyparis lawsoniana.* A cedar common to the coastal belt of western Oregon and extreme northern California, having limited markets, but its light-colored wood is greatly desired in Japan for exposed use in houses.

position (1) A trader's open contracts in the futures market. (2) A reference to a shipping period, as in "Feb/March position."

positioned weld A weld on a joint that as been oriented to facilitate the welding.

position indicator A device that shows the position of an elevator in its hoistway. Also called a "hall position indicator" if at a landing, or a "cab position indicator" if in the cab.

position trading An approach to futures trading in which the trader either buys or sells contracts and holds them for an extended period, as distinguished from a day trader, who will normally initiate and offset his position in a single trading day.

positive cutoff A below-ground wall that extends to an impervious lower stratum to block subsurface seepage.

positive displacement Wet-mix shotcrete delivery equipment in which the material is pushed through the material hose in a solid mass by a piston or auger.

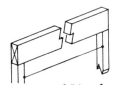

Post-and-Lintel Construction

Post Shore

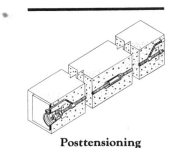

Posttensioning

positive moment A condition of flexure in which, for a horizontally simply supported member, the deflected shape is normally considered to be concave downward and the top fibers subjected to compression stresses. For other members and other conditions, consider positive and negative as relative terms. See also *negative moment*. (Note: For structural design and analysis, moments may be designated as positive or negative with satisfactory results as long as the sign convention adopted is used consistently.)

positive reinforcement Reinforcement for positive moment.

possum-trot plan Plan of a house with two areas separated by a breezeway, all sections having a common roof.

post (1) A member used in a vertical position to support a beam or other structural member in a building, or as part of a fence. In lumber, 4x4s are often referred to as posts. Most grading rules define a post as having dimensions of five inches by five or more inches in width, with the width not more than two inches greater than the thickness. (2) Vertical formwork member used as a brace. Also called a "shore", "prop", "jack".

post-and-beam construction See *post-and-lintel construction.*

post-and-beam framing Framing in which the horizontal members are supported by a distinct column, as opposed to a wall.

post-and-lintel construction A construction that uses posts or columns and a horizontal beam to span an opening, as opposed to a construction using arches or vaults.

post and pane A type of construction in which timber framings are filled in with brick or plaster panels, leaving the timbers exposed.

postbuckling strength The load that can be carried by a structural member after it has been subjected to buckling.

post-completion services See *post-construction services*.

post-construction services (1) Under traditional forms of agreement, additional services rendered after issuance of the final certificate for payment, or in the absence of the final certificate for payment, more than sixty days after the date of substantial completion of the work. (2) Under designated services forms of agreement, services necessary to assist the owner in the use and occupancy of the facility.

postern (1) Any small, often inconspicuous, door. (2) A smaller door, for pedestrian passage, located next to a large door that is used by vehicles.

post pole See *post shore*.

post shore, pole shore Individual vertical member used to support loads.

  adjustable timber single-post shore Individual timber used with a fabricated clamp to obtain adjustment and not normally manufactured as a complete unit.

  fabricated single-post shore (1) Type I: Single all-metal post, with a fine-adjustment screw or device in combination with pin-and-hole adjustment or clamp. (2) Type II: Single or double wooden post members adjustable by a metal clamp or screw and usually manufactured as a complete unit.

  timber single-post shore Timber used as a structural member for shoring support.

posttentioning A method of prestressing reinforced concrete in which tendons are tensioned after the concrete has hardened.

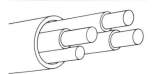

Power Cable

Power Panelboard

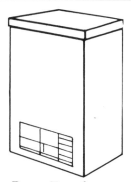

Power Transformer

potable water Water that satisfies the standards of the responsible health authorities as drinking water.

pot floor A floor surface of structural clay tiles.

pot life Time interval, after preparation, during which a liquid or plastic mixture is usable.

pound-calorie The amount of heat required to raise one pound of water one degree centigrade.

pour coat, top mop The top coating of asphalt on a built-up roof, sometimes including embedded gravel or slag.

pouring of concrete See **placement** and **placing**.

pouring rope See **asbestos joint runner**.

powdered asphalt Hard asphalt crushed or ground to fine particles, and softened by adding flux oil.

powder molding A method of manufacturing objects by melting polyethylene powder in a mold.

powder post A condition in which wood has decayed to powder or been eaten by borers that leave holes full of powder.

power The rate of performing work or the rate of transforming, transferring, or consuming energy. Power is usually measured in watts, BTU/hour, or horsepower.

power buggy A wheelbarrow-sized machine powered by a gasoline engine or an electric motor.

power cable A usually heavy cable, consisting of one or more conductors with insulation and jackets, for conducting electric power.

power consumption The rate at which power is consumed by a device or unit (such as a building), usually expressed in kilowatt-hours, BTU/hour, or horsepower-hours.

power drill An electric-powered, hand-held drill, activated by pressing a trigger-like switch.

power float See **rotary float**.

power of attorney An instrument authorizing another to act as one's agent. See also **attorney-in-fact**.

power panelboard A panel board used for circuits supplying motors and other heavy power-consuming devices, as opposed to a panel board used for lighting circuits.

power sander An electric-powered hand tool used for smoothing and/or polishing.

power shovel A self-propelled, power-operated machine used to excavate and/or load soils or debris.

powers spacing factor See **spacing factor**.

power take-off A place in a transmission or engine to which a shaft can be attached to drive an external device.

power transformer A device in an alternating-current electrical system that transfers electric energy between circuits, usually changing the voltage in the process.

power trowel See **mechanical trowel**.

power wrench See **impact wrench**.

pozzolan A siliceous, or siliceous and aluminous, material which, in itself, possesses little or no cementitious value but will, in finely divided form and in the presence of moisture, chemically react with calcium hydroxide at ordinary temperatures to form compounds possessing cementitious properties.

pozzolan cement A natural cement, used in ancient times, made by grinding pozzolan with lime.

pozzolanic Of or pertaining to a pozzolan.

pozzolanic reaction See **pozzolan**.

Preaction Sprinkler System

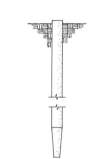

Precast Concrete Pile

Precast Concrete Wall Panel

Pratt truss A type of truss with parallel chords, all vertical members in compression, and all diagonal members in tension. The diagonals slant toward the center.

preaction sprinkler system A dry pipe sprinkler system in which water is supplied to the piping when a smoke or heat detector is activated.

preassembled lock A factory assembled lock, requiring little or no alterations on installation.

preboring (1) Drilling a pilot hole. (2) Boring a hole, for a timber pile, through a hard stratum that would damage the pile if driven.

precast (1) A concrete member that is cast and cured in other than its final position. (2) The process of placing and finishing precast concrete.

precast concrete pile A concrete pile, either reinforced or prestressed, cast elsewhere than its final position.

precast concrete wall panel A concrete wall panel, either reinforced or prestressed, cast elsewhere than its final position.

precast pile A reinforced concrete pile manufactured in a casting plant or at the site but not in its final position.

precipitator See **electrostatic precipitator**.

precise level An instrument similar to an ordinary surveyor's level but capable of finer readings and including a prism arrangement that permits simultaneous observation of the rod and the leveling bubble.

precise leveling rod A leveling rod with fine graduation on an insert of metal with a low thermal expansion coefficient.

precoating See **tinning**.

precompressed zone The area of a flexural member which is compressed by the prestressing tendons.

preconsolidation pressure The greatest effective pressure a soil has experienced.

precure The process of curing a glued joint prior to pressing or clamping.

precured period See **presteaming period**.

precuring The premature curing of an adhesive due to press temperatures being too high, a too-rapid resin-curing speed, or a malfunctioning press. Precuring can result in plywood delamination or a poor-quality surface in particleboard.

precut A lumber item, usually a stud, that is cut to a precise length at the time of manufacture, so that it may be used in construction without further trimming at the job site.

predesign services Additional services of the architect provided prior to and preceding the customary basic services, including services to assist the owner in establishing the program, financial and time requirements, and limitations for the project. See also **programming phase**.

predrilled Lumber, such as roof decking, that has been drilled at the mill to accommodate bolts or other hardware.

prefab Abbreviation for prefabricated.

prefabricate To fabricate units or components at a mill or plant for assembly at another location.

prefabricated construction A construction method which uses standard prefabricated units that are assembled at a site along with site fabrication of some minor parts.

prefabricated flue A metal vent for fuel-fired equipment that is assembled from factory-made parts.

prefabricated joint filler A compressible material used to fill control, expansion, and

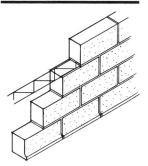

Prefabricated Tie

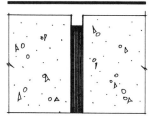

Preformed Asphalt Joint Filler

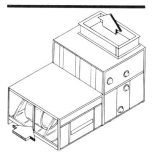

Preheater

contraction joints and may be used alone or as a backing for a joint sealant.

prefabricated masonry panel A wall panel of masonry units constructed at an assembly site and moved to a job site for erection.

prefabricated pipe conduit system Prefabricated units consisting of insulated piping for one or more utilities ready to be installed either above or below ground.

prefabricated tie A manufactured assembly consisting of two heavy parallel wires tied together by welded lighter wires. The tie is laid in masonry joints to tie two wythes together.

prefabricated wall Same as demountable partition.

preferred angle (1) Any angle of pitch of a stair between 30 degrees and 35 degrees. (2) Any angle of pitch of a ramp 15 degrees or less.

prefilled A particleboard panel whose surface has been made smooth by the application of a solvent-based filler before being shipped. Such panels have decorative overlays or laminates applied to them.

prefilter In an air conditioning system, a filter placed before the main filter(s). A prefilter is coarser and is used to remove larger particles.

prefinished Lumber, plywood, molding, or other wood product with a finish coating of paint, stain, vinyl, or other material applied before it is taken to the job site.

prefinished door A standard-sized door with both faces factory-finished and cuts and recesses provided for hardware.

prefiring Raising the temperature of refractory concrete under controlled conditions prior to placing it in service.

preformed asphalt joint filler Premolded strips of asphalt, vegetable or mineral filler, and fibers for use as a joint filler.

preformed foam Foam produced in a foam generator prior to introduction of the foam into a mixer with other ingredients to produce cellular concrete.

preformed joint sealant See **preformed sealant**.

preformed sealant A factory-shaped sealant that requires little field fabrication prior to installation.

preframed A construction term for wall, floor, or roof components assembled at a factory.

preheat coil A coil, in an air conditioning system, used to preheat air which is below 32 degrees F (0 degrees C).

preheater (1) A heat exchanger used to heat air that is to be used in the combustion chamber of a large boiler or furnace. (2) See **preheat coil**.

preheat fluorescent lamp A fluorescent lamp, the electrodes of which must be preheated before the arc can be started. The preheating can be manual or automatic.

prehung door A packaged unit consisting of a finished door on a frame with all necessary hardware and trim.

preliminary drawings Drawings prepared during the early stages of the design of a project. See also **schematic design phase** and **design development phase**.

preliminary estimate See **statement of probable construction cost**.

premature stiffening See **false set** and **flash set**.

premium (1) In commodity futures trading, a sum above the value of the item in the cash market. (2) A better quality than another product.

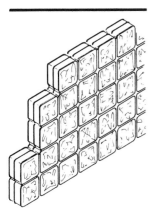

Pressed Glass

Pressure Connector

premium grade A general term describing the quality of one item as superior to another.

premolded asphalt panel A panel with a core of asphalt, minerals, and fibers, covered on each side with asphalt-impregnated felt or fabric and pressure bonded. The outside is then coated with hot asphalt.

prepacked concrete See **concrete, preplaced-aggregate**.

prepared roofing See **asphalt prepared roofing**.

prepayment meter A coin-operated water or gas meter that passes a fixed amount of fluid for each coin.

preplaced-aggregate concrete See **concrete, preplaced-aggregate** and **colloidal concrete**.

preposttensioning A method of fabricating prestressed concrete in which some of the tendons are pretensioned and a portion of the tendons are posttensioned.

prepreg In reinforced plastic, the reinforcing with applied resin before molding.

prequalification of bidders The process of investigating the qualifications of prospective bidders on the basis of their experience, availability, and capability for the contemplated project, and approving qualified bidders.

preservationist A term applied to one who objects to the use of natural resources because of a belief that such use will destroy basic values of the resource. The term is often used to refer to a member of various groups opposed to the expansion of industrial/commercial uses of public lands.

preservative Any substance applied to wood that helps it resist decay, rotting, or harmful insects.

preset period See **presteaming period**.

preshrunk (1) Concrete which has been mixed for a short period in a stationary mixer before being transferred to a transit mixer. (2) Grout, mortar, or concrete that has been mixed 1 to 3 hours before placing to reduce shrinkage during hardening.

pressed brick Brick that is molded under mechanical pressure. The resulting product is sharp-edged and smooth and is used for exposed surfaces.

pressed edge Edge of a footing along which the greatest soil pressure occurs under conditions of overturning.

pressed glass Glass units, such as pavement lights or glass block, that are pressed into shape.

pressed steel Die-stamped building components.

pressure (1) The force per unit area exerted by a homogenous liquid or gas on the walls of a container. (2) The force per unit area transferred between surfaces.

pressure bulb The zone in a loaded soil mass that is bounded by a selected stress isobar.

pressure cell An instrument used to measure the pressure within a soil mass or the pressure of the soil against a rigid wall.

pressure connector A mechanical device which forms a conductive connection between two or more electric conductors, or between one or more conductors and a terminal, without the use of solder.

pressure creosoting The process of forcing creosote, by use of pressure chambers, into timber.

pressure drop (1) The drop in pressure between two ends of a pipe or duct, between two points in a system, or across valves, fittings, etc., caused by friction losses. (2) In a water system, the drop caused by a difference in elevation.

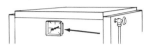

Pressure Gauge

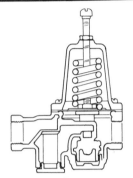

Pressure-reducing Valve

Prestress

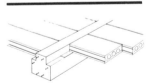

Prestressed Concrete

pressure forming A thermoforming process for plastics in which pressure is used to force a sheet against a mold, as opposed to vacuum forming.

pressure gauge An instrument for measuring fluid pressure.

pressure gun See **caulking gun**.

pressure line Locus of force points within a structure, resulting from combined prestressing force and externally applied load.

pressure-locked grating Metal grating in which cross bars and bearing bars are locked together at their intersections by deforming or swaging the metal.

pressure preserved Wood that has been treated with a preservative under pressure in a closed container.

pressure process The process of treating wood under pressure in a closed container. Pressure is usually preceded or followed by a vacuum.

pressure-reducing valve A valve which maintains a uniform fluid pressure on its outlet side as long as pressure on the inlet side is at or above a design pressure.

pressure-relief damper A damper which will open when pressure on the inside exceeds a design pressure.

pressure-relief device A disk or seal that is designed to open or rupture when pressure on a designated side exceeds a design value.

pressure-relief hatch A roof hatch designed to open or blow off under pressure from an explosion in a building. Some smoke-and-heat vents are also designed as pressure relief hatches.

pressure-relieving joint A horizontal expansion joint in panel-wall masonry, usually below supporting hangers at each floor. The joints prevent the weight of higher panels from being transmitted to the masonry below.

pressure-sensitive Capable of adhering to a surface when pressed against it.

pressure-sensitive adhesive An adhesive material that remains tacky after the solvents evaporate and will adhere to most solid surfaces with the application of light pressure.

pressure treating A process of impregnating lumber or other wood products with various chemicals, such as preservatives and fire retardants, by forcing the chemicals into the structure of the wood using high pressure.

pressure weather stripping Weather stripping designed to provide a seal by means of spring tension.

presteaming period In the manufacture of concrete products, the time between molding of a concrete product and start of the temperature-rise period.

prestress (1) To place a hardened concrete member or an assembly of units in a state of compression prior to application of service loads. (2) The stress developed by prestressing, such as by pretensioning or posttensioning. See also **prestressed concrete, prestressing steel, pretensioning,** and **post-tensioning**.

prestressed concrete Concrete in which internal stresses of such magnitude and distribution are introduced that the tensile stresses resulting from the service loads are counteracted to a desired degree. In reinforced concrete, the prestress is commonly introduced by tensioning the tendons.

prestressed concrete wire Steel wire with a very high tensile strength, used in prestressed concrete. The wire is initially stressed close to its tensile strength. Then some of this load is transferred to the concrete, by chemical bond or mechanical

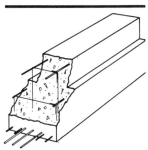

Prestressing Steel

Pretensioned Concrete

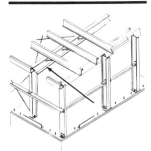

Primary Member

anchors, to compress the concrete.

prestressing Applying a load to a structural element to increase its effectiveness in resisting working loads. Prestressed concrete is a common example.

prestressing cable A cable or tendon made of prestressing wires.

prestressing steel High-strength steel used to prestress concrete, commonly seven-wire strands, single wires, bars, rods, or groups of wires or strands. See also **prestress, prestressed concrete, pretensioning,** and **post-tensioning.**

pretensioned concrete Tendons stressed in form before placing concrete and tendons released to provide load transfer where concrete has acheived strength.

primary consolidation Soil compaction caused by the application of sustained loads, principally due to the squeezing out of water in the voids of the affected soil.

pretensioning A method of prestressing reinforced concrete in which the tendons are tensioned before the concrete has hardened.

pretensioning bed (or bench) The casting bed on which pretensioned members are manufactured and which resists the pretensioning force prior to release.

pricking up Scoring the first coat of plaster on lath.

pricking-up coat The first, or base, coat of plaster on lath.

prick post An intermediate post in a truss. Theoretically, no loads are placed on it.

prick punch A pointed steel hand punch used to mark metal.

primacord A detonating fuse for high explosives, consisting of an explosive core in a strong, waterproof covering.

primary air Air which is fed to a burner to be mixed with gas.

primary battery A battery of two or more primary cells.

primary blasting The blasting operation in which a natural rock formation is dislodged from its original location.

primary branch (1) A drain between the base of a soil or waste stack and a building drain. (2) The largest single branch of a water-supply line or an air supply duct in a building.

primary cell A cell that generates electric current by electrochemical action. In the process, one of the electrodes is consumed and the process can not be effectively reversed. The cell cannot be recharged from an external source of electric power.

primary crusher A heavy crusher suitable for the first stage in a process of size reduction of rock, slag, or the like.

primary distribution feeder A feeder which operates at primary voltage supplying a distribution circuit.

primary excavation Excavating undisturbed soil.

primary light source (1) A source of light in which the light is produced by a transformation of energy. (2) The most obvious source of light when several sources are present.

primary member One of the main load-carrying members of a structural system, generally columns or posts.

primary nuclear vessel Interior container in a nuclear reactor designed for sustained loads and for working conditions.

prime (1) A grade of finish lumber, ranking below superior, the highest grade, and above E, the lowest grade of finish. Finish graded prime must present a fine appearance and is designed for application where finishing requirements are less exacting.

Principal Post

Processed Shake

(2) To supply water to a pump to enable it to start pumping. (3) In blasting, to place the detonator in a cartridge or charge of explosive.

prime coat (1) An application of low-viscosity liquid asphalt to an absorbent surface. (2) The first or preparatory coat in a paint system.

prime contract Contract between owner and contractor for construction of the project or portion thereof.

prime contractor Any contractor on a project having a contract directly with the owner.

prime mover (1) A powerful tractor or other vehicle used to move other machines. (2) A machine used to convert a fuel or steam into mechanical energy.

prime professional Any person or entity having a contract directly with the owner for professional services.

primer (1) (OSHA) A cartridge or container of explosives into which a detonator or detonating chord is inserted or attached. (2) A bituminous covering over compacted soil for waterproofing. (3) See **prime coat, def. 1 and 2**.

priming (1) The application of a prime coat. (2) Filling a pump or siphon with fluid to enable flow. (3) The first or annual filling of a canal or reservoir with water.

Prince Albert Fir Another name for West Coast Hemlock.

Princess Pine Another name for Jackpine.

princess post Subsidiary verticals, between queen posts and walls or the king post, used to stiffen a roof truss.

principal (in professional practice) Any person legally responsible for the activities of a professional practice.

principal beam The main beam in a structural frame.

principal-in-charge The architect or engineer in a professional practice firm charged with the

responsibility for the firm's services in connection with a given project.

principal planes See **principal stress**.

principal post A corner post in a framed building or a door post in a framed partition.

principal rafter One of the diagonals in a roof truss which support the purlins on which rafters are laid.

principal stress Maximum and minimum stresses, at any point on a structural member, acting at right angles to the mutually perpendicular planes, at zero shearing stress, designated as the principal planes.

prismatic glass Glass with parallel prisms rolled into one face. The prisms refract light rays and change their direction.

prismatic rustication Rusticated masonry with a diamond-shaped projections worked into the face of each stone.

prism glass See **prismatic glass**.

private sewer A sewer that is not in the public sewer system and subject only to the provisions of the local code.

private stairway A stairway intended to serve only one tenant.

privy An outhouse serving as a toilet.

probabilistic design Method of design of structures using the principles of statistics (probability) as a basis for evaluation of structural safety.

probable construction cost See **statement of probable construction cost**.

processed shake A sawn cedar shingle which is textured on one surface to resemble a split shingle.

Proctor compaction test A test to determine the moisture content of a soil at which maximum compaction can be obtained. The test establishes the density-moisture relationship in a soil.

	DESCRIPTION	COST $	%	MARCH 26 27 28 29 30 2 3 4 5
12	CLEAR & GRUB	2154	1.34	
14	EXC. FTG'S & UTILITIES	680	.42	
16	FORM FOOTINGS	2023	1.36	
24	UNDERGROUND PIPING	4875	3.03	
18	REINF. & CONC. FTG'S	1755	1.09	
20	FOUNDATION BLOCK	4521	2.81	
24	PERIMETER INSULATION	971	.60	
28	GRAVEL FILL	613	.38	
26	BACKFILL	1711	1.06	
30	MASONRY BR'G WALLS	45107	28.04	
30	ROUGH PLUMBING FIX.	3326	2.07	
36	PRSTRS'D CONC. SLABS	1813	1.13	
38	FINE GR. FORMS, & W.W.F.	2834	1.76	
42	JOISTS & STEEL DECK	9677	6.02	
42	CONC. FLOOR SLABS	5679	3.53	
44	RF'G, S. M., & SKYLIGHTS	14435	8.97	
43	BLOCK PARTITIONS	1422	.88	
54	TOILET FIXTURES	3653	2.27	
62	ROOF HEATING UNITS	11825	7.35	
62	OVERHEAD DOORS	8260	5.13	
54	WIRE MESH PARTITIONS	1488	.93	
54	STEEL STAIRS	2068	1.29	
54	ROUGH & FIN. CARP.	2331	1.39	

Progress Chart

Proctor penetration needle An instrument that measures the resistance of a fine-grained soil to penetration by a standard needle at a standard rate.

producer Manufacturer, processor, or assembler of building materials or equipment.

product data Illustrations, standard schedules, performance charts, instructions, brochures, diagrams, and other information furnished by the manufacturer to illustrate a material, product, or system for some portion of the work.

products liability insurance Insurance for liability imposed for damages caused by an occurrence arising out of goods or products manufactured, sold, handled, or distributed by the insured or others trading under the insured's name. Occurrence must occur after product has been relinquished to others and away from premises of the insured. See also **completed operations insurance**.

product standard A published standard that establishes: (1) dimensional requirements for standard sizes and types of various products; (2) technical requirements for the product; and (3) methods of testing, grading, and marking the product. The objective of product standards is to define requirements for specific products in accordance with the principal demands of the trade. Product standards are published by the National Bureau of Standards of the U.S. Department of Commerce, as well as by private organizations of manufacturers, distributors, and users.

professional advisor An architect engaged by the owner to direct a design competition for the selection of an architect.

professional engineer Designation reserved, usually by law, for a person professionally qualified and duly licensed to perform engineering services such as structural, mechanical, electrical, sanitary, and civil.

professional fee See **fee**.

professional liability insurance Insurance coverage for the insured professional's legal liability for claims for damages sustained by others, allegedly as a result of negligent acts, errors, or omissions in the performance of professional services. See also **negligence**.

professional practice The practice of one of the environmental design professions in which services are rendered within the framework of recognized professional ethics and standards and applicable legal requirements. See **environmental design**.

profile (1) A drawing showing a vertical section of ground, usually taken along the center line of a highway or other construction project. (2) A template used for shaping plaster. (3) A guide used in masonry work. (4) A British term for batter board.

pro forma invoice An invoice sent before the order has been shipped in order to obtain payment before shipment.

program A written statement setting forth design objectives, constraints, and criteria for a project, including space requirements and relationships, flexibility and expandability, special equipment, and systems and site requirements.

programming phase That phase of the environmental design process in which the owner provides full information regarding requirements for the project, including a program. See also **predesign services**.

progress chart (1) A chart that shows various operations in a construction project, such as excavating and foundations, along

389

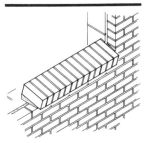

Progress Schedule

Projected Window

Projection

with planned starting and finish dates in the form of horizontal bars. Progress is indicated by filling in the bars. (2) A similar chart for the design phase of a project. The bars usually identify specific drawings.

progressive kiln A dry kiln in which green lumber enters one end and is dried progressively as it moves to the other end, where it is removed.

progressive scaling The progressive disintegration of materials, such as concrete, which first appears as surface scaling but continues in deeper layers.

progress payment Partial payment made during progress of the work on account of work completed and/or materials suitably stored.

progress schedule A diagram, graph, or other pictorial or written schedule showing proposed and actual times of starting and completion of the various elements of the work. See also **critical path method (CPM)** and **PERT schedule**.

project The total construction of which the work performed under the contract documents may be the whole or a part.

project application for payment Certified requests for payment from individual contractors on a construction management project, assembled for certification by the architect and submittal to the owner. See also **application for payment**.

project architect The architect designated by the principal-in-charge to manage the firm's services related to a given project. See also **project manager**.

project certificate for payment A statement from the architect to the owner confirming the amounts due individual contractors, where multiple contractors have separate direct agreements with the owner.

See also **certificate for payment**.

project cost Total cost of the project, including construction cost, professional compensation, land costs, furnishings and equipment, financing, and other charges.

project designer (1) (architect's office) The individual designated by the principal-in-charge to be responsible for setting the overall direction of the architectural design of a given project. (2) (consultant's office) The individual who is responsible for the design of a specific portion of a project, such as structural, mechanical, electrical, sanitary, civil, acoustical, food service, and the like. See also **project engineer**.

projected window A window with one or more sashes which swing either inward or outward.

project engineer The engineer, either in the architect's office or the consultant's office, as the case may be, designated to be responsible for the design and management of specific engineering portions of a project.

projecting belt course A course of masonry which projects beyond the face of the wall to form a decorative shelf.

projecting brick One of a number of bricks that project from a wall to form a pattern.

projecting scaffold A work platform which is cantilevered from the face of a building by means of brackets.

projecting sign A sign attached to the face of a building and extending outward.

projection Any component member or part which extends out from a building for a relatively short distance.

projection booth (1) A booth, usually at the rear of a room or

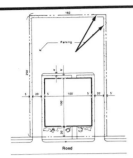

Property Line

hall, used for the operation of still or moving projectors or spotlights.

project manager A term frequently used to identify the individual designated by the principal-in-charge to manage the firm's services related to a given project. Normally, these services include administrative responsibilities as well as technical responsibilities.

project manual The volume(s) of document(s), prepared by the architect or project engineers for a project, which may include the bidding requirements, sample forms and conditions of the contract, and the specifications.

projector (1) A lighting unit which concentrates light within a limited solid angle by means of lenses or mirrors. (2) A line dropped perpendicularly from a point to a plane surface.

project representative The architect's representative at the project site who assists in the administration of the construction contract.

project site See **site**.

promoter See **catalyst**.

proof stress (1) Stress applied to materials sufficient to produce a specified permanent strain. (2) A specific stress to which some types of tendons are subjected in the manufacturing process as a means of reducing the deformation of anchorage, reducing the creep of steel, or insuring that the tendon is sufficiently strong.

prop See **post** and **shore**.

propeller twist Twisted grain in a Red Cedar shake that prevents it from lying flat and causes it to take the shape roughly resembling an airplane propeller.

property damage insurance Insurance coverage for the insured's legal liability for claims for injury to or destruction of tangible property, including loss of use resulting therefrom.

Property damage insurance usually does not include coverage for injury to or destruction of property which is in the care, custody, and control of the insured. See also **care, custody, and control.**

property insurance Coverage for loss or damage to the work at the site caused by the perils of fire, lightning, extended coverage perils, vandalism and malicious mischief, and additional perils (as otherwise provided or requested). Property insurance may be written on (1) the completed value form in which the policy is written at the start of a project in a predetermined amount representing the insurable value of the work (consisting of the contract sum less the cost of specified exclusions) and adjusted to the final insurable cost on completion of the work, or (2) the reporting form in which the property values fluctuate during the policy term, requiring monthly statements showing the increase in value of work in place over the previous month. See also **builder's risk insurance, extended coverage insurance, insurable value, and special hazards insurance.**

property line A recorded boundary of a plot.

proportional dividers An instrument for enlarging or reducing lengths of lines that consists of two pivoted bars pointed at each end. The pivot is moveable to adjust the relative lengths between the two pairs of points.

proportional limit The greatest stress which a material is capable of developing without any deviation from proportionality of stress to strain (Hooke's Law).

proportioning Selection of proportions of ingredients for mortar or concrete to make the most economical use of available

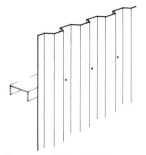

Protected Metal Sheeting

materials to produce mortar or concrete of the required properties.

proposal (contractor's) See **bid**.

proposal form See **bid form**.

proscription The acquisition of title to real property by one who openly and continuously is in adverse possession of it for a period sufficiently long that the statute of limitation bars the previous owner from reclaiming it (usually 20 years).

protected corner Corner of a slab with adequate provision for load transfer, such that at least 20 percent of the load from one slab corner to the corner of an adjacent slab is transferred by mechanical means or aggregate interlock.

protected metal sheeting Sheet metal which is coated with zinc, paint and/or a thin coating of asphalt for corrosion protection.

protected noncombustible construction Noncombustible construction in which bearing walls (or bearing portions of walls), whether interior or exterior, have a minimum fire resistance rating of 2 hours and are stable under fire conditions. Roofs and floors, and their supports, have minimum fire resistance ratings of 1 hour. Stairways and other openings through floors are enclosed with partitions having minimum fire resistance ratings of 1 hour.

protected opening An opening, in a rated wall or partition, which is fitted with a door, window, or shutter having a fire resistance rating appropriate to the use of the wall.

protected ordinary construction Construction in which roofs and floors and their supports have a minimum fire resistance rating of 1 hour, and stairways and other openings through floors are enclosed with partitions that have minimum fire resistance ratings of

1 hour. Such construction must also meet all the requirement of ordinary construction.

protected paste volume The portion of hardened cement paste that is protected from the effects of freezing by proximity to an entrained air void.

protected waste pipe A waste pipe from a fixture that is not directly connected to a drain, soil, vent, or waste pipe.

protected wood-frame construction Construction in which roofs and floors and their supports have minimum fire resistance ratings of 1 hour, and stairways and other openings through floors are enclosed with partitions with minimum fire resistance ratings of 1 hour. Such construction must also meet all the requirements of wood frame construction.

protection screen A screen woven of moderately heavy stainless steel wires and set in a steel frame, used to protect windows at a psychiatric unit.

protective covenant (1) An agreement in writing which restricts the use of real property. (2) A restriction, in the legal document conveying title to real property, that restricts the use of the property.

protective lighting Lighting that is provided to facilitate the night time policing of a property or area.

protractor A graduated drafting instrument used to lay out angles.

proving ring A device for calibrating load indicators of testing machines, consisting of a calibrated elastic ring and a mechanism or device for indicating the magnitude of deformation under load.

proximate cause The cause of an injury or of damages which, in natural and continuous sequence, unbroken by any legally

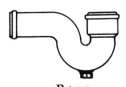

P-trap

recognized intervening cause, produces the injury, and without which the result would not have occurred. Existence of proximate cause involves both (1) causation in fact, i.e., that the wrongdoer actually produced an injury or damages, and (2) a public policy determination that the wrongdoer should be held responsible.

proximity switch A sensor which is activated by the intrusion of objects into an area.

proximo, prox A credit term meaning next month.

pry bar A heavy steel bar, shaped like a chisel at one end, used for prying.

psychrometer An instrument for measuring water vapor in the atmosphere, utilizing both wet- and-dry-bulb thermometers. A wet-bulb thermometer is kept moistened and is cooled by evaporation, giving a slightly lower reading than the dry-bulb thermometer. Because evaporation is greater in dry air, the difference between the two thermometer readings is greater in a dry atmosphere.

P-trap A P-shaped trap that provides a water seal in a waste or soil pipe, used mostly at sinks and lavatories.

public area An area that is open to the public.

public garage Garage for temporary parking or storage of small- to medium-size motor vehicles, usually for a fee.

public housing Low-cost housing owned, maintained, and administered by a municipal or other government agency.

public liability insurance Insurance covering liability of the insured for negligent acts resulting in bodily injury, disease, or death of persons other than employees of the insured, and/or property damage. See also **comprehensive general liability insurance** and **contractor's liability insurance**.

public sewer A common sewer controlled completely by a public authority.

public space (1) An area within a building to which the public has free access, such as a foyer or lobby. (2) An area or piece of land legally designated for public use.

public system A water or sewer system owned and operated by a governmental authority or by a utility company that is controlled by a government authority.

public utility A service for the public such as water, sewers, telephone, electricity, or gas.

public water main A water supply pipe controlled by public authority.

public way A street, alley, or other parcel of land open to the outside air and leading to a public street. A public way is deeded or otherwise permanently appropriated for public use. A minimum width is usually specified by code.

puddle (1) To settle loose soil by flooding and turning it over. (2) To vibrate and/or work concrete to eliminate honeycomb. (3) Clay which has been worked with some water to make it homogeneous and increase its plasticity so it can be used to seal against the passage of water.

puddle weld A weld used to join two sheets of light-gauge metal. A hole is burned in the upper sheet and filled with weld metal to fasten the two together.

puddling See **puddle**.

puff pipe A short vent pipe on the outlet side of a trap, used to prevent siphoning.

Puget Sound Pine An archaic term for Douglas Fir.

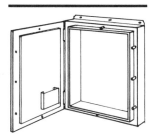

Pull Box

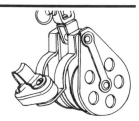

Pulley

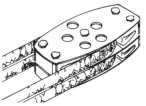

Pulley Block

Pumped Concrete

pug mill (1) A machine for mixing and tempering clay. (2) The part of an asphaltic concrete plant where the concrete is mixed.

pull (1) A handle used for opening a door, drawer, etc. (2) To loosen rock at the bottom of a hole by blasting.

pull box (1) A box, with a removable cover, placed in electric raceway to facilitate the pulling of conductors through the raceway. (2) A manual activator for a fire alarm system.

pull-chain operator A chain or control used to open or close a device such as a damper.

pulldown handle A handle fixed to the bottom rail of the upper sash of some double hung windows.

pulley (1) A wheel, with a grooved rim, that carries a rope or chain and turns in a frame. Also called a "pulley sheave". (2) A pulley block.

pulley block A frame that contains one or more pulleys.

pulley mortise See **chase mortise**.

pulley sheave See **pulley, def. 1**.

pulley stile The upright in a window frame on which the sash pulleys are supported and along which the sash slides.

pull hardware A handle or grip on a door for opening the door.

pulling (1) The drag on a paint brush caused by high paint viscosity. (2) Installing and connecting wires in an electric system.

pulling over Smoothing a lacquer on wood by rubbing with a solvent-soaked cloth.

pulling up The softening of a coat of paint as the next coat is applied.

pull scraper A hand scraper, consisting of a steel blade at approximately right angles to the handle, that is used to remove old finishes or for smoothing wood.

pull shovel Same as backhoe.

pull switch Same as chain-pull switch.

pulpboard A solid board composed of wood pulp. Also see **fiberboard**.

pulverized fuel ash See **fly ash**.

pumice A highly porous lava, usually of relatively high silica content, composed largely of glass drawn into approximately parallel or loosely entwined fibers.

pumice concrete A lightweight concrete in which pumice is used as the coarse aggregate and which has good thermal insulation value.

pumice stone A solid block of pumice used to rub or polish surfaces.

pumicite Naturally occurring finely divided pumice.

pump A machine, operated by hand or a prime mover, used to compress and/or move a fluid.

pumped concrete Concrete which is transported through hose or pipe by means of a pump.

pumping (of pavements) The ejection of water, or water and solid materials, such as clay or silt, along transverse or longitudinal joints and cracks, and along pavement edges. Pumping is caused by downward slab movement activated by the passage of loads over the pavement after the accumulation of free water on or in the base course, subgrade, or subbase.

punch (1) A small pointed tool which is struck with a hammer and used for centering and starting holes. (2) A steel tool, usually cylindrical, with sharpened edges and used in a hydraulic machine to make holes through metal.

puncheon (1) Roughly dressed, heavy timber used as a flooring, or as a footing for a foundation. (2) Short timbers supporting horizontal members in a coffer dam.

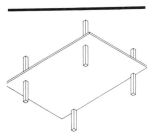

Punching Shear

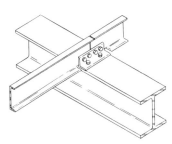

Purlin Cleat

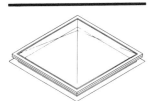

Pyramidal Light

punching shear (1) Shear stress calculated by dividing the load on a column by the product of its perimeter and the thickness of the base or cap, or by the product of the perimeter taken at one half the slab thickness away from the column and the thickness of the base or cap. (2) Failure of a base when a heavily loaded column punches a hole through it.

punch list See **inspection list**.

punning An obsolete term designating a light form of ramming. See also **ramming** and **tamping**.

purchase order A document, sent to a seller by the buyer, listing details of the order, such as stock descriptions, price, and shipping instructions.

purchaser A person or company that buys or contracts to buy real property.

purge To remove unwanted air or gas from a ductline, pipeline, container, space, or furnace, often by injecting an inert gas.

purge valve See **air purge valve**.

purlin In roofs, a horizontal member supporting the common rafters.

purlin cleat A shaped metal fastener used to secure a purlin to its support.

purlin plate A purlin, in a curb roof, located at the curb and supporting the ends of the upper rafters.

purlin post A strut which supports a purlin to reduce sag.

purlin roof A roof in which purlins are supported directly on walls rather than rafters.

push bar A heavy bar across a glazed door, screen door, or horizontally pivoted window sash, used to open or close the door or window.

push button A device in an electric circuit with a button that is pressed to activate or disconnect the circuit.

push drill A hand drill which is operated by pushing on its handle. A spiral ratchet rotates the bit.

push hardware A fixed bar or plate operated by pushing.

push plate A metal plate used to protect a door while it is pushed open.

push-pull rule A thin steel rule which coils into a case when not in use.

putlog Short pieces of timber that support the planks of a scaffold. One end of the timber is supported by the scaffold; the other is inserted in a temporary hole left in the masonry.

putlog hole The temporary hole in masonry that support one end of a putlog.

puttied split A split in a wood product, such as a panel surface, that has been filled with putty, usually an epoxy, then sanded.

putty A dough-like mixture of pigment and vehicle, used to set glass in window frames and fill nail holes and cracks.

putty coat Final smooth coat of plaster.

putty knife A knife, with a broad flexible blade with a flat end, used to apply putty.

pycnometer A vessel for determining the specific gravity of liquids or solids.

pylon (1) A steel tower used to support electrical high-tension lines. (2) A movable tower for carrying lights. (3) A truncated pyramidal form used in gateways to Egyptain monuments.

pyramidal light A skylight shaped like a pyramid.

pyramid roof A roof with four slopes terminating at a peak.

pyrometric cone A small, slender, three-sided oblique pyramid made of ceramic or refractory material for use in determining the time-temperature effect of heating and in obtaining the pyrometric cone equivalent of refractory material.

pyrometric cone equivalent The number of that cone whose tip would touch the supporting plaque simultaneously with that of a cone of the refractory material being investigated, when tested in accordance with a specified procedure such as ASTM C 24.

 ABBREVIATIONS

The abbreviations listed below are those most commonly used in the construction industry. Alternative forms (usually nonstandard) are shown in parentheses.

q quart

qda quantity discount agreement

QF quick firing

qr quarter

QR quarter-round

qs quarter-sawn

qt quart

QTR (1) quarry-tile roof. (2) On drawings, abbr. for quarter.

quad. quadrant

QUAD. quadrangle

QUAL quality

quar quarterly

Q DEFINITIONS

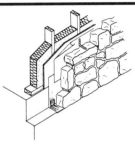

Quarry-faced

Quarter Bend

Quarter-turn

quadrangle, quad An open rectangular courtyard surrounded on all sides by buildings.

quadrant (1) One quarter of the circumference of a circle; an arc of 90 degrees. (2) An angle measuring instrument.

quadripartite An assembly involving four parts; usually describing groined vaulting.

quaking concrete A method of vibrating mass concrete.

quality A grade of Idaho White Pine equivalent to D Select in other species.

quality assurance A system of procedures for selecting the levels of quality required for a project or portion thereof to perform the functions intended, and assuring that these levels are obtained.

quality control A system of procedures and standards by which a constructor, product manufacturer, materials processor, or the like, monitors the properties of the finished work.

quantity overrun/underrun The difference between estimated quantities and the actual quantities in the completed work.

quantity survey Detailed listing and quantities of all items of material and equipment necessary to construct a project.

quarrel A small diamond or square-shaped tile or piece of glass, often set diagonally.

quarry An open excavation in the surface of the earth for mining stone.

quarry-faced Freshly split ashlar squared off at the joints only, used for facing a masonry wall.

quarry run Indiscriminate building stone as it comes from the pit without regard to color or structure.

quarrystone bond In masonry, a term applied to the manner in which stone is arranged in rubblework.

quarter bend A 90 degree bend; as in piping.

quarter closer, quarter closure A brick cut to 1/4 of its normal length, used either as a spacer or to complete a course.

quarter-cut See **quartersawn**.

quartered See **quartersawn**.

quartered lumber Lumber that has been quarter-sawn approximately radially from the log.

quartered veneer Veneer that has been sliced in a radial direction; that is, at right angles to the growth rings. The term quartered comes from the use of blocks that have been cut into quarters before slicing. Quarter slicing brings out the presence of medullary rays; quarter-sliced veneer appears striped.

quarter hollow A concave molding formed by a 90 degree arc; the opposite of a quarter round.

quartering (1) A method of obtaining a representative sample by dividing a circular pile of a larger sample into four equal parts and discarding opposite quarters successively until the desired size of sample is obtained. (2) Small timbers used as studs in a framed partition. (3) Quarter sawing.

quarter round A type of molding used as a base shoe; presenting the profile of a quarter circle.

quartersawn, rift-sawn Lumber sawn so that the annual rings form angles of 45 to 90 degrees with the surface of the piece.

quarter-turn Descriptive of a stair which turns 90 degrees as it progresses from top to bottom.

quartz glass, silica glass Glass consisting of pure, or nearly pure, amorphous silica. Of all glass, quartz glass has the highest heat resistance and ultraviolet transmittance.

quartzite A sandstone composition consisting primarily of quartz and a siliceous cement.

Quebec Pine *Pinus resinosa*. Red Pine.

Quebec Spruce *Picea alba, P. glauca*. White Spruce.

queen closer A half brick; of normal thickness but half normal width. Used in a course of brick masonry to prevent vertical joints from falling above one another.

queen post One of the two vertical members in a queen-post truss.

queen-post truss, queen truss A pitched roof support using two vertical tie posts connected between the tie beam and the rafters.

queen rod, queen bolt A metal rod used as a queen post.

quetta bond Vertical voids in brickwork in which reinforcing rods are installed. The voids are then filled with mortar.

quick condition Soil which is weakened by the upward flow of water. Minute channels are created, which significantly reduces the bearing capacity of the soil.

quick-leveling head A ball and socket attachment under the head of a surveyor's level or transit.

quicklime Calcium oxide (CaO). See also **lime**.

quicksand Fine sand in a quick condition; having virtually no bearing capacity.

quick set See **flash set** and **false set**.

quilt insulation A thermal barrier with paper faces that are stitched or woven.

quirk A narrow groove or bead at or near the intersection of two

surfaces or next to a molding, to reduce the possibility of uncontrolled cracking.

quitclaim deed A document which transfers the sellers interest in property to another party.

quoin, coign, coin (1) A right angle stone in the corner of a masonry wall to strengthen and tie the corner together.

quoin bonding In masonry, the interlocking of stones in the corner of a wall with staggered stretchers and headers.

Quonset hut A prefabricated building with a semicircular cross section; usually built with corrugated steel and thermal insulation.

quotation A price quoted by a contractor, subcontractor, material supplier, or vendor to furnish materials, labor, or both.

R

# R ABBREVIATIONS

The abbreviations listed below are those most commonly used in the construction industry. Alternative forms (usually nonstandard) are shown in parentheses.

r rain, range, rare, red, river, roentgen, run
R radius, right
Ra radium
R.A. registered architect
rab rabbeted
RAB rabbet
rad radiator
raft. rafter
RBM reinforced brick masonry
RC, R/C reinforced concrete
RC asphalt rapid-curing asphalt
RCP reinforced concrete pipe
1/4 RD quarter-round
1/2 RD half-round
rd road, rod, round
RD roof drain, round
rebar reinforcing bar
recap. recapitulation
recd received
recip reciprocal
RECP receptacle
rec. room recreation room
rect rectangle, rectified
red. reduce, reduction
ref reference, refining
REF refer, reference
REFR refractory, refrigerate
refrig refrigeration
reg registered
Reg regular
REG register, regulator
rein., reinf reinforced
REINF reinforce, reinforcing
REM removable
remod remodel
rent. rental
rep, REP repair
repl, REPL replace, replacement
REPRO reproduce
reqd, REQD required
res resawn
ret retain, retainage

RET. return
rev revenue, reverse, revised
REV revise
rf roof
RF roof, radio frequency
Rfg roofing
RFP request for proposal
rgh, Rgh rough
Rh rhodium, Rockwell hardness
RH relative humidity
RHN Rockwell hardness number
RI refractive index
R.I.B.A. Royal Institute of British Architects
rib. gl. ribbed glass
riv river
RJ road junction
R/L random lengths
rm ream, room
RM room
r. mld. raised mold
rms root mean square
rnd round
ROP record of production
ROPS roll-over protection system
rot. rotating, rotation
rpm revolutions per minute
RRGCP reinforced rubber gasket concrete pipe
RRS railroad siding
RSJ rolled steel joist
rt right
RT raintight
Rub., rub. Ruberoid, rubble
r.w. redwood, roadway, right-of-way
R/W right-of-way
R/W&L random widths and lengths
rwy, ry railway

R DEFINITIONS

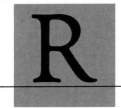

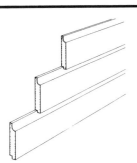

Rabbeted Siding

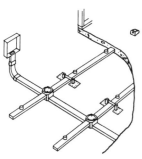

Raceway

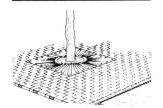

Radial Grating

rabbet (1) A cut or groove along or near the edge of a piece of wood that allows another piece to fit into it to form a joint. (2) A rabbet joint.

rabbet bead A bead placed in the reentry corner of a rabbet.

rabbet depth The depth of a glazing rabbet.

rabbeted doorjamb A doorjamb with a rabbet to receive a door.

rabbeted lock, rebated lock A lock which is fitted into and flush with the rabbet on a rabbeted doorjamb.

rabbeted siding See **drop siding**.

rabbeted stop A stop that is an integral part of a door or window frame.

rabbet joint A longitudinal edge joint formed by fitting together rabbeted boards.

rabbet plane A plane for cutting a rabbet or fillister in wood, open on one side with the blade extending to the open side.

rabbet size The actual size of a rabbeted glass opening, equal to the glass size plus two-edged clearances.

race (1) A channel intended to contain rapidly moving water. (2) A groove in a machine part in which an object moves.

raceway Any furrow or channel constructed to loosely house electrical conductors. These conduits may be flexible or rigid, metallic or non metallic and are designed to protect the cables they enclose.

rack A framework of metal bars used to prevent water born trash from entering a waterway.

rack-and-pinion elevator A platform or enclosure in which linear motion is created by the use of a rotating gear pinion mated to

and revolving on a vertical stationary rack.

racked A temporary support used to brace and prevent deformation.

racking, racking back A method entailing stepping back successive courses of masonry in an unfinished wall.

rack saw A saw with teeth that are wide.

radial Radiating from or converging to a common center.

radial-arm saw, radial saw A circular saw suspended above the saw table on a cantilevered arm. The material remains stationary while the saw is free to move along its projecting beam.

radial grating Grating in which the bearing bars extend radially from a common center and the cross bars form concentric circles.

radially-cut grating Rectangular grating cut into panels shaped as segments of a circle to fit annular or circular openings.

radiant glass Glass with embedded heating elements.

radiant heating system A heating system which transfers heat in the form of waves or particles to be naturally circulated throughout the space.

radiant panel test An ASTM method using radiant heat to test the surface flammability of different materials.

radiation The transmission of energy by means of electromagnetic waves of very long wave length. The energy travels in a straight line at the speed of light and is not affected by the temperature or currents of the air through which it passes.

radiation-shielding concrete A high density concrete used primarily for enclosing nuclear installations.

Rafter

Rail

radiator A visually exposed heat exchanger consisting of a series of pipes that allows the circulation of steam or hot water. The heat from the steam or hot water is given up to the air surrounding the pipes.

radius of gyration An imaginary distance from an axis to a point such that, if an objects mass were concentrated at the point, the moment of inertia would not change.

radius rod A long arm with a marker at one end and an adjustable pointer at the other, used to draw large radius circles or curves.

rafter One of a series of sloping parallel beams used to support a roof covering.

rafter plate A plate used to support the lower end of rafters and to which they are fastened.

rafter table A carpenters square with a table of values for determining the lengths and angles of cut for roof rafters.

rafter tail That part of a rafter overhanging the wall.

raft foundation A continuous slab of concrete, usually reinforced, laid over soft ground, or where heavy loads must be supported to form a foundation. See **mat foundation**.

rag bolt See **lewis bolt**.

rag felt An asphaltic felt consisting of rags saturated with asphalt, creating a waterproof composition for use on roofs, i.e. roofing paper and asphalt shingles.

raggle, reglet, raglin A narrow channel or furrow in building stone for which metal flashing is fitted.

rag-rolled finish A decorative paint finish made by rolling a twisted rag over a wet paint surface so that portions of the undercoat show through. The effect can be simulated with a special paint roller.

ragstone Building stone quarried in thin blocks or slabs.

ragwork Courses of irregular stone masonry laid in a random pattern without parallel surfaces. The stones are roughly shaped with nonuniform joints.

rail (1) A horizontal member supported by vertical posts, i.e. a handrail along a stairway. (2) A horizontal piece of wood, framed into vertical stiles, such as a panelled door.

rail fence A barrier or boundary constructed of rails and their supporting posts.

rail head (1) A loading point, or the nearest point of access to a railroad. (2) The farthest point on a railroad to which rails have been laid.

railing (1) A solid wood band around one or more edges of a plywood panel. (2) A balustrade.

rails (1) The horizontal members that form the outside frame of a door, including pieces used as a cross-bracing between the top and bottom rails. (2) The horizontal members of a fence between posts. (3) The side pieces of a ladder to which rungs or steps are attached.

rail steel reinforcement Reinforcing bars hot-rolled from standard T-section rails.

Raimann patch A patch, elliptical in shape, used to fill voids caused by defects in the veneers of plywood panels. The Raimann patch is sometimes called a "football patch" and is one of the two patch designs allowed in A-grade faces.

rain leader See **downspout**.

rainwet (1) Lumber that has excess moisture content because of exposure to rain after it was dried. (2) Surfaced lumber that has been stained or weathered by exposure to rain.

Rake (2)

raised flooring system A floor constructed of removable panels supported by stringers allowing easy access to the space below.

raised girt, flush girt, raised girth A stiffening member, parallel to and level with the floor joists, used to strengthen and protect.

raised grain A roughened condition on the surface of dressed lumber whereby the hard summerwood is raised above the softer springwood but not torn loose from it.

raised molding A molding not on the same level or plane as the wood member or assembly to which it is applied.

raising piece A piece of timber laid on top of a brick wall or across posts of a timber framed house, to support a beam or beams.

raising plate A horizontal timber carrying the feet of other structural members.

rake (1) To slant or incline from the vertical or horizontal. (2) A board or molding that is placed along the sloping edge of a frame gable to cover the edges of the siding. (3) A tool used to remove mortar from the face of a wall.

rake classifier Machine for separating course and fine particles of granular material temporarily suspended in water. The course particles settle to the bottom of a vessel and are scraped up an incline by a set of blades. The fine particles remain in suspension to be carried over the edge of the classifier.

raked joint A joint in a masonry wall which has the mortar raked out to a specified depth while it is still soft.

rake-out, raking out An action for removing mortar an even distance from the face of a wall.

raker (1) A sloping brace for a shore head. (2) A tool for raking out decayed mortar from the joints of brickwork.

raking bond A bricklaying pattern which is created by arranging face brick diagonally or vertically i.e., herringbone, basket weave, or diagonal.

raking course A course of bricks laid diagonally across faces of a brick wall for added strength.

raking flashing A parallel flashing used at the intersection of a chimney or other projection and a sloping roof.

raking pile See **batter pile**.

raking riser A riser that is not vertical but slopes away from the nosing for added foot room on the lower riser.

raking stretcher bond A bond in brickwork in which the vertical joint of each brick is displaced a small, fixed distance from the vertical joint of the brick below.

rammed earth A mixture of aggregate and water which has been compressed and dried.

ramming A form of heavy tamping of concrete, grout, or the like by means of a blunt tool forcibly applied. See **dry pack, punning, and tamping**.

ramp A sloping surface to provide an easy connection between floors.

rampant vault A type of barrel vault in which the two quarters have different radii. The quarters share a common crown but the abutments are at different levels.

randle bar A rod horizontally assembled into the walls of an open chimney for which hooks are attached for hanging cooking pots and utensils.

random ashlar, random bond Ashlar masonry in which stones are set without continuous joints and appear to follow a random pattern, although a large pattern may be repeated.

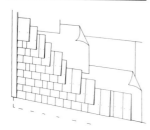

Random Shingle

Ratchet

random course An ashlar or squared stone masonry course of irregular height.

random lengths (RL) (1) Lumber of various lengths, usually in even two-foot increments. A random length loading is presumed to contain a fair representation of the lengths being produced by a specific manufacturer. (2) An American corporation that regularly publishes information about wood products markets in North America.

random line A surveyor's reference line laid between two points which are not visible from each other.

random paneling Board paneling of varying widths of the same grade and pattern. The term refers to plywood paneling grooved to represent random width paneling.

random range ashlar See **random work**.

random rubble Stone masonry, built of rubble.

random shingle One of a number of shingles of the same length but varying widths.

random widths Boards, lumber, and shingles of varying widths.

random work (1) A wall built of odd sized stones. (2) Masonry of rectangular stone laid in broken courses by use of stones of different heights and widths.

range (1) A row or course of masonry. (2) A straight line of objects such as columns.

range closet A multi-seated latrine.

range-in, wiggling-in The term used by a surveyor for the trial and error procedure of finding a previously established line.

ranger See **wale**.

range rod, range pole A device that consists of a pole seven to eight feet long, equipped with a metal point. It is painted alternately red and white in bands one foot wide. The pole is used by surveyors for locating points of reference.

ranging bond Small strips of wood at the face of a masonry wall, usually laid in joints and projecting slightly, used to provide a nailing surface for furring and battens.

rapid-curing asphalt A liquid composed of asphalt cement and a highly volatile diluent such as naphtha or a gasoline.

rapid-curing cutback See **rapid-curing asphalt**.

rapid-hardening cement A high-early-strength cement.

rapid-start fluorescent lamp A fluorescent lamp designed for operation with a ballast that provides a low-voltage winding to preheat the electrodes and initiate the arc without a starting switch or the application of high voltage.

rasp A file with projecting teeth for rough work.

ratchet A mechanism with a hinged catch, or pawl, that slides over and locks behind sloped teeth on a gear or rod, allowing motion in one direction only.

ratchet brace A carpenter's clamp used in confined spaces where a full turn of the brace cannot be achieved. It is fitted with a pawl mechanism allowing the bit to be rotated while in the hole.

ratchet drill A hand operated drill that works with a hinged catch engaging on a wheel or chuck.

rated lamp life The average life of a particular lamp fixture.

rated load The overall weight that a piece of machinery is designed to carry.

rated speed The speed at which a piece of machinery is designed to operate with an appropriate load.

ratio The relative size of two quantities expressed as one divided by the other.

rat stop A barrier in a masonry wall to prevent rats from

Rawl Plug

burrowing down along the exterior of a masonry wall.

rat-tail file A file with a circular cross section which tapers to a small diameter at the end opposite the handle.

rattrap bond A bond in brickwork in which headers and stretchers are alternated and laid on edge on each face, creating a wall of two brick wythes thick with a series of cavities between strechers.

raummeter A German term for one cubic meter of stacked wood. A raummeter is also called a "stere".

raveling A term used for the progressive deterioration of asphalt pavement. The aggregate dislodges and becomes fragmented.

raw linseed oil Linseed oil which has been refined but has not received further processing such as boiling, blowing, or bodying.

rawl plug A short fiber cylinder with a lead lining, driven into a hole in wood, masonry, glass, plaster, tile, concrete, or other materials to receive and hold a screw.

raw mix Blend of raw materials, ground to desired fineness, correctly proportioned, and blended ready for burning, such as that used in the manufacture of cement clinker.

raw water (1) Water that requires treatment before it can be used, such as water for steam generation. (2) Any water used in ice making except distilled water.

raze To tear down, demolish, or level to ground.

reactive aggregate Aggregate containing substances capable of reacting chemically with the products of solution or hydration of the portland cement in concrete or mortar, under ordinary conditions of exposure, resulting in harmful expansion, cracking, or staining.

reactive silica material Several types of materials which react at high temperatures with portland cement or lime during autoclaving, including pulverized silica, natural pozzolan, and fly ash.

ready-mixed concrete Concrete manufactured for delivery to a purchaser in a plastic and unhardened state. See **central-mixed concrete, shrink-mixed concrete,** and **transit-mixed concrete**.

real estate Property in the form of land and all improvements such as buildings and paving.

realignment A change in the horizontal layout of a highway, may also affect vertical alignment.

real property Land including everything on it and beneath it with some rights to the airspace directly above.

ream To enlarge or smooth a hole using a reamer.

reamer A bit with sharp, spiral fluted cutting edges along the shaft, may be slightly tapered, used to enlarge drilled holes or remove burrs from the inside of pipe.

rear arch An arch forming a true opening and faced by a larger arch for decoration.

reasonable care and skill See **due care**.

rebar Short for reinforcing bar.

rebate See **rabbet**.

rebound Aggregate and cement or wet shotcrete which bounces away from a surface against which it is being projected.

rebound hammer An apparatus providing a rapid indication of the strength of concrete based on the distance of rebound of a spring-driven missile.

rebutted and rejointed (R&R) Shingles with edges machine-trimmed to be exactly parallel and butts retrimmed at precisely right

Receptacle

Recessed Fixture

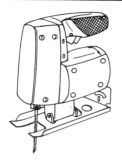

Reciprocating Saw

angles for use primarily in sidewall applications.

receptacle A contact device installed in an electric outlet box for the connection of portable equipment or appliances.

receptacle outlet An electric outlet with one or more receptacles.

receptacle plug A device, usually attached to a flexible chord, which is inserted in a receptacle to connect portable equipment or appliances to an electric circuit.

receptor The shallow basin in the floor of a shower.

recessed fixture A lamp fixture which has its bottom edge flush with the ceiling.

recessed head A specially constructed mechanical fastener which is designed to fit flush into the surface.

recessed pointing A type of joint where the mortar is kept back approximately one quarter of an inch from the face of the wall. This particular joint protects the mortar from peeling.

recharge (1) The replenishment of ground water through direct injection or infiltration from trenches outside the area. (2) The replenishment of electric energy in a storage battery.

reciprocal levelling A surveying technique used in leveling across streams, gullies, and other obstructions. To eliminate instrumental errors, levels are taken from two set-ups; one near each point.

reciprocating drill See **push drill**.

reciprocating engine A steam or internal-combustion engine with cylinders and pistons as opposed to a turbine.

reciprocating saw A saw operated in a back-and-forth or up-and-down motion extending from an engine or other power source, i.e., a sabre saw.

recirculated air Return air which is reconditioned and distributed once again, as opposed to make-up air.

reconditioned wood Hardwood lumber steam-dried to correct defects that occurred during the original curing, such as collapse or warp.

record drawings Construction drawings revised to show significant changes made during the construction process, usually based on marked-up prints, drawings, and other data furnished by the contractor to the architect. Record drawings are preferable to as-built drawings.

record sheet A sheet or form which preserves the memory of events.

recovery peg A temporary surveying marker of known location and elevation, used to reestablish a permanent marker that is to be replaced.

rectangular tie A piece of bent heavy wire in the shape of a rectangle used as a wall tie.

Red Alder See *Alnus rubra*, **Alder**.

Red Cedar See **Western Red Cedar, Incense Cedar**, and **Eastern Red Cedar**.

Red Fir (1) An archaic description of Douglas Fir lumber made from smaller, younger timber principally from the coast range and southern Oregon. (2) See **California Red Fir**.

red knot A slang term for a knot caused by cutting through a live pine branch.

red label A grade of shingle between blue label #1 and black label #5, graded by the Red Cedar Shingle and Handsplit Shake Bureau.

Reducing Joint

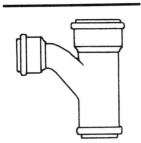

Reducing Pipe Fitting

red lead A compound of lead used in paints to improve the anticorrosive properties of the paint used on steel and iron.

Red Meranti *Shorea* species. Any of a number of tropical hardwoods, mainly used as veneers, such as lauan.

Red Oak *Quercus rubra* and others, such as Southern Red Oak.

red oxide A pigment used in paints to prevent rusting and corrosion.

Red Pine *Pinus resinosa.* This species, commonly called Norway Pine, is found in the Great Lakes states, the northeastern U.S., and eastern Canada. The wood is used for general construction and for remanufacturing into such items as sash, molding, flooring, and crating.

redress (1) Financial compensation for a loss caused by the actions of another. (2) To set right, remedy, or rectify.

red rot An early stage of decay.

redry To turn material to a dry kiln or veneer dryer for additional drying when the material is found to have a higher moisture content level than desired.

Red Silver Fir *Abies amabilis*, Pacific Silver Fir.

Red Spruce *Picea rubens.* This species, also called "Eastern Spruce", is found in southeastern Canada, the New England states, and along the Appalachian Mountains as far south as North Carolina. In addition to general construction applications, the wood is often used for such products as ladder rails and as a source of pulpwood.

red stripe Fungi that have entered cracks in wood and caused decay.

reducer (1) A solvent used in paints to reduce the viscosity of the paint. (2) A pipe fitting, larger at one end than at the other, a reducing coupling.

reducing joint The connecting joint of unequally sized electrical conductors.

reducing pipe A pipe fitting with ends of unequal diameters for connecting different sizes of pipes.

reducing pipe fitting Fitting that is constructed to facilitate the joining of unequally sized pipes.

reduct A small piece of material cut from a larger piece to make the larger piece uniform or symmetrical.

reduction of area The difference between the original cross-sectional area of a tensile specimen and the smallest area after rupture, expressed as a percentage of the original area.

Redwood *Sequoia sempervirens.* This species is found only in limited areas of northern California and southern Oregon. It is resistant to decay and is used for many of the same purposes as cedar, especially siding and paneling. Another species of Redwood, *Sequoia gigantea*, grows in the Sierra Mountains of central California. It is protected from harvest.

reeving (1) Threading a rope through the blocks to assemble a block and tackle. (2) Threading or placing a pulling or control line.

reference line A series of two or more points in line to serve as a reference for measurements.

reference mark A supplementary mark close to a survey station. One or more such marks are located and recorded with sufficient accuracy so that the original station can be reestablished from the references.

reflected plan A plan of an upper surface, such as a ceiling projected downward.

reflective glass Glass with a special metallic coating used to repel light and radiant heat.

reflective insulation A thermal material having one or both faces metallically coated to reflect the

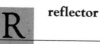

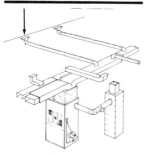

Register

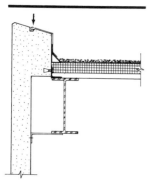

Reglet

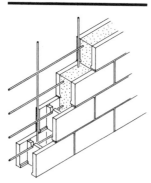

Reinforced Blockwork

radiant energy that strikes its surface.

reflector A device used to redirect light or sound energy by the process of reflection.

reflector lamp An incandescent lamp in which a part of the bulb acts as a reflector.

refractories Materials, usually nonmetallic, used to withstand high temperatures.

refractoriness In refractories, the property of being resistant to softening or deformation at high temperatures.

refractory Resistant to high temperatures.

refractory aggregate Materials having refractory properties which, when bound together into a conglomerate mass by a matrix, form a refractory body.

refractory brick A brick manufactured for high temperature use.

refractory cement A high temperature bonding material used in furnaces, consisting of crushed brick and diatomaceous earth.

refractory concrete Concrete having refractory properties and suitable for use at high temperatures, generally about 315 degrees to 1315 degrees C, in which the binding agent is a hydraulic cement.

refractory insulating concrete Refractory concrete having low thermal conductivity.

refrigerant The medium used to absorb heat in a cooling cycle.

refrigerant compressor unit A unit consisting of a pump with various controls for regulating refrigerant flow.

refrigeration cycle A thermodynamic process whereby a refrigerant accepts and rejects heat in a repetitive sequence.

refrigeration system A system in which a refrigerant is compressed,

condensed, and expanded as a means of removing heat from a cold reservoir. The heat is rejected elsewhere at a higher temperature.

refrigerator A container or space and the machinery for cooling it.

register An opening to a room or space for the passage of conditioned air. The register has a grill and a damper for flow regulation.

registered architect See **architect**.

reglet A groove in a wall to receive flashing.

regrating The cleaning of masonry by removing the outer layer, leaving a fresh stone surface.

regressed luminaire A luminaire mounted above a ceiling with its opening above the ceiling line.

regulated-set cement A portland cement with admixtures for the purpose of controlling its set and early strength.

reheat coil A coil in an air supply duct, used to control the temperature of air being supplied to individual spaces or a group of spaces.

reheating The heating of return air in an air-conditioning system for reuse.

reimbursement expenses Amounts expended for or on account of the project which, in accordance with the terms of the appropriate agreement, are to be reimbursed by the owner.

reinforced bitumen felt A light roofing felt saturated with bitumen and reinforced with jute cloth.

reinforced blockwork Masonry blockwork with reinforcing steel and grout placed in the voids to resist tensile, compressive, or shear stresses.

reinforced brick masonry Brick masonry in which steel reinforcement is placed.

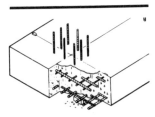

Reinforcement

Helical

Hoop

Mesh

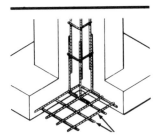

Two-Way

reinforced cames Lead bars with a steel core for reinforcing for use in leaded lights.

reinforced concrete Concrete containing adequate reinforcement, prestressed or not prestressed, and designed on the assumption that the two materials act together in resisting forces. See **plain concrete**.

reinforced concrete masonry Concrete masonry construction in which steel reinforcement is so embedded that the materials act together in resisting tensile, compressive, and/or shear stresses.

reinforced masonry Unit masonry construction in which steel reinforcement is so embedded that the materials act together in resisting tensile, compressive, and/or shear stresses.

reinforced T-beam A concrete T-beam strengthened internally with steel rods to resist tensile and/or shear stresses.

reinforcement Bars, wires, strands, and other slender members embedded in concrete in such a manner that the reinforcement and the concrete act together in resisting forces.

cold-drawn wire Steel wire made from rods that have been hot rolled from billets and cold-drawn through a die used for concrete reinforcement of small diameter such as in gauges not less than 2 mm nor greater than 16 mm.

cold-worked steel Steel bars or wires which have been rolled, twisted, or drawn at normal ambient temperatures.

distribution-bar Small-diameter bars, usually at right angles to the main reinforcement, intended to spread a concentrated load on a slab and to prevent cracking.

dowel-bar See *dowel*.

expanded metal fabric See *expanded metal*.

four-way A system of reinforcement in flat slab construction comprising bands of bars parallel to two adjacent edges and also to both diagonals of a rectangular slab.

helical Steel reinforcement of hot rolled bar or cold-drawn wire fabricated into a helix, more commonly known as spiral reinforcement.

high-strength Concrete reinforcing bars having a minimum yield of 60,000 psi or 414 MPa.

hoop A one-piece closed tie or continuously wound tie not less than #3 in size, the ends of which have a standard 135-degree bend with a ten-bar-diameter extension, that encloses the longitudinal reinforcement.

lateral Usually applied to ties, hoops, and spirals in columns or column-like members.

mesh See *welded-wire fabric* and *welded-wire fabric reinforcement*.

principal Elements or configurations of reinforcement that provide the main resistance of reinforced concrete to loads borne by structures. See *reinforcement, secondary*.

secondary Reinforcement other than main reinforcement.

spiral See *spiral reinforcement*.

transverse Reinforcement at right angles to the longitudinal reinforcement. Transverse may be main or secondary reinforcement.

twin-twisted bar Two bars of the same nominal diameter twisted together.

two-way Reinforcement arranged in bands of bars at right angles to each other.

411

Relief Valve

welded Reinforcement joined together by welding.

reinforcement displacement Movement of reinforcing steel from is specified position in the forms.

reinforcement ratio Ratio of the effective area of the reinforcement to the effective area of the concrete at any section of a structural member. See **percentage of reinforcement**.

reinforcing bar A steel bar, usually with manufactured deformations, used in concrete and masonry construction to provide additional strength.

reinforcing plate An added plate used to strengthen a member or part of a member.

reinforcing rod See **reinforcing bar**.

reinforcing unit A box-shaped reinforcement in a hollow metal door housing a bored lock and providing supports for the latch.

rejection of work The act of rejecting work that is defective or does not conform to the requirements of the contract documents.

relaskop A patented instrument used by foresters to measure tree heights and/or slope gradients by trigonometric means.

related trades The different building trades required to complete a project.

relative compaction The dry density of soil expressed as a percentage of the density of the soil after a standard compaction test.

relative humidity The ratio of the quantity of water vapor actually present to amount present in a saturated atmosphere at a given temperature, expressed as a percentage.

relative settlement See **differential settlement**.

relaxation of steel Decrease in stress in steel as a result of creep within the steel under prolonged strain or as a result of decreased strain of the steel, such as results from shrinkage and creep of the concrete in a prestressed concrete unit.

relay An electro magnetic or electromechanical device using small currents and voltages to activate switches or other secondary devices.

release agent Material used to prevent bonding of concrete to a surface. See **bond breaker**.

release of lien Instrument executed by a person or entity supplying labor, materials, or professional services on a project which releases that person's or entity's mechanic's lien against the project property. See **mechanic's lien** and **waiver of lien**.

release paper A protective film, coated lightly on one side with adhesive, used to protect a surface during shipment and installation.

relief (1) Carved or embossed decoration raised above a background plane. (2) The representation of elevations of the earth's surface, usually referenced to a mean-sea-level datum plane, by means of contour lines.

relief damper A damper in an air-conditioning system which opens automatically at a set pressure differential to balance pressures in a building.

relief map A map showing relief by means of contour lines, shading, tinting, or relief models.

relief valve A pressure activated valve held closed by spring tension and designed to automatically relieve pressure in excess of its setting.

relief vent An auxiliary vent, supplementary to regular vent pipes, the primary purpose being to provide supplementary circulation of air between drainage and vent pipes.

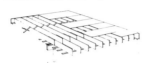

Repetitive Member

relish In carpentry, the shoulder around or at the side of a tenon.

remodeling See **alterations**.

remoldability The readiness with which freshly mixed concrete responds to a remolding effort such as jigging or vibration, causing it to reshape its mass around reinforcement and to conform to the shape of the form. See **flow**.

remolding test A test to measure remoldability.

removable mullion A door mullion which can be removed from a door frame to allow passage of oversized objects.

removable stop A removable molding or trim used to stop motion and permit the installation of window panes or doors.

render (1) To shade an architectural or machine drawing to give it a more real life appearance. (2) To apply plaster, especially the first coat, directly to brickwork or other masonry.

rendered brickwork Brickwork that has a waterproof material applied to the face.

rendering (1) The application, by means of a trowel or float, of a coat or mortar. (2) A drawing of a project or portion thereof with an artistic delineation of materials, shades, and shadows.

repeatability Variability among replicate test results obtained on the same material within a single laboratory by one operator. Repeatability is a quantity that will be exceeding in only about 5 percent of the repetitions by the difference, taken in absolute value, or two randomly selected test results obtained in the same laboratory on a given material. In use of the term, all variable factors should be specified.

repeating theodolite A theodolite designed so that an angle can be turned several times and successive angle measurement accumulated with the final reading representative of the accumulations.

repetitive member One of a series of framing or supporting members such as joists, studs, planks, or decking that are continuous, or spaced not more than 24 inches apart, and are joined by floor, roof, or other load-distributing elements. In repetitive-member framing, each member is connected to, and receives some shared support from, the others.

replacement value The estimated cost to replace an existing building based on current construction costs.

repointing See **pointing**.

reporting form property insurance See **property insurance**.

reposting See **reshoring**.

reproducibility Variability among replicate test results obtained on the same material in different laboratories. Variability is a quantity that will be exceeding in only about 5 percent of the repetitions by the difference, taken in absolute value, of two single test results made on the same material in two different, randomly selected laboratories. In use of the term, all variable factors should be specified.

request for payment See **application for payment**.

resaw (1) To saw a piece of lumber along its horizontal axis. (2) A bandsaw that performs such an operation.

resawn board A piece of lumber, most commonly 11/16-inch thick, obtained by resawing a piece of 6/4 common. Used mostly for sheathing and industrial applications. Produced most often from Ponderosa Pine and White Fir.

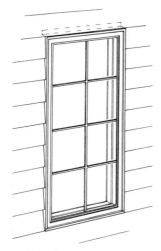

Residence Casement

Resilient Flooring

resawn lumber Lumber that has been sawn on a horizontal axis to produce two thinner pieces. See **resawn board**.

resealing trap A trap connected to a plumbing fixture drain pipe so constructed as to allow the rate of flow to seal the trap without causing self-siphonage.

reservoir A tank or receptacle used for the accumulation and retention of a fluid.

resetting (of forms) Setting of forms separately for each successive lift of a wall to avoid offsets at construction joints.

reshoring The construction operation whereby the original shoring or posting is removed and replaced in such a manner as to avoid deflection of the shored element or damage to partially cured concrete.

residence casement A window sash that opens outwards used in homes, hotels, and commercial buildings.

resident engineer An engineer employed by the owner to represent the owner's interests at the project site during the construction phase. The term is frequently used on projects in which a governmental agency is involved. See **owner's inspector** and **clerk of the works**.

residential occupancy Occupancy of a building in which sleeping accommodations are provided for normal residential purposes. The term excludes institutional occupancy.

resident inspector See **owner's inspector, resident engineer,** and **project representative**.

residual deflection The amount of deviation that remains after an applied load has been removed.

residual soil Soil formed by the aging of the subsurface mineral material.

residual stress The stress remaining in a member after an applied force has been removed.

resilience The work done per unit volume of a material in producing strain.

resilient channel A mounting device with flexible connectors used for fastening gypsum board to studs or joists and helps to reduce the transmission of vibrations.

resilient clip A flexible metal device for mounting gypsum board to studs or joists used to reduce noise and vibrations.

resilient flooring A floor covering that is very durable and has the ability to resume its original shape, such as linoleum.

resin (1) A natural or synthetic, solid or semisolid organic material of indefinite and often high molecular weight having a tendency to flow under stress, usually has a softening or melting range and fractures conchoidally. (2) A natural vegetable substance occurring in various plants and trees, especially the coniferous species, used in varnishes, inks, medicines, plastic products, and adhesives.

resin chipboard A particleboard which uses a resin to ensure a uniform consistency.

resin-emulsion paint Paint with a vehicle consisting of an oil, oleo-resinous varnish, or resin binder dispersed in water.

resin mortar (or concrete) See **polymer concrete**.

resistance See **electrical resistance**.

resistance brazing A process in which coalescence is produced by the heat obtained from the resistance of the work to the flow of an electrical current.

resistance welding A process creating a metallurgical bond between metal by producing heat obtained from the resistance offered by the work to the flow of

Retaining Wall

electric current, and by the application of an external pressure.

resistor A device included in an electric circuit to introduce a desired resistance.

resonance A condition reached in an electrical or mechanical system, when the applied load frequency coincides and is equal to the natural undamped frequency of the system.

resorcinol adhesive A bonding material that is workable and water soluble for two to four hours. After this time span it becomes insoluble and chemically resistant.

resorcinol resin adhesive A synthetic resin adhesive made with resorcinol-formaldehyde resin using a hardener, usually composed of wood flour and paraform. The adhesive provides an exterior quality glueline and will cure at room temperature. It is used in the manufacture of laminated beams or scarf joints.

respond A support, usually a pilaster or corbel, attached to a wall and supporting one end of an arch, groin, or vault rib.

responsible bidder See **lowest responsible bidder**.

rest bend A right angle pipe fitting with an integral seat for mounting on a support See **Base elbow**

restoration A series of actions that tends to bring an object or building back to its original condition.

restraint (of concrete) Restriction of free movement of fresh or hardened concrete following completion of placing in formwork or molds or within an otherwise confined space. This restraint can be internal or external and may act in one or more directions.

restricted list of bidders See **invited bidders**.

restrictive covenant An agreement between individuals, incorporated within a deed, which stipulates how land can be used. The constraints may include: the specific use to which property may be put, setback of buildings, size of yards, locations and dimensions of fences, and type of architecture. Racial and religious restrictions on inhabitants are not legally enforceable.

resurfacing The placing of a new surface on an existing pavement to improve its conformation or increase its strength.

retainage A sum withheld from progress payments to the contractor in accordance with the terms of the owner-contractor agreement.

retained percentage See **retainage**.

retaining wall (1) A structure used to sustain the pressure of the earth behind it. (2) Any wall subjected to lateral pressure other than wind pressure.

retardation Reduction in the rate of hardening or setting, i.e., an increase in the time required to reach initial and final set or to develop early strength of fresh concrete, mortar, or grout. See **retarder**.

retarder An admixture which delays the setting of cement paste, and hence of mixtures such as mortar or concrete containing cement.

retempering Addition of water and remixing of concrete or mortar that has lost its workability and become too stiff. This is not a good practice as some of the strength is lost. See **tempering**.

retention (1) A percentage, usually 10%, withheld from a periodic payment to a contractor, in accordance with the owner-

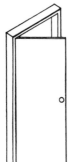

Reversed Door

contractor agreement, for work completed. The retention is held until all terms of the contract have been fulfilled. (2) The amount of preservative, fire-retardant treatment, or resin retained by treated or impregnated wood.

reticulated Covered with crossing lines, netted.

reticuline bar Sinuously bent bars that interconnect adjacent *bearing bars* in a grating.

retort A vessel used to treat wood by applying various chemicals under pressure.

return The continuation, in a different direction, of a molding or projection, usually at right angles.

return air Air returned from a conditioned room or space for processing and recirculation.

return-air intake An opening, usually with a control damper, through which return air reenters an air conditioning system.

return bend A 180 degree bend in a pipe.

return corner block A concrete masonry unit with a plane surface at one end as well as the front face, used at outside corners.

return grille The grille, usually with a damper, on a return air intake.

return mains Pipes or ducts which reroute fluid or air back to the supply source.

return pipe The drain line that returns the water from condensed heating steam to the boiler for reuse.

return system A series of ducts, pipes, or passages that returns a substance, whether it be air or water, to the source for reuse.

return wall A short wall placed perpendicular to the end of a wall to increase its stability.

reveal The side of an opening in a wall for a window or door. Reveal is the depth of exposure of

aggregate in an exposed aggregate finish. See **exposed-aggregate finish**.

reveal pin, reveal tie A screw or clamp located inside a window opening used to secure scaffolding to the opening.

revent pipe An individual vent that connects directly to a waste fixture and the main or branch vent pipe.

reverse In plastering, a template with the reverse shape of the molding it is intended to match.

reverse-acting diaphragm valve A valve which opens on application of pressure on a diaphragm and closes when the pressure is released.

reverse-acting thermostat A thermostat that activates a circuit when a preset high temperature is detected.

reverse bevel A latch bolt that is inclined or sloped in the opposite direction of a regular bolt.

reverse board & batten A siding pattern in which the wider boards are nailed over the battens, producing a narrow inset.

reversed door See **reverse-swing door**.

reverse-swing door, reversed door A door which swings outward from a room.

reversible grating A framework of bars which can be installed with either side exposed without affecting the load capabilities or appearance.

reversible lock A lock which can be adapted to fit a door of either hand.

reversible siding Resawn board siding that may be installed with either the surfaced or sawn side exposed.

reversion (1) The tendency of a processed material to return to its original state. (2) Chemical reaction leading to deterioration

Rheostat

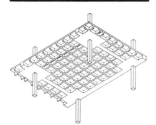

Ribbed Slab

of a sealant due to moisture trapped behind the sealant.

revertible flue A flue or chimney that momentarily redirects the gas flow path downward during its normal upward ascent.

revet To face a foundation or embankment with a layer of stone, concrete, or other suitable material.

revetment A facing, such as masonry, used to support an embankment.

revibration One or more applications of vibration to concrete after completion of placing and initial compaction but preceding initial setting of the concrete.

revolving door An exterior door consisting of four leaves set at right angles to each other and revolving on a central pivot, used to limit the passage of air through the opening and eliminate drafts.

revolving shelf See **lazy susan**.

rheology The science dealing with flow of materials, including studies of deformation of hardened concrete, the handling and placing of freshly mixed concrete, and the behavior of slurries, pastes, and the like.

rheostat An electric resistor, so constructed that its resistance may be varied without opening the circuit in which it is installed, used to control the flow of electric current as in a light dimmer.

rib (1) One of a number of parallel structural members backing sheathing. (2) The portion of a T-beam which projects below the slab. (3) In deformed reinforcing bars, the deformations or the longitudinal parting ridge.

riband, ribband See **ribbon strip**.

ribbed panel A panel composed of a thin slab reinforced by a system of ribs in one or two directions, usually orthogonal.

ribbed slab See **ribbed panel**.

ribbed vault A vault with ribs which support or appear to support the web of the vault.

ribbing A more-or-less regular corrugation of the surface of wood, caused by differential shrinkage. Ribbing is also called "crimping" or "washboarding".

ribbing up Circular joinery made by gluing several layers of veneer with a parallel grain structure.

ribbon A narrow strip of wood or other material used in formwork.

ribbon board (1) A horizontal brace used in balloon framing where the board is applied to notches in the studs. (2) A let-in brace.

ribbon course A rugged roof texture which can be achieved by applying a triple thickness of asphalt shingles on alternate courses. This feature adds visual appeal and is especially suited for large roof areas of a long one story building.

ribbon loading Method of batching concrete whereby the solid ingredients, and sometimes the water, enter the mixer simultaneously.

ribbon strip A horizontal board set into the studs to help support the ends of rafters or joists.

ribbon stripe A pattern on quartered veneer that has the appearance of a ribbon.

rich concrete Concrete of high cement content. See **lean concrete**.

rich lime A pure lime, when mixed with mortar it improves the plasticity or workability of the mortar.

rich mixture A concrete mixture containing a high proportion of cement.

rich mortar See **fat mortar**.

riddle A coarse sieve for separating and grading granular material.

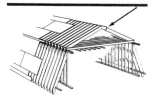

**Ridgeboard
Ridge roof**

Ridge cap

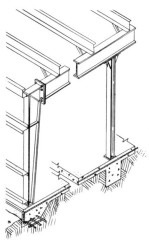

Rigid Frame

rider cap See **pile cap**.

ridge The horizontal line formed by the upper edges of two sloping roof surfaces.

ridge beam A horizontal timber to which the tops of rafters are fastened.

ridgeboard The longitudinal board set on edge used to support the upper ends of the rafters.

ridgecap, ridge cap A layer of wood or metal topping the ridge of a roof.

ridgepole See **ridgeboard**.

ridge rib A projecting structural element following the ridge of a vault.

ridge roll (1) A round wooden strip with the underside "V" cut, used to finish the upper edges of the roof surface. (2) A flexible metal covering used to cap the ridge of a roof.

ridge roof A roof with two slopes meeting at a central crest.

ridge stop Descriptive of the flashing that forms a watertight barrier between the edge of the roofing membrane and the vertical wall or chimney rising above it.

ridging Long narrow blisters in the surface of built-up roofing.

rift A narrow fissure in rock.

rift sawn See **quarter sawn**.

rig (1) To provide with equipment or gear for a special purpose. (2) To assemble in a makeshift manner. (3) An assembled piece of equipment, such as an oil-drilling rig.

rigger (1) A long-haired, slender brush used for precision painting. (2) A worker who prepares heavy equipment or loads of materials for lifting.

rigging Lines or cables used to lift heavy loads.

rigging line A line or cable used with others to lift heavy loads.

riggot A trough or gutter for draining off rainwater.

right bank The bank located on the right as one looks downstream.

right-hand door See **hand**.

right-hand lock A lock to be used on a right-hand door.

right-hand reverse door See **hand**.

right hand rule A rule for determining polarity produced by flowing electric current.

right-hand stairway A stairway having the outside hand rail on the right as one ascends.

right-of-way A strip of land, including the surface and overhead or underground space, which is granted by deed or easement for the construction and maintenance of specific linear elements such as power and telephone lines, roadways and driveways, and gas or water lines.

rigid connection A connection between two structural members which prevents end rotation of one relative to the other.

rigid foam See **cellular plastic**.

rigid frame A frame depending on moment in joints for stability.

rigidized The term used for the strengthening and increased rigidity of sheet metal by embossing the steel in a rolling process.

rigid metal conduit A raceway constructed for the pulling in or withdrawing of wires or cables after the conduit is in place and made of standard weight metal pipe permitting the cutting of standard threads.

rigid pavement Pavement that will provide high bending resistance and distribute loads to the foundation over a comparatively large area.

rim latch A surface-mounted latch.

rim lock A face-mounted door lock.

Riser

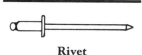

Rivet

Riveting

Ringelmann chart A table of values used as a standard for determining the density of smoke emitting from a chimney.

ring scratch awl A pointed tool made especially for use on sheet metal.

ring-shank nail A nail with ring-like grooves around the shank to improve its grip.

rip To saw a piece of lumber parallel to the grain.

riparian right A landowners right to the use of a body of water from which his land borders.

ripper A device with protruding teeth that is pulled by a tractor. Used to penetrate and disrupt the earths surface up to 3 feet in depth.

ripping See **ripsawing**.

ripping bar See **pinch bar**.

riprap An irregularly broken, large piece of rock, used along stream banks and ocean fronts as protection against erosion.

ripsaw A coarse-toothed saw used for cutting wood in the direction of the grain.

ripsawing Sawing lumber parallel to the grain.

rise Defined by OSHA as: (1) the vertical distance from the top of a tread to the top of the next higher tread, (2) the height of an arch from springing to the crown, and (3) the vertical height from the supports to the ridge of a roof.

rise and run The angle of inclination or slope of a member or structure, expressed as the ratio of the vertical rise to the horizontal run.

riser A vertical member between two stair treads.

riser board The member forming the vertical face of a step.

riser height The vertical distance between the tops of two successive treads.

rising hinge, rising butt hinge A door hinge fabricated with a slope on the knuckle which causes the door to rise when opened. A rising hinge is used when there are obstructions along the doors bottom edge i. e., carpeting.

rivet A metal cylinder or rod with a head at one end which is inserted through holes in the materials to be fastened. The protruding end is flattened to tie the two pieces together.

riveted grating A grating assembled from straight bearing bars and bent connecting bars, which are joined at contact points by rivets.

riveting The act of fastening or securing two or more parts with rivets.

road forms Wood or steel forms set on edge to form the side of a road slab, also used as screeds.

road heater A traveling machine which prepares a road surface for treatment by blowing a flame or hot air on it.

road oil A heavy petroleum oil, usually a slow-curing asphalt.

rock drill A high powered pnuematic or electrically driven device for boring holes into rock.

Rock Elm *Ulmus racemosa*. A lighter colored and finer textured wood than common elm, used primarily for veneer.

rock pocket A porous, mortar-deficient portion of hardened concrete, consisting primarily of coarse aggregate and open voids, caused by leakage of mortar from form, separation or segregation during placement, or insufficient consolidation. See **honeycomb**.

Rockwell hardness A measure of the resistance of a material to indentation, expressed as an index number obtained by pressing a standard steel ball or diamond into the material under controlled conditions. A high index number indicates a hard material.

rock wool A type of mineral wool made by forming fibers from

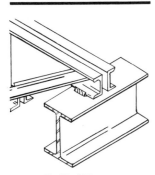

Rolled Beam

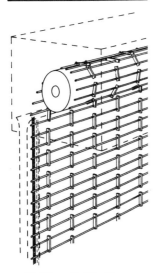

Rolling Grille Door

Roof Deck

molten rock and slag, used as insulation in walls and ceilings.

rod Sharp-edged cutting screed used to trim shotcrete to forms or ground wires. See **screed**.

rod bender A power assisted device with adjustable rollers and supports, used for bending steel reinforcing rods into usable shapes.

rod buster The colloquial term for one who installs reinforcement for concrete.

rod cutter A bench type hydraulically operated wedge-like shear, used to cut steel reinforcing rods.

roddability The susceptibility of fresh concrete or mortar to compaction by means of a tamping rod.

rodding Compaction of concrete or the like by means of a tamping rod. See **rod, tamping,** and **roddability**.

rod dowel See **dowel**.

rod level A device used to assure a leveling rod or stadia rod is in the vertical position before taking any instrument readings.

rod tamping A round, straight, steel rod having one or both ends rounded to hemispherical tip.

rod target A metal disc that slides upon a track on a leveling rod, used for taking sights in surveying.

roll (1) A quantity of sheet material wound in cylindrical form. (2) A rounded strip of roofing fastened to and running along the ridge of a roof. (3) Any type of rounded molding. (4) Any heavy, metal cylinder used to flatten, smooth, or form material.

rolled beam, rolled steel beam A beam that is fabricated of steel and passed through a hot-rolling mill.

roller (1) A heavy, self-propelled or towed device used to compact granular fill. (2) A small hand tool used to smooth wall covering and

flooring. (3) See **paint roller**.

rolling The use of heavy metal or stone rollers on terrazzo topping to extract excess matrix.

rolling doors See **roll-up door**.

rolling grille door A device similar to a roll-up door but with an open grille rather than slats, used as security protection.

roll joint A point of connection formed by rolling two edges of sheet-metal together and then compressing the roll.

roll-up door, rolling shutters A device consisting of horizontal interlocking metal slats that ride along wall guides. When the door is opened the slats coil around a barrel assembly located above the door.

Roman cement A misnomer for a hydraulic cement made by calcining a natural mixture of calcium carbonate and clay, such as argillaceous limestone, to a temperature below that required to sinter the material but high enough to decarbonate the calcium carbonate, followed by grinding. The term is so named because its brownish color resembles ancient Roman cements produced by use of lime-pozzolan mixtures.

roof The outer cover and its supporting structures on the top of a building.

roof covering The covering material installed in a building over the roof deck. The type of covering used depends on the roofing system specified to weatherproof the structure properly.

roof-deck (1) The foundation or base for which the entire roofing system is dependent upon. Types of decks include steel, concrete, cement and wood. (2) A flat open portion atop a roof, such as a terrace or sundeck.

roof decking Prefabricated sections of lightweight insulated

Roof Hatch

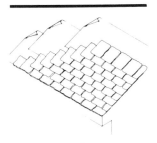

Roofing

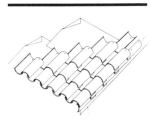

Roofing Tile

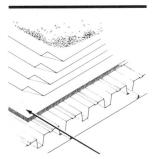

Roof Insulation

waterproof panels. This type of roof deck is quickly completed and consequently inexpensive. Roof decking is made of different roofing materials, such as plywood, aluminum, steel, or timber.

roof drain A collection point for rainwater runoff which is then discharged to a downspout.

roofers Lumber used as backing for shingles or shakes on a roof, usually 1x4s.

roof flange A collar that fits around a pipe that penetrates through the roof making the opening watertight.

roof framing A group of members fitted or joined together to provide support for the roof covering.

roof hatch A weather-tight assembly with a hinged cover, used to provide access to a roof.

roofing Any material that acts as a *roof covering* making it impervious to the weather such as shingles, tile, or slate.

roofing bond A guarantee by a surety company that a roof installed by a specified roofer in accordance with specific specifications will be repaired if it fails within a certain period of time. Failure must be due to normal weathering.

roofing bracket Defined by OSHA as a bracket used in slope roof construction, having provisions for fastening to the roof or supported by ropes fastened over the ridge and secured to some suitable object.

roofing nail A special purpose short threaded nail with a large head, usually galvanized or aluminum with a neoprene or plastic washer to aid in fastening roof coverings.

roofing square (1) A steel square used by carpenters. (2) A measure of roofing material. (3) See **square**.

roofing tile A preformed slab of baked clay, concrete, cement, or plastic laid in rows as a roofing cover. Tiles have a variety of patterns, but fall into two classifications, roll and flat.

roof insulation Lightweight concrete used primarily as insulating material over structural roof systems.

roof live load Any external loads that may be applied to a roof deck, such as rain, snow, construction equipment, and personnel.

roof pitch The slope of a roof expressed as the ratio of the rise of the roof to the horizontal span. More roofing material is required to cover the roof when the slope or pitch is great.

roof plate A wall plate which supports the lower end of rafters.

roof principal A roof truss.

roof purlin See **purlin**.

roof saddle See **saddle**.

roof scuttle See **roof hatch**.

roof sheathing Any sheet material, such as plywood or particleboard, connected to the roof rafters to act as a base for shingles or other roof coverings.

roof space The space between the roof and the ceiling of the highest room.

roof span (1) The shortest distance between the seats of opposite common rafters. (2) The distance across a roof.

roof structure Any structure on or above the roof of a building.

roof tree The ridgeboard of a roof.

roof truss A truss used in the structural system of a roof.

roof valley See **valley**.

roof vent A device used to ventilate an attic or roof cavity.

root (1) The part of a tenon which widens at the shoulders. (2) The point where the back or bottom of the weld meets the base metal.

Rotary Float

rope (1) Twisted strands of fiber made into strong flexible cord. (2) Strands of wire braided or twisted together, used for heavy hoisting or hauling.

rope caulk A preformed, rope-like bead of caulking compound, which may contain twine reinforcement to facilitate handling.

roped hydraulic elevator A car connected by cables to an operating ram. A motor driven positive displacement pump discharges oil into the pressure cylinder which extends the ram. The ram is retracted by bleeding oil off the pressure cylinder and into a storage tank.

rope diameter The largest diameter of the cross section of a wire rope. Fiber ropes are usually measured by circumference.

rope lay The direction in which the wires or strands are twisted during manufacture.

rope suspension equalizer A device that equalizes the tensions in hoisting cables.

rose (1) The metal plate or escutcheon between a doorknob and the door. (2) See **rosette**.

rose bit A bit for countersinking holes in wood.

rose nail A wrought nail with a cone shaped head.

rosette (1) A round pattern with a floral motif. (2) A circular or oval ornamental wood plaque, used to terminate a wood piece such as a stair rail at a wall. (3) A decorative nailhead or screwhead.

rosewood See **bubinga**.

rosin See **resin**.

rot Decay in wood caused by fungi and other microorganisms. Rot reduces the strength, hardness and density of the wood.

rotary cutting A method of obtaining wood veneers by rotating logs against a flat knife

and peeling the veneer off in a long continuous sheet. Peeling provides a greater volume and a more rapid production then does sawing or slicing.

rotary drill A machine for making holes in rock or earth by a cutting bit at the end of a metal rod, usually turned by a hydraulically or pneumatically driven motor.

rotary float, power float Motor-driven revolving blades that smooth, flatten, and compact the surface of concrete slabs or floor toppings.

rotary kiln A long steel cylinder with a refractory lining, supported on rollers so that it can rotate about its own axis, and erected with a slight inclination from the horizontal so that prepared raw materials fed into the higher end move to the lower end, where fuel is blown in by air blast. See **kiln** and **cement**.

rotary pump Any pump using gears, vaned-wheels, or a screw mechanism to displace liquid, usually delivering large volumes at low pressure.

rotary trowel See **rotary float**.

roto operator A mechanical device consisting of a crank driven gear which operates a hinged lever, used to open and close jalousies, casement windows, and awning windows.

rottenstone A porous, lightweight, siliceous limestone used for polishing soft metals and wood.

rotunda A circular building or hall that is round inside and out and usually domed.

rough ashlar A block of stone before dressing as delivered from a quarry.

roughback (1) The concealed end of bondstone in a masonry wall. (2) A slab of stone with one side rough and the other sawn, cut from a block fed through a gang saw.

Rough Floor

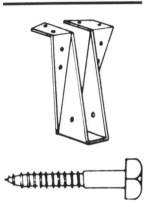

Rough Hardware

Router

rough-cut joint, flat joint, flush joint, hick joint A mortar joint that is flush with the face of the brickwork.

rough floor A base or sub floor, consisting of a layer of boards or plywood nailed to the floor joists.

rough flooring Any materials used to construct an unfinished floor.

rough grading To smooth or level the earth for preparation of finish work.

rough grind The initial operation in which coarse abrasives are used to cut the projecting chips in hardened terrazzo down to a level surface.

rough hardware Any fittings, such as screws, bolts, or nails which should be concealed for a finished product.

roughing-in (1) The base coat in three coat plasterwork. (2) Any unfinished work in a construction job.

roughing-out A preliminary shaping operation in carpentry.

rough opening An opening in a wall or framework into which a doorframe or window frame, subframe or rough buck is fitted.

rough rendering Applying a coat of plaster without removing the irregularities.

rough sill A horizontal member laid across the bottom of an unfinished opening to act as a base during construction of a window frame.

rough work The framing, boxing, and sheeting for a wood framed building.

round (1) A molding which may be semicircular to full round, as in a closed rod. (2) A turn of wire rope around a drum.

roundel (1) A semicircular panel, window, or recess. (2) A small, semicircular molding or astragal.

round molding, round A fairly large molding with a circular or nearly circular cross section.

round timber Felled trees which have not been converted to lumber.

rout To deepen and widen a crack to prepare it for patching or sealing.

router An electrically driven device with various bits for cutting grooves or channels in wood.

router gauge A carpenter's tool consisting of a guide, a bar with a scale, and a narrow chisel as a cutter, used in inlaid work to cut out narrow channels in which colored strips are laid.

router patch A wood patch in plywood with straight sides and rounded ends, similar to a tongue depressor, used to fill voids caused by defects. The router patch is one of two patch designs allowed in A grade faces and is also known as a "Davis Patch".

router plane, plough, plow A woodworker's plane for cutting grooves, or channels in wood.

row house A house in a row of houses separated by party walls and covered by a continuous roof.

row spacing The measured distance between the centers of mechanical fasteners in a row.

royals Shingles with 24-inch edges and a thickness of 1/2-inch at the butt.

rubbed finish (1) On woodwork, a dull finish obtained from hand rubbing with a rag or pad saturated with water or oil and pumice. (2) A finish obtained by using an abrasive, often a carborundum stone, to remove surface irregularities from concrete. See **sack rub**.

rubbed joint A process for joining two narrow boards. Both boards are planed smooth and then coated with glue and rubbed together until all air pockets and excess glue are expelled from the joint. No clamping is necessary and the joint is extremely strong.

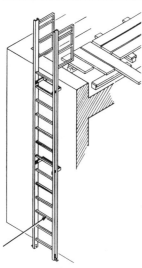

Rung

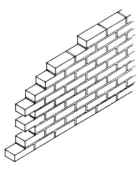

Running Bond

rubbed work Masonry work having a rubbed finish.

rubber (1) A highly resilient natural material manufactured from the juice of rubber trees and other plants. (2) Any of various synthetically manufactured materials with properties similar to natural rubber. (3) A cushioned backing for a carpet.

rubber-emulsion paint See **latex paint**.

rubber set See **false set**.

rubber silencer, bumper A small round rubber device that attaches to a rabbeted doorjamb which silences the noise caused by a slamming door.

rubber tile A soft and yielding floor covering that reduces the transmission of impact noises produced by walking or from other causes.

rubber-tired roller A machine for compacting and kneading soil using pneumatic tired rollers.

rubbing brick A silicon-carbide brick used to smooth and remove irregularities from surfaces of hardened concrete.

rubble Rough stones of irregular shape and size, broken from larger masses by geological processes or by quarrying.

rubble concrete (1) Concrete similar to cyclopean concrete, except that small stones, such as one man can handle, are used. (2) Concrete made with rubble from demolished structures. See **cyclopean concrete**.

rub brick See **rubbing brick**.

ruff sawn A designation for plywood paneling or siding which has been saw-textured to provide a decorative rough sawn appearance.

rule (1) A straight edge with graduations used for measuring, laying out lengths, or drawing straight lines. (2) A straight edge for working plaster to a plane surface.

rule joint A pivoted joint connecting two flat strips and allowing relative rotation around the pivot.

rule of thumb A statement or formula that is not exact but is close enough for practical work.

ruling pen A pen used by draftsmen to draw ink lines of uniform thickness.

run (1) In plumbing, a pipe or fitting that continues in the same straight line as the direction of flow. (2) In roofing, the horizontal distance between the outer face of the wall and the roof ridge. (3) In stairs, the horizontal distance from the face of the first riser to the face of the last riser.

rung A bar, usually of circular cross section, used as a step in a ladder.

run molding A formed molding of plaster or similar material formed by passing a template over the plastic material.

runner (1) As defined by OSHA, the lengthwise horizontal bracing or bearing members. (2) A cold rolled channel used to support steel studs in a partition or ceiling tile. (3) Tile as ledger.

running (1) Descriptive of a repeating design in a band having a smooth progression. (2) Forming a cornice, of plaster or similar material, in place with a running mold. (3) Operating a powered hand tool, particularly a drill.

running bond See **stretcher bond**.

running foot A linear foot. The term is a measurement of the actual length of a piece of lumber, without regard to the thickness or width of the piece.

running ground Earth in a semi-plastic state that will not stand without support.

running inch A linear inch. The term is a measurement of the

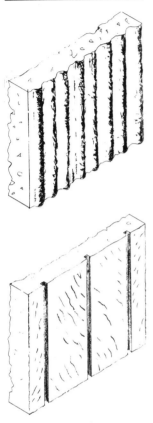

Rustication

actual length of a piece of lumber. See **running foot**.

running mold A template shaped to a desired cornice and mounted on a wood frame, used by plasterers to form a molding or cornice in place by applying plaster as the mold is moved along the ceiling line.

running screed A narrow strip of plaster used in place of a rule to guide a running mold.

running shoe A metal guide on a running mold to prevent wear and allow it to slide freely on a rule.

running trap A U-shaped pipe fitting installed in a drain line to prevent the backflow of sewer gases.

run-of-bank gravel See **bank-run gravel**.

runway Decking over area of concrete placement, usually of movable panels and supports, on which buggies of concrete travel to points of placement.

ruptured grain Breaks in veneer that have resulted from cutting or irregular grain.

rupture disk A pressure activated safety device consisting of a breakable disk that bursts at a predetermined pressure.

rust Any of various powdery or scaly reddish-brown or reddish-yellow hydrated ferric oxides formed on the surface of iron or steel which is exposed to moisture and air. Rust eventually will weaken or destroy the material if allowed to progress.

rusticated A formed or cut reveal in concrete masonry or stone used to highlight or conceal joints in concrete, masonry or stone fascia.

rustication A groove in a concrete or masonry surface.

rustication strip A strip of wood or other material attached to a form surface to produce a groove

or rustication in the concrete.

rustic brick A brick that has a rough textured finish produced by covering it with sand, wire brushing, or impressing it with a pattern. These bricks are often a variety of colors.

rustic joint A deeply sunk mortar joint that is accented by beveling the edges of the adjacent stone.

rustic or washed finish A type of terrazzo topping in which the matrix is recessed by washing prior to setting so as to expose the chips without destroying the bond between chip and matrix. A retarder is sometimes applied to the surface to facilitate this operation. See **exposed aggregate finish**.

rustic stone Any rough broken stone suitable for masonry with a rough appearance, most commonly limestone or sandstone, usually set with the longest dimension horizontal.

rustic woodwork Decorative or structural woodwork made of unpeeled logs and saplings.

rust-inhibiting paint See **anticorrosive paint**.

rust joint A watertight pipe connection that uses iron filings as a catalyst to induce rusting in iron pipe joints.

rust pocket An area in the bottom of a ventilating pipe for the collection and removal of rust and debris.

rutile A reddish brown to black natural mineral, a form of titanium dioxide used in paints and fillers.

"R" Value A measure of a material's resistance to heat flow given a thickness of material. The term is the reciprocal of the "U" value. The higher the "R" value the more effective the particular insulation.

ABBREVIATIONS

S side, south, southern, seamless, subject, sulphur

S&E surfaced one side and edge

S&G studs and girts

S&H staple and hasp

S&M surfaced and matched

S/A shipped assembled

SAE Society of Automotive Engineers

SAF safety

SAN sanitary

sanit sanitation

sat. saturate, saturation

sch school

SCH schedule

scp spherical candlepower

SD sea-damaged, standard deviation

S/D shop drawings

SDA specific dynamic action

Sdg siding

Se selenium

S/E square-edged

SE&S square edge and sound

sec second

SECT section

sed sediment, sedimentation

sel select, selected

Sel select

sep, SEP separate

SERV service

SE Sdg, S.E. Sdg. square-edge siding

SEW. sewer

Sftwd. softwood

sf surface foot

sfu supply fixture unit

SGD sliding glass door

sh shingles

SH sheet, shower, single-hung

shf superhigh frequency

shp shaft horsepower

sht sheet, sheath

Si silicon

SIC Standard Industrial Classification

sid siding

SIM similar

sk sack

SK sketch

sky. skylights

S/L, S/LAP shiplap

SL&C shipper's load and count

slid. sliding

SM standard matched, surface measure

s. mld. stuck mold

SMS sheet-metal screw

so. south

SO. seller's option

soln solution

SOV shutoff valve

sp specific, specimen, spirit, single pitch (roof)

SP soil pipe, standpipe, self-propelled, single pole

SPEC specification

sp. gr., SP GR specific gravity

sp. ht. specific heat

SPKR loudspeaker

spl spline

SPL special

spr spruce

SPT Standard Penetration Test

sp. vol. specific volume

sq. square

sq. e. square edge

sq. E&S square edge and sound

sq. ft. square foot

sq. in. square inch

sq. yd. square yard

SR sedimentation rate

ss single strength (glass)

SS, S/S stainless steel

sst standing seam tin (roof)

SST stainless steel

st stairs, stone, street
ST steam, street
STC sound transmission class
std, STD standard
Std. M standard matched
STG storage
STK stock
STL steel
STP standard temperature and pressure
Stpg stepping
str stringers
Str. structural
STR. strike
Struc structural
st. sash. steel sash
ST W storm water
sty. story
sty. hgt. story height
SUB. substitute
sub. fl. subfloor
subpar subparagraph
subsec subsection
sup supplementary, supplement
SUP supply
supp supplement
SUPSD supersede
supt, SUPT superintendent
SUPV supervise
supvr supervisor
sur, SUR surface
surv survey, surveying, surveyor
svc service
sw switch
SW switch, seawater, southwest
SWBD switchboard
SWG, S.W.G. standard wire gauge
sy jet syphon jet (water closet)
SYM symmetrical
SYN synthetic
SYS system
syst system
S1E surfaced one edge
S1S surfaced one side
S1S1E surfaced one side and one edge

S1S2E surfaced one side and two edges
S2E surfaced two edges
S2S surfaced two sides
S2S&CM surfaced two sides and center matched
S2S&SL surfaced two sides and shiplapped
S2S1E surfaced two sides and one edge
S4S surfaced four sides
S4S&CS surfaced four sides and caulking seam

Definitions

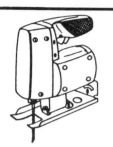

Saber saw

Saddle

saber saw A hand-held power saw with a reciprocating blade extending through the base of the saw.

sack See **bag**.

sack joint A flush masonry joint that has been wiped or rubbed with a rag or an object such as a rubber heel or a burlap sack.

sack rub, sack finish A finish for formed concrete surfaces, designed to produce even texture and fill all pits and air holes (see **bug holes**). After dampening the surface, mortar is rubbed over the surface. Then, before it dries, a mixture of dry cement and sand is rubbed over it with a wad of burlap or a sponge-rubber float to remove surplus mortar and fill the voids.

sacrificial protection The use of a metallic coating, such as zinc-rich paint, to protect steel. In the presence of an electrolyte, such as salt water, the metallic coating dissolves instead of the steel.

saddle (1) A fitted device used with hangers to support a pipe. (2) A series of bends in a pipe over an obstruction. (3) A short horizontal member set on top of a post as a seat for a girder. (4) Any hollow-backed structure with a shape suggesting a saddle, as a ridge connected to two higher elevations or a saddle roof. (5) See **threshold**. (6) See **cricket**.

saddleback (1) A coping stone with its top surface sloped from a high line in the center to either edge. (2) See **saddle joint**.

saddleback board See **threshold**.

saddle-backed coping See **saddleback, def. 2**.

saddleback joint See **saddle joint**.

saddleback roof See **saddle roof**.

saddle bar One of the horizontal iron bars across a window opening to secure leaded lights.

saddle bend A bend made in a conduit to provide clearance where it crosses another conduit.

saddle block The boom swivel block through which the stick of a dipper shovel slides.

saddle board A board used to cover the joint at the ridge of a pitched roof.

saddle coping See **saddleback, def. 2**.

saddle fitting A type of gasketed fitting clamped around the exterior of a pipe, used when a connection to a previously installed pipe is required.

saddle flange A flange which is curved to conform to the surface of the boiler or tank to which it is welded, riveted, or otherwise attached, and designed to accept a threaded pipe.

saddle flashing Flashing installed over a cricket.

saddle joint (1) A joint in sheet-metal roofing, in which one end of one sheet is folded downward over the turned-up edge of the adjacent sheet. (2) A stepped joint in a projecting masonry course to prevent the penetration of water.

saddle piece A metal cricket used in sheet-metal roofing.

saddle roof A roof with two gables and one ridge, resembling a saddle.

saddle scaffold A scaffold erected so as to bridge the ridge of a roof, usually used during chimney repair.

safe (1) A built-in or portable chamber used to protect materials or documents from fire and/or theft. (2) A pan or other collector placed beneath a pipe or fixture to collect leakage or overflow.

Safety Nosing

Safety Switch

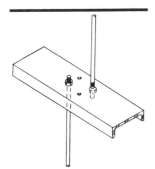

Sag Rod

safe-edge A strip-form detector mounted on the leading edge of an elevator door. If an object is in the way of the door, it will reopen or close slowly depending on how the activating mechanism is programmed. The safe-edge is vertically mounted to extend from the bottom to the top of a door panel.

safe leg load The load which can safely be directly imposed on the frame leg of a scaffold. See also **allowable load**.

safe load The maximum load on a structure that does not produce stresses greater than those allowable.

safety belt A belt-like device worn around the waist and attached to a life-line or structure to prevent a worker from falling.

safety can (OSHA) An approved closed container, of not more than five gallons capacity, having a flash-arresting screen, spring-closing lid, and spout cover, and so designed that it will safely relieve internal pressure when subjected to fire exposure.

safety curtain See **asbestos curtain**.

safety factor See **factor of safety**.

safety fuse A cord containing black powder or other burning medium encased in flexible wrapping and used to convey fire at a predetermined and uniform rate for firing blasting caps.

safety glass (1) See **wire glass**. (2) See **tempered glass**. (3) See **laminated glass**.

safety lintel A load-carrying lintel positioned behind a more decorative but somewhat less functional lintel, as in the aperture of a window or door.

safety nosing An abrasive, nonslip stair nosing whose surface is flush with the tread against which it is placed.

safety shoe (1) A workman's shoe with a steel-protected toe and low-slip sole and heal. (2) See **safe-edge**.

safety shutoff device A device in a gas burner that will shut off the supply if the gas flame is extinguished.

safety switch In an interior electric wiring system, a switch enclosed within a metal box but having a handle which protrudes through the box to allow switching to be accomplished from outside the box.

safety tread A tread on a stair which has a roughened surface or roughened inserts to prevent a foot from slipping.

safety valve See **pressure-reducing valve**.

safe working pressure The maximum working pressure at which a vessel, boiler, flask, or cylinder is allowed to operate, as determined by the American Society of Mechanical Engineers Boiler Code, and usually so identified on each individual unit.

safflower oil An oil, obtained from safflower seeds, that has properties similar to linseed oil and is used in paints.

safing (1) Non-combustible material used as a fire barrier around the perimeter of a floor or around protrusions or penetrations. (2) In ductwork, a type of barrier or similar device installed around a component to ensure that air flows through that component and not around it.

sagging (1) Subsidence of shotcrete material from a sloping, vertical, or overhead placement. (2) The condition of a horizontal structural member bending downward under load. See also **sloughing**.

sag rod A tension member used to limit the deflection of a girt or purlin in the direction of its weak axis or to limit the sag in angle bracing.

sailing course See **string course**.

431

Saltbox

salamander A portable source of heat, customarily oil-burning, used to heat an enclosure around or over newly placed concrete to prevent the concrete from freezing.

sal ammoniac A material used in soldering flux and iron cement. Also called "ammonium chloride".

salient Descriptive of a projecting part of an object or member, as a salient corner.

sally A projection, such as the end of a rafter beyond the notch which has been cut to fit a plate or beam.

salmon brick, chuff brick, place brick Comparatively soft and underburnt brick, usually because of its high placement in the kiln, and so named because of its color.

saltbox A wood-framed house, common in colonial New England, with a short pitched roof in front and a roof which sweeps close to the ground in back.

salt-glazed brick, brown-glazed brick Brick whose surface faces have a lustrous glazed finish resulting from the thermochemical reaction between silicates of clay and vapors of salt or other chemicals during firing.

salt-glazed tile Facing tile with a lustrous glazed finish, obtained by thermochemical reaction between silicates of clay and vapors of salt or other chemicals during firing.

salt treatment One method of preserving wood, using any of various waterborne salts to impregnate the wood. Among the more widely used systems are Wolman Salts and Osmose Salts.

samples Physical examples which illustrate materials, equipment, or workmanship and establish standards by which the work will be judged.

sand (1) Granular material passing the 3/8-inch sieve and almost entirely passing the No.4 (4.75-millimeter) sieve and predominantly retained on the No. 200 (75-micrometer) sieve, and resulting from natural disintegration and abrasion of rock or processing of completely friable sandstone. (2) That portion of an aggregate passing the No. 4 (4.75-millimeter) sieve, and resulting from natural disintegration and abrasion or rock or processing of completely friable sandstone. See also **fine aggregate**. *Note*: The definitions are alternatives to be applied under differing circumstances. Definition (1) is applied to an entire aggregate either in a natural condition or after processing. Definition (2) is applied to a portion of an aggregate. Requirements for properties and grading should be stated in the specifications. Fine aggregate produced by crushing rock, gravel, or slag commonly is known as manufactured sand.

sand asphalt A mixture of ungraded sand and liquid asphalt used for an economical base or wearing surface for pavements.

sandbag A canvas bag which is filled with sand and used as a counterweight or for emergency damming for water flow.

sandblast A system of cutting or abrading a surface such as concrete by a stream of sand ejected from a nozzle at high speed by compressed air. Sandblasting is often used for cleanup of horizontal construction joints or for exposure of aggregate in architectural concrete.

sand box (or sand jack) A tight box filled with clean, dry sand, on which rests a tight-fitting timber plunger that supports the bottom of posts used in centering. Removal of a plug from a hole near the bottom of the box permits the sand to run out when it is necessary to lower the centering.

Sander

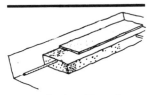

Sand Filter Trenches

sand clay A naturally occurring sand that contains about 10% clay, or just enough to make the material bind tightly when compacted.

sand-coarse aggregate ratio Ratio of fine to coarse aggregate in a batch of concrete, by weight or volume.

sand-dry Descriptive of a stage in the drying process of paint where sand will not adhere to the surface.

sanded Panel products that have been processed through a machine sander to provide a smooth surface on one or both sides. In sanded plywood, A- or B-grade veneers are used for one side of the panel.

sanded-face shingle A shingle with retrimmed edges and butts that has been sanded to remove saw marks, etc., and is to be applied to a wall as part of a decorative effect.

sanded fluxed-pitch felt A felt that is saturated with a fluxed coal tar, coated both sides with the same material, sanded and rolled for handling.

sanded grout See **sand grout**.

sand equivalent A measure of the relative proportions of detrimental fine dust or claylike material in soils or fine aggregate.

sander A machine designed to smooth wood and remove saw or lathe marks and other imperfections. Sanders range in size from hand-held to large drums or belts capable of surfacing a full-size panel.

sand-faced brick A brick formed in a mold that has been sprinkled with sand to facilitate removal.

sand filter A bed of sand laid over graded gravel, used as a filter for a water supply.

sand filter trenches A network of sewage effluent filtering trenches incorporating perforated pipe or drain tiles surrounded by fine sand sandwiched between coarse aggregate, and equipped with an underdrain to remove whatever material has passed through. The trenches are used to remove solid or colloidal material that cannot be removed by sedimentation.

sand finish (1) In plastering, a textured final coat, usually containing sand, lime putty, and Keene's cement. (2) A smooth finish derived from rubbing and sanding the final coat.

sand-float finish A rough plaster finish obtained by using a wood float.

sand grout Any grout in which fine aggregate is incorporated into the mixture. Also called "sanded grout".

sanding block A device to hold a piece of sandpaper while sanding by hand.

sanding machine A stationary machine having a moving belt, disk, or spindle with an abrasive surface, usually a sandpaper, used for smoothing surfaces.

sand jack See **sand box**.

sand-lime brick See **calcium-silicate brick**.

sand, Ottawa See **Ottawa sand**.

sandpaper Strong, tough paper coated on one side with glue or other adhesive material, into which an abrasive such as flint, silica, or aluminum oxide has been embedded. The resulting product is used primarily for resurfacing or cutting wood, metal, plastic, or glass and is available in an extensive range of coarseness or grit.

sandpile A foundation formed by ramming sand into a hole left by a pile that was driven and removed.

Sand Pine *Pinus clausa*. This pine is found almost exclusively in north central Florida, where it grows in very sandy soil. The wood is used mostly for pulp.

sand plate A flat steel plate or strip welded to the legs of bar supports

Sandwich Panel

Sanitary Tee

for use on compacted soil.

sand pocket A zone in concrete or mortar containing sand without cement.

sandstone A cemented or otherwise compacted sedimentary rock composed predominantly of sand grains.

sand streak A streak of exposed fine aggregate in the surface of formed concrete caused by bleeding.

sand trap See **sand interceptor**.

sandwich beam See **flitch beam**.

sandwich construction Composite construction usually incorporating thin layers of a strong material bonded to a thicker, weaker, and lighter core material, such as rigid foam or paper honeycomb, to create a product which has high strength-to-weight and stiffness-to-weight ratios.

sandwich panel (1) A panel formed by bonding two thin facings to a thick, and usually lightweight, core. Typical facing materials include plywood, single veneers, hardboard, plastics, laminates, and various metals, such as aluminum or stainless steel. Typical core materials include plastic foam sheets, rubber, and formed honeycombs of paper, metal, or cloth. (2) A prefabricated panel which is a layered composite, formed by attaching two thin facings to a thicker core, such as a precast concrete panel, consisting of two layers of concrete separated by a nonstructural insulating core.

sanitary cove A piece of metal used in a stair between the surface of the tread and the face of the riser to facilitate cleaning.

sanitary cross In a soil pipe system, a cross pipe, all of whose 90-degree transitions are curved to direct the flow from the branches toward the direction of the main flow.

sanitary engineering That part of civil engineering related to public health and the environment, such as water supply, sewage, and industrial waste.

sanitary sewage, domestic sewage Sewage containing human excrement and/or household wastes, which originates from sanitary conveniences of a dwelling, business, building, factory, or institution. Sanitary sewage does not include storm water.

sanitary sewer The conduit or pipe which carries sanitary sewage.

sanitary tee A T-fitting for pipe, having a slight curve in the 90-degree transition so as to channel flow from a branch line toward the main flow.

sanitary ware Devices of porcelain enamel, stainless steel, or other material, such as bathtubs, sewer pipes, toilet bowls, and washbasins.

Santorin earth A volcanic tuff originating on the Grecian Island of Santorin and used as a pozzolan.

sap grade (1) A grade of Southern Yellow Pine export lumber. (2) KD saps.

sapwood The wood just beneath the bark of a tree, normally lighter in color than the rest of the wood, but usually not as strong as the rest of the wood.

sarking, sarking board A thin board employed in sheathing applications, as under the tiles or slates of a roof.

sash, window sash The framework of a window that holds the glass.

sash adjuster See **casement stay**.

sash and frame A preassembled unit consisting of a cased frame and a double-hung window.

sash balance A spring-loaded device, usually a spring balance or tape balance, used as a counterbalance for a sash in a double-hung window. A sash balance replaces sash weights, cords, and pulleys.

sash bar See **glazing bar**.

sash block See **jamb block**.

sash center The support for a horizontally pivoting sash or transom, consisting of a socket which is secured to the jamb or frame, and a pin on which the sash or transom actually pivots. Also called a "sash plate".

sash chain A metal chain used in place of a sash cord to connect a vertically hung sash with its counterweight.

sash chisel A chisel having a wide blade honed on both sides, used for deep cutting, such as cutting the mortises in pulley stiles.

sash cord A rope connecting a sash with its counterweight in a double-hung window.

sash-cord iron A metal holder used to connect a sash cord or chain to the window.

sash counterweight See **sash weight**.

sash door See **glazed door**.

sash fast, sash fastener, sash holder Any fastening device which holds two window sashes together to prevent their opening or rattling. A sash fast is usually attached to the meeting rails of a double-hung window.

sash fillister (1) A rabbet cut in a glazing bar to receive the glass and glazing compound. (2) A special plane for cutting such rabbets. See **fillister plane**.

sash hardware All accessories used to balance a vertically hung sash, including chains or cords, weights, and pulleys.

sash holder See **sash fast**.

sash lift See **window lift**.

sash lift and hook A sash lift having a locking lever that holds the window fixed by contact with a strike in the frame. Raising the sash automatically releases the strike.

sash line See **sash cord**.

sash lock A sash fast that is controlled by a key. See **sash fast**.

sash plane A carpenter's plane having a notched cutting blade for trimming the inside of a doorframe or window frame.

sash plate One of the pair of plates constituting the pivoting mechanism for a horizontally pivoting sash or transom. Also called a "sash center".

sash pocket See **pocket**.

sash pull A plate, with a recess for fingers, set in a sash rail, or a handle attached to a rail to use in raising or lowering a window.

sash pulley A pulley mortised into the side of the frame of a double-hung window. The sash chord or chain passes over the pulley to the sash weight.

sash ribbon A metal tape used in place of a sash chord or chain.

sash saw A small miter saw used to cut the tenons of sashes.

sash spring bolt See **window spring bolt**.

sash stop A small strip fastened to a cased frame to hold a sash of a double-hung window in place. Also called a "window stop".

sash stuff Wood cut to standard sizes and shapes for use in making window frames.

sash tool A round brush used in painting items such as window frames and glazing bars.

sash weight A weight, usually cast iron, used to balance a vertically hung window.

sash window Any window, but usually a double-hung window, having a vertically or horizontally sliding sash.

satin sheen Descriptive of a paint film with a subdued gloss resembling satin.

satinwood The hard, fine-grained, pale to golden yellow wood of the

Saucer Dome

gum arabic tree, which is especially used in cabinetwork and decorative paneling.

satisfaction The cancellation of an encumbrance on a piece of real property, most often resulting from the payment of whatever debt had been secured by it.

satisfaction piece The document which records and acknowledges the payment of an indebtedness secured by a mortgage.

saturant A bituminous material with a low softening point, used for impregnating felt in asphalt prepared roofing.

saturated surface dry Condition of an aggregate particle or other porous solid when the permeable voids are filled with water and no water is on the exposed surfaces.

saturation line A line used on a cross-sectional drawing to indicate the ground water level.

saturation temperature The air temperature at which the air is saturated for any given water vapor content.

saucer dome A dome that has a rise less than its radius.

sauna A room in which a person bathes in steam produced when water is sprayed or poured over heated rocks. Some contemporary units employ heated surfaces other than rocks.

saw To cut by means of a hand or powered tool, having a thin, flat metal blade, band, or stiff plate with cutting teeth along the edge.

saw bench A bench on which a circular saw is mounted.

sawbuck See **sawhorse**.

saw cut A cut made in hardened concrete by diamond or silicone-carbide blades or discs.

sawdust concrete Concrete in which the aggregate consists mainly of sawdust from wood.

sawed finish Descriptive of the surface of any stone which has been sawn.

sawed joint A joint cut in hardened concrete by special equipment to less than the full depth of the member.

sawhorse, sawbuck A four-legged bench, usually used in pairs, made primarily to hold wood while being sawed.

saw kerf A slot or kerf that is cut into wood with a saw.

sawmill A plant in which logs are converted to lumber by running them through a series of saws.

sawn face See **sawed finish**.

sawn veneer Veneer that has been cut from a block with a saw, rather than peeled on a lather or sliced off by a blade. Sawn veneer is sometimes said to be more solid than sliced or peeled veneer. Because of saw-kerf waste, it is more costly to produce.

saws Toothed steel devices used to cut construction materials. The principal saws used in cutting construction materials include:

band An endless ribbon, toothed on one or both edges, held in tension on two pulley wheels, and powered by one or both of them.

chain Usually portable, a gas-powered saw having an articulated chain with cutting teeth running around a bar of flat steel. A chain saw is used in felling trees and cross-cutting logs.

circular A circular steel blade fitted with cutting teeth and mounted on an arbor.

drag A powered reciprocating saw used to cross-cut logs.

gang A number of toothed steel ribbons, fitted in a sash or frame, operating together.

rock A circular saw that removes a wide kerf on the upper surface of a log. A rock saw is used to remove stones or debris before a log enters the headrig.

sash A saw fitted in a frame that moves vertically.

Scaffolding

slashers A set of circular saws operated in combination for quick cross-cutting of lengths of wood before chipping or grinding into pulp fibers, or as fuel.

swing A circular saw, suspended on a pendulum, used in cross-cutting.

twin Two circular saws mounted one above the other to cut in the same plane. Also called "double arbor saws".

saw set (1) The angle at which the teeth of a saw are set. (2) A tool used to set the teeth of a saw at a desired angle.

saw table The table or platform of a power saw.

saw tex See **saw-textured**.

saw textured A texture put on a piece of siding or paneling by a saw or knurled drum to give it a textured, rough, and/or resawn appearance.

saw-tooth roof, sawtooth roof A roof with a profile similar to the teeth in a saw, composed of a series of single-pitch roofs, whose shorter or vertical side has windows for light and air.

sawyer In a sawmill, a worker who operates the head rig, or main saw, to make the initial cuts on a log.

scab A short piece of wood fastened to two formwork members to secure a butt joint.

scabbing hammer See **scabbling hammer**.

scabble To dress stone with a pick, scabbling hammer, or broad chisel, leaving prominent toolmarks so that a rough surface is left. Finer dressing usually follows.

scabbled rubble Rubble with only the roughest irregularities removed.

scabbling (1) A chip or fragment of stone produced during rough dressing. (2) Dressing down rough stone.

scabbling hammer A hammer, used for rough-dressing stone, with one end pointed for picking. Also called "scabbing hammer".

scaffold (1) (OSHA) Any temporary elevated platform and its supporting structure used for supporting workmen and/or materials. (2) Any raised platform.

scaffold board A board used in forming a work platform on a scaffold.

scaffold height That height of a wall under construction which necessitates the addition of another section of scaffold so that construction of the wall can continue.

scaffolding A temporary structure for the support of deck forms, cartways, and/or workmen, such as an elevated platform for supporting workmen, tools, and materials. Adjustable metal scaffolding is frequently adapted for shoring in concrete work.

scagliola Plasterwork in imitation of ornamental marble, consisting of ground gypsum and glue colored with marble or granite dust.

scale (1) A draftman's tool with proportioned graduated spaces. (2) A system of proportioned drawing in which lengths on a drawing represent larger or smaller lengths on a real object or surface. (3) The flakey material resulting from corrosion of metals, especially iron or steel. (4) A heavy oxide coating on copper or copper alloys resulting from exposure to high temperatures and an oxidizer. (5) Any device for measuring weight.

scale drawing A drawing in which all dimensions are reduced proportionately according to a predetermined scale, such as 1 inch = 40 feet.

Schedule

scaling Local flaking or peeling away of the near-surface portion of hardened concrete or mortar, or of a layer from metal. See also **peeling, spalling**, and **mill scale**.

light Does not expose coarse aggregate.

medium Involves loss of surface mortar to 5 to 10 mm in depth and exposure of coarse aggregate.

severe Involves loss of surface mortar to 5 to 10 mm in depth with some loss of mortar surrounding aggregate particles 10 to 20 mm in depth.

very severe Involves loss of coarse aggregate particles, as well as mortar, generally to a depth greater than 20 mm.

scallop One of a series of segments of a circle, used in decorative patterns.

scalper A sieve for removing oversized particles.

scalping The removal of particles larger than a specified size by sieving.

scalp rock The waste material from a rock-grading operation.

scant Less than standard or required size.

scantling (1) A small piece of lumber, ordinarily yard lumber, two inches thick and less than eight inches wide, or lumber not more than five inches square. (2) The dimensions, especially width and thickness, of construction materials such as stone or timber. (3) A stud or similar upright framing timber. (4) Any hardwood that has been squared but is not of standard dimensions.

scarf The end of one of the pieces of a scarf joint.

scarf connection A connection made by precasting, beveling, halving, or notching two pieces to fit together. After overlapping, the pieces are secured by bolts or other means.

scarf joint A joint made by chamfering, or beveling, the ends of two pieces of lumber or plywood to be joined. The angled cut on each piece is made to correspond to the other so that the surfaces of the two pieces being joined are flush. See **scarf connection**.

scarifier (1) An attachment, on a machine, having a series of long teeth which are used to tear up a pavement. (2) A plastering tool with long wirelike teeth, used to scratch a plaster coat to form a surface for another coat.

scarify To break up or scratch a surface.

scarp A steep slope, natural or man-made.

SC asphalt Slow-curing asphalt.

schedule A supplemental listing, usually in chart form, of a project system, subsystem, or drawings. See **progress schedule**.

schedule of values A statement furnished by the contractor to the architect reflecting the portions of the contract sum allocated to the various portions of the work and used as the basis for reviewing the contractor's applications for payment.

schematic design documents Drawings and other documents illustrating the scale and relationship of project components.

schematic design phase, schematic drawing The phase of the architect's services in which the architect consults with the owner to ascertain the requirements of the project and prepares schematic design studies, consisting of drawings and other documents illustrating the scale and relationship of the project components to the owner. The architect also submits to the owner a statement of probable construction cost based on current area, volume, or other unit costs.

Scissors Truss

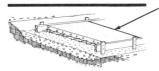

Score

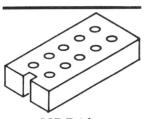

SCR Brick

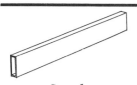

Screed

schist A metamorphic rock, the constituent minerals of which have assumed roughly parallel beds, used principally for flagging.

scissors truss A roof truss with tension members extending from the foot of each principal rafter to the upper half of its opposite member.

scollop See **scallop**.

score (1) In concrete work, to modify the top surface of one pour, as by roughening, so as to improve the mechanical bond with the succeeding pour. (2) To tool grooves in a freshly placed concrete surface to reduce cracking from shrinkage. (3) To scratch or otherwise roughen a surface to enhance the bond of plaster, mortar, or stucco which will be applied to it. (4) To groove, notch, or mark a surface for practical or decorative purposes.

scoria Vesicular volcanic ejecta of larger size, usually of basic composition and characterized by dark color. The material is relatively heavy and partly glassy, partly crystalline. The vesicles do not generally interconnect. See also **aggregate, lightweight**.

Scotch Pine *Pinus sylvestris*. An imported European conifer widely used in reforestation.

scotia A deep, concave molding more than 1/4 round in section, especially as found in Classical architecture at the base of a column.

Scots Fir See **Scotch Pine**.

Scots Pine See **Scotch Pine**.

scour Erosion of a concrete surface, exposing the aggregate.

scouring Smoothing freshly applied mortar or plaster by working it in circular motions with a cross-grained wooden float.

scragg mill An arrangement of four fixed saws used to break down logs or peeler cores of fairly uniform sizes. The scragg

produces two slabs, two 2-inch cants, and a 4-inch cant in a single operation. These can then be reduced to 2x4s and 4x4s in a resaw.

scraper A digging, hauling, and grading machine having a cutting edge, a carrying bowl, a movable front wall or apron, and a dumping or ejecting mechanism.

scratch To score or groove a coat of plaster to provide a better bonding surface for a successive coat.

scratch awl An awl used for scribing surfaces of wood, plastic, etc.

scratch-brushed finish, satin finish A surface finish rendered by mechanical wire brushing or abrasive buffing.

scratch coat The first coat of plaster or stucco applied to a surface in three-coat work and usually cross-raked or scratched to form a mechanical key with the brown coat.

scratch tool Any hand tool for scratching a plaster surface to increase bonding of the successive coat, such as a devil float, drag, or scarifier.

SCR brick Brick whose nominal dimensions are 2-2/3 inches by 6 inches by 12 inches, as designated, or classified, by the Structural Clay Research (trademark of the Structural Clay Products Institute). It will render three courses in eight inches and will render a wall whose nominal thickness is eight inches.

screed (1) To strike off concrete lying above the desired plane or shape. (2) A tool for striking off the concrete surface, sometimes referred to as a strikeoff.

screed coat The plaster coat made flush with the screeds.

screed guide Firmly established grade strips or side forms for unformed concrete which will

Screen Door

Screwdriver

guide the strikeoff in producing the desired plane or shape.

screeding The operation of forming a surface by the use of screed guides and a strikeoff. See also **strikeoff**.

screed rail Grade strips or side forms for concrete which will also guide the strikeoff in screeding.

screed strip One of a series of long narrow strips of plaster, carefully leveled to serve as guides for the application of plaster to a specified thickness.

screen See **sieve**.

screen analysis See **sieve analysis**.

screen door A lightweight exterior door, with a wood or aluminium frame and small mesh screening in place of panels, which permits ventilation but bars insects.

screen-door latch A light locking or latching device for use on screen doors and operated by a knob, a lever handle, or a handle and push button.

screen facade An architectural facing used to disguise the shape or size of a building.

screenings That portion of granular material which is retained on a sieve.

screen mold A molding, originally used in the construction of screens and now used extensively in cabinet work and finished carpentry, where a clear strip is required, as on the edge of a shelf made of plywood or particleboard.

screw A fastener with an external thread.

screw anchor A type of molly whose metal, plastic, or fiber shell is inserted into a hole in masonry, plaster, or concrete, and expanded when the screw is driven in.

screw clamp A woodworking tool consisting of a pair of opposing jaws that can hold pieces of wood and are adjusted by two screws.

screw dowel A dowel pin with a straight or tapered thread.

screwdriver A hand tool with a handle and a shank with a tip shaped to fit the recess in the head of a screw, used to drive or remove the screw. The most common tip is wedged-shaped for slotted screws, but star and Phillip's-head-shaped tips are available.

screw eye A screw with a loop or eye for its head.

screw jack See **jackscrew**.

screwless knob A door knob that is attached to the spindle using a special wrench, rather than the more common side screw.

screwless rose A rose with a concealed method of attachment.

screwnail See **drivescrew**.

scriber A pointed tool used such as to mark guide lines on wood, metal, and plastic.

scrim A coarse, meshed material, such as wire, cloth, or fiberglass, that spans and reinforces a joint over which plaster will be applied.

scroll saw A handsaw consisting of a thin blade in a deep U-shaped frame and a handle, used for cutting thin boards, veneers or plates. A scroll saw is especially good for cutting curves.

scrubboard See **baseboard**.

scrub plane A wood plane having a blade with a convex cutting edge, used in rough carpentry work.

scrub sink A plumbing fixture equipped to enable medical personnel to scrub their hands prior to a surgical procedure. The hot and cold water supply is activated by a knee-action mixing valve or by wrist or foot control.

scum (1) A deposit, sometimes formed on the surface of clay bricks, caused by soluble salts in the clay which accumulate on the surface as moisture escapes during drying, or by the formation of deposits during kiln firing. (2) A mass of organic matter that floats

Scupper

Seam

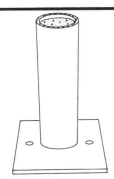

Seamless Pipe

on the surface of sewage. (3) A film of impure matter which forms on the surface of a body of water.

scumbling The process of lightly rubbing a paint brush containing a small amount of opaque or semi-opaque color over a surface to soften and blend bright tints, or to produce a special effect. The deposited coat may be so thin as to be semi-transparent.

scupper Any opening in a wall, parapet, bridge curb, or slab that provides an outlet through which excess water can drain.

scutch, scotch A tool used by bricklayers to cut, trim, and dress masonry units.

scutching Dressing stone with a special hammer whose head contains several steel points.

scuttle See **roof hatch**.

s-dry A description of lumber seasoned to a moisture content of 19% or less prior to surfacing.

seal (1) An embossing device, stamp, or other device used by a design professional on drawings and specifications as evidence of registration in the state where the work is to be performed. (2) A device formerly consisting of an impression upon wax or paper, or a wafer, which is used in the execution of a formal legal document, such as a deed or contract. The statute of limitations applicable to a contract under seal may be longer than for a contract not under seal.

sealant See **joint sealant** and **membrane curing**.

seal coat See **sealer**.

sealed bearing A conventional bearing which has been provided with seals on its sides so that the bearing can be used for longer periods without greasing.

sealed bid A bid, based on contract documents, that is submitted sealed for opening at a designated time and place.

sealer (1) Any liquid applied to the surface of wood, paper, or plaster to prevent it from absorbing moisture, paint, or varnish. (2) A liquid coating applied over bitumen or creosote to restrict it from bleeding through other paints. (3) A final application of asphalt or concrete to protect against moisture. (4) Any liquid coating used to seal the pores of the surface to which it is applied.

sealing compound See **joint sealant** and **membrane curing**.

seal weld A weld used primarily to seal a joint against leakage.

seam A joint between two sheets of material, such as metal.

seamer A hand tool used in the making of seams in sheet metal.

seaming The process of joining metal sheets by bending over or doubling the edges and pinching them together.

seamless door (1) A hollow-metal door having no visible seams on its faces. (2) A door constructed from sheet-steel bonded to a solid, structural mineral core so as not to have any edge seams.

seamless flooring Fluid- or trowel-applied floor surfaces that do not contain aggregates.

seamless pipe An extruded seamless tube having certain standardized sizes of outside diameter and wall thickness commonly designated by "Nominal Pipe Sizes" and American National Standards Institute's "Schedule Numbers".

seamless tubing Tubing manufactured with no visible seams.

seam weld A resistance weld made in overlapping parts.

season crack A crack occurring in metals that have undergone rolling or some other process that causes internal stress.

seasoned (1) Timber that is not green, having a moisture content of 19% or less, and is air or kiln-

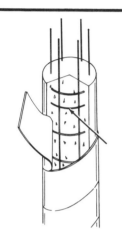

Seat Angle

Secondary Reinforcement

dried. (2) Cured or hardened concrete.

seasoning (1) The process by which lumber is dried, either by air or in a kiln. (2) The curing or hardening process which occurs in concrete.

seasoning check A small split, or check, which occurs in the grain of wood when moisture is extracted too rapidly.

seat angle A short length on a steel angle fastened to a column or girder to support the end of a beam.

seat cut The shaped cut in the end of a rafter where it rests on, and is connected to, a plate or beam.

secant modulus See **modulus of elasticity**.

second (1) A piece of secondary quality or one not meeting specified dimensions. (2) A unit measure of time.

secondary air (1) Air which is introduced into a burner above or around the flames to promote combustion, in addition to the primary air which is premixed with the fuel or forced as a blast under a stoker. (2) Air already in an air-conditioned space, in contrast to primary air which is supplied to the space.

secondary beam A flexural member that is not a portion of the principal structural frame of a building.

secondary blasting Blasting used to reduce the dimensions of oversized material to make it suitable for handling.

secondary branch In the plumbing of a building drain or water-supply main, any branch that is not the primary branch.

secondary consolidation, secondary compression, secondary time effect The reduction in volume of a soil mass caused by the application of a sustained load to the mass and due principally to the adjustment

of the internal structure of the soil mass after most of the load has been transferred from the soil water to the soil solids.

secondary crusher A crusher used for the second stage in a process of size reduction. See also **primary crusher**.

secondary light source A light source which is not itself a luminaire or otherwise intrinsically light-producing or light-emitting but which, instead, receives light from another source and simply serves to redirect it, as by reflection or transmission.

secondary moment In statically indeterminate structures, the additional moments caused by deformation of the structure due to the applied forces. In statically indeterminate prestressed concrete structures, the additional moments caused by the use of a nonconcordant prestressing tendon.

secondary nuclear vessel Exterior container or safety container in a nuclear reactor subjected to a design load only once in its lifetime, if at all.

secondary reinforcement Reinforcing steel in reinforced concrete, such as stirrups, ties, or temperature steel. Secondary steel is any steel reinforcement other than main reinforcement.

secondary sheave A supplementary sheave or block used to enable double wrapping of hoisting ropes and increase traction and/or lifting power.

secondary winding The winding on the output side of a transformer.

second coat The second coat of plaster, which is the brown coat in three-coat work or the finish coat in two-coat work.

second-growth timber Wood from trees grown after a virgin forest has been cut down.

Sectional Overhead Doors

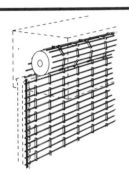

Security Screen

seconds A grade of lumber used in the hardwood industry and sometimes in the overseas trade, as in firsts and seconds.

secret dovetail, miter dovetail A joint whose external appearance implies a simple miter joint, but having dovetailing concealed within it.

secret gutter A concealed gutter.

secret nailing See **blind nailing**.

section (1) A topographical measure of land area, equal to one mile square, or 640 acres. One of the 36 divisions in a township. (2) The most desired pieces of veneer, clipped to standard widths of 54 and 27 inches, because of the ease of using them in assembling a panel. The actual width may vary from 48-54 inches, or 24-27 inches. (3) A part of a flat log raft. (4) A drawing of a surface revealed by an imaginary plane cut through the project, or portion thereof, in such a manner as to show the composition of the surface as it would appear if the part intervening between the cut plane and the eye of the observer were removed. See also **drawings**. (5) A subdivision of a division of the specifications which covers a unit of work.

sectional insulation Insulation that is manufactured to be assembled in the field, such as pipe insulation molded in two parts to fit around a pipe in the field.

sectional overhead doors Doors made of horizontally hinged panels that roll into an overhead position on tracks, usually spring assisted.

section modulus A term pertaining to the cross section of a flexural member. The section modulus with respect to either principal axis is the moment of inertia with respect to that axis divided by the distance from that axis to the most remote point of

the tension or compression area of the section, as required. The section modulus is used to determine the flexural stress in a beam.

sectorwood A method of wood processing, patented by Weyerhaeuser Company, in which round logs are first quartered and then cut into pie-shaped pieces called sectors. These pieces are then glued together to form wood two inches thick and up to three or four feet in width. This wood can then be ripped to the desired widths. The process is designed to greatly increase the yield from small logs.

security glass (1) See **bullet resisting glass**. (2) See **laminated glass**.

security screen A heavy screen in a special frame used as a barrier against escapes or break-ins. See **detention screen**, **protection screen**.

security window A steel window used in stores, warehouses, and similar commercial buildings to provide protection against burglary.

sediment Material that settles to the bottom of a liquid. Often called "silt".

sedimentary rock Rock such as limestone or sandstone that is formed from deposits of sediment consolidated by cementitious material and/or the weight of overlaying layers of sediment.

seedy Descriptive of a rough finish on paint caused by a dirty brush or by undispersed particles of pigment or insoluble gel particles in the paint.

seepage (1) The slow percolation of water through a soil. (2) The quantity of water which has moved slowly through a porous material.

seepage bed A trench at least a yard wide into which coarse aggregate and a system of distribution pipes are placed so as to allow the treated sewage which passes through them to seep into the surrounding soil.

seepage force The force which is transferred to soil particles by seepage.

seepage pit A lined excavation in the ground which receives the discharge from a septic tank and is designed to permit the effluent from the septic tank to seep through its bottom and sides.

segmental arch An arch where the curve through the length of the arch is less than a semicircle.

segmental member A structural member made up of individual elements prestressed together to act as a monolithic unit under service loads.

segment saw A large-diameter circular saw consisting of pie-shaped sections and whose narrow kerf makes it especially suitable for cutting veneer.

segregation The differential concentration of the components of mixed concrete, aggregate, or the like, resulting in nonuniform proportions in the mass. See also **separation**.

seismic design Construction designed to withstand earthquake force.

seismic load, earthquake load The assumed lateral load which an earthquake might cause to act upon a structural system in any horizontal direction.

seizing (1) Friction damage to a metal surface. (2) Restriction or prevention of motion of a metal component caused by fusion or coherence with another component.

select A high-quality piece of lumber graded for appearance. Select lumber is used in interior and exterior trim, and cabinetry. It is most often sold S4S in a 4/4 thickness, but may also be produced S2S in a variety of

thicknesses, usually for remanufacturing.

selected bidder The bidder selected by the owner for discussions relative to the possible award of a construction contract.

selected list of bidders See **invited bidders**.

selective digging Separating two or more types of soil while excavating, such as loam from sandy soil.

select material Excavated soil suitable for use as a foundation for a granular base course of a road or for bedding around pipes.

select merchantable (1) A grade of boards intended for use where knotty-type lumber of fine appearance is required. (2) An export grade of sound wood with tight knots and close grain, suitable for high quality construction and remanufacture.

select structural The highest grade of structural joists and planks. This grade is applied to lumber of high quality in terms of appearance, strength, and stiffness.

select tight knot (STK) A grade term frequently used for cedar lumber. Lumber designated STK is selected from mill run for the tight knots in each piece, as differentiated from lumber which may contain loose knots or knotholes.

selenite Gypsum in a transparent, foliated, crystalline state, usually used as ornamental building stone.

selenitic cement, selenitic lime A type of lime cement whose hardening properties have been improved by the addition of 5-10% plaster of paris.

self-centering lath Expanded-metal rib lath used as lathing in solid plaster or pneumatic placed concrete partitions and walls, or as form work for concrete slabs.

self-closing fire door A fire door equipped with a closing device.

Self-furring

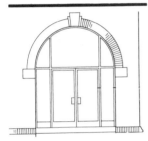

Semicircular Arch

Semidetached House

self-contained air conditioning See **package air conditioner.**

self-desiccation The removal of free water by chemical reaction so as to leave insufficient water to cover the solid surfaces and to cause a decrease in the relative humidity of the system. The term is applied to an effect occurring in sealed concretes, mortars, and pastes.

self-faced stone An undressed stone such as a flagstone.

self-furring Metal lath or welded-wire fabric formed in the manufacturing process to include means by which the material is held away from the supporting surface, thus creating a space for "keying" of the insulating concrete, plaster, or stucco.

self-furring nail A nail with a flat head and a washer or spacer on the shank, used for fastening reinforcing wire mesh and spacing it from the nailing member.

self-reading leveling rod A leveling rod designed so that the instrument man sights on a target and the rod man takes the reading off the rod.

self-sealing paint A type of paint that can be applied over a surface of inconsistent porosity to seal it and still dry to a uniform color and sheen.

self-spacing tile Ceramic tile with protuberances on the sides which space the tiles for grout joints.

self-stressing concrete (mortar or grout) Expansive-cement concrete (mortar or grout) in which expansion, if restrained, induces persistent compressive stresses in the concrete (mortar or grout). Also known as "chemically prestressed concrete"

self-supporting wall A non-load-bearing wall.

self-tapping screw A screw designed to cut its own threads in a predrilled hole.

seller's market A condition in which demand for goods is greater than supply, giving sellers the upper hand in negotiations.

selvage, selvedge (1) An edge or edging that differs from the main part of a fabric or granule-surfaced roll roofing. (2) The finished lengthwise edge of woven carpet that will not unravel or require binding.

selvage joint In roofing, a lapped joint used with mineral-surfaced cap sheets. A small part of the longitudinal edge of the sheet below contains no mineral surfacing so as to improve the bond between the lapped top sheet surface and the bituminous adhesive.

semiautomatic arc welding Arc welding with equipment that feeds the filler metal at a preset rate. The advance of the welding is controlled manually.

semiautomatic batcher See **batch mixer.**

semibasement A basement that is only partly below ground level.

semicircular arch A round arch whose intrados is a complete semicircle.

semicircular dome A dome constructed in the shape of a half sphere.

semiconductor (1) An electric conducting material, with resistivity in the range between metals and insulators, such as germanium or silicon. (2) Miniature electric devices manufactured from semiconductor materials.

semidetached dwelling One of a pair of dwellings with a party wall between them.

semidetached house One of a pair of houses with a party wall between them.

semidirect lighting Lighting from luminaires that direct 60 degrees to 90 degrees of the emitted light

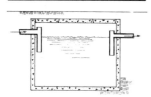

Septic Tank

downward and the balance upward.

semidome A half-dome equal to 1/4 of a hollow sphere, as might be found above a semicircular niche or apse.

semiengineering brick Brick whose crushing strength is somewhere between that of a building brick and an engineering brick.

semiflexible joint In reinforced concrete construction, a connection in which the reinforcement is arranged to permit some rotation of the joint.

semigloss (1) The degree of surface reflectance midway between glass and eggshell. (2) Paints and coatings displaying these properties.

semihoused stair A stair which has a wall on one side only.

semihydraulic lime A type of lime categorized intermediately between high-calcium lime and eminently hydraulic limes and which, when run to putty, is almost as workable as non-hydraulic lime, even though its hydraulic characteristics can be substantially reduced by soaking in water for several hours.

semi-indirect lighting Lighting from luminaires that direct 60 to 90 degrees of the emitted light upward and the balance downward.

semirigid frame A type of structural framework construction in which some flexibility is allowed at the joints of columns and beams.

semi-split A trade name used by the Red Cedar Shingle and Handsplit Shake Bureau for a product with a partially sawn and split face.

semisteel Low-carbon cast iron produced by adding steel scrap to molten pig iron.

semitrailer A trailer that has a set or sets of wheels at the rear only.

The front part is supported on a towing vehicle.

sensible heat Heat that alters the temperature of a material without causing a change in state in that material.

sensor A device designed to detect an abnormal ambient condition, such as smoke or high temperature, and to sound an alarm or operate a device.

separate-application adhesive An adhesive consisting of two parts, each part being applied to a different surface. The surfaces are brought together to form a joint.

separate contract One of several prime contracts on a job. A separate contract is not a subcontract.

separate contractor On a given project, a contractor who has a separate contract with the owner.

separated aggregate (1) A quantity of aggregate that has been separated into two or more batches based on size. (2) Fine and coarse aggregate considered separately before mixing.

separation (1) The tendency, as concrete passes from the unconfined ends of chutes or conveyor belts, or similar arrangements, for coarse aggregate to separate from the concrete and accumulate at one side. (2) The tendency, as processed aggregate leaves the ends of conveyor belts, chutes, or similar devices with confining sides, for the larger aggregate to separate from the mass and accumulate at one side. (3) The tendency for the solids to separate from the water by gravitational settlement. See also **bleeding** and **segregation**.

septic tank A watertight receptacle that receives the discharge from a drainage system, or part thereof, and is designed and constructed to separate solids from the liquid, digesting organic matter through a period of detention.

Service Conductors

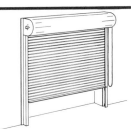

**Service Door,
Service Entrance**

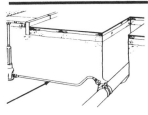

Service Pipe

sequence-stressing loss In post-tensioning, the elastic loss in a stressed tendon resulting from the shortening of the member when additional tendons are stressed.

serial distribution A group of absorption trenches, seepage pits, or seepage beds arranged in series so that the total effective absorption area of one is used before flow enters the next.

series circuit A circuit supplying electric energy to a number of devices connected so that the same current passes through each device.

serpentine Rock largely composed of hydrous magnesium silicate and commonly occurring in greenish shades. Serpentine is the main constituent in some marbles.

service box Within a building, a metal box located at that point where the electric service conductors enter the building.

service conductors Those conductors (wires) which carry electric current from the street mains, or transformer, to the service equipment of the building being supplied.

service dead load The dead weight supported by a member.

service door, service entrance An exterior door in a building, intended primarily for deliveries, for removal of waste, or for the use of service personnel.

service drop The overhead conductors which connect the electric supply or communication line to the building being served.

service elevator An elevator intended for combined passenger and freight use.

service ell, street ell A fitting for threaded pipe with a 45-degree or 90-degree bend, a male thread on one end, and a female thread on the other.

service entrance switch The circuit breaker or switch, with fuses and accessories, located near the point of entrance of supply conductors to a building and intended to be the main control and cut-off for the supply to that building.

service lateral Those underground electrical service conductors (wires) between the street main, including any risers at a pole or other structure or from transformers, and the point of initial connection with the service entrance conductors in a terminal box or other enclosure either inside or outside a wall of the building being served. In the absence of such a box or enclosure, the service lateral connect the street main and the point at which the service conductors enter into the building.

service live load The live load specified by the general building code or bridge specification, or the actual non-permanent load applied in service.

service load See **service dead load** and **service live load**.

service period In lighting, the number of hours per day for which natural daylight provides a specified amount of illumination, often expressed as a monthly average.

service pipe (1) The water or gas pipe which leads from a supply source, usually public distribution mains in the street, to the particular building(s) being served. (2) The pipe or conduit through which underground service conductors are run from the outside supply wires to the customer's property.

service refrigerator Any refrigerated display case or reach-in refrigerator from which an attendant serves a customer.

service road (1) A lesser road parallel to the main road, used primarily by local traffic. (2) A road or drive in a complex which

447

is intended for vehicles making deliveries or collecting waste.

service systems The heating, ventilating, air conditioning, water, and electric distribution systems in a building.

service tee A tee fitting used for threaded pipe and having one end threaded on the outside, and the other end and the branch threaded on the inside.

service valve In a piping system, any valve which isolates a device or apparatus from the rest of the system.

set (1) The condition reached by a cement paste, mortar, or concrete when it has lost plasticity to an arbitrary degree, usually measured in terms of resistance to penetration or deformation. Initial set refers to first stiffening; final set refers to attainment of significant rigidity. See **permanent set**. (2) The rehydration and consequential hardening of gypsum plaster. (3) The strain which remains in a member after the removal of the load that initially produced the deformation. (4) To transform a resin or adhesive from its initial liquid or plastic state to a hardened state by physical or chemical action, such as condensation, polymerization, oxidation, vulcanization, gelation, hydration, or the evaporation of volatile ingredients. (5) To drive a nail so far that its head is below the surface into which it has been driven. (6) To apply a finish coat of plaster. (7) The permanent distortion produced in a spring which has been stressed beyond the elastic limit of its constituent material. (8) The overhang given to the points of sawteeth to result in a kerf slightly wider than the saw as to facilitate sawing motion.

setback The minimum distance required by code or ordinance between a building and a property line or other reference.

setback buttress A buttress set slightly back from the angle or corner of a building.

setscrew (1) An often headless screw used to secure two separate parts in a position relative to each other, preventing the independent motion of either part. (2) A screw used to adjust the tension of a spring.

setting bed The mortar subsurface to which terrazzo is applied.

setting block In glazing, a small block of wood, lead, neoprene, or other suitable material placed under the bottom edge of a light or panel to support it within the frame and prevent it from settling down onto the lower rabbet or channel.

setting shrinkage A reduction in volume of concrete prior to the final set of cement, caused by settling of the solids and by the decrease in volume due to the chemical combination of water with cement.

setting space The distance between the finished face of a veneer, such as a brick panel, and the outside of the main wall.

setting temperature The temperature to which an adhesive or resin must be subjected in order for setting to occur.

setting time See **initial setting time** and **final setting time**.

setting-up That point in the initial drying of a paint or other liquid coating at which it is no longer able to flow.

settlement Sinking of solid particles in grout, mortar, or fresh concrete, after placement and before initial set. See also **bleeding**.

settling The lowering in elevation of sections of pavement or structures due to their mass, the loads imposed on them, or shrinkage or displacement of the support.

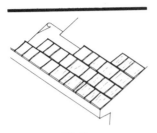

Shake

settling basin An enlargement or basin within a water conduit, which provides for the settling of suspended matter, such as sand, and is usually equipped with some means of removing the accumulated material.

set up (1) The stationing of a surveying instrument, such as a transit. (2) Descriptive of concrete or similar firm material. (3) In plumbing, to bend up the edge of a sheet of lead lining material. (4) To caulk a pipe joint with lead by driving it in with a blunt chisel.

sewage Any liquid-borne waste which contains animal or vegetable matter in suspension or solution. Sewage may include chemicals in solution, and ground, surface, or storm water may be added as it is admitted to or passes through the sewers.

sewage disposal The treatment and dispersal of sewage.

sewage ejector A plumbing device used to raise sewage to a higher elevation.

sewage gas The mixture of gases, odors, and vapors, sometimes including poisonous and combustible gases, found in a sewer.

sewage treatment Any process to which sewage is subjected to remove or alter its objectionable constituents by reduction in the organic and bacterial content, rendering it less offensive and dangerous.

sewage treatment plant Structures and appurtenances that receive raw sewage and bring about a reduction in organic and bacterial content of the waste so as to render it less dangerous and less odorous.

sewer (1) Generally, an underground conduit in which waste matter is carried in a liquid medium. (2) A pipeline in which sewage is conveyed.

sewerage The entire works required to collect, treat, and dispose of sewage, including the sewer system, pumping stations, and treatment plant.

sewer appurtenances Manholes, sewer inlets, and other devices, constructions, or accessories related to a sewer system but exclusive of the actual pipe or conduit.

sewer brick Low-absorption, abrasive-resistant brick intended for use in drainage structures.

sewer gas See **sewage gas**.

sewer pipe The piping used in a sewer.

sewer plank Timbers, mostly in the sizes of 3x8 and 3x10, 18 and 20-foot lengths, which are used to repair or construct drainageways, especially in older cities.

sewer tile Impervious clay tile pipe intended to carry water or sewage.

sewer trap See **building trap**.

s-green A description of lumber surfaced while at a moisture content of more than 19%.

shack See **shanty**.

Shade Pine See **Sugar Pine.**

shaft (1) That portion of a column between the base and the capital. (2) An elevator well. (3) A pit dug from the ground surface to a tunnel to furnish access and ventilation. (4) Any enclosed vertical space in a building used for utilities or ventilation. (5) Any cylindrical rod connecting moving parts in a machine.

shake (1) Roofing material produced from wood, usually cedar, with at least one surface with a grain split face. (2) A crack in lumber due to natural causes.

shale A laminated and fissile sedimentary rock, the constituent particles of which are principally clay and silt.

shank (1) The main body of a nail, screw, bolt, or similar fastener extending between the head and

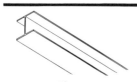

Shape

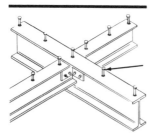

Shear Connector

the point. (2) The usually metal part of a drill or other tool that connects the working head to the handle. (3) In a Doric frieze, a plain space between channels of a triglyph.

shanty A small wooden building, usually temporary and haphazardly constructed.

shape (1) A solid section, other than flat product, rod, or wire sections, furnished in straight lengths and usually made by extrusion but sometimes fabricated by drawing. (2) A solid section other than regular rod, bar, plate, sheet, strip, or flat wire, and may be oval, half oval, half round, triangular, pentagonal, or of any special cross section furnished in straight lengths. (3) A wrought product that is long in relation to the dimensions of its cross section, which is of a form other than that of sheet, plate rod, bar, tube, or wire. (4) To give a profile or detail to a piece of work, such as providing a board with a head or rounded edge. (5) To work on a piece of wood, or other material to make it conform to a predetermined desired or required pattern, or to render from its surface a specific texture or degree of smoothness.

shaper (1) A woodworking machine with a vertically revolving cutter for cutting irregular outlines, moldings, etc., in wood placed on a table below the cutter. (2) A metalworking machine similar to a planer, except that the cutting tool is moved back and forth across the surface.

shaping machine See **shaper**.

sharp sand Coarse sand made up of particles of angular shape.

Shasta Red Fir See **California Red Fir**.

shatterproof glass See **laminated glass**.

shaving A very thin slice of wood

removed in dressing and used in some types of panels.

shear (1) An internal force tangential to the plane on which it acts. See also **shearing force**. (2) The relative displacement of adjacent planes in a single member. (3) To cut metal with two opposing passing blades or with one blade passing a fixed edge. (4) The tool used for the operation in definition 3.

shear connector (1) A welded stud, spiral bar, short length of channel, or any other similar connector which resists horizontal shear between components of a composite beam. (2) A timber connector.

shear diaphragms Members in a structure to resist shear forces, such as those caused by wind load. See also **shear wall**.

shear failure, failure by rupture Failure in which movement caused by shearing stresses in a soil or rock mass is of sufficient magnitude to destroy or seriously endanger a structure.

shearhead Assembled unit in the top of the columns of flat slab or flat plate construction to transmit loads from slab to column.

shearing force The algebraic sum of all the tangential forces acting on either side of the section at a particular location in a flexural member.

shearing machine (1) An apparatus having a moveable blade that passes a fixed cutting edge to cut metal. (2) A machine used in carpet finishing to remove stray fibers and fuzzy loop pile and produce a smooth, level surface on cut pile.

shear legs A hoisting device from two or more poles fastened together near their apex, from which a pulley is hung to lift heavy loads.

Shear Plate (2)

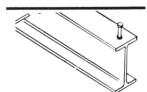

Shear Stud

Sheathed Cable

Sheathing

She Bolt

shear plate (1) A shear-resisting plate used to reinforce the web of a steel beam. (2) In heavy timber construction, a round steel plate usually inserted in the face of a timber to provide shear resistance in joints between wood and non-wood.

shear-plate connector A type of timber connector employed in wood-to-wood or wood-to-steel applications.

shear reinforcement Reinforcement designed to resist shear or diagonal tension stresses.

shears A metal-cutting tool which has two opposing pivoted blades whose beveled edges face each other.

shear slide A type of landslide where a section of earth slides away from the material beneath it in a single consolidated mass.

shear splice A type of splice designed to distribute the shear between the two members which it joins.

shear strain, shearing strain The angular displacement or deformation of a member caused by a force perpendicular to its length.

shear strength The maximum shearing stress which a material or structural member is capable of developing, based on the original area of cross section.

shear stress The shear-producing force per unit area of cross section, usually expressed in pounds per square inch.

shear stud A short unthreaded bolt welded to the top flange of a steel beam. The shear studs are embedded in a concrete slab to form a composite beam and stud.

shear wall, shearwall A wall portion of a structural frame intended to resist lateral forces, such as earthquake, wind, and blast, acting in or parallel to the plane of the wall.

sheath An enclosure in which post-tensioning tendons are encased to prevent bonding during concrete placement. See **duct**.

sheathed cable Electric cable protected by non-conductive covering, such as vinyl. See **non-metallic sheathed cable**.

sheathing (1) The material forming the contact face of forms. Also called "lagging" or "sheeting". (2) Plywood, waferboard, oriented strand board, or lumber used to close up side walls, floors, or roofs preparatory to the installation of finish materials on the surface. The sheathing grades are also commonly used for pallets, crates, and certain industrial products.

sheathing felt Roofing felt which has been saturated or impregnated, usually with some type of bitumen of low softening point (100 degrees-160 degrees F).

sheathing paper See **building paper**.

sheave (1) The grooved wheel of a pulley, or block. (2) The entire assembly over which a rope or cable is passed, including not only the pulley wheel but also its shaft bearings, and side plates.

sheave block A pulley with a housing and bail.

she bolt A type of form tie and spreader bolt in which the end fastenings are threaded into the end of the bolt, eliminating cones and reducing the size of holes left in the concrete surface.

shed A small, usually roughly constructed shelter or storage building, sometimes having one or more open sides, and sometimes built as a lean-to.

shed dormer A dormer window having vertical framing projecting from a sloping roof, and an eave line parallel to the eave line of the principle roof. A shed dormer is designed to provide more space

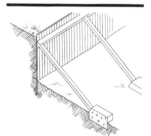

Sheeting

Sheet-metal Screw

under a roof than a gabled dormer would provide.

shed roof A roof having a single sloping plane.

sheen On a flat paint finish, the luster or gloss which can be seen only from certain angles.

sheepsfoot roller A powered or towed earth-tamping or compacting machine consisting of a large drum with projecting studs or feet with enlarged outer ends.

sheer legs See **shear legs**.

sheet A thin piece of material, such as glass, veneer, plywood, or rolled metal. See **panel**.

sheet asphalt A plant-mixed asphalt paving material containing sand which has passed through a 10-mesh sieve, and some type of mineral filler.

sheet glass Flat glass made by continuous drawing. Also known as "common window glass".

sheeting (1) Planks used to line the sides of an excavation, such as for shoring and bracing. See also **sheathing**. (2) 7/8-inch tongue-and-groove board. (3) Sheet piling. (4) A form of plastic in which the thickness is very small in proportion to length and width and in which the plastic is present as a continuous phase throughout, with or without filler.

sheeting clip A metal clip designed to fasten different thicknesses of plasterboard, plywood paneling, and the like.

sheeting driver An air hammer attachment that fits onto the ends of planks to allow their being driven without splintering.

sheeting jacks Push-type turnbuckles used to support sheeting in a trench.

sheet lath Heavier and stiffer than expanded-metal lath, lath fabricated by punching geometrical perforations in copper alloy steel sheet.

sheet lead Lead which has been

cold-rolled into a sheet and whose designation is determined by the weight of one square foot of the finished product.

sheet metal Metal, usually galvanized steel but also including aluminum, copper, and stainless steel, which has been rolled to any given thickness between 0.06 and 0.249 inches and cut into rectangular (usually 4 feet by 8 feet) sections which are then used such as in the fabrication of ductwork, pipe, and gutters.

sheet-metal screw Usually a self-tapping screw, either round, flat, or pan-headed, threaded from tip to head, and used without a nut for fastening lapped sheet metal and some other materials. When equipped with a pointed tip, no tap hole is required.

sheet-metal work The fabrication, installation, and/or final product, such as the ductwork of a heating or cooling system, as performed or produced by a worker skilled in that trade.

sheet pile A pile in the form of a plank driven in close contact or interlocking with others to provide a tight wall to resist the lateral pressure of water, adjacent earth, or other materials. A sheet pile may be tongued and grooved if made of timber or concrete, or interlocking if made of metal.

sheetpiling A barrier or diaphragm formed of sheet piles which is used to prevent the movement of soil or keep out water during excavation and construction. Sheet piles are constructed of timber, sheet steel, or concrete.

sheetrock screwdriver A specialized power tool designed to accept the flat Phillips head of a sheetrock screw and drive the screw when the point is placed on the sheetrock and the tool is pushed toward it.

sheet-roofing nail See **roofing nail**.

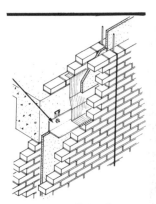

Shelf Angles

Shielded Conductor

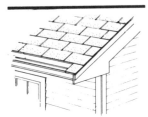

Shingle

shelf (1) Any horizontally mounted board, slab, or other flat-surfaced device upon which objects can be stored, supported, or displayed. (2) A ledge, as of rock, of a setback.

shelf angle A section of angle iron or steel which is welded or otherwise secured to an I-beam or channel section to provide support for the formwork or the hollow tiles of a concrete slab, or which, when similarly attached to a girder, carries the ends of joists.

shelf angles Structural angles with holes or slots in one leg for bolting to the structure to support brick work, stone, or terra cotta.

shelf bracket Any structural member secured to a wall or upright, from which it also projects to support a shelf.

shelf cleat A wood (or other) strip which supports a shelf along one edge.

shelf life Maximum duration during which a material can be stored and still remain in a usable condition.

shelf nog A shelf support consisting of a piece of wood which has been built into a wall but protrudes outward from it.

shelf rest, shelf pin, shelf support A type of angle bracket through whose vertical portion a pin is passed and inserted into one of several position-adjusting holes such as in a wall, or cabinet.

shelf strip See **shelf cleat**.

shell (1) Structural framework. (2) In stressed-skin construction, the outer skin applied over the frame members. (3) Any hollow construction when accomplished with a very thin curved plate or slab. (4) The outer portion of a hollow masonry unit when laid.

shellac A transparent coating produced by dissolving lac, a resinous secretion of the lac bug, in denatured alcohol.

shell aggregate A type of aggregate made up of fine sands and the crushed shells of mollusks, such as clams, oysters, and scallops.

shell construction (1) A type of reinforced concrete construction in which thin curved slabs are primary elements. (2) Construction in which a curved exterior surface has been obtained by using shaped steel and hardboard or curved plywood panels.

shelling See **checking**.

shelly structure See **perlitic structure**.

shielded conductor An insulated electric conductor enclosed in a metal sheath or envelope.

shielded metal-arc welding An arc welding process wherein coalescence is produced by heating with an arc between a covered metal electrode and the work. Shielding is obtained from decomposition of the electrode covering. Pressure is not used, and filler metal is obtained from the electrode.

shielding concrete Concrete, used as a biological shield to attenuate or absorb nuclear radiation, usually characterized by high specific gravity or high hydrogen (water) content or boron content. See also **biological shielding**.

shift (1) The lateral movement of a faulted seam. (2) A work period.

shim A strip of metal, wood, or other material used to set base plates or structural members at the proper level for placement of grout, or to maintain the elongation in some types of post-tensioning anchorages.

shingle A roof-covering unit made of asphalt, wood, slate, asbestos, cement, or other material cut into stock sizes and applied on sloping roofs in an overlapping pattern.

shingle backer A specially fabricated sheathing-grade structural insulating board used as

Shoe (2)

a backer strip in coursed shingle construction.

shingle lap A type of lap joint whereby the thinner of two tapered surfaces is lapped over the thicker.

shingle nail A galvanized steel, aluminum alloy, or coated steel nail, often having a threaded shank. There are two types of shingle nails: one for wood shingles and one for asphalt shingles, the latter having a larger head.

shingle ridge finish See **Boston hip**.

shingle stain A low-viscosity, pigmented finish that penetrates wood shingles to provide moisture protection as well as color.

shingle tile A flat clay tile used primarily for roofing and laid so as to overlap.

Shinglewood Another name for "Western Red Cedar"

shingling hatchet, claw hatchet Similar to a lath hammer, this tool consists of a hammer head, a hatchet, and a notch or nail claw.

ship decking Full-sawn vertical grain pieces, knot-free on the faces, used in the construction of ship decks.

shiplap (1) Lumber that has been worked to make a rabbeted joint on each edge so that pieces may be fitted together snugly for increased strength and stability. (2) A similar pattern cut into plywood or other wood panels used as siding to assure a tight joint.

shipping position A seller's estimate of the time required after an order is placed until it can be shipped.

shivering The splintering which critical compressive stresses can precipitate in fired glazes or other ceramic coatings. Also called "peeling".

shock See **thermal shock**.

shock hazard (OSHA) Considered to exist between an accessible part in a circuit and the ground or other accessible parts, if the potential is more than 42.4 volts peak and the current through a 1,500-ohm load is more than 5 milliamperes.

shock load The impact load of material such as aggregate or concrete as it is released or dumped during placement.

shock mount See **vibration isolator**.

shoddy work Any work which has been performed carelessly and/or unprofessionally.

shoe (1) Any piece of timber, metal, or stone receiving the lower end of virtually any member. Also called a "soleplate". (2) A metal device protecting the foot, or point, of a pile. (3) A metal plate used at the base of an arch or truss to resist horizontal thrust. (4) A ground plate forming a link of a track, or bolted to a track line. (5) A support for a bulldozer blade or other digging edge to prevent cutting down. (6) A cleanup device following the buckets of a ditching machine. (7) A short section used at the base of a downspout to direct the flow of water away from a wall.

shoe molding A base shoe used at the bottom of a baseboard to cover the space between the finished flooring and the baseboard.

shoe rail The molding on the top of a stair string to support the balusters.

shoot To straighten the edge of a board with a plane.

shooting The placing of shotcrete.

shooting board A jig to hold a board while its edge is being squared or beveled.

shooting plane A light plane used to square or bevel the edge of a board with a shooting board.

Shore

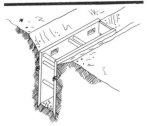

Shoring

Shoring, Horizontal

Short Nipple

shop drawings Drawings, diagrams, schedules, and other data specially prepared for the work by the contractor or any subcontractor, manufacturer, supplier, or distributor to illustrate some portion of the work.

shop lumber, factory lumber Lumber graded, in a shop or factory, according to the number of pieces, of designed size and quality, into which it may be cut.

shop painting The application of paint, usually one coat, to metals in the shop before shipment to a job site.

shopping center A collection of stores, shops, and service centers, along with parking facilities.

shop rivet A rivet driven in the shop.

shopwork Work manufactured or assembled in the shop.

shore A temporary support for formwork and fresh concrete or for recently-built structures that have not developed full design strength. Also called a prop, tom, post, strut. See also **T-head**.

shore hardness number A number representing the relative hardness of materials. It is the height of rebound observed when a standard hammer strikes the material being tested.

shore head A wood or metal horizontal member placed on and fastened to a vertical shoring member. See also **raker**.

shore up To support with shores.

shoring (1) Props or posts of timber or other material in compression used for the temporary support of excavations, formwork, or unsafe structures. (2) The process of erecting shores.

shoring, horizontal A metal or wood load-carrying strut, beam, or trussed section used to carry a shoring load from one bearing point, column, frame, post, or wall to another. Horizontal shoring may be adjustable.

shoring layout A drawing prepared prior to erection showing the arrangement of equipment for shoring.

short circuit An accidental electric connection of relatively low resistance between two points of different potential in an electric circuit, causing a high current flow between the two points.

short face A sales term used when describing tongue-and-groove siding that includes the tongue within the stated width. Also called narrow face.

short grain Wood in which part of the grain runs diagonally to the length of the piece and which is subject to failure under load.

Shortleaf Pine *Pinus echinata*. This species, one of the Southern Yellow Pine group, is found in a broad range from Texas to as far north as Pennsylvania and New Jersey.

short-length (1) A piece of stock lumber usually less than 8 ft (2.44m) long. (2)(Brit.) A piece of sawn hardwood usually less than 6 ft (1.83 m) long.

short nipple A pipe nipple with a short unthreaded length between the threaded ends.

short-oil varnish A varnish having a low vehicle content, containing less than 15 gallons of oil per 100 pounds of resin.

shorts Short pieces of lumber. The lengths described as shorts vary widely by species, products, and regions. Generally, a dimension of 12 feet or less is described as short, while boards of 6 feet or less are also shorts.

short ton A unit of measurement of weight in the English system equal to 2,000 pounds.

short working plaster Plaster that is difficult to work and from which the sand separates, due to

**Shower Bath,
Shower, Shower Stall**

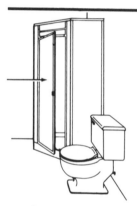

Shower Stall Door

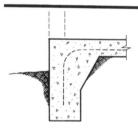

Shrinkage Reinforcement

using plaster that has been affected by moisture over a long storage period.

shotblasting A process similar to sand blasting, except that metal shot is used.

shotcrete (1) Mortar or concrete pneumatically projected at high velocity onto a surface. Also known as air-blown mortar. (2) Pneumatically applied mortar or concrete, sprayed mortar, and gunned concrete. See also **dry-mix shotcrete, pneumatic feed, positive displacement,** and **wet-mix shotcrete**.

shotcrete gun A pneumatic device used to deliver shotcrete or to move concrete.

shot tower A tower used in the manufacture of metal shot. Molten lead is dropped from a height into water.

shoulder An unintentional offset in a formed concrete surface usually caused by bulging or movement of formwork.

shouldering Placing mortar under the top edge of a slate so the lower edge will fit closer to the slate beneath.

shoved joint A vertical joint in brickwork made by laying the brick in a bed of mortar and shoving the brick toward the last brick laid.

shower bath, shower, shower stall The compartment and plumbing provided for bathing by overhead spray.

shower-bath drain A floor drain in a shower-bath.

shower head A nozzle used to spray water in a shower bath.

shower mixer A valve in a shower bath used to mix hot and cold water supplies to obtain a desired water temperature.

shower pan A pan of concrete, terrazzo, concrete and tile, or metal used as a floor in a shower bath.

shower partition A partition used around a shower bath for privacy.

shower stall door A door in a shower partition.

show rafter A rafter, often ornamental, which is visible below a cornice.

show window A window used to display merchandise.

shrinkage Volume decrease caused by drying and/or chemical changes, such as of concrete or wood.

shrinkage-compensating A characteristic of grout, mortar, or concrete made using an expansive cement in which volume increase, if restrained, induces compressive stresses intended to approximately offset the tendency of drying shrinkage to induce tensile stresses. See also **expansive cement**.

shrinkage crack A crack due to restraint of shrinkage.

shrinkage cracking The cracking of a structure or member due to failure in tension caused by external or internal restraints from carbonation and/or reduction in moisture content.

shrinkage, drying See **drying shrinkage** and **shrinkage**.

shrinkage joint See **contraction joint**.

shrinkage limit The water content at which a reduction in water content will not cause a decrease in volume of the soil mass but an increase in water will increase the volume. See **Atterberg limits**.

shrinkage loss Reduction of stress in prestressing steel resulting from shrinkage of concrete.

shrinkage reinforcement Reinforcement designed to resist shrinkage stresses in concrete.

shrink-mixed concrete Ready-mixed concrete mixed partially in a stationary mixer and then mixed in a truck mixer. See also **preshrunk**.

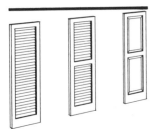

Shutter

Siamese Connection

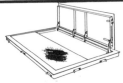

Sidewalk Door

Siding, Weatherboarding

Siding Shingle

shrunk joint A joint made by placing a piece of heated pipe over the ends of two cool pipes and allowing it to contract.

shunt An electrical device with a low resistance or impedance connected in parallel across another electrical device to divert current from the second device.

shutter A movable cover or screen used to cover an opening, especially a window opening.

shutter bar A hinged bar used to fasten a pair of shutters together in the closed position.

shutter blind An exterior adjustable louver used at a window.

shutter butt A narrow hinge used on shutters and small doors.

shuttering See **formwork**.

shutting shoe A receptacle of metal or stone set in the paving or ground beneath a gate with two leaves to receive the vertical bolt.

siamese connection A wye connection on the outside of a building with two inlet connections, used by the fire department to supply water to a sprinkler and/or standpipe system.

sidecast During the construction of a logging road, the pushing of waste soil and debris over the downhill side.

sidecasting Piling soil beside the excavation.

side-construction tile Tile designed to be used with the cells set horizontally.

side-dump loader An earth loader with its bucket mounted on a pivot so it can be dumped to either side as well as forward.

side-entrance manhole A deep manhole with the access shaft built into the side of the inspection chamber.

side gutter A small gutter along a side of a chimney or dormer.

sidehill A slope crossing the line of work.

sidehill cut An excavation for a highway through a sidehill, leaving a bank on one side only.

side hook See **bench hook**.

side-hung window See **casement window**.

side jamb A vertical member forming a side of a door opening.

side knob screw A setscrew used to secure a door knob to a spindle.

side lap The distance that a piece of material, such as steel roof deck, overlaps an adjacent piece.

side light A fixed frame of glass beside a window or door.

side outlet An ell or tee fitting with an outlet at right angles to the line of run.

side post One of a pair of posts in a roof truss, each set at an equal distance from the center of the truss.

side vent A vent connected to the drain pipe at an angle of 45 degrees or less from vertical.

sidewalk door A cellar door opening directly onto a sidewalk. The door is flush with the sidewalk when closed.

sidewalk elevator (1) An elevator opening onto a sidewalk. (2) An elevator platform without a cab which rises to level flush with a sidewalk.

sidewall The exterior wall of a building.

side yard The open space between a side of a dwelling and property line.

siding, weatherboarding Lumber or panel products intended for use as the exterior wall covering on a house or other building.

siding gauge See **clapboard gauge**

siding shingle A thin piece of material, such as wood, cement, asbestos, or plastic, used as an exterior wall finish over sheathing.

sieve A metallic plate or sheet, a woven wire cloth, or other similar device, with regularly spaced apertures of uniform size, mounted in a suitable frame or holder, for use in separating material according to size. In mechanical analysis, an apparatus with square openings is a sieve; one with circular apertures is a "screen".

sieve analysis The determination of the proportions of particles within certain size ranges in a granular material by separation on sieves of different size openings. See also **grading**.

sieve correction Correction of a sieve analysis to adjust for deviation of sieve performance from that of standard calibrated sieves.

sieve number A number used to designate the size of a sieve, usually the approximate number of openings per linear inch. The number is applied to sieves with openings smaller than 6.3 mm (1/4 in.).

sieve size The nominal size of openings usually between cross wires of a testing sieve.

sight draft An instrument of payment negotiated through banks. Negotiable documents are attached, such as an order bill of lading, thus ensuring payment by the consignee to the negotiating bank in exchange for the documents and prior to the delivery of the goods.

sight glass A glass tube used to indicate the liquid level in boilers, tanks, etc.

sighting rod (1) See **leveling rod**. (2) See **range rod**.

sight rail A series of rails set using a surveying instrument and used to check the vertical alignment of a pipe in a trench.

sight rod (1) See **leveling rod**. (2) See **range rod**.

sign, signboard (1) A board or surface displaying directions, instructions, identification, or advertising. (2) (OSHA) A warning of hazard, temporarily or permanently affixed or placed at a location where a hazard exists.

signal sash fastener A sash fastener beyond reach of a person. The fastener consists of a catch operated by a ring moved by a fitting on a long pole.

silencer See **rubber silencer**.

silica Silicon dioxide (SiO_2).

silica brick A refractory brick made from quartzite containing approximately 96% silica with alumina and lime.

silica flour Very finely divided silica, a siliceous binder component that reacts with lime under autoclave curing conditions. The flour is prepared by grinding silica, such as quartz, to a fine powder. Also known as "silica powder".

silica gel, synthetic silica A drying agent made from a form of silica.

silica powder See **silica flour**.

silicate Salt of a silicic acid.

silicate paint A paint in which sodium silicate is used as the binding agent.

silicon A metallic element used mainly as an alloying agent but used in pure form in electrical rectifiers.

silicon bronze A copper alloy used in hardware and other applications where a high resistance to oxidation is desirable.

silicon carbide An artificial product (SiC), granules of which may be embedded in concrete surfaces to increase resistance to wear or as a means of reducing skidding or slipping on stair treads or pavements. Silicon carbide is also used as an abrasive in saws and drills for cutting concrete and masonry.

Sill

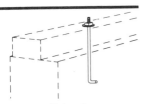

Sill Anchor

Sill Course

Simple Span

silicone A resin, characterized by water-repellent properties, in which the main polymer chain consists of alternating silicon and oxygen atoms, with carbon-containing side groups. Silicones may be used in caulking or coating compounds or admixtures for concrete.

silicone-carbide paper A tough, black, water-resistant sandpaper used in wet sanding and for fine finishing.

silicone paint A heat-and-chemical-resistant paint used on chimneys, stoves, and heaters, which requires heat to cure.

sill (1) The horizontal member forming the bottom of a window or exterior doorframe. (2) As applied to general construction, the lowest member of the frame of the structure, resting on the foundation and supporting the frame. See **mud sill**.

sill anchor An anchor bolt used to fasten a sill to its foundation.

sill block A solid concrete masonry unit used as a sill for an opening.

sill cock A water faucet located on the exterior of a building roughly at the top of the sill. A sill cock is usually threaded to provide a connection for a hose.

sill course A course of brickwork at a window sill, usually projected to shed water.

sill high (1) Located at the height of a sill above the floor. (2) Located at the height of a sill above ground level.

silo (1) A tower-like structure, usually cylindrical, used to store items such as grain, coal, or minerals. (2) A structure built in the ground to house a military missile.

silt, inorganic silt, rock flour (1) A granular material resulting from the disintegration of rock, with grains largely passing a No. 200 (75-micrometer) sieve. (2)

Such particles in the range from 2 to 50 micrometers in diameter.

silt box A steel box in a catch basin that can be removed to clean out the silt.

silt grade Descriptive of sediment having particles of silt size.

Silver Fir See **Pacific Silver Fir**.

silver solder A solder containing silver and having a high melting point, used for high-strength solder joints.

Silver Spruce Sitka Spruce.

silver white (1) Any white pigment used in paints. (2) A very pure white lead.

simple beam A beam without restraint or continuity at its supports.

simple span The support-to-support distance of a simple beam.

simplex casement A swing-out window with no mechanical device for opening and closing.

single-acting door, single-swing door A swing door that swings in one direction only.

single bridging Single pieces of wood fixed between joists, as opposed to diagonal bridging.

single-cleat ladder A ladder made of two side rails with single treads between them. See **double-cleat ladder**.

single contract A contract for construction of the project under which a single prime contractor is responsible for all of the work.

single-cut file A file having serrations cut in one direction only.

single-duct system An air conditioning system using one duct to serve a number of different areas.

single Flemish bond A bond in a brick wall using Flemish bond for the facework and English bond for the body.

Single-hub Pipe

Sink

Siphon Trap

single floor A floor of joists and flooring only, without intermediate support.

single-framed roof A roof-framing system in which the rafters are tied together by boards or the framing of the floor or ceiling below.

single-hub pipe A pipe with a hub at one end only.

single-hung window A window with a movable sash, vertically hung, and a fixed sash.

single-lap tile A curved roofing tile laid so as to overlap only the tile directly below it.

single-package refrigation system A factory-assembled and tested refrigeration system shipped in one section. No refrigerant-containing parts are connected in the field.

single-point adjustable suspension scaffold A manual or power-operated platform supported by a single wire attached to its frame, used for light work only.

single-pole scaffold A platform resting on putlogs or cross beams supported on ledger beams and posts on the outer side and by the wall on the inner side.

single-pole switch An electric switch with one movable and one fixed contact.

single-roller catch A catch consisting of a spring-loaded roller on the door and a strike plate on the jamb. The door catches automatically and is opened by pushing or pulling.

single roof A roof supported only by common rafters.

single-sized aggregate Aggregate in which a major portion of the particles are of sizes between narrow limits.

single-stage curing An autoclave curing process in which precast concrete products are put on metal pallets for autoclaving and

remain there until stacked for delivery or yard storage.

single-swing frame A frame in which is mounted one swinging door.

single-throw switch An electric switch opened or closed by the operation of a single pair of contacts.

sink A plumbing fixture consisting of a water supply, a basin, and a drain connection.

sinker nail A wire nail with a small, slightly depressed head, driven below the surface of a wood member as a hidden fastener.

sinking (1) A shallow depression in an object. (2) The process of removing wood in a jamb so that the hinges can be installed flush.

sinking in The penetration of a paint binder into an unprimed, porous surface, resulting in a finish coat with low gloss.

sinter The process of forming a material by heating powder to a temperature just below its melting point so that it fuses together.

sintering grate A grate on which material is sintered.

siphonage The removal of fluid from a device, such as a trap, caused by suction produced by fluid flow.

siphon trap An S-shaped plumbing trap containing water as a gas seal.

S-iron A retaining plate at each end of a tie rod used with a turnbuckle to tie two masonry walls together.

sisal A fiber from the leaves of the sisal plant, used in making rope and cord and as reinforcing in plaster.

site The geographical location of the project, usually defined by legal boundary lines.

site analysis services (of the architect) Those services described in the schedule of designated services necessary to

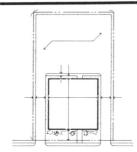

Site Plan

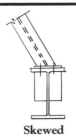

Skewed

establish site related limitations and requirements for the project.

site built The construction of a structure at the site where it is to remain.

site drainage (1) An underground system of piping carrying rainwater or other wastes to a public sewer. (2) The water so drained.

site exploration See **site investigation**.

site-foamed insulation Thermal insulation foamed in place.

site investigation A complete examination, investigation, and testing of surface and subsurface soil and conditions. The report resulting from the investigation is used in design of the structure.

site plan A plan of the area at the proposed construction showing the building outline, parking, work areas, and/or property lines.

Sitka Spruce *Picea sitchensis.* A lightweight, but particularly strong, species still used in aircraft construction and for other special uses. The range of Sitka Spruce is a narrow belt extending along the Pacific Coast from Alaska to Northern California.

sitzbath A bathtub designed to support a person in the sitting position, used for therapeutic treatment.

size (1) To bring a piece to specified dimensions. (2) See **sizing**.

sized dry Surfaced or sawn to a specific size after being dried.

sized green Surfaced or sawn to a specific size while still green, and subject to further shrinkage. The National Grading Rule for dimension lumber sets slightly larger sizes for green lumber than for dry to reflect shrinkage.

sized lumber Lumber uniformly manufactured to net surfaced sizes. Sized lumber may be rough, surfaced, or partly surfaced on one or more faces.

sizing, size The application of diluted glue or adhesive to hardwood veneer to prepare the wood for application of standard concentration of glue. Sizing reduces the amount of standard glue that will be absorbed. Sizing is often used when woods of different densities are glued together.

sizing saw A saw that trims large timbers to the desired dimensions.

skeleton construction A type of construction in which loads are carried on frames made of beams and columns. Wall of upper floors are supported on the frames.

skew back, skewback A sloping surface against which the end of an arch rests, such as a concrete thrust block supporting the thrust of an arch bridge. See also **chamfer strip**.

skew chisel A woodworking chisel with a sharp cutting edge cut at an oblique angle and beveled sides.

skew corbel A specially shaped stone at the bottom of a stone gable forming an abutment for the coping, eave gutters, or wall cornices.

skewed Forming an oblique angle with a main center line.

skew fillet A fillet-shaped piece nailed along the gable coping under the slate to divert water from the edge.

skew nailing See **toe nailing**.

skew plane A woodworking plane with the blade at an oblique angle across the face.

skid resistance A measure of the frictional characteristics of a surface.

skim coat A thin coat of plaster, usually either the finish coat or the leveling coat.

skin (1) The materials, such as steel, aluminum and/or glass that make up a curtain wall. (2) The

Skylight

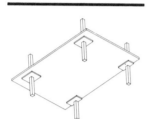

Slab Floor

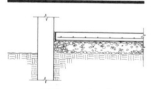

Slab-on-Grade

outer veneer or ply of a lamination or built-up piece. (3) The thin face of a hollow-core door. (4) A tough layer formed on the surface of paint in a container. (5) A dense layer on the surface of a cellular material.

skin drying The rapid drying of the surface of a paint film while the material underneath remains wet.

skin friction The friction between soil and a structure, such as a retaining wall, or between soil and a pile.

skinner saw A saw that trims and sizes the eight-foot side of a plywood panel as it leaves the press. The four-foot side is trimmed by a cut-off saw.

skintled brickwork Brick work laid so the resulting face is irregular.

skirting block (1) A corner block where a base and vertical framing meet. (2) A concealed block to which a baseboard is attached.

skylight A glazed opening in a roof to admit light.

slab (1) A flat, horizontal or nearly so, molded layer of plain or reinforced concrete, usually of uniform but sometimes of variable thickness, either on the ground or supported by beams, columns, walls, or other framework. (2) The outside, lengthwise cut on a log. See also **flat slab** and **flat plate.**

slab bar bolster A wire form used to support reinforcing while placing concrete.

slab board A rough board cut from the side of a log with bark and sapwood intact.

slab bolster See **bolster, slab.**

slab floor A floor of reinforced concrete.

slab form The formwork used in placing a concrete slab.

slab on grade A concrete slab placed on grade, sometimes having

insulation board or an impervious membrane beneath it.

slab-on-grade construction A type of construction in which the floor is a concrete slab poured after plumbing and other equipment is installed.

slab spacer Bar support and spacer for slab reinforcement, similar to slab bolster but without corrugations in top wire, no longer in general use. See also **bolster, slab.**

slab strip See **middle strip.**

slack-rope switch A safety device that shuts off electric power to the drive of an elevator if the supporting cables become slack.

slag See **blast-furnace slag.**

slag block A concrete masonry unit with blast furnace slag as the coarse aggregate.

slag concrete Portland cement concrete with blast furnace slag as the coarse aggregate.

slag sand A sand-like material made by crushing and grading blast furnace slag.

slake (1) To add water to quicklime to make putty. (2) To crumble or disintegrate on exposure to moist air.

slaked lime See **hydrated lime.**

slaking box A wooden box used to hold materials while slaking lime.

slamming strip An inlay along the edge of the lockstile of a flush wood door.

slant A sewer pipe connecting a house sewer to a common sewer.

Slash Pine *Pinus elliottii.* One of several pine species grouped under the designation of Southern Yellow Pine. Slash Pine is native to the Southeastern and Gulf Coast states. It is fast growing and matures early. Its wood closely resembles that of the Longleaf Pine, another of the SYP group.

slat A thin, narrow strip of wood, metal, or plastic.

Slating

Sliding Sash

slate A fine-grained metamorphic rock possessing a well-developed fissility (slaty cleavage) usually not parallel to the bedding planes of the rock.

slate-and-a-half slate Slate 1-1/2 times as wide as other slates on a roof but of the same length.

slate batten A batten fastened across rafters to support slates on a roof.

slate boarding Close boarding used to support slates or tiles.

slate powder A powder made from slate, used as an extender in paint.

slate roll, slate ridge A cylindrical rod formed from slate, having a V-shaped notch cut on its underside, and used to form the ridge on a roof.

slating (1) The installation of slates on a roof or wall. (2) The slate shingles on a roof or wall, taken collectively.

slave unit A device or machine controlled or activated through another unit.

sledgehammer, sledge A heavy hammer, up to 100 pounds (45kg), having two striking faces and swung with both hands.

sleeper (1) One of many strips of wood fastened to the top of a concrete slab to support a wood floor. (2) Any horizontal timber laid on the ground to distribute load from a post.

sleeper clip A metal clip used to fasten sleepers to a concrete subfloor.

sleeper joist A joist supported on sleepers.

sleeper wall Any short wall that supports floor joists.

sleepiness A film defect consisting of an area of lower gloss in a high-gloss film.

sleeve See **pipe sleeve**.

sleeve fence A short, decorative fence extending out from a dwelling.

slender beam A beam which, if loaded to failure without lateral bracing of the compression flange, would fail by buckling rather than in flexure.

slenderness ratio The ratio of effective length or height of a wall, column, or pier to the radius of gyration. The ratio is used as a means of assessing the stability of the element.

slewing Rotating the jib of a crane so the load moves through a horizontal arc.

sliced veneer Veneer sliced from a face of a squared-off log.

slicker See **darby**.

slick line The end section of a pipe line used in placing concrete by pump which is immersed in the placed concrete and moved as the work progresses.

slide rule A device used to make certain calculations rapidly, consisting of a base of wood, metal, plastic, or cardboard with various scales and a scaled sliding insert. The calculations are made by offsetting appropriate scales.

sliding bearing A support for a structure that slides relative to a base structure.

sliding fire door A fire door hung from an overhead track. The track may be sloped for direct gravity operation or horizontal with operation by weights, cables and pulleys.

sliding form See **slipform**.

sliding sash Any window that moves horizontally in grooves.

sliding window See **sliding sash**.

slimline lamp A type of instant-starting fluorescent lamp with a single pin base.

slip Movement occurring between steel reinforcement and concrete in stressed reinforced concrete indicating anchorage breakdown.

slip form, slipform, sliding form A form that is pulled or raised as concrete is placed. The form may

Slip-joint Conduit

Slop Sink

move in a generally horizontal direction to lay concrete evenly for highway paving or on slopes and inverts of canals, tunnels, and siphons; or vertically to form walls, bins, or silos.

slip joint (1) A vertical joint between an old and a new brick wall, made by cutting a slot in the old wall and filling it with brick projecting from the new wall. (2) A joint in plumbing in which one pipe slips within another and the seal is made by a pressure device fitting over the joint, often threaded to the larger pipe.

slip-joint conduit Electric conduit connected by short couplings that slip over the ends of the conduit.

slip-joint pliers Pliers having jaws connected by a rivet through a slot in one part, allowing two settings for the jaws.

slip mortise See **slot mortise.**

slip newel A newel that either has a hollowed base to slip over a peg or is grooved to fit a wall.

slip-on flange A flange slipped over the end of a pipe and welded in place.

slippage The lateral movement of adjacent roofing plies.

slip-resistant tile Ceramic tiles with abrasive particles or grooves in the surface.

slip sheet Protective paper placed over the faces of prefinished plywood paneling to protect them during transport.

slip sill A sill cut to fit between jambs and installed after the walls are in place.

slip tongue See **spline.**

slogging chisel A heavy chisel used to cut off the heads of bolts.

slope (1) See **grade.** (2) See **pitch.** (3) See **incline.** (4) See **grain slope.**

slope correction See **grade correction.**

sloped footing A footing having a sloping top or sloping side faces.

slope map A map displaying the topography of an area, along with a discussion of topographic features.

slope ratio Relation of the horizontal projection of a surface to its rise. For example, 2 feet horizontal to 1 foot rise is shown as 2:1 or 2 to 1.

slop sink A deep sink set low on a wall, used to clean mops and to empty and clean pails.

slot mortise, open mortise, slip mortise A mortise that does not extend the full length of the member in which it is cut.

slot outlet An air supply outlet with a length-to-width ratio greater than 10:1.

slot weld A weld between two members made by welding within a slot in one of the members.

sloughing Subsidence of shotcrete, due generally to excessive water in mixture. See also **sagging.**

slow-burning A misleading term implying that a material is fire-safe. The term must be related to a particular test for interpretation.

slow-burning construction Heavy timber construction with large flat surfaces, as opposed to joisted construction.

slow-burning insulation Insulation that burns or chars without a flame. Also see **slow-burning.**

slow-curing asphalt Liquid asphalt made of asphalt cement and oils of low volatility.

slow-evaporating solvent A solvent with a high boiling point, used to lengthen, and thus improve, the drying time of paint.

sludge (1) Waste material composed of wet fines produced from grinding a terrazzo floor. (2) Accumulated solids in the wash water reservoir of paint spray booths. (3) The semiliquid settled solids from treated sewage.

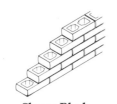

Slump Block

Smoke and Fire Vent

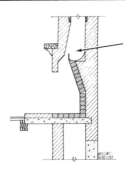

Smoke Chamber

slugging The pulsating and intermittent flow of shotcrete material due to improper use of delivery equipment and materials.

sluice (1) A steep, narrow passage for water. (2) See **sluicing**.

sluicing Moving soil by use of rapidly flowing water, for excavation or particle grading.

slump A measure of consistency of freshly mixed concrete, mortar, or stucco equal to the subsidence measured to the nearest 1/4 in. (6 mm) of the molded specimen immediately after removal of the slump cone.

slump block A concrete masonry unit intentionally removed early from a mold so that it slumps slightly.

slump cone A mold in the form of the lateral surface of the frustum of a cone with a base diameter of 8 in. (203 mm), top diameter 4 in. (102 mm), and height 12 in. (305 mm), used to fabricate a specimen of freshly mixed concrete for the slump test. A cone 6 in. (152 mm) high is used for tests of freshly mixed mortar and stucco.

slump loss The amount by which the slump of freshly mixed concrete changes during a period of time after an initial slump test was made on a sample or samples thereof.

slump test The procedure for measuring slump.

slurry A mixture of water and any finely divided insoluble material, such as portland cement, slag, or clay in suspension.

slurry seal machine A self-propelled machine used for mixing and delivering a mixture for use as an asphalt emulsion slurry seal.

slushed joint A vertical joint made by slushing mortar into the joint after the unit is laid.

slush grouting Distribution of a grout, with or without fine aggregate, as required, over a rock or concrete surface that is subsequently to be covered with concrete. The grouting is usually accomplished by brooming it into place to fill surface voids and fissures.

smoke (1) A suspension in air of particles which are usually solid. (2) Carbon or soot particles less than 0.1 micron in size, resulting from incomplete combustion, such as of oil or coal.

smoke and fire vent A vent cover, installed on a roof, which opens automatically when activated by a heat-sensitive device, such as a fusible link.

smoke chamber The transition portion of a chimney between the fireplace and the flue.

smoke damper A damper arranged to close and stop flow automatically when smoke is detected.

smoke-developed rating A relative number indicating the smoke produced when the surface of a material burns, as measured during an ASTM E-119 flame spread test.

smoke door A smoke and fire vent over a stage which opens either automatically or manually by cutting a line.

smoke-dried lumber Lumber seasoned by exposing the material to the heat and smoke from a fire.

smoke hatch See **smoke door**.

smoke shelf A concave shelf at the back of a smoke chamber to redirect downdrafts up the chimney.

smokestop A partition intended to retard the spread of smoke. Any opening in a smokestop should be protected by a door with an automatic closer.

smoke test A test using a nontoxic, visible smoke to determine the routes taken by air currents and/or to detect leaks.

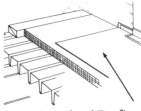

Smooth-surfaced Roofing

smooth ashlar A rectangular stone block with smooth faces used in masonry construction.

smoothing iron A hand-held hot iron with a long handle used for smoothing asphaltic pavement joints.

smoothing plane A small, fine carpenter's plane used for finishing.

smooth-surfaced roofing A built-up roofing membrane, the top surface of which is either hot mopped asphalt, an asphalt emulsion of a cutback coating, or an inorganic top felt.

smudge (1) An accidental mark or smear on a surface. (2) A paint primer made from the scraping of paint pots. (3) A mixture of glue and lamp black spread on a surface to prevent adhesion of solder.

snake (1) A long, resilient wire used by electricians in running wires through conduit. The snake is pushed through and then used to pull the wires. (2) A flexible metal wire used to clear clogged plumbing fixtures.

snake hole A hole driven into rock at an angle.

snakeholing Drilling blast holes under a rock or surface.

snap head See **button head**.

snap header A half bat.

snapped work Masonry laid using more snap headers than full headers.

snapping line A layout line made by stretching a chalked line across a surface and snapping the line.

snap switch A manually operated switch used to control low-power indoor circuits.

snap tie A proprietary concrete wall-form tie, the end of which can be twisted or snapped off after the forms have been removed.

snatch block A pulley or block with a side that can be opened to

receive a rope or line.

snips See **tin snips**.

snow fence A fence of wood strips connected by wire and used to catch drifting snow.

snow load The live load allowed by local code, used to design roofs in areas subjected to snowfall.

soaker A piece of flashing used on a slate roof at a hip or valley, or at the intersection of a roof and a vertical wall.

soaking period In high-pressure and low-pressure steam curing, the time during which the live steam supply to the kiln or autoclave is shut off and the concrete products are exposed to the residual heat and moisture.

soap A brick or tile of normal face dimensions but with a nominal thickness of 2 inches.

soapstone A soft rock containing a high proportion of talc, used for such items as sinks, bench tops, and carved ornaments. Also see **steatite** .

socket (1) British term for the enlarged end of bell-and-spigot pipe. (2) A mechanical device for supporting a lamp or plug fuse and completing the electric circuit. (3) See coupling.

socket fuse See **plug fuse**.

socketing A connection of two timber pieces by fitting one into a mating cavity in the other.

socket plug A plug for the end of an interior threaded pipe fitting consisting of a threaded piece with a recess into which a tool is placed to turn it.

socket wrench A box wrench with a recessed socket that fits over a nut for tightening or removing the nut.

sod The upper layer of soil containing roots of grass.

soda-acid fire extinguisher A fire extinguisher that discharges water under pressure. The pressure is

Soffit

produced by mixing soda and acid to generate carbon dioxide.

sod house A house with walls made of layers of sod. The roof is usually logs covered with earth and sod.

sodium light The orange-yellow light from a low-pressure sodium-vapor lamp.

soffit The underside of a part or member of a structure, such as a beam, stairway, or arch.

soffit block A special concrete masonry unit used under a beam and slab concrete floor to conceal the beam soffits and provide a flat ceiling.

soffit board A board that forms the soffit of a cornice.

soffit bracket A bracket used to mount an exposed exterior door closer to a doorframe head or transom bar.

softening point The temperature at which bitumen softens or melts, used as an index of fluidity.

soft glass Glass susceptible to thermal shock because it has a high coefficient of thermal expansion and a low softening point.

soft light Dispersed light producing soft shadows.

Soft Maple Acer rubrum, A. saccharinum. Red Maple and River Maple, softer than Rock or Sugar Maple.

soft-mud brick Brick produced by molding clay with a high moisture content (20% to 30%).

soft particle An aggregate particle possessing less than an established degree of hardness or strength as determined by a specific testing procedure.

soft rot Decay in wood, where the residue is chiefly cellulose.

softwood (1) A general term referring to any of a variety of trees having narrow, needle-like or scale-like leaves, generally coniferous. (2) The wood from

such trees. The term has nothing to do with the actual softness of the wood; some "softwoods" are harder than certain of the "hardwood" species.

soil A generic term for unconsolidated natural surface material above bedrock.

soil absorption system A disposal system, such as an absorption trench, seepage bog, or seepage pit, that utilizes the soil for subsequent absorption of treated sewage.

soil analysis See **mechanical analysis**.

soil binder Soil that just passes through a No. 40 (40 squares per inch) sieve.

soil branch A branch line of a soil pipe.

soil-cement Soil, portland cement, and water mixed and compacted in place to make a hard surface for sidewalks, pool linings, and reservoirs, or for a base course for roads.

soil class A classification of soil by particle size, used by the U.S. Department of Agriculture: (1) gravel, (2) sand, (3) clay, (4) loam, (5) loam with some sand, (6) silt-loam, and (7) clay-loam.

soil classification test A series of tests combined with sensory observations used to classify a soil. The tests may include such aspects as grain size, distribution, plasticity index, liquid limit, and density.

soil cover See **ground cover**.

soil creep The slow movement of a mass of soil down a slope, caused by gravity and aggravated by pore water.

soil drain A horizontal soil pipe.

soil mechanics The application of the laws and principles of mechanics and hydraulics to engineering problems dealing with soil as a building material.

Solar Flat Plate Collector

Solarium

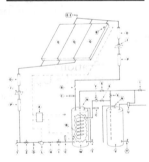

Solar Heating System

soil pipe A pipe that conveys the discharge from water closets or similar fixtures to the sanitary sewer system.

soil pressure See **contact pressure**.

soil profile A vertical section through a site showing the nature and sequence of layers of soil.

soil sample A representative sample of soil from a specific location or elevation of a construction site, usually extracted to determine bearing capacity.

soil stabilizer (1) A machine that mixes in place soil and an added stabilizer, such as cement or lime, in order to stiffen the soil. (2) A chemical added to soil to stiffen it and increase the stability of a soil mass.

soil stack A vertical soil pipe that carries the discharge from water closet fixtures.

soil structure The pattern in which the soil particles are arranged in the aggregate.

soil survey See **subsurface investigation**.

soil test See **soil classification test**.

solar collector Any device intended to collect solar radiation and convert it to energy.

solar constant The average rate at which radiation energy from the sun is received at the surface of the earth, equal to 430 BTUs per hour per sq. ft. (1.94 cal per min per sq cm).

solar dryer A wood dryer that uses solar energy to raise the dry-bulb temperature of the air being circulated through the wood.

solar energy The radiant energy from the sun.

solar flat plate collector A solar collector in the shape of a flat plate consisting either of a series of photo voltaic cells or a sandwich panel made up of a black surface, a film of circulating water or air, and a transparent cover.

solar fraction The percentage of the seasonal heating requirement of a building that is provided by a solar system.

solar gain That part of a building's heat load produced by solar radiation striking the building or passing through windows.

solar heat Heat supplied by radiation from the sun.

solar heat exchanger A means of transferring the heat in storage, in a solar heating system, to the area to be heated.

solar heating system An assembly of components, including collectors, heat exchangers, piping, storage system, controls, and supplemental heat source, used to provide heat and/or hot water to a building, with the sun as the main source of energy.

solar house A house designed and located so as to use the sun's rays to maximum advantage in heating the house.

solarium A room or porch with a great deal of glass and located for maximum exposure to the sun.

solar orientation The alignment of a building relative to the sun, initially set either for maximum or minimum heat gain, depending on the local climate.

solar screen (1) An openwork or louvered panel of a building, so placed as to act as a sunshade. (2) A perforated wall used as a sun shade.

solar storage Fluid and/or rocks used to hold some of the heat energy collected by a solar heat collector.

solder (1) An alloy, usually lead-tin, with a melting point below 800 degrees F (427 degrees C), used to join metals or seal joints. (2) The process of joining metals or sealing joints using solder and heat.

soldered joint A gastight pipe joint made by applying solder to a heated joint.

soldering gun A tool with a pistol grip and a small electrically heated bit that reaches operating temperatures rapidly, used to solder electric components.

soldering nipple A nipple with one end threaded and the other plain. The plain end is used for a soldered joint and the threaded end for a mechanical joint.

solderless connector See **pressure connector**.

soldier (1) A vertical wale used to strengthen or align formwork or excavations. (2) A masonry unit set on end so its long, narrow face is vertical on the face of the wall.

soldier arch A flat, brick arch in which the stretchers are set vertically.

soldier beam A rolled-steel section driven into the ground to support a horizontally sheeted earth bank.

soldier course A course of brick units set on end with the long, narrow face vertical on the wall face.

soldier pile, soldier (1) In an excavation, a vertical member that supports horizontal sheeting and is supported by struts across the excavation and by embedment below the excavation. (2) A vertical member used to support formwork and held in place by struts, bolts, or wires.

sole (1) See **solepiece**. (2) See **soleplate**.

solenoid valve A valve opened by a plunger in which movement is controlled by an electromagnet.

solepiece (1) Any horizontal member used to distribute the loads from one or more uprights or struts. (2) A member that supports the foot of a raking shore.

soleplate (1) A solepiece or shoe that serves as a base for studs in a partition. (2) A plate welded or bolted to the underside of a plate girder that bears on a pad. (3) A sill.

solid block A masonry block meeting the specifications for a solid masonry unit.

solid brick A brick meeting the specifications for a solid masonry unit.

solid bridging See **block bridging**.

solid core (1) The inner layers of a plywood panel which contain no open irregularities, such as gaps or open knotholes, and in which the grain runs perpendicular to the outer plies. Solid core is primarily used as underlayment for resilient floor covering. (2) A flush door, used in entries and as fire-resistant doors, in which particleboard or wood blocks completely fill the area between the door skins.

solid-core door A door having a core of solid wood or mineral composition, as opposed to a hollow-core door.

solid cross (X) band A cross band in plywood consisting of plugged veneer allowing limited open defects or splits. The term is no longer used by softwood plywood manufacturers, who have substituted "plugged innerplies" or "plugged crossbands" as a more accurate description.

solid door See **solid-core door**.

solid frame The frame of a door or window having jambs, head, and sill made of solid pieces of timber rather than parts.

solid glass door A door in which the glass essentially provides all the structural strength.

solid loading Filling a drill hole with explosive, except for a stemming space at the top.

solid masonry unit (1) (U.S.A.) A masonry unit whose minimum net cross-sectional area parallel to its bearing surface is 75% or more of its gross cross-sectional area.

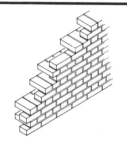

Solid Masonry Wall

Solid-web Steel Joist

(2) A masonry unit having holes less then 3/4 in. (2 cm) wide or less than 0.75 sq. in. (5 sq cm) passing through it, or having frogs that do not exceed 20% of its volume. In either type, up to 3 handling holes, not exceeding 5 sq in (32.5 sq cm) each, are allowed. The total area of all through holes may not exceed 25% of the gross area.

solid masonry wall A wall built of solid masonry units with all joints filled with mortar and no hollow wythes.

solid mopping The application of hot asphalt over an entire roof surface.

solid molding A molding produced from a single piece of wood, as distinguished from finger-jointed moldings produced from two or more pieces of wood joined together end to end.

solid panel A solid slab, usually of constant thickness.

solid partition A partition with no cavity.

solid plasterwork Solid core plaster formed in place.

solid punch A steel rod used to drive bolts out of holes.

solid rock Rock that can not be moved or processed without being blasted.

solids That part of paint, varnish, or lacquer that does not evaporate but stays on the surface to form the film.

solid-state welding A welding process in which coalescence takes place at temperatures below the melting point of the metals being joined and without use of a brazing filler metal. Sometimes this is accomplished through the use of pressure.

solid wall (1) A wall of solid concrete. (2) See **solid masonry wall**.

solid-web steel joist A steel truss,

or light beam, having a solid web, usually cold-formed from sheet and having a channel shape.

solution A liquid solvent in which one or more substances are dissolved.

solvent A liquid in which another substance may be dissolved.

solvent adhesive An adhesive having a volatile organic liquid as a vehicle.

sonic modulus See **dynamic modulus of elasticity**.

sonic pile driver A machine used to drive piles or sheet piling into soil using a head vibrated at a high frequency, usually less than 6,000 times per minute.

Sonotube[*] A product consisting of a preformed casing made of laminated paper used to form cylindrical piers or columns.

soot door An access door to a flue for cleaning or repairing the flue.

soot pocket A place at the foot of a chimney where soot collects, usually fitted with a door so the soot can be removed easily.

sound (1) A vibratory disturbance, with the frequency in the approximate range between 20 to 20,000 cycles per second, capable of being detected by a human ear. (2) Wave motion in the air.

sound absorption (1) The process of dissipating sound energy. (2) The measure of the absorptive ability of a material or object, expressed in sabins or metric sabins.

sound absorption coefficient The fraction of sound energy striking a surface which is absorbed or otherwise not reflected from the surface.

sound attenuating door See **sound-rated door**.

sound attenuator An assembly installed in a duct system to absorb sound.

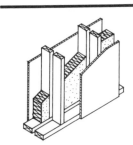

Sound Insulation

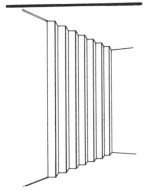

Sound-rated Door

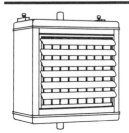

Space Heater

sound deadening board A board with good sound absorption qualities, used in sound-control construction.

sound door See **sound-rated door**.

sounding well A vertical conduit used to determine the elevation of grout being placed in preplaced aggregate concrete.

sound-insulating glass A glazing unit consisting of two or more lights fixed in resilient mountings and sealed so as to provide one or more dead air spaces.

sound insulation, sound isolation (1) The use of materials and assemblies to reduce sound transmission from one area to another or within an area. (2) The degree to which sound transmission is reduced.

sound isolation See **sound insulation**.

sound knot A dead knot in wood which is undecayed, at least as hard as the surrounding wood, and held firmly in place.

sound level The reading of a sound level meter, expressed in decibels and based on one of three weighting networks, A, B, or C.

soundness The freedom of a solid from cracks, flaws, fissures, or variations from an accepted standard. In the case of a cement, soundness is freedom from excessive volume change after setting. In the case of aggregate, soundness is the ability to withstand the aggressive action to which concrete containing it might be exposed, particularly that due to weather.

soundproofing (1) The design and construction of a building or unit to reduce sound transmission. (2) The materials and assemblies used in a building or unit to reduce sound transmission.

sound-rated door A door constructed to provide greater sound attenuation than that provided by a normal door, usually carrying a rating in terms of its sound transmission class (STC).

sound transmission The passage of sound from one point to another, as from one room to another, or from a street to a room within a building.

sound transmission class, STC A single number indicating the sound insulation value of a partition, floor-ceiling assembly, door, or window, as derived from a curve of insulation value as a function of frequency. The higher the number is, the greater the insulation value.

sound trap See **sound attenuator**.

Southern Balsam Fir See **Balsam Fir**.

Southern Pine See **Southern Yellow Pine**.

Southern Red Cedar *Juniperus silicicola*. A species very similar to Eastern Red Cedar and found in wet lowlands and swampy areas along the southern Atlantic and Gulf coasts.

Southern Yellow Pine A species group composed primarily of Loblolly, Longleaf, Shortleaf, and Slash Pines. Various subspecies also are included in the group. The Southern Yellow Pine region refers to the southeastern United States, from Texas to Virginia.

space frame Any three-dimensional structural frame capable of transmitting loads in the three dimensions to supports. A space frame is usually an interconnected system of trusses or rigid frames.

space heater A small heating unit, usually equipped with a fan, intended to supply heat to a room or portion of a room. The source of heat energy may be electricity or a fluid fuel.

spacer (1) A device that maintains reinforcement in proper position,

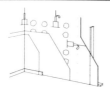

Spackle, Spackling

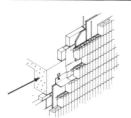

Spandrel Beam

or wall forms at a given distance apart before and during concreting. (2) A small block of wood or other material placed during installation on the edges of a pane of glass to center it in the channel and maintain uniform width of sealant beads to prevent excessive sealant distortion. See also **spreader**.

spacing factor An index related to the maximum distance of any point in a cement paste or in the cement paste fraction of mortar or concrete from the periphery of an air void. Also known as "Powers spacing factor".

spackle, spackling, sparkling A paste, or a dry mixture mixed with water to form a paste, used to fill holes and cracks in plaster, wallboard, or wood.

spade A sturdy digging tool having a thick handle and a flat blade that can be pressed into the ground with a foot.

spading Consolidating mortar or concrete by repeatedly inserting and withdrawing a flat, spade-like tool.

spall A fragment, usually in the shape of a flake, detached from a larger mass by a blow, by the action of weather, by pressure, or by expansion within the larger mass; a small spall involves a roughly circular depression not greater than 20 mm in depth nor 150 mm in any dimension; a large spall may be roughly circular or oval or, in some cases, elongated, more than 20 mm in depth, and 150 mm in greatest dimension.

spalling The development of spalls.

span (1) The distance between supports of a member. (2) The measure of distance between two supporting members.

spandrel, spandril That part of a wall between the head of a window and the sill of the window above it.

spandrel beam A beam in the perimeter of a building, spanning between columns and usually supporting floor or roof loads.

spandrel frame A triangular-shaped frame.

spandrel glass Opaque glass used in curtain wall construction to conceal structural elements.

span length See **effective span**.

spanner (1) A horizontal cross brace. (2) A collar beam.

span roof A pitched roof with the same slope on either side.

spar (1) A common rafter. (2) A heavy, round timber. (3) A bar used as a gate latch.

sparge pipe A perforated pipe used to distribute flushing water for a urinal.

spark arrester A device at the top of a chimney or stack to catch sparks, embers, or other ignited material over a given size. Also called a "bonnet".

sparrow peck (1) A texture given to finish plaster by dabbing it with a stiff brush. (2) A texture given to stone by pitting it with a pick.

spar varnish A varnish with superior weather-resisting qualities, used on exterior wood.

spat A protective sheet, usually stainless steel, installed at the bottom of a door frame.

spatter dash, spatterdash A rich mixture of portland cement and coarse sand, thrown onto a background by a trowel, scoop, or other appliance so as to form a thin, coarse-textured, continuous coating. As a preliminary treatment before rendering, it assists bond of the undercoat to the background, improves resistance to rain penetration, and evens out the suction of variable backgrounds. See also **parge** and **dash-bond coat**.

special assessment A charge imposed by a government on a particular class of properties to defray the cost of a specific improvement or service, presumably of benefit to the public but of special benefit to the owners of the charged properties.

special conditions A section of the conditions of the contract, other than general conditions and supplementary conditions, which may be prepared to describe conditions unique to a particular project.

special hazards insurance Insurance coverage for damage caused by additional perils or risks to be included in the property insurance (at the request of the contractor or at the option of the owner). Examples often included are sprinkler leakage, collapse, water damage, and coverage for materials in transit to the site or stored off the site. See **property insurance.**

special matrix terrazzo A finish flooring consisting of colored aggregate and organic matrix.

special-purpose industrial occupancy Industrial occupancy for particular operations characterized by a relatively low density of employee population with much of the area occupied by machinery or equipment. Hazardous usage is excluded.

special-quality brick A brick that is durable when used under conditions of exposure.

species A category of biological classification. A species is a class of individuals having common attributes and designated by a common name. "Species" is always properly used with the "s" when referring to trees or other biological classifications; "specie" refers only to money in coin.

specification A detailed and exact statement of particulars, especially a statement prescribing materials, dimensions, and workmanship for something to be built or installed.

specifications A part of the contract documents contained in the project manual consisting of written requirements for materials, equipment, construction systems, standards, and workmanship. Under the Uniform Construction Index, the specifications comprise sixteen divisions.

specific gravity (1) The ratio of the mass of a unit volume of a material at a stated temperature to the mass of the same volume of gas-free distilled water at a stated temperature. (2) The ratio of the density of one substance to another when used as the standard. Water is the standard for determining specific gravity of solids and liquids; hydrogen is used for gases.

apparent specific gravity The ratio of the mass in air of a unit volume of a material at a stated temperature to the mass in air of equal density of an equal volume of gas-free distilled water at a stated temperature. If the material is a solid, the volume is that of the impermeable portion.

bulk specific gravity The ratio of the mass in air of a unit volume of a permeable material, including both permeable and impermeable voids normal to the material, at a stated temperature to the mass in air of equal density of an equal volume of gas-free distilled water at a stated temperature.

bulk (saturated-surface-dry basis) specific gravity Same as bulk specific gravity, except that the mass includes the water in the permeable voids.

specific gravity factor The ratio of the weight of aggregates, including all moisture, as introduced into the mixer, to the effective volume displaced by the aggregates.

specific heat A quantity that describes the ability of a body to absorb heat and increase its

Spiral Reinforcement

Spiral Stair

temperature. The specific heat of a substance is the ratio of the amount of heat that must be added to a unit mass to raise its temperature through one degree to the amount of heat that is required to raise the temperature of an equal mass of water one degree.

specific retention The ratio of the volume of water retained by rock or soil to the gross volume of rock or soil, after the sample has been saturated and drip-drained.

specific surface The surface area of the contained particles in a unit weight of material.

specific yield The ratio of the volume of water obtained by draining a sample of saturated rock or soil to the gross volume of the sample.

spectral power distribution The distribution of radiant power, usually expressed in watts per nanometer, with respect to wavelength.

spectrophotometer An instrument for measuring the intensity of radiant energy of desired frequencies absorbed by atoms of molecules. Substances are analyzed by converting the absorbed energy to electrical signals, proportional to the intensity of radiation.

speculative builder One who develops and constructs building projects for subsequent sale or lease.

spigot (1) The end of a pipe that fits into the bell, or upset, end to form a joint after caulking. (2) A faucet.

spigot joint See **bell-and-spigot joint**.

spike A very heavy nail, usually square in cross section, and 3 to 12 inches (7.6 to 30.5 centimeters) long.

spike-and-ferrule installation A type of gutter installation in which the gutter is secured by long nails in metal sleeves.

spiked-and-linked chain An iron chain consisting of spikes linked with a length of chain and attached to posts to enclose a planted area.

spile (1) A peg or plug of wood. (2) A spout for directing the flow of sap from Sugar Maples. (3) A heavy wooden stake.

spillway A passage to convey overflow water from a dam or similar structure.

spindle (1) A small axle on which an object turns. (2) The bar in a lock that connects the knob(s) or handle(s) with the latching mechanism. (3) A short, turned-wood part, such as a baluster.

spiral A continuously wound length of reinforcing steel in the form of a cylindrical helix.

spirally reinforced column A column in which the vertical bars are enveloped by spiral reinforcement; i.e., closely spaced continuous hooping.

spiral ratchet screwdriver A screwdriver with a ratchet stem and a spring-loaded handle that permits a screw to be driven or withdrawn by pushing on the handle.

spiral reinforcement Continuously wound reinforcement in the form of a cylindrical helix.

spiral stair A flight of stairs whose treads wind around a central newel in a spiral or helix shape.

spire Any long, slender, pointed construction on top of a building. A spire is often a narrow, octagonal pyramid set on a short, square tower.

spirit level A device used to set an instrument to true horizontal or true vertical, consisting of a glass tube nearly filled with liquid so a

Splashboard

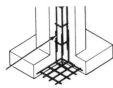

Splice

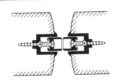

Split Astragal

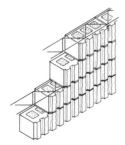

Split Face

traveling air bubble is formed. Cheap levels are bent concave; better levels are ground internally to an overall concave shape.

spirit stain An alcohol-soluble dye with good penetrating properties, used to stain wood.

spirit varnish A varnish with a highly volatile liquid as the solvent for the oil or resin.

spit The depth of one hand-shovel blade.

splashboard (1) A board placed on a wall at a sink to protect the wall. (2) A weather molding placed on the bottom of an outside door. (3) A board set against a wall at a scaffold to protect the wall.

splash brush A brush used to apply water to a finish coat of plaster while it is being troweled smooth.

splash lap That part of the overlap of a seam in sheet-metal roofing that extends onto the flat surface of the next sheet.

splay (1) To form with an oblique angle, or to bevel. (2) To spread out or extend.

splay brick, cant brick A brick with one side beveled at about 45 degrees.

splayed edge A bevel across the full thickness of a piece of wood.

splayed heading joint A joint between the ends of two adjacent floor boards. Their ends are cut at an angle of 45 degrees, rather than 90 degrees as in a butt joint.

splayed joint A joint between the ends of two adjacent members in which the ends of each are beveled to form an overlap.

splice Connection of two similar materials to another by lapping, welding, gluing, mechanical couplers, or other means, such as the connection of welded wire fabric by lapping and the connection of piles by mechanical couplers.

splice plate A plate laid over a joint and fastened to the pieces being joined to provide stiffness.

spline, false tongue, feather, slip feather, slip tongue A piece of metal or wood used to join two pieces of wood, such as decking, together.

spline joint A joint formed by inserting a spline into slots formed in the two pieces to be joined.

splinter (1) A small, thin, sharp piece of wood broken from a larger piece. (2) To split or break something into splinters.

split (1) A crack extending completely through a piece of wood or veneer. (2) A tear in a built-up membrane resulting from tensile stresses. (3) A masonry unit one-half the height of a standard unit.

split astragal Two pieces of molding attached to the meeting edges of two leaves of a pair of swing doors, allowing both leaves to be active.

split-batch charging A method of charging a mixer in which the solid ingredients do not all enter the mixer together. Cement, and sometimes different sizes of aggregate, may be added separately.

split block See **split-face block**.

split-conductor cable A cable in which each conductor consists of two or more insulated wires normally in parallel.

split course A course made of bricks cut so they are of less than normal thickness.

split face An exposed rough face of a masonry unit or stone created by splitting rather than forming or sawing the face.

split-face block Concrete masonry unit with one or more faces produced by purposeful fracturing of the unit, to provide architectural effects in masonry wall construction.

Split-level

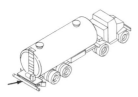

Spray Bar

split-face finish A rough-faced building stone cut from stratified rock so that the split face exposes the bedding. Usually the stone is cut so the bedding is set horizontally, but sometimes the cut is made to set the bedding vertically.

split fitting A section of electric conduit that is split longitudinally. After conductors have been placed in one half, the unit is assembled and secured with screws.

split frame, split jamb A doorframe with the jambs split in two or more pieces to allow the use of a pocket-type sliding door.

split-level A type of house in which the floor levels on one side of the house are one-half floor above or below those on the other side.

split pin A pin or spike that spreads during insertion or that can be split after insertion.

split-ring connector A timber connector consisting of a metal ring set in circular grooves in two pieces and the assembly held by bolts.

splitting tensile strength Tensile strength of concrete determined by a splitting tensile test.

splitting tensile test (diametral compression test) A test for tensile strength in which a cylindrical specimen is loaded to failure in diametral compression applied along the entire length.

spoil Dirt or rock excavated and removed from its original location.

spoil area A site where excavated material is to be dumped.

spokeshave A carpenter's tool for shaping curved edges, consisting of a blade set between two handles and used as a drawing knife or planing tool.

sponge rubber Expanded rubber having interconnected cells, used as a resilient padding and as

thermal insulation.

spoon (1) A small steel plasterer's tool used in finishing moldings. (2) A recovery tool used in soil sampling.

spoon bit See **dowel bit**.

spot board See **mortarboard**.

spot elevation A point on a map or plan whose existing or proposed elevation is noted.

spot ground A piece of wood attached to a plaster base and used as a gauge for thickness of applied plaster.

spotter (1) The person who directs a truck driver into a loading or unloading position. (2) The horizontal framework between the machinery deck of a pile driver and the leads.

spotting (1) Directing trucks for loading or unloading. (2) Spots of adhesive material used to fasten a veneer to its backing. (3) A defect in a painted surface consisting of spots of a different color, shade, or gloss than the rest.

spot-weld A small circular weld between two metal pieces made by applying heat and pressure.

spout A short tube used to direct water from gutters, balconies, etc., so the water will run down the building wall.

spray bar A pipe with ports used to apply binder to a road surface.

spray booth An enclosed or partly enclosed area used for the spray-painting of objects, usually equipped with a waterfall system to catch overspray and/or be furnished with a filtered air supply.

spray drying A method of evaporating the liquid from a solution by spraying it into a heated gas.

sprayed acoustical plaster An acoustical plaster applied with a special spray gun. The plaster usually has a rough surface and may be perforated with hand tools before hardening.

Sprayed Fireproofing

Spread Footing

sprayed asbestos A fire-resistive and sound-absorbing coating of asbestos fibers and an adhesive, applied to surfaces with a spray gun. Sprayed asbestos is no longer used in the U.S.A.

sprayed fireproofing An insulating material sprayed directly onto steel structural members with or without wire mesh reinforcing to provide a fire-endurance rating.

sprayed insulation See **spray-on insulation**.

sprayed mortar See **shotcrete**.

spray-on insulation Any of a number of lightweight concretes or mineral fibers and adhesives applied to surfaces and/or structural members for fire resistance, thermal insulation, or acoustic absorption.

spray painting Applying paint, lacquer, or a similar material with a special tool activated by air pressure.

spread The mobile power equipment, such as a paving spread or earth-moving spread, under the direction of a spread superintendent.

spreader (1) A piece of wood or metal used to hold the sides of a form apart until the concrete is placed. (2) A brace between two wales. (3) A device for spreading gravel or crushed stone for a pavement base course. (4) A stiffening member used to keep door or window frames in proper alignment during shipment and installation.

spreader bar A temporary member at the base of a preassembled door frame to maintain alignment during shipment and installation.

spread footing A generally rectangular prism of concrete larger in lateral dimensions than the column or wall it supports, to distribute the load of a column or wall to the subgrade.

spreading rate (1) The number of square yards of surface covered by a gallon of paint. (2) The rate at which bitumen or other material is applied to the surface of a roof.

spread-of-flame test A fire test for roof coverings in which a specified flame source is applied to a sample while a specified air stream is directed at the sample. The sample may be mounted horizontally or at an angle, depending on the intended end use of the product.

spring An elastic body or shape, such as a spirally wound metal coil, which stores energy by distorting and imparts that energy when it returns to its original shape. (2) The line or surface from which an arch rises.

spring-bolt, cabinet lock A spring loaded, beveled bolt that self-latches when a door or drawer is closed.

spring buffer An assembly containing a spring designed to absorb and dissipate kinetic energy, such as that from a descending elevator car or counter weight.

spring clamp Any clamp with pressure exerted by a spring. One variety is used to hold parts during gluing; another is used as a temporary electric connection.

springer, skewback, summer (1) The stone from which an arch springs. (2) The bottom stone of the coping of a gable. (3) The rib of a groined vault.

spring hinge A spring with one or more springs mounted in its barrel to return a door to the closed position. The hinge may be single-acting or double-acting for a swing door.

springing, spring (1) The support point from which an arch rises. (2) The initial angle of rise of an arch.

springing line The horizontal line connecting the points from which

Sprinklered

Spud Vibrator

an arch or arches rise.

sprinklered Said of an area of a building that is protected from fire by an automatic sprinkler system.

sprinkler head A distribution nozzle used in a fire-protection sprinkler system. The head may be closed by a plug held in place by a heat-sensitive device.

sprinkler system See **fire-protection sprinkler system**.

sprocked eaves Eaves having been raised by a sprocket.

sprocket, cocking piece, sprocket piece A wedge-shaped piece of wood attached to the upper side of a rafter at the eave to form a break in the roof line.

Spruce *Picea* species. Yellow to reddish-white wood, usually straight-grained, light, and soft. See **White Spruce, Red Spruce**, etc.

Spruce Fir See **Red Spruce**.

Spruce Pine *Pinus glabra*. A minor pine species found in the south bearing a close resemblance to spruce. The wood is considerably weaker than the major Southern Yellow Pine species. Lumber from this species must be stamped "Spruce Pine".

Spruce-Pine-Fir (SPF) Canadian woods of similar characteristics that have been grouped for production and marketing. The SPF species have moderate strength, are worked easily, take paint readily, and hold nails well. They are white to pale yellow. The largest volume comes from western Canada (British Columbia and Alberta), where the principal species in the group are White Spruce, Engelmann Spruce, Lodgepole Pine, and Alpine Fir. The principal species in the group originating in eastern Canada are Red Spruce, Black Spruce, Jack Pine, and Balsam Fir. Some lumber production in the New England states also is marketed as

Spruce-Pine-Fir and includes these species.

sprung molding A molding that has its interior corner beveled off to better fit a right angle joint.

spud (1) A hand tool used to strip bark from logs. (2) A sharp, narrow bar or spade used for removing gravel and roofing from a built-up roof.

spud vibrator A vibrator used for consolidating concrete, having a vibrating casing or head, that is inserted into freshly placed concrete.

spud wrench A tool, used by ironworkers to align holes and tighten bolts, that has a long tapering steel handle with an open-end wrench on one end.

spun concrete Concrete compacted by centrifugal action, such as in the manufacture of pipes.

spur (1) An appendage to a supporting member such as a buttress, shore or prop. (2) A decorative stone base that makes the transition from a round column to a square or polygonal plinth. (3) A carpenter's tool with a sharp point used for cutting veneer. (4) A rock ridge left projecting from a side wall after a blast. (5) A short length of railroad track usually parallel to a main track and used for loading, unloading or storage.

square (1) A quantity of shingles, shakes, or other roofing or siding materials sufficient to cover 100 square feet when applied in a standard manner; the basic sales units of shingles and shakes. (2) See **carpenter's square**.

square bolt A door bolt with a square cross section.

squared rubble Wall construction using various-sized square stones to make patterns.

square-framed Joinery framing with all angles of stiles, rails, and mountings cut square.

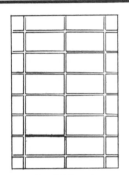

Stack Bond

square-headed Being cut off at right angles at the top, such as a doorway with a horizontal lintel as opposed to an arched opening.

square roof A roof with two sloping surfaces, each at an angle of 45 degrees from the horizontal.

square staff A narrow wood strip used as an angle bead for plastering.

square up To trim a timber or piece of wood, using a plane, so that its cross section is rectangular.

squaring Aligning or constructing an item or assembly so that all angles are 90 degrees.

squatter's right The right to acquire the ownership of land through long-continued occupancy.

squeezed joint A joint formed by coating surfaces of two pieces with cement and squeezing them together.

squinch A small arch built across an interior corner of a room to support a superimposed load, such as a dome or the spire of a tower.

squint A small oblique opening in an interior wall of a medieval church to allow a view of the high altar from the aisles.

stability A measure of the ability of a structure to withstand overturning, sliding, buckling, or collapsing.

stabilizer A substance that makes a solution or suspension more stable, usually by keeping particles from precipitating.

stack (1) A vertical structure containing one or more flues for the discharge of hot gas. (2) A chimney. (3) A vertical supply duct in a warm air heating system. (4) Any vertical plumbing pipe, such as soil pipe, waste pipe, vent, or leader pipe. (5) A collection of vertical plumbing pipes. (6) A tier of shelves for books.

stack bond, stacked bond A masonry pattern bond in which all vertical and horizontal joints are continuous and aligned.

stack cap See **chimney cap**

stack effect See **chimney effect**.

stack vent (1) The extension of a soil or waste stack above the highest horizontal drain or fixture connected to the stack. (2) A device installed through a built-up roof covering to allow entrapped water vapor to escape from the insulation.

stack venting A method of venting a single fixture connected directly into a soil or waste stack. The term is applicable only to the top fixture or fixtures of a stack.

stadia rod, stadia A graduated surveyor's rod used with a transit or similar instrument to determine distances. An observed intercept on the rod, as defined by two lines in the reticle of a telescope, is converted into the distance between the instrument and the rod by use of similar triangles.

staff man A person who sets the leveling rod for a surveyor in leveling or stadia work. Also called a "rodman".

stage grouting Sequential grouting of a hole in separate steps or stages, in lieu of grouting the entire length at once.

staggered Descriptive of fasteners, joints, or members arranged in two or more rows so that the beginning of each row is offset from the adjacent one.

staggered course A course, such as of shingles or tiles, where the butts do not form a continuous horizontal line.

staggered riveting Rows of rivets installed in a staggered pattern.

staggered-stud partition A partition made of two rows of studs with alternating studs supporting opposite faces of the partition and each stud making contact with only one wall.

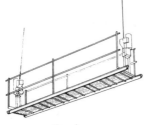

Staging

Stanchion

Staggered-stud partitions often have an interwoven fiberglass blanket to improve the sound-insulation value of a partition.

staging (1) A temporary working platform against or within a building for construction, repairs, or demolition. Also called "scaffolding". (2) A temporary working platform supported by the temporary timbers in a trench.

stain (1) Color in a dissolving vehicle. When spread on wood or similar material, the stain penetrates and gives color to the material. (2) A discoloration in the surface of a material, such as wood or plastic.

stained glass Glass given a desired color by use of an additive to the molten material, or by firing a stain into the surface of the formed glass. Stained glass is used for decorative windows or mosaics.

stainless steel Any of a number of steels alloyed with chromium and nickel. Depending on the alloy, the metal may possess good corrosion resistance, high heat tolerance, or high strength.

stair (1) A single step. (2) A series of steps or flights of steps connected by landings, used for passage from one level to another.

stair bolt See **handrail bolt**.

stair bracket A stiffener, often decorative, fixed under the nosing of a tread on an open string stair.

stairbuilder's truss Crossed beams used to stiffen a landing of a stair.

staircase (1) A single flight or multiple flights of stairs including supports, frame works, and handrails. (2) The structure containing one or more flights of stairs.

stairhead The first stair at the top of a flight of stairs.

stair headroom The least clear vertical distance measured from a nosing of a tread to an overhead obstruction.

stair rod A metal rod used under a stair nosing to hold a carpet in place.

stair turret (1) A structure containing only a winding stair. (2) A housing projecting above a roof and containing a stair.

stairwell A vertical shaft enclosing a stair.

stake (1) A short, pointed piece of wood or metal driven into the ground as a marker or an anchor. (2) The process of marking a work or survey site with stakes. (3) A small anvil, used in working sheet metal, supported by a sharp vertical leg inserted into a hole in a work surface.

staking out The process of driving stakes for batter boards to locate the limits of an excavation.

stalactite A downward-pointing deposit formed as an accretion of mineral matter produced by evaporation of dripping water from the surface of rock or of concrete, commonly shaped like an icicle. See also **stalagmite**.

stalagmite An upward pointing deposit formed as an accretion of mineral matter produced by evaporation of dripping water, projecting from the surface of rock or concrete, commonly conical in shape. See also **stalactite**.

stanchion (1) A vertical post or prop supporting a roof, window, etc. (2) An upright bar or post, as in a window, screen, or railing.

standard (1) A grade of lumber suitable for general construction and characterized by generally good strength and serviceability. In light framing rules, the standard grade applies to lumber that is two to four inches thick and two to four inches wide. It falls between the construction and utility grades. (2) A grade of Idaho White Pine boards equivalent to #3 common in other species. (3) In the British timber trade, a quantity of lumber that equals 1,980 board feet.

standard absorption trench An absorption trench 12 to 36 inches (30 to 90 centimeters) wide, containing 12 inches (30 centimeters) of clean coarse aggregate and a distribution pipe covered with a minimum of 12 inches (30 centimeters) of earth.

standard air Air with a density of 0.075 lb. per cu. ft. (0.0012 gm per cc) which is close to air at 68 degrees F. (20 degrees C.) dry bulb and 50% relative humidity at a barometric pressure of 29.9 in. (76 cm) of mercury.

standard and better A mix of lumber grades suitable for general construction. The "and better" signifies that a portion of the lumber is actually of a higher grade than standard (but not necessarily of the highest grade). The proportion of higher grades included is a factor in determining market value.

standard atmosphere A pressure equal to 14.7 lb. per sq. in. (1.01 x 10 dynes per sq cm).

standard curing Exposure of test specimens of concrete to specified conditions of moisture, humidity, and temperature. See also **fog curing**.

standard deviation A statistic, used as a measure of dispersion in a distribution, that is equal to the square root of the arithmetic average of the squares of the deviations from the mean.

standard hook A hook at the end of a reinforcing bar made in accordance with a standard.

standard knot Any knot in wood 1-1/2 inches (3.8 centimeters) or less in diameter.

standard matched Tongue-and-groove lumber with the tongue and groove offset, rather than centered as in center-matched lumber. See also **center matched**.

standard net assignable area That portion of the area of a project that is available for assignment or rental to an occupant, computed in accordance with AIA Document D101.

standard penetration resistance, Proctor penetration resistance The load required to produce a standard penetration of a standard needle into a soil sample at a standard rate.

standard penetration test A test to estimate the degree of compactness of soil in place by counting the number of blows required to drive a standard sampling spoon 1 foot (0.3 meter) using a 140 pound (64 kilogram) weight falling 30 inches (0.8 meter).

standard pile See **guide pile**.

standard pressure See **standard atmosphere**.

standard sand Ottawa sand accurately graded to pass a U.S. Standard No. 20 (850-micrometer) sieve and be retained on a U.S. Standard No. 30 (600-micrometer) sieve, for use in the testing of cements. See also **Ottawa sand** and **graded standard sand**.

standard source A, light source A A tungsten filament lamp operating at a color temperature of 2856 degrees K. (2583 degrees C.).

standard source B, light source B A light source having a correlated color temperature of 4874 degrees K. (4601 degrees C.), approximating sunlight at noontime.

standard source C, light source C A light source having a correlated color temperature of 6774 degrees K. (6501 degrees C.), approximating a combination of direct sunlight and a clear sky.

standard tolerance A generally accepted tolerance for a specific product.

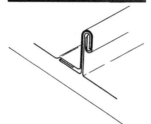

Standing Seam

Starter

standard wire gauge The legal standard wire gauge in Great Britain and Canada.

standby lighting Lighting provided to supply illumination in event of failure of the usual lighting system so that normal activities can continue.

standing bevel A bevel that forms an obtuse angle.

standing finish Those items of interior finish that are permanent and fixed, as distinguished from such items as doors and moveable windows.

standing leaf An inactive leaf of a double door bolted in the closed position.

standing panel A panel with the longer dimension vertical.

standing seam A seam, in sheet metal and roofing, made by turning up two adjacent edges and folding the upstanding parts over on themselves.

standing timber Trees that have not been cut but are of merchantable size.

standing waste A vertical overflow pipe connected to piping at the bottom of a water tank to control the height of storage.

standpipe (1) A pipe or tank connected to a water system and used to absorb surges that can occur. (2) A pipe or tank used to store water for emergency use, such as fire fighting.

standpipe system A system of tanks, pumps, fire department connections, piping, hose connections, connections to an automatic sprinkler system, and an adequate supply of water used in fire protection.

staple A double-pointed, U-shaped piece of metal used to attach wire mesh, insulation batts, building paper, etc., usually driven with a stapling gun.

staple gun A spring-driven gun used to drive staples used for fastening materials such as building paper and batt insulation.

staple hammer A hand tool that holds a magazine of staples and drives a staple when the face strikes a surface.

stapler See **staple gun**.

star drill See **plugging chisel**.

star expansion bolt A fastener used in concrete, consisting of two semicircular shields that are forced apart, to bear on the walls of a hole, as a bolt is driven into the shields.

starter (1) A device used with a ballast to start an electric-discharge lamp. (2) An electric controller used to accelerate a motor from rest to running speed and to stop the motor.

starter board A 6- or 8-inch board used at the eave of a roof to provide a solid nailing surface for the first courses. A starter board is also used in reroofing to replace the old shingles at the eaves.

starter frame Shallow form work projecting above floor level, used to locate a column or wall.

starter strip, starting strip A strip of composition roofing material applied along the eaves before the first row of shingles is laid.

starting board The first board nailed in place at the base of a form for concrete.

starting course The first course of shingles applied along the eaves.

starting strip The first row of composition roofing laid along the eaves.

starved joint A poorly bonded glue joint resulting from the use of too little glue.

statement of probable construction cost Cost forecasts prepared by the architect during the schematic design, design development, and construction document phases of basic services for the guidance of the owner.

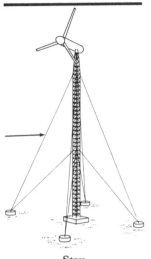

Stay

statements of ethical principles Statements of ethical principles promulgated by professional societies to guide their members in the conduct of professional practice.

static head The static pressure of a fluid expressed as the height of a column of the fluid that the pressure could support.

static load The weight of a single stationary body or the combined weights of all stationary bodies in a structure, such as the load of a stationary vehicle on a roadway. During construction, the static load is the combined weight of forms, stringers, joists, reinforcing bars, and the actual concrete to be placed. See also **dead load**.

static modulus of elasticity The value of Young's modulus of elasticity obtained by arbitrary criteria from measured stress-strain relationships derived from other than dynamic loading. See also **modulus of elasticity**.

static pressure (1) The pressure exerted by a fluid on a surface at rest with respect to the fluid. (2) The pressure a fan must supply to overcome resistance to air flow in an air distribution system.

statics That branch of mechanics dealing with forces acting on bodies at rest. Statics is the basis of structural engineering.

static test (1) A test that subjects a curtain wall to a pressure differential equal to a specified wind pressure. (2) A test that simulates the flow of water over a curtain wall during a hurricane. (3) A test to determine the pressure exerted by a fluid while the system is at rest.

station (1) A point on a survey traverse, particularly a roadway. (2) An identified point on the earth's surface, the location of which has been determined by surveying.

stationary hopper A container used to receive and temporarily store freshly mixed concrete.

station roof (1) A roof shaped like an umbrella and supported by a single column. (2) A long roof supported on a row of columns and cantilevering off one or both sides of the column line.

station yards of haul The product of the number of cubic yards in a haul and the number of 100-foot (30.48-meter) stations through which it is hauled.

statute of frauds A statute specifying that certain kinds of contracts, such as for the sale or lease of real property, are unenforceable unless signed and in writing or unless there is a written memorandum of terms signed by the party to be charged. The statute varies by state.

statute of limitations A statute specifying the period of time within which legal action must be brought for alleged damage or injury, or other legal relief. The lengths of the periods vary from state to state and depend upon the type of legal action. Ordinarily, the period commences with the occurrence of the damage or injury, or discovery of the act resulting in the alleged damage or injury. In construction industry cases, many jurisdictions define the period as commencing with completion of the work or services performed in connection therewith.

statutory bond A bond, the form or content of which is prescribed by statute.

stave A narrow strip of wood used in the construction of barrels or buckets.

stave bolt The bolt from which wooden staves are cut.

stay Anything that stiffens a frame or stabilizes some structural component.

stay bolt A long metal rod with a threaded end and used as a stay.

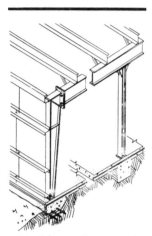

Steel-frame Construction

Steel Sheet Piling

stay rod Any rod-shaped tie used temporarily or permanently to prevent spreading of the parts it connects.

steam box An enclosure for steam curing concrete products. See also **steam curing room**.

steam chest A container in which wood is steamed for bending or forming.

steam cleaner A machine that provides pressurized steam to a nozzle for the purpose of cleaning grease or dirt from a surface. Detergents or chemicals are sometimes added.

steam curing Curing of concrete or mortar in water vapor at atmospheric or higher pressures and at temperatures between about 100 and 420 F (40 and 215 C).

steam-curing cycle (1) The time interval between the start of the temperature-rise period and the end of the soaking period or the cooling-off period. (2) A schedule of the time and temperature of periods that make up the cycle.

steam-curing room A chamber for steam curing of concrete products at atmospheric pressure.

steamed wood Wood that has been softened in preparation for bending or for producing veneer.

steam hammer A pile hammer activated by steam pressure.

steam heating system A heating system in which heat is transferred from a boiler or other source, through pipes, to a heat exchanger. The steam can be below, at, or above atmospheric pressure.

steaming A process in which logs are heated with steam or hot water in special vats prior to peeling them into veneer. Steaming results in smoother veneer and improved recovery from the log. A similar

process is used to prepare wood for bending or shaping.

steam kiln See **steam-curing room**.

steam pipe Any pipe for the conveyance of steam.

steam shovel A power shovel operated by steam from an integral boiler. Steam shovels have largely been replaced by diesel-powered shovels.

steam table A table or section of counter equipped with removable food containers, the contents of which are kept warm by steam, hot water, or hot air circulating beneath the containers.

steam vat A container used in steaming to soften logs or flitches before peeling or slicing veneers. Sometimes referred to as a "steam chest".

steel Any of a number of alloys of iron and carbon, with small amounts of other metals added to achieve special properties. The alloys are generally hard, strong, durable, and malleable.

steel casement A casement generally made from hot-rolled steel sections and classified as a residence, intermediate, or heavy-intermediate steel casement.

steel erector A contractor who undertakes to place, plumb, and secure a steel structure.

steel-frame construction Construction in which steel columns, girders, and beams comprise the structural supporting elements.

steel sheet Cold-formed sheet or strip steel shaped as a structural member for the purpose of carrying the live and dead loads in lightweight concrete roof construction.

steel sheet piling Interlocking rolled-steel sections driven vertically into the ground to serve

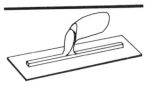

Steel Trowel

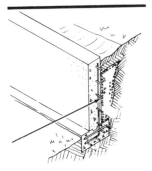

Stem Bars

as sheeting in an excavation or to cut off the flow of ground water.

steel square A steel carpenter's square.

steel stud anchor A steel clip fastened to a doorframe and used to secure the frame to a steel stud.

steel trowel See **trowel**.

steel troweling A steel hand tool or machine used to create a dense, smooth finish on a concrete surface.

steel wool A matted mass of fine steel fibers, used principally for cleaning and polishing wood or metal surfaces.

steelworker A craftsman skilled in assembling structural steel or concrete reinforcing steel. Also called an "ironworker".

steep asphalt Roofing asphalt having a high softening point, used on surfaces with a steep slope.

steeplejack A craftsman who builds and/or repairs steeples, chimneys, and other tall masonry structures.

steering brake A brake to slow or stop one side of a tractor.

steering clutch A clutch that can disconnect power from one side of a tractor.

stem bars Bars used in the wall section of a cantilevered retaining wall or in the webs of a box. When a cantilevered retaining wall and its footing are considered as an integral unit, the wall is often referred to as the stem of the unit.

stemming (OSHA) A suitable inert, incombustible material or device used to confine or separate explosives in a drill hole, or to cover explosives in mudcapping.

step brazing A method of brazing in which filler metals of successively lower brazing temperatures are used to prevent softening of previously brazed joints.

step down transformer An electric transformer with a lower voltage at the secondary winding terminals than at the primary winding terminals.

step joint (1) A notched joint used to connect two wood members meeting at an angle, such as a tie beam and a rafter. (2) A joint between two rails with different heights and/or cross sections.

stepladder A ladder with an integral supporting frame and narrow treads instead of rungs, usually built in the form of an inverted "V".

stepped flashing Flashing at the intersection of a sloped roof and a wall or chimney. The upper edge of each sheet is stepped with relation to adjacent sheets to maintain a safe distance from the sloped roof surface.

stepped floor A floor on a stage or platform that rises in steps, as opposed to a ramped floor.

stepped footing (1) A wall footing with horizontal steps to accommodate a sloping grade or bearing stratum. (2) A column or wall footing composed of two or more steps on top of one another to distribute the load.

stepped foundation A foundation constructed in a series of steps that approximate the slope of the bearing stratum. The purpose is to avoid horizontal force vectors that might cause sliding.

stepping Lumber designed to be used for stair treads. Stepping is vertical-grained and is customarily shipped kiln-dried, surfaced three sides, and bull-nosed on one edge. Besides the C & Better and D grades of solid wood, stepping is also made from particleboard for use where it will be covered.

stepping off Laying off the required length of a rafter by use of a carpenter's square.

step-plank Any hardwood lumber used to make steps, usually 1-1/4 to 2 in. (3.2 to 5.1 cm) thick.

Stick Built

Stile

step soldering A method of soldering in which solders with successively lower application temperatures are used to prevent the disturbing of previously soldered joints.

stereobate The foundation, substructure, or platform upon which a building is constructed.

sterling A grade of Idaho White Pine boards, equivalent to #2 common in other species.

stick (1) Any long, slender piece of wood. (2) A waxed-paper cartridge containing an explosive, usually 1-1/8 X 8 in. (3 X 20 cm). (3) A rigid bar fastened to the bucket and hinged at the boom of a power shovel.

stick built A term describing frame houses assembled piece by piece from lumber delivered to the site with little or no previous assembly into components. The more typical type of residential construction is stick built.

sticky cement Finished cement that develops low or zero flowability during or after storage in silos, or after transportation in bulk containers, hopper-bottom cars, etc. Sticky cement may be caused by: (a) interlocking of particles; (b) mechanical compaction; (c) electrostatic attraction between particles. See also **warehouse set**.

stiffback See **strongback**.

stiffened compression element A compression element that has been stiffened on its weak axis in order to resist buckling on that axis.

stiffener (1) A bar, angle, channel, or other shape attached to a metal plate or sheet to increase its resistance to buckling. (2) Internal reinforcement, usually light-gauge channels, for a hollow-metal door.

stiff frame See **rigid frame**.

stiff-mud brick Brick formed by extruding a stiff, but plastic, clay (12% to 15% moisture) and

cutting the extruded material to length with wires.

stiffness Resistance to deformation.

stiffness factor A measure of the stiffness of a structural member. For a prismatic member, the stiffness factor is equal to the ratio of the product of the moment of inertia of the cross section and the modulus of elasticity for the material to the length of the member.

S-tile A roofing tile with an S-shaped cross section.

stile The vertical members forming the outside framework of a door or window.

stilt (1) A post used to raise a structure above ground or water level. (2) A compression member placed above or below a similar member to gain additional height.

stipple To make dots or short dashes on a surface as a decorative effect.

stippled finish A dotted or pebbly-textured finish on the surface coat of paint, plaster, or porcelain enamel, induced by punching the unset surface with a stiff brush.

stipulated sum agreement A contract in which a specific amount is set forth as the total payment for performance of the contract.

stirrup (1) A reinforcement used to resist shear and diagonal tension stresses in a structural member. (2) A steel bar bent into a "U" or box shape and installed perpendicular to or at an angle to the longitudinal reinforcement, and properly anchored. (3) Lateral reinforcement formed of individual units, open or closed, or of continuously wound reinforcement. The term "stirrups" is usually applied to lateral reinforcement in flexural members and the term "ties" to lateral reinforcement in vertical

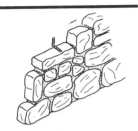

Stone

Stonework

Stool

compression members. See also **tie.**

stitched veneer Veneer sheets composed of random width pieces of veneer sewn together with heavy thread. Several stitch lines run across the width of each piece of stitched veneer. In veneer production, the random width pieces are run through a large sewing machine and reclipped in desired (usually standard 48-54-inch) sizes. Stitched veneer is usually used in the core plies of plywood.

stitch rivet One of a number of rivets, installed at regular intervals and in a straight line to connect two parts and provide lateral stiffness.

stitch welding The joining of two or more parts with a line of short welds equally-spaced.

St. John's Spruce Canadian Spruce.

stock (1) Material or devices readily available from suppliers. (2) The body or handle of a tool. (3) A frame to hold a die when cutting external threads on a pipe.

stockhouse set See **sticky cement** and **warehouse set.**

stock lumber Lumber cut to standard sizes and readily available from suppliers.

stock millwork Millwork manufactured in standard shapes and sizes and readily available from suppliers.

stockpile Material excavated and stored for later use.

stock size A standard size of an item or piece that is readily available from suppliers.

stone (1) Individual blocks of rock processed by shaping, cutting, or sizing, for use in masonry work. (2) Fragments of rock excavated, usually by blasting, from natural deposits and further processed by crushing and sizing, for use as aggregate. (3) A carborundum or other natural or artificial hone

used to sharpen cutting edges of tools.

stone bolt A bolt cemented in a hole in a unit of masonry for connection of another object.

stone chip A small, angular chip of stone that has been washed of dust.

stone dust Pulverized stone used in the construction of walkways or other stable surfaces. The dust is mixed with soil and compacted or used with gravel to fill spaces between irregular stones.

stone mason A craftsman skilled in constructing stone masonry, including any preparation at the site.

stone sand Fine aggregate resulting from the mechanical crushing and processing of rock. See also **sand** and **fine aggregate.**

stone seal An asphaltic wearing surface made by rolling aggregate into a heavy application of bituminous material. Application is usually repeated several times.

stone-setter's adjustable multiple-point suspension scaffold A swinging-type scaffold having a platform suspended at the four corners by hoisting devices to permit lowering or raising the platform to any desired working level.

stone slate Slate-like slabbing or flagging obtained from limestone or sandstone that has been separated along its bedding, used as rough shingling on a roof.

stonework (1) Masonry construction using stone. (2) The preparation or setting of stone for building or paving.

stool (1) A narrow shelf, across the lower part of a window opening, that butts against the sill. (2) A framed support.

stoop A small platform at the entrance to a house, often consisting of several wide steps.

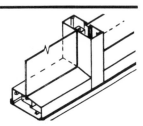

Storefront Sash

Storm Door, Weather Door

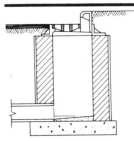

Storm Drain

Storm Window

stop A type of molding nailed to the face of a door frame to prevent the door from swinging through. A stop is also used to hold the bottom sash of a double-hung window in place.

stop-and-check valve See **nonreturn valve**.

stop-and-waste cock A stopcock with a drain plug in the valve. When the valve is closed, water in the downstream piping can be drained through the plug. A stop-and-waste cock is often used at connections to outdoor faucets so the exterior piping can be drained for the winter.

stopcock A valve to shut off the flow of fluid in a branch of a distribution system in a building.

stop glazing Either the lip at the back of a rabbet or the molding applied at the front. These serve to hold a light in a sash or frame with the help of spacers.

stop stone A stone in the ground against which the meeting stiles of a pair of gates close.

stop valve Any valve in a piping distribution system that is used to stop flow.

stopwork A mechanism on a lock to hold the bolt in the latched position so the lock can not be opened from the outside. A stopwork can usually be set by a sliding or push button.

stop work order An order issued by the owner's representative to stop work on a project. Reasons for the order include failure to conform to specifications, unsatisfied liens, labor disputes, and inclement weather.

storage hopper See **stationary hopper**.

storage life, shelf life The length of time that a product, such as a package adhesive or sealant, can be stored at a specified temperature range and remain usable.

storage tank Any tank that receives a fluid and holds it for later distribution.

storefront sash An assembly of light metal members that form a frame for a fixed-glass storefront.

storm clip A clip on the exterior of a glazing bar that holds the pane in place.

storm door, weather door A door, usually fully glazed, fitted to the outside of the frame of an exterior door to provide additional protection against cold and wind.

storm drain A drain used to convey rain water, subsurface water, condensate, or similar discharge, but not sewage or industrial waste.

storm porch A porch, or portion thereof, intended to protect an entrance to a house from the weather.

storm sash See **storm window**.

storm sewage That material flowing in combined sewers or storm sewers as a result of rainfall.

storm sewer A sewer used for conveying rainwater and/or similar discharges, but not sewage or industrial waste, to a point of disposal.

storm-sewer system A sewer system consisting of only storm sewers.

storm sheet A sheet of roofing curved down over an eave to protect the eave from rain.

storm water That portion of rainfall or other precipitation that runs over the ground surface for the short period after a storm as the flow exceeds the normal runoff.

storm window, storm sash An exterior window placed over an existing window to provide additional protection against the weather.

story (1) That part of a building between the upper surface of a

Story-and-a-half

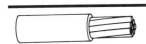

Stranded Wire

floor and the upper surface of the floor above. Building codes differ in designations applied when a part of a "story" is below grade. (2) A major division in the height of a building even when created by architectural features, such as windows, rather than horizontal divisions. For example an area may be "two stories high".

story-and-a-half The designation of a building in which the second story rooms have low headroom at the eaves.

story height See **floor height**.

story post An upright post that helps support a beam on which a floor rests.

story rod A wooden rod of a length equal to one story height, sometimes divided in parts each equal to one riser in a stair, for use in stair construction.

stove bolt A common bolt with a round or flat head and a slot for a screwdriver.

straddle pole One of two poles laid along a roof line from an upright at the eaves and connected to the other pole at the ridge, used in a saddle scaffold.

straddle scaffold See **saddle scaffold**.

straightedge, rod A rigid, straight piece of wood or metal used to strikeoff or screed a concrete surface to proper grade, or to check the planeness of a finished grade. See also **rod, screed,** and **strikeoff**.

straight grain A piece of wood in which the principal cells run parallel to its length.

straight jacket A stiff timber attached to a wall to reinforce and increase the wall's rigidity.

straight joint (1) A continuous joint in a wood floor formed by the butt ends of parallel boards. (2) A butt joint between two parallel timbers. (3) Vertical joints in masonry that form a continuous straight line.

straight-joint tile A term describing single-lap tiles designed to be laid so that the edges of each course run in a straight line from eave to ridge.

straight-line edger A mechanically fed saw used to dress and straighten edges of lumber or veneer. Also called a "straight-line ripsaw".

straight-line theory An assumption in the reinforced-concrete analysis according to which the strains and stresses in a member under flexure vary in proportion to the distance from the neutral axis.

straight lock A lock mounted on the inside face of a door. The only preparation is cutting a key hole.

strain Deformation of a material resulting from external loading. The measurement for strain is the change in length per unit of length.

strain gauge A sensitive electrical or mechanical instrument used to measure strain in loaded members or objects, usually for research or development.

straining beam, straining piece, strutting piece (1) A horizontal strut in a truss, placed above the tie beam or the bottom of the rafters, usually midheight. (2) The chord between the upper ends of the posts in a queen post truss.

strake (1) A run of clapboard on the side of a house. (2) A row of steel plates in a steel chimney.

strand A prestressing tendon composed of a number of wires twisted about a center wire or core.

stranded wire A group of fine wires used as a single electric conductor.

strand grip A device used to anchor a prestressing tendon.

strap (1) A metal plate fastened across the intersection of two or more timbers. (2) See **pipe strap**.

strap bolt (1) A bolt in which the middle portion of its shank is flattened, so the unit can be bent in a U-shape. (2) See **lug bolt**.

straphanger A hanger made from a thin, narrow strip of material.

strap hinge A surface-mounted hinge with one leaf or both leaves elongated to accommodate additional fasteners.

strap joint A butt joint between two pieces secured by a strap, or straps, and fasteners.

strapping (1) Flexible metal bands used to bind units for ease of handling and storage. (2) Another name for "furring".

stratification (1) The separation of overwet or overvibrated concrete into horizontal layers with increasingly lighter material toward the top. Water, laitance, mortar, and coarse aggregate will tend to occupy successively lower positions in that order. (2) A layered structure in concrete resulting from placing of successive batches that differ in appearance. (3) The occurrence in aggregate stockpiles of layers of differing grading or composition. (4) A layered structure in a rock formation.

stratified rock Rock that has been formed by compaction, cementation, or crystallization of successive beds of deposited material.

Stratling's compound Dicalcium aluminate monosilicate-8-hydrate, a compound that has been found in reacted lime-pozzolan and cement-pozzolan mixtures.

stratum A bed or layer of rock or soil.

streamline flow Fluid flow in which the velocity at every point is equal in magnitude and direction.

stream shingle Flat pieces of thin rock having a sloped, overlapping pattern resembling shingling, usually found in small, fast streams.

street elbow A pipe elbow with male threads on one end and female threads on the other. Also called a "service ell".

street ell See **street elbow**.

street floor That floor of a building nearest to street level. According to some building codes, the street floor is a floor level not more than 21 in. above or 12 in. below grade level at the main entrance.

street lighting luminaire A complete lighting unit intended to be set on a pole or post with a bracket.

street lighting unit An assembly consisting of a pole or post, bracket, and luminaire.

street line (1) A line dividing a lot or other area from a street. (2) A side boundary of a street, as legally defined.

strength See **compressive strength, fatigue strength, flexural strength, shear strength, splitting tensile strength, tensile strength, ultimate strength,** and **yield strength**.

strength design method, ultimate strength method A design method in which service loads are increased sufficiently by factors, often referred to as "load factors", to obtain the ultimate design load. The structure or structural element is then proportioned to provide the desired ultimate strength.

stress Intensity of internal force (i.e., force per unit area) exerted by either of two adjacent parts of a body on the other across an imagined plane of separation. When the forces are parallel to the plane, the stress is called "shear stress"; when the forces are normal to the plane the stress is called "normal stress"; when the normal stress is directed toward the part on which it acts it is

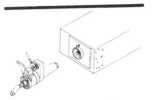

Stressing End

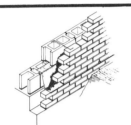

Stretcher Bond

Striated

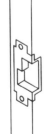

Strike Plate, Strike

called "compressive stress"; when it is directed away from the part on which it acts it is called "tensile stress".

stress corrosion Corrosion of a metal accelerated by stress.

stress corrosion cracking Cracking of metal that occurs under the combined influence of certain corrosive environments and applied or residual stresses.

stress crack An external or internal crack in a plastic as a result of applied tensile stresses.

stress diagram See **stress-strain diagram**.

stressed-skin construction Construction in which a thin material on the surface of a building is used to carry loads.

stressed-skin panel A panel assembled by fastening plywood or similar sheets over a frame or core, on both sides. The result is a composite structural member.

stress-graded lumber Lumber graded for strength according to the rate of growth, slope of grain, and the number of knots, shakes, and other defects.

stressing end In prestressed concrete, the end of the tendon from which the load is applied when tendons are stressed from one end only.

stress relaxation Stress loss developed from strain when a constant length is maintained under stress.

stress-strain diagram A diagram for a particular material with values of stress plotted against corresponding values of strain, usually with stress plotted as the ordinate.

stretcher A masonry unit laid with its length horizontal and parallel with the face of the wall. See also **header**.

stretcher block A concrete masonry unit used as a stretcher.

stretcher bond, running bond, stretching bond A masonry bond with all courses laid as stretchers and with the vertical joint of one course falling midway between the joints of the courses above and below.

striated Having parallel grooves in the face, as in a fluted column.

striated face The face of a plywood panel that has been given closely spaced, shallow grooves to provide a vertical pattern.

strike (1) In masonry, to cut off the excess mortar at the face of a joint with a trowel stroke. (2) To remove formwork. (3) A work stoppage by a body of workers.

strike edge See **leading edge.**

strike jamb The jamb on which the strike plate is mounted.

strike off, strikeoff (1) To remove concrete in excess of that which is required to fill the form evenly or bring the surface to grade, performed with a straightedged piece of wood or metal by means of a forward sawing movement or by a power-operated tool appropriate for this purpose. (2) The name applied to the tool. See also **screed** and **screeding.**

strike plate, strike, striking plate A plate or box, mounted in a jamb, with a hole or recess shaped to receive and hold a bolt or latch from a lock on the door.

striker A slightly beveled metal plate attached to a strike plate to guide a door latch to its socket.

strike reinforcement A metal piece welded inside a hollow metal frame to which the strike plate is attached and which also serves to strengthen the frame.

strike stile See **strike plate.**

striking The removal of temporary supports from a structure.

striking plate See **strike plate.**

striking stile See **lock stile.**

stringcourse, belt course A horizontal band of masonry,

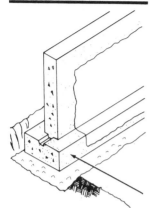

Strip Footing

Strongback

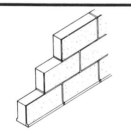

Structural Clay Tile

usually narrower than the other courses, which may be flush or projecting, and plain or ornamented.

stringer (1) A secondary flexural member parallel to the longitudinal axis of a bridge or other structure. See also **beam**. (2) A horizontal timber used to support joists or other cross members.

stringer bead A continuous bead of weld metal, made by moving the electrode in a direction parallel to the bead without much transverse oscillation.

stringing mortar Spreading a long enough mortar bed to lay several masonry units at one time.

strip (1) Board lumber one inch in nominal thickness and less than four inches in width, frequently the product of ripping a wider piece of lumber. The most common sizes are 1x2 and 1x3. See **furring**. (2) To remove formwork or a mold. (3) To remove an old finish with paint removers. (4) To damage the threads on a nut or bolt.

strip core, blockboard A composite board whose core is made of strips of wood, laid loose or glued together. Veneer is glued to both sides of the core with its grain at right angles to the grain of the core pieces.

strip flooring Hardwood finish flooring in the form of narrow tongue-and-groove strips, commonly made of mahogany, maple, or oak.

strip footing See **continuous footing**.

strip foundation A continuous foundation of which the length considerably exceeds the breadth.

striplight A lighting assembly used to flood all or part of a stage, consisting of a row of lamps mounted in a trough with a reflecting hood and color frames.

strip mopping A method of applying hot bitumen to a roof deck in parallel strips.

stripper A liquid compound formulated to remove coatings by chemical and/or solvent action.

stripping See **strip**.

stripping agent See **release agent**.

stripping felt A narrow strip of roofing felt used to cover a flange of metal flashing.

stripping piece A splayed narrow member used to facilitate removal of formwork in a confined space.

strip taping See **taping strip**.

strongback A frame attached to the back of a form to stiffen or reinforce it during concrete placing or handling operations.

struck joint A masonry joint in which excess mortar is removed by a stroke of a trowel.

structural A term applied to those members in a structure that carry an imposed load in addition to their own weight.

structural analysis The determination of stresses in members in a structure due to imposed loads from gravity, wind, earthquake, thermal effects, etc.

structural clay facing tile A structural clay tile with a finished ceramic face.

structural clay tile A hollow masonry unit molded from clay, shale, fireclay, or a mixture of such materials.

structural engineering That branch of engineering concerned with the design of the load-supporting members of a proposed structure, and also with the investigation of existing structures which are suspect.

structural failure (1) The inability of a structure or structural member to perform its intended function, perhaps caused by collapse or excessive deformation.

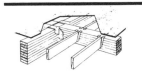

Structural Glued-laminated Timber

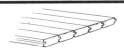

Structural Planks

Structural Timbers

(2) A marked increase in strain without an increase in load.

structural frame All the members of a building or other structure used to transfer imposed loads to the ground.

structural glass Rectangular tiles or panels of glass used as finish for walls.

structural glued-laminated timber A wooden structural member made from selected boards strongly glued together.

structural joists and planks Lumber two, three, or four inches thick and six inches or wider, graded for its strengh properties. Such planks are used primarily for joists in residential construction and graded, in descending order, select structural, #1, #2, and #3. The #1 and #2 grades are usually marketed in combination as #2 & Better.

structural light framing A category of dimension lumber up to four inches in width which provides higher bending strength ratios for use in engineered applications, such as roof trusses. The lumber is often referred to by its fiber strength class, such as 175f for #1 & Better Douglas Fir, or as stress-rated stock.

structural lumber Any lumber with nominal dimensions of 2 inches or more in thickness and 4 inches or more in width and which is intended for use where working stresses are required. The working stress is based on the strength of the piece and the use for which it is intended, such as beams, stringers, joists, planks, posts, and girders.

structural shape A hot-rolled or cold-formed member of standardized cross section and strength, generally used in a structural frame. Common structural shapes are angle irons, channels, tees, H-sections, and wide flange beams.

structural steel Steel rolled in a variety of shapes and manufactured for use as load-bearing structural members.

structural steel fastener Any fastener used to connect structural steel members to each other, or to supporting elements, or with concrete to make a composite section.

structural tee (1) A structural shape made by cutting a wide-flange beam or I-beam in half. (2) A hot-rolled steel member shaped like the letter "T".

structural timber connector See **timber connector**.

structural timbers Structural lumber with a nominal dimension of 5 inches or more on each side, used mainly as posts or columns.

structural veneer Veneers used in the construction of structural plywood panels, as opposed to decorative veneers.

structure A combination of units fabricated and interconnected in accordance with a design and intended to support vertical and horizontal loads.

structure height The vertical distance from grade to the top of the structure.

strut See **shore**.

stub A short projecting element.

stub mortise A mortise that does not project entirely through the piece in which it is cut. See **blind mortise**.

stub mortise and tenon See **blind mortise and tenon joint**.

stub pile A short pile.

stub tenon A short tenon cut to fit a stub mortise.

stub wall A low wall, usually 4 to 8 in. (100 to 200 mm) high, placed monolithically with a

Stucco

Stud Partition

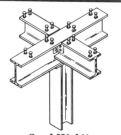

Stud Welding

Stuffing Box

concrete floor or other members to provide for control and attachment of wall forms.

stucco A cement plaster used for coating exterior walls and other exterior surfaces of buildings. See also **plaster**.

stud (1) A vertical member of appropriate size (2x4 to 4x10 in.) or (50x100 to 100x250 mm) and spacing (16 to 30 in.) or (400 to 750 mm) to support sheathing or concrete forms. (2) A framing member, usually cut to a precise length at the mill, designed to be used in framing building walls with little or no trimming before it is set in place. Studs are most often 2x4's, but 2x3's, 2x6's and other sizes are also included in the stud category. Studs may be of wood, steel, or composite material. (3) A bolt having one end firmly attached.

stud anchor A floor anchor for studs in a wall.

stud bolt A bolt firmly anchored in, and projecting from, a structure, such as a concrete pad, and used to secure another member, as in bolting a sill plate to concrete.

studding (1) The material from which studs are cut. (2) Same as stud, def. 2.

stud driver A device for driving a hardened steel fastener into concrete or other hard material, consisting of a hand held driver that positions the fastener. A blow on the head of the driver forces the fastener into the material.

stud grade A grade of framing lumber under the National Grading Rule established by the American Lumber Standards Committee. Lumber of this grade has strength and stiffness values that make it suitable for use as a vertical member of a wall, including use in load-bearing walls.

stud gun A stud driver in which the impact is from a contained blank cartridge.

stud opening A rough opening in wood framing.

stud partition A partition in which studs are used as the structural base. A wallboard is usually applied over the studs.

stud shooting Installing fasteners with a stud gun.

stud welding Attaching a special metal shear stud into a steel member by resistance welding. A special gun is used to hold the stud and provide electric current for the welding process. Shear studs are used for composite construction in which steel and concrete form a composite beam.

studwork Brick masonry interspaced with studs.

study (1) A preliminary sketch or drawing to facilitate the development of a design. (2) A room of a house or apartment intended or equipped for reading, writing, and studying.

stuff Sawn timber.

stuffing box A packing gland surrounding a shaft to prevent leakage, usually filled with soft packing to prevent fluids or gases from leaking.

stump veneer Veneer that has been produced from the roots of a tree, used in hardwood plywood manufacturing because of its grain configurations.

Sturd-I-Floor A trade name registered by the American Plywood Association for a panel designed specifically for use as combined subfloor/underlayment in residential floor applications. It is available in several thicknesses, each keyed to a recommended spacing of floor joists from 16 to 48 inches.

Sturd-I-Wall APA 303 siding panels intended to be attached directly to studs, as combined

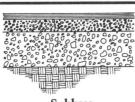

Subbase

Submersible Pump

sheathing/siding, or over non-structural wall sheathing. The name is a registered grade-trademark of the American Plywood Association.

Subalpine Fir *Abies lasiocarpa.* One of the true firs, this species is grouped for commercial purposes under the White Woods designation.

subbase (1) A layer in a pavement system between the subgrade and base course or between the subgrade and the concrete pavement. (2) The bottom front strip or molding of a baseboard.

subbasement (1) A level, or levels, of a building below the basement. (2) A story immediately below a basement.

sub-bidder, subbidder A person or entity who submits a bid to a bidder for materials or labor for a portion of the work.

subcontract An agreement between a prime contractor and a subcontractor for a portion of the work at the site.

subcontractor A person or entity who has a direct contract with the contractor to perform any of the work at the site. See also **supplier** and **vendor**.

subdivided truss A truss with additional framing members in the panels to shorten the effective length of the panels.

subdivision regulations Local municipal ordinances specifying the conditions under which a tract of land can be subdivided. The ordinances may include layout and construction, street lighting and signs, side walks, sewage and storm water systems, water supply systems, and dedication of land for schools, parks, etc.

subfeeder An electric feeder that originates at a distribution center other than the main distribution centers and supplies one or more branch-circuit distribution centers.

subfloor, blind floor, counterfloor A rough floor laid on floor joists and serving as a base for the finish floor. A subfloor may also be used as a structural diaphram to resist lateral loads.

subflooring Plywood sheets or construction grade lumber used to construct a subfloor.

subframe, rough buck, subbuck (1) A structural frame of wood members or channel-shaped metal members that support the finish frame of a door or window. (2) A framework supporting wall siding or sheets.

subgrade (1) The soil prepared and compacted to support a structure or a pavement system. (2) The elevation of the bottom of a trench in which a sewer or pipeline is laid.

subgrade reaction See **contact pressure**.

subheading A subdivision of a heading used in the filing system.

subject to mortgage A legal subjection of a property to an existing mortgage if the purchaser has actual or constructive notice of the mortgage, for example if a mortgage of real property has been recorded. The new owner is not liable for mortgage payments unless he has agreed to that liability. However, the mortgagee may foreclose if there is a default in payments.

sublease A lease by a tenant of part or all of leased premises, for part or all of the terms of the original lease.

submerged arc welding A type of arc welding in which the arc and weld are shielded by a blanket of powdered metal called "flux".

submersible pump A type of pump with the motor and liquid handling unit in a watertight case that can be lowered directly into the liquid to be pumped.

Sub-purlin, subpurlin

subordinate lien Any mortgage lien subsequent to the first. In event of foreclosure, holders of such liens may resort to the property for payment only to the extent of any surplus after prior liens have been paid, with priority usually determined by the chronological sequence in which the mortgages were initiated.

subparagraph In the AIA Documents, the first subdivision of a paragraph, identified by three numerals (e.g.2.2.2). A subparagraph may be subdivided into clauses.

sub-purlin, subpurlin A light structural section used as a secondary structural member, and used in lightweight concrete roof construction to support the formboards over which the lightweight concrete is placed.

subrogation The substitution of one person or entity for another with respect to legal rights such as a right of recovery. Subrogation occurs when a third person, such as an insurance company, has paid a debt of another or claim against another and succeeds to all legal rights which the debtor or person against whom the claim was asserted may have against other persons.

subsealing The placing of a waterproofing material under an existing pavement, to waterproof the pavement and to fill voids in the subsoil.

subsidence Settlement over a large area as opposed to settlement of a single structure.

subsill (1) An additional sill fitted to the outside of a window as a stop for screens and to increase the shedding distance. (2) A rough doorsill fixed to the groundsill.

subsoil The bed or stratum of soil lying immediately below the surface soil and which is usually devoid of humus or organic matter.

subsoil drain A drain installed to collect subsurface or seepage water and convey it to a point of disposal.

substantial completion See **date of substantial completion**.

substitution A material, product, or item of equipment offered in lieu of that specified.

substrate An underlying material that supports or is bonded to another material on its surface.

substrate failure A surface failure in a concrete wall at a joint, caused by a sealant of high tensile strength that tears weak concrete or mortar from the face of the joint.

substructure The foundation of a building that supports the superstructure.

sub-subcontractor A person or entity who has a direct or indirect contract with a subcontractor to perform any of the work at the site.

subsurface course The top course of a pavement that acts as a wearing surface.

subsurface investigation The soil boring and sampling program, together with the associated laboratory tests, necessary to establish subsurface profiles and the relative strengths, compressibility, and other characteristics of the various strata encountered within depths likely to have an influence on the design of the project. Sometimes called "geotechnical investigation". Preferable to soil survey.

subsurface sand filter A sewage filtering system consisting of a number of lines of perforated pipe or drain tile surrounded by coarse aggregate, an intermediate layer of sand as filtering material, and a system of under drains to carry off the filtered liquid.

subsurface sewage disposal system A system for treating and disposing of domestic sewage,

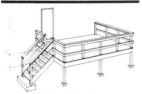

Sun Deck

usually from a single residence, by means of a septic tank and a soil absorption system.

successful bidder The bidder chosen by the owner for the award of a construction contract. See also **selected bidder**.

sucker A shoot rising from an underground root or stem.

suction (1) The absorption of water from a plaster finish coat by the base coat, gypsum block, or gypsum lath, functioning to increase bond and promote better adhesion to the base. (2) The adhesion of mortar to brick.

suction pump A pump that draws water from a reservoir at a level lower than the pump, or from a pipe operating at a lower hydraulic gradient than the pump.

Sugar Pine *Pinus lambertiana*. A pine species, found in northern California and southern Oregon, that is light, smooth, and easily worked. It is widely used in millwork, patternwork, and various interior applications. The Sugar Pine takes its name from the sugary, sweet tasting deposits of resin exuded from its bark after injury to the tree.

sulfate attack A chemical and/or physical reaction between sulfates usually in soil or ground water and concrete or mortar, primarily with calcium aluminate hydrates in the cement-paste matrix, often causing deterioration.

sulfate resistance The ability of concrete or mortar to withstand sulfate attack. See also **sulfate attack**.

sulfur cement A cement of clay or a similar substance, usually with additives such as sulfur, metallic oxides, silica, or carbon, used for sealing joints and coatings in high-temperature areas, such as furnace fire boxes.

sullage (1) Drainage, sewage, or other water-borne waste.

(2) Sediment or slit transported by flowing water.

summer (1) A horizontal beam supporting floor joists or a wall of a superstructure. (2) Any heavy timber that serves as a bearing surface. (3) A lintel of a door or window. (4) A stone set on a column as a support for construction above, such as a base for a column-supported arch.

summertree See **summer**.

sump (1) A pit, tank, or basin that receives sewage, liquid waste, or seepage or overflow water and is located below the normal grade of the disposal system. A sump must be emptied by mechanical means. (2) A depression in a roof deck at a drain.

sump pump A small pump used to remove accumulated waste or liquid from a sump.

sun deck A deck or flat roof intended for sunbathing.

sunken joint A small, narrow depression on the face of a piece of plywood. Sunken joints occur over joint gaps in the core plies of plywood.

sunk face A building stone with material removed from a portion of the face to give the appearance of a sunken panel.

sunk fillet A fillet formed by a groove in a flat surface.

sunk molding A molding that is slightly depressed below the surface it frames.

sunk panel A panel carved into solid masonry or timber, or recessed below the surrounding molding.

superheated steam Steam at a temperature higher than the saturation temperature corresponding to the pressure.

superimposed drainage (1) A natural drainage system developed by erosion and having little relation to the area's geological structure. (2) A man-made

Supply Grille

Supply System

drainage system developed against the existing geological structure.

superimposed load The load other, than its own weight, that is resisted by a structural member or system.

superintendent The contractor's representative at the site who is responsible for continuous field supervision, coordination, completion of the work, and, unless another person is designated in writing by the contractor to the owner and the architect, for the prevention of accidents.

superstructure (1) That part of a building or other structure above the foundation. (2) That part of a bridge above the beam seats or the spring line of an arch.

supervision Direction of the work by the contractor's personnel. Supervision is neither a duty nor a responsibility of the architect as part of professional services.

supplemental authorization In professional services, a written agreement authorizing a modification to a professional services agreement.

supplemental conditions See **supplementary conditions**.

supplemental services Concerning the architect those services described in the, schedule of designated services which are in addition to the generally sequential services, from predesign through post-construction, of the architect, including such items of service as renderings, value analyses, energy studies, project promotion, expert testimony, and the like.

supplementary conditions A part of the contract documents which supplements and may also modify, change, add to, or delete from provisions of the general conditions. Preferable to

supplemental conditions.

supplementary lighting Lighting used to provide additional quality and quantity to the general lighting system, usually local and for specific work requirements.

supplier A person or entity who supplies materials or equipment for the work, including that fabricated to a special design, but who does not perform labor at the site. See also **vendor**.

supply air The conditioned air delivered to a space or spaces.

supply fixture unit A measure of the probable demand on a water supply by a particular type of plumbing fixture. The value depends on the volume of water supplied, the average duration of a single use, and the number of uses per unit time.

supply grille A grille through which conditioned air is delivered to a space.

supply mains (1) The pipes that bring water, gas, and the like into a building. (2) The pipes through which the heating or cooling fluid flows from the source of heating or cooling to the laterals or risers leading to heating or cooling units.

supply system The connected ducts, plenums, and fittings through which conditioned air is transferred from a heat exchanger to the space or spaces to be conditioned.

Suppressed Pine Small Ponderosa Pine that has grown under adverse conditions. It usually has dark bark, as opposed to the reddish bark found on most mature Ponderosa Pine. Also called "Blackjack Pine" or "Bull Pine".

supreme A grade in Idaho White Pine equivalent to B&Btr - 1&2 clear, the highest grade of select lumber.

surety A person or entity who promises in writing to make good

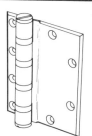

Surface Hinge

the debt or default of another.

surety bond A legal instrument under which one party agrees to answer to another party for the debt, default, or failure to perform of a third party.

surface, specific See **specific surface**.

surface active Having the ability to modify surface energy and to facilitate wetting, penetrating, emulsifying, dispersing, solubilizing, foaming, and frothing of other substances.

surface-active agent An additive to a concrete mix to reduce the surface tension of the mixing water and facilitate wetting, penetrating, emulsifying, dispersing, solubilizing, foaming, and frothing of other additives.

surface area See **specific surface**.

surface hardware preparation The reinforcement of a hollow-metal door or frame to receive surface hardware.

surface hinge A hinge, often ornamental, mounted on a face of a door rather than the edge.

surface latch A latch mounted on a face of a door rather than inset.

surface moisture, free water, surface water Free water retained on surfaces of aggregate particles and considered to be part of the mixing water in concrete, as distinguished from absorbed moisture.

surface planer A machine used to plane and smooth the surface of materials such as wood, stone or metal.

surface retarder A retarder used by application to a form or to the surface of freshly mixed concrete to delay setting of the cement to facilitate construction joint cleanup or to facilitate production of exposed aggregate finish.

surface tension That property, due to molecular forces, in the surface film of all liquids that tends to prevent the liquid from spreading.

surface texture Degree of roughness or irregularity of aggregate particles or of the exterior surfaces of hardened concrete.

surface vibrator A vibrator used for consolidating concrete by application to the top surface of a mass of freshly mixed concrete. Four principal types exist: vibrating screeds, pan vibrators, plate or grid vibratory tampers, and vibratory roller screeds.

surface voids Cavities visible on the surface of a solid. See also **bugholes**.

surface water See **surface moisture**.

surface wiring switch An electric switch intended for mounting on a surface, such as of a wall post, with most of the body of the switch exposed.

surfacing (1) The upper wearing or protective layer of materials, such as in a roof or road. (2) See **surfacing weld**.

surfacing weld A deposit of weld metal on the surface of a material to provide desired dimensions or properties.

surform tool A hard tool similar to a file with a surface of sharp cutting teeth pitched at an angle of about 45 degrees which removes surface wood like a plane.

surge tank A tank in a water supply system used to absorb water during a sharp pressure rise or to supply water in a sudden pressure drop.

survey (1) To do boundary, topographic, and/or utility mapping of a site. (2) To measure an existing building. (3) To analyze a building for the use of space. (4) To determine the owner's requirements for a project. (5) To investigate and report on required data for a project.

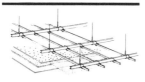

Suspended Acoustical Ceiling

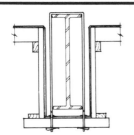

Suspended Ceiling

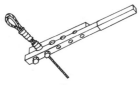

Suspended Formwork

Swage

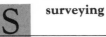

Sweep Fitting

surveying The measurement of distances, elevations, angles, etc., of the earth's surface, including natural and man made features, usually for the preparation of a plan or map.

surveyor An engineer or technician skilled in surveying.

surveyor's arrow See **chaining pin**.

surveyor's compass A large, graduated compass, with a sighting device, used by surveyors to determine magnetic bearings.

surveyor's level See **level, def. 1**.

survey stakes Small pieces of wood, usually 1x2 or 2x2, that have been cut and pointed for driving into the ground to mark a survey line or some boundary of construction.

survey traverse A sequence of lengths and angles between points on the earth established and measured by a surveyor. A survey traverse is used as a reference in making a detailed survey.

suspended absorber A sound-absorbing assembly designed for overhead suspension in an area.

suspended acoustical ceiling A ceiling designed to be sound absorbent and to be hung from the structural slab or beams in an area.

suspended ceiling, dropped ceiling A finished ceiling suspended from a framework below the structural framework.

suspended floor A floor system that spans the entire distance between end supports without intermediate supports.

suspended formwork Formwork suspended by hangers from a structural system.

suspended metal lath A system in which metal lath is attached to a light framing system that, in turn, is hung from a structural slab or beams.

suspended span A span supported by two cantilevers or a cantilever and a column or pier, used mainly in bridges and some roofs.

suspended-type furnace A unit furnace designed to be suspended from a floor or roof and to supply heated air through ducts to areas other than that in which the furnace is located.

suspension roof A roof system supported by cables attached to a frame.

swage (1) A tool or die used to bend or form cold metal. (2) A tool used to set the teeth of a saw by bending each one to the proper transverse angle.

swale (1) A shallow depression, in a flat area of land, that may be artificial and used in a storm water drainage system. (2) A low tract of land, usually marshy.

swan-neck (1) The connector between a gutter outlet and the downspout. (2) The curved portion of the handrail of a stair connecting the rail to a newel-post.

swaybrace A diagonal brace used to resist wind or other lateral forces. See also **bracing** and **X-brace**.

sweating (1) A soldering technique for joining metal parts. (2) A gloss that develops on a dull, or matte, paint or varnish film and is caused by rubbing the dry film. (3) A collection of condensation moisture on a surface that is below the dew point of the air.

sweat-out A defective condition in gypsum plaster, characterized by a soft damp area surrounded by dry plaster, and usually caused by poor air circulation.

sweep (1) The curvature or bend in a log, pole, or piling, classified as a defect. (2) A bend in an electrical conduit.

sweep fitting Any plumbing fitting with a large radius curve.

sweep lock A device used to secure two sections of sash. One part is a

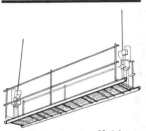

Swinging Scaffold

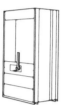

Switchboard

Switchgear

flat lever attached to the upper sash; the other is a receiving catch attached to the lower sash.

sweep strip A flexible weather stripping used on the top and the bottom edges of a revolving door.

swell factor The ratio of the weight or volume of loose excavation material to the weight or volume of the same material in place.

swelling A volume increase caused by wetting and/or chemical changes.

swift A reel or turntable on which prestressing tendons are placed to facilitate handling and placing.

swing (1) The movement of a door on hinges or a pivot. (2) The rotation of a power shovel on its base.

swing check valve A check valve with a hinged gate to permit fluid to flow in one direction only, used mostly where fluid velocities are low.

swinging door See **double-acting door**.

swinging latch bolt A latch bolt hinged to a lock or door and operated by swinging rather than sliding.

swinging scaffold A scaffold suspended from a roof by hooks, cables, and a block and tackle.

swing joint A type of joint in threaded pipe that allows for expansion and contraction without stressing the piping.

swing leaf (1) The active leaf in a double door. (2) A hinged sash in a casement window.

swing offset In surveying, a method of referencing a point by swinging bisecting arcs, with a tape, from established points nearby.

swing scaffold See **swinging scaffold**.

swing-up garage door A rigid overhead door that opens as one unit.

swirl The irregular, curving pattern in wood at knots or crotches.

swirl finish, sweat finish A nonskid texture imparted to a concrete surface during final troweling by keeping the trowel flat and using a rotary motion.

Swiss hammer See **rebound hammer**.

switch A device used to open, close, or change the connection of an electric circuit.

switchboard A large panel, frame, or assembly with switches, overcurrent and other protective devices, fuses, and instruments mounted on the face and/or back. Switchboards are usually accessible from front or rear and are not intended to be mounted in cabinets.

switchgear Any switching and interrupting devices combined with associated control, regulating, metering, and protective devices, used primarily in connection with the generation, transmission, distribution, and conversion of electric power.

switch plate A flush plate used to cover an electric switch.

switch-start fluorescent lamp See **preheat fluorescent lamp**.

swivel spindle A spindle in a door latch unit with a joint in the middle of its length to allow one knob to be fixed by a stop works while the other knob operates the latch.

sycamore A tough, yellowish wood with a close firm texture, that takes a fine polish and is used for flooring and veneer.

symmetrical construction A plywood panel in which the plies on one side of the center ply are balanced in thickness with those on the other side. Also called "balanced construction".

synchronous motor A constant-speed electric motor, usually direct current. The operational speed is equal to the frequency of the supply voltage divided by one-half the number of poles or windings on the machine.

syneresis The contraction of a gel, usually evidenced by the separation from the gel of small amounts of liquid, a process possibly significant in the bleeding and cracking of fresh portland cement mixtures.

synthetic resin Any of a large number of products similar to natural resins and made either by polymerization or condensation or by modifying a natural material.

synthetic rubber A chemically manufactured elastomer, rubber-like in its degree of elasticity.

systems A process of combining prefabricated assemblies, components, and parts into single integrated units utilizing industrialized production, assembly, and methods.

T ABBREVIATIONS

The abbreviations listed below are those most commonly used in the construction industry. Alternative forms (usually nonstandard) are shown in parentheses.

t temperature, time, ton
T tee, township, true, thermostat
Ta tantalum
t.b. turnbuckle
TB through-bolt
TC terra-cotta
Te tellurium
TE table of equipment, trailing edge
tech technical
TEL telephone
T.E.M. Total Energy Management
temp temperature, temporary
TEMP temperature
TER terrazzo
t.f. tar felt
T&G, T and G tongue-and-groove
TG&B tongued, grooved, and beaded
t.g.&d. tongued, grooved, and dressed
TH true heading
therm thermometer
THERMO thermostat
thou thousand
THK thick
thp thrust horsepower
THRU through
Ti titanium
TL transmission loss
tlr trailer
TM technical manual
tn ton, town, train
TN true north
tnpk/tpk turnpike
t.o. take off (estimate)
TOL tolerance
tonn tonnage
topog/topo topography
TOT. total
tp title page, township, tar paper
tps townships
tr tread
trans transom

TRANS transformer
transp transportation
trf tuned radio frequency
trib tributary
trib. ar tributary area
ts tensile strength
TU trade union, transmission unit
TUB. tubing
TV terminal velocity
twp township
TYP typical

T DEFINITIONS

Tail

tabby A type of concrete consisting of a mixture of lime and water with shells, gravel, or stones used to make blocks for masonry.

table joint Any of various joints in which the fitted surfaces are parallel to the edges of the pieces being joined, with a vertical break in the middle. This break is termed the table. Such joints are used in lengthening structural members.

tabled joint A joint in stone masonry in which a projection cut into a stone fits into a channel cut into the stone below.

Table-Mountain Pine *Pinus pungens*. This pine is found primarily in the Appalachian states. It is not a commercial species.

table saw A power saw, with a circular blade, set below a table. The blade projects above the table through a slot with the projection height adjustable to suit the work.

tack (1) A short sharp pointed nail with a large head used in laying linoleum and carpets. (2) A strip of metal, usually lead or copper, used to secure edges of sheet metal in roofing. (3) The property of an adhesive that allows it to form a strong bond as soon as the parts and adhesive are placed in contact. (4) To glue, weld, or otherwise fasten in spots.

tack coat A coat of emulsion which enhances the bonding of two layers of asphalt.

tack dry The state of an adhesive at which it will adhere to itself although it seems dry to the touch.

tack-free dry The point where paint or varnish no longer feels sticky.

tackless strip A metal or wood strip, with many small upstanding hooks, fastened to a floor or stair. A carpet is stretched beyond the strip and the hooks engage the backing and hold the carpet in place.

tack rag A rag treated with slow-drying or nondrying varnish or resin and used to clean dirt and foreign matter from the surfaces of articles before they are painted.

tack weld (1) A temporary weld to position parts. (2) One of a series of short welds used where a continuous weld is unnecessary.

tag A strip of sheet metal folded on itself and used as a wedge to hold metal flashing in a masonry joint.

tagline (1) A line that runs from a crane boom to a clamshell bucket and keeps the bucket firmly in position during operations. (2) A safety line used by workers performing a job on high elevation or in another dangerous location where a fall could be injurious.

tail The bottom part of a slate shingle.

tail bay (1) The area between a wall and the nearest column line. (2) In a framed floor or roof the area between an end wall and the nearest girder.

tailgate The hinged moving rear wall of a dump truck through which materials are dumped.

tailing in (1) To secure or fasten one end of a timber, such as a floor joist, at a wall. (2) To secure one end or edge of a projecting unit of masonry, such as a cornice.

tailing iron A steel member, built into a wall, to take the upward thrust of a cantilevered member which projects from the wall.

tailings (1) Stones left on a screen used to grade material, as in a crushing operation. (2) The waste material or residue of a product.

Tandem roller

Tap

tailpiece (1) A subordinate joist, rafter, or the like supported by a header joist at one end and a wall or sill at the other. (2) A handle on the bar end of a two-man power saw.

take-up block A pulley block rigged so that its weight or a spring will prevent slack in the lines passing through it.

talc A mineral with a greasy or soapy feel, very soft, having the composition Mg Si O (OH) . See **masonry cement**.

Tamarack *larix laricina*. This species, also called "Eastern Tamarack", is common in the northeastern states. However, its range extends as far as southern Alaska. As a member of the Larch family, it is one of the few conifers that sheds its needles in the fall.

tamp To press a loose material such as soil or fresh concrete into à firm, compact mass by pounding repeatedly on it.

tamper (1) An implement used to consolidate concrete or mortar in molds or forms. (2) A hand-operated device for compacting floor topping or other unformed concrete by impact from the dropped device in preparation for strikeoff and finishing. Contact surface often consists of a screen or a grid of bars to force coarse aggregates below the surface to prevent interference with floating or trowelling. See **jitterbug**.

tamping The operation of compacting freshly placed concrete by repeated blows or penetrations with a tamping device.

tamping rod See **rod, tamping**.

tandem (1) A pair in which one moving part follows the action of the other, such as tandem rollers. (2) A vehicle made up of two units attached to one another, as the cab and trailer of a truck.

tandem drive A vehicle with three axles, two of which are driving axles.

tandem roller A unit of two rollers of equal diameter sat one behind the other on the same track.

tang The narrow extending tongue or prong on an object by which it is affixed to another piece, such as the tongue that secures a chisel to a handle.

tangent A straight line or curve that touches another curve at a single point without crossing the curve.

tangent modulus See **modulus of elasticity**.

tap (1) A connection to a water supply. (2) A faucet. (3) A tool used to cut internal threads.

tap bolt A machine bolt that screws into a hole in material without requiring a nut or internal threads.

tap borer A hand tool used to bore tapered holes, as in a lead pipe for a connection.

tape (1) A flexible measuring strip of fabric or steel marked off with lines similar to the scale of a carpenter's rule, usually contained in a case to allow rewinding or retracting after use. (2) One of a number of adhesive-backed fabric or paper strips of assorted width used for various purposes within construction systems.

tape balance A balancing device on a sash which counterbalances the weight of the sash by force of a metal tape coiled around a spring-loaded reel.

tape correction A correction that is figured into a distance measured by a tape in order to compensate for errors resulting from the condition of the tape or the manner in which it was handled.

tape measure, tapeline A steel strip used by builders and surveyors to measure distances, usually graduated in feet, tenths,

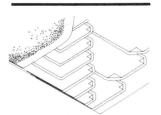

Tar-And-Gravel Roofing

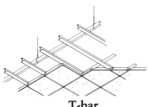

T-bar

and hundredths of a foot for use in surveying and engineering, or in feet, inches, and fractions of an inch for use in the building trades.

taper A gradual diminution of thickness, diameter, or width in an object.

tapered tenon A tenon decreasing in width from the root to the projecting end.

taper file A triangular file equipped with fine teeth, used to sharpen saws.

tapersawn A trade name used by the Red Cedar Shingle and Handsplit Shake Bureau for shakes fully sawn on both faces.

tapersplit Shakes produced by using a mallet and froe, a sharp steel blade, to obtain split faces on both sides. The taper of the shake is a result of reversing the block and splitting the shake from a different end after each split.

taper thread A screw thread developed on a frustum of a cone, used in piping systems to ensure a tight joint, also used on some fasteners such as a screw or plug for a hole with worn threads.

taping The process of measuring the distance across ground with a chain or tape.

taping pin See **chaining pin**.

taping strip (1) A strip of roofing felt laid over the joint between adjacent precast concrete units before roofing operations. (2) A strip of tape used to cover the joint between adjacent roof insulation boards.

tapped fitting Any fitting which has one or more tapped internal threads to receive threaded pipe.

tapped tee A cast iron soil pipe tee with a tapped outlet.

tapping machine A machine producing a series of uniform impacts on a floor surface, used to measure sound transmission of floor ceiling assembly.

tapping screw See **sheet metal screw**.

tap screw See **tap bolt**.

tar A dark, glutinous oil distilled from coal, peat, shale, and resinous woods, used as surface binder in road construction and as a coating in roof installation. See **coal-tar pitch**.

tar-and-gravel roofing Built-up roofing made up of gravel or sand, poured over a heavy coating of coal-tar pitch applied to an underlayer of felt.

tar cement Heavier grades of tar for use in construction and maintenance of bituminous concrete pavements.

tare The weight of a rail car, truck, or other conveyance when empty. The tare is deducted from the gross weight of the vehicle and its contents to determine the weight of the freight in figuring freight rates.

target A red and white target mounted on a surveyor's leveling rod, used to facilitate the sighting of the markings on the rod by the transit man.

target leveling rod A leveling rod fitted with a target to facilitate setting or reading.

target rod See **leveling rod**.

tarmac, tarmacadam See **macadam**.

tar paper See **asphalt prepared roofing**.

tarpaulin A waterproof cloth, generally used in large sheets to cover and to protect construction materials or other goods stored out-of-doors.

tarred felt See **asphalt felt**.

tax abatement A reduction in taxes on real property, usually accomplished by reducing the assessed value.

T-bar A light-gauge, T-shaped member used to support panels in a suspended acoustical ceiling.

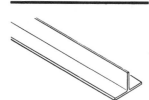

T-beam

Tee

Telescope

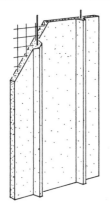

**Temperature
Reinforcement**

T-beam A beam composed of a stem and a flange in the form of a "T", usually of reinforced concrete or rolled metal.

teak Wood from trees in southeastern Asia, India, and Burma which is dark, heavy, and particularly resistant to attack of insects, used for exterior construction, plywood, flooring and decorative paneling.

teazel An angle post used in the construction of a timber-frame building.

technical trader A futures trader who bases his buying and selling decisions primarily on conditions that develop within the futures market rather than on developments in the market for the physical commodity.

tee (1) An elaborately turned finial in the shape of an umbrella, used as finishing ornamental on pagodas, stupas, and topes. (2) See **pipe tee**. (3) A metal structural member with a T-shaped cross section.

tee beam See **T-beam**.

tee bevel See **bevel square**.

tee handle A T-shaped handle used in place of a doorknob to operate the bolt on a door lock.

tee head See **T-head**.

tee hinge See **T-hinge**.

tee iron (1) A piece of flat heavy sheet metal, shaped into a "T" and equipped with drilled countersunk holes, used to reinforce joints in wood construction. (2) A steel T-beam section.

tee joint A joint between two members that intersect at right angles to form a T-shaped joint.

tee square See **T-square**.

telescope To insert or slide one piece inside another.

telltale Any device designed to indicate movement of formwork

or a point along the length of a pile under load.

tempera A fast-drying, water soluble paint composed of eggs, gum, pigment, and water, used chiefly for painting murals.

temperature cracking Cracking due to tensile failure, caused by temperature drop in members subjected to external restraints or temperature differential in members subjected to internal restraints.

temperature reinforcement Reinforcement designed to carry stresses resulting from temperature changes.

temperature rise The increase of temperature caused by absorption of heat or internal generation of heat, as by hydration of cement in concrete.

temperature rise period The time interval during which the temperature of a concrete product rises at a controlled rate to the desired maximum in autoclave or atmospheric-pressure steam curing.

temperature steel A steel reinforcement used within concrete slabs and other units of masonry to reduce the chances of cracking caused by temperature changes.

temperature stress Stress in a structure or a member due to changes or differentials in temperature in the structure or member.

temperature stress rod Concrete reinforcing that is provided to control cracks that are caused by shrinkage and temperature stresses.

tempered board A strong, durable composition board as wood fiber that has been impregnated with drying oil or an oxidizing resin and then cured with heat to improve its hardness and moisture resistance.

Template

Temporary Shoring

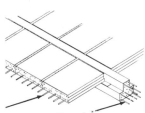

Tendon

tempered glass Glass that is prestressed by heating and then rapidly cooled, a process that makes it two to four times stronger than ordinary glass.

tempering The addition of water and mixing of concrete or mortar as necessary to bring it to the desired consistency during the prescribed mixing period. For truck-mixed concrete this will include any addition of water as may be necessary to bring the load to the correct slump on arrival at the work site but not after a period of waiting to discharge the concrete.

template, templet (1) A thin plate or board frame used as a guide in positioning or spacing form parts, reinforcement, or anchors. (2) A full-size mold, pattern or frame, shaped to serve as a guide in forming or testing contour or shape.

template hardware Hardware that is an exact match for a master template drawing, conforming to the spacing of all holes and dimensions.

temporary shoring Shoring installed to support a structure while it is being built and removed when construction is finished.

temporary stress A stress which may be produced in a precast concrete member of component or a precast concrete member during fabrication or erection, or in cast-in-place concrete structures due to construction or test loadings.

tenancy Occupation of property by one who has less than a fee interest, whether a life tenancy arranged by agreement with the owner or heirs, or a tenancy created by lease for a stated term of years.

tenant A person or firm occupying a building or a part thereof to which they hold tenancy rights by lease or title.

tenant's improvement Improvements made to a house or parcel of land by a tenant at his own expense which commonly become a part of the property and cannot be removed by a departing tenant unless the owner gives his consent.

tender (1) An offer showing a willingness to buy or sell at specific price and conditions. (2) In futures, the act on the part of a seller of a contract of giving notice to the clearinghouse that he intends to deliver the physical commodity in satisfaction of a futures contract. (3) A hooktender. (4) A railroad car attached to a steam locomotive for carrying fuel and water or a ship providing similar services in a fleet.

tendon A steel element such as a wire, cable, bar, rod, or strand used to impart prestress to concrete when the element is tensioned.

tendon profile The path or trajectory of the prestressing tendon.

tenement, tenement house Any dwelling, but usually an apartment house that is situated in a poorer section of a city and has fallen into disrepair.

Tennessee Red Cedar *Juniperus virginiana*. A pioneer species found in most of the Eastern U.S. It is more commonly known as "Eastern Red Cedar".

tenon A projecting, tongue-like part of a wood member designed to be inserted into a slot or mortise of another member to form a mortise and tenon joint.

tenon-and-slot mortise A glued wood joint formed by a tenon and mortise, usually the two pieces join at a right angle to each other.

tenoner A machine used to form tenons.

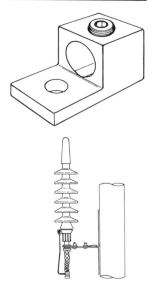

Terminal

Terne Metal

tenon saw A small backsaw used for cuttings tenons, also called a "mitre saw".

tenpenny A particular size of nail, generally 3 inches long with 69 to a pound. The term is believed to have originated in England, where nails of the same size once sold for tenpence per hundred.

tensile strain Elongation of a material that has been subjected to tension by pulling or stretching.

tensile strength Maximum unit stress which a material is capable of resisting under axial tensile loading, based on the cross-sectional area of the specimen before loading.

tensile stress Stress resulting from tension.

tension The state or condition imposed on a material or structural member by pulling or stretching.

tension member A tie or another structural member subjected to tension.

tension reinforcement Reinforcement designed to carry tensile stresses such as those in the bottom of a simple beam.

tension wood Defective wood usually cut from the upper side of hardwood branches or leaning trunks. The processed lumber exhibits high longitudinal shrinkage, causing warping and splitting.

ten wheeler A tandem truck having ten wheels, commonly called a bogie. In construction, the term refers to a tandem dump truck.

teredo A marine borer that damages untreated wood by boring holes in it.

terminal (1) An element attached to the end of a conductor or to a piece of electric equipment to serve as a connection for an external conductor. (2) A decorative element forming the end of an item of construction. (3) A point of departure or arrival such as a railway or airport terminal.

terminal box A box, on a piece of electrical equipment, that contains leads from the equipment, ready for connection to a power source. The box is usually provided with a removable cover.

terminal unit A unit, at the end of a duct in an air conditioning system, through which air is delivered to the conditioned space.

terminal expense An expense incurred by one or both parties to a contract at the time of its termination.

terminal velocity The average speed of an airstream in an air-conditioning system as it reaches the end of its throw, used as an indicator of comfort level and degree of draftiness.

termination expenses In professional services, expenses directly attributable to the termination of a professional service agreement, including an amount allowing for compensation earned to the time of termination.

termites Insects that destroy wood by eating the wood fiber. Termites are social insects that exist in most parts of the U.S., but they are most destructive in the coastal states and in the southwest. Termites can enter wood through the ground or above the ground, although the subterranean type is most common in the U.S. They eat the softer springwood first and prefer sapwood over heartwood.

termite shield A sheet of metal used on a foundation wall or pier as a projecting shield to prevent the passage of termites from the ground to a structure.

terne metal An alloy of lead, composed of up to 20 percent tin.

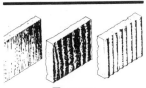

Texture

Texture 1-11

terneplate Sheet steel coated with an alloy of lead and tin, used chiefly in roofing.

terrace (1) An embankment with a level top surface and stabilized side slopes. It may be used for agriculture or, paved and/or planted for recreational use. (2) A platform or paved embankment adjoining a building and used for recreational purposes.

terra-cotta Units of hard, unglazed fired clay, used for ornamental masonry.

tertiary treatment An advanced stage of the wastewater treatment process, employing methods such as ion exchange, carbon absorption, reverse osmosis, and demineralization of residual solids.

tessera A small square of glass, stone, tile or marble used in geometric and figurative mosaic work in pavements or floors.

test A trial, examination, observation, or evaluation used as a means of measuring a physical or chemical characteristic of a material, or a physical characteristic of a structural element or a structure.

test cylinder A sample of a concrete mix, cast in a standard cylindrical shape, cured under controlled or job conditions and used to determine the compressive strength of the mix after a specified time interval.

testing machine A device for applying test conditions and accurately measuring results.

test piling A foundation piling that is installed on the site of a proposed construction project and used to conduct load tests to determine the size and quantity of pilings needed for the actual structures.

test pit An excavation made to examine the subsurface conditions on a potential construction site, including in a soil, also, a pit excavated to inspect the condition of existing foundations.

test plug A device that contains water test pressure in drainage pipes as part of a process to check plumbing systems for leakage.

test pressure Air pressure or water pressure used in plumbing to test pipes and fittings for strength and watertightness.

test tee A pipe tee that is inserted into a drainage system to test it for leaks by subjecting it to water pressure.

tetracalcium aluminoferrite A compound in the calcium aluminoferrite series, having the composition $4CaO.Al2O3.Fe2O3$, abbreviated C4AF, which is usually assumed to be the aluminoferrite present when compound calculations are made from the results of chemical analysis of portland cement. See **brown-millerite**.

texture The pattern or configuration apparent in an exposed surface, as of concrete or mortar, including roughness, streaking, striation, or departure from flatness.

Texture 1-11 A registered trade name of the American Plywood Association for siding panels with special surface treatment, such as saw textured, and having grooves spaced regularly across the face.

texture brick A rough fired-clay brick, often multicolored, commonly used for facing work. See **rustic brick**.

texturing The process of producing a special texture on unhardened or hardened concrete.

T&G joint See **tongue-and-groove joint**.

thatch A roof covering made of straw, red palm leaves, or similar material fastened together to shed

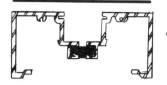

Thermal Break

water and provide thermal insulation.

T-head (1) In precast framing, a segment of girder crossing the top of an interior column. (2) The top of a shore formed with a braced horizontal member projecting on two sides forming a T-shaped assembly.

theodolite An instrument used in precise surveying consisting of an alidade which is equipped with a telescope, a leveling device, and an accurately graduated horizontal circle. A theodolite may also have an accurately graduated vertical circle.

therm A quantity of heat equal to 100,000 Btu's.

thermal break, thermal barrier An element of low conductivity placed between two conductive materials to limit heat flow, for use in metal windows or curtain walls which are to be used in cold climates.

thermal conductance The rate at which heat flows from one surface of a material to the other, usually measured or specified at the rate over a unit area and under a unit temperature differential.

thermal conduction The process of heat transfer through a material by internal molecular action.

thermal conductivity A property of a homogeneous body measured by the ratio of the steady state heat flux, time rate of heat flow per unit area, to the temperature gradient, temperature difference per unit length of heat flow path, in the direction perpendicular to the area.

thermal cutout An overcurrent protective device containing a heater element and a renewable fusible member which opens the circuit. The thermal cutout is not designed to interrupt short-circuit currents.

thermal diffusivity Thermal conductivity divided by the product of specific heat and unit weight. The term is an index of the facility with which a material undergoes temperature change.

thermal expansion The change in length or volume experienced by a material or mass when subjected to a change in temperature.

thermal insulating cement A dry mixture of cement and granular flaky, fibrous, or powdery materials of low conductivity. When mixed with water and applied to a surface, it dries to provide an insulating covering.

thermal insulation, heat insulation A material which provides a high resistance to heat flow. Examples are foamed plastics, mineral or glass fibers, cork, and foamed glass. The material is used in the forms of blankets, boards, blocks and poured or granular fill.

thermal process A type of wood preservation treatment whereby the wood is heated in the preservative for several hours and then submerged in cold preservative for several more hours. Creosote or penta are the preservatives most commonly used with this treatment method, which is also called the hot and cold bath treatment.

thermal protector A protective device, installed in an electric machine, to guard the machine against overheating, caused by failure to start or overload.

thermal shock The subjection of a material or body, such as partially hardened concrete, to a rapid change in temperature which may be expected to have a potentially deleterious effect.

thermal stress, temperature stress Stress induced in an object or structural member by restraint against movement required to accommodate temperature changes.

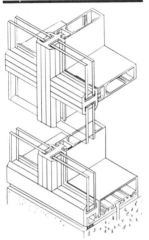

Thermopane

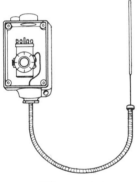

Thermostat

thermal transmittance, U-value The measure of the rate of heat flow per unit area under steady conditions from the fluid on the warm side of a barrier to the fluid on the cold side, per unit temperature difference between the fluids.

thermal unit A unit of heat energy, usually the British Thermal Unit in the English system or the calorie in the metric system.

thermal valve A valve with an activating element that responds to temperature or rate of temperature change.

thermomechanical pulping (TMP) A process in which wood chips are heated and softened by steam before being ground into fibers.

thermometer well A specially designed enclosure connected into a piping system and into which a thermometer can be placed to measure the temperature of a fluid in the piping.

Thermopane Trade name of a double-glazed panel for doors or windows, used for thermal insulation.

thermoplastic Becoming soft when heated and hard when cooled.

thermosetting Becoming rigid by chemical reaction and not remeltable.

thermostat An electric switch controlled by an element that responds to temperature, used in heating and/or cooling systems.

thermostatic expansion valve A valve regulating the flow of volatile refrigerant to a cooling unit, activated by changes in cooling unit pressure and superheat of the refrigerant leaving the cooling unit.

thermostatic trap A steam trap using a thermally actuated element to expand and close a discharge port when a designed amount of steam flows through it, and to

contract and allow condensate to flow through as the temperature drops; usually used on steam radiators.

thickness gauge See **feeler gauge**.

thick panels Plywood panels 5/8-inch and thicker.

thimble (1) A protective sleeve in a part intended to hold an item supported by or passing through the part. (2) A protective sleeve of metal in the wall of a chimney used to hold the end of a stovepipe or smoke pipe.

thin butt A defect in Western Red Cedar shakes in which the butt end of the shake fails to meet minimum thickness requirements.

thin edge A defect in shakes in which the thickness of 24-inch shakes, within 10 inches of the butt, is less than half the minimum specified thickness.

T-hinge, tee-hinge A T-shaped, surface mounted hinge. The cross arm of the T is fixed to a door frame or post while the length of the T supports a door or gate.

thinner Any volatile liquid used to lower the viscosity of a paint, adhesive, or other like material.

thin-set Descriptive of bonding, materials for tile which are applied in a layer approximately 1/8 inch or 3 mm thick.

thin-shell concrete Reinforced or prestressed concrete used to form a large shell. The thickness of the concrete is small relative to the span of the shell.

thin-shell precast Precast concrete characterized by thin slabs and web sections. See **shell construction**.

thin-wall conduit Electric conduit with a wall thickness that will not support threads. Sections are joined by couplings held in place by set screws.

Third Clear The highest grade of shop lumber. Officially called Factory Select or #3 Clear, this grade will yield a high percentage

Three-wire System

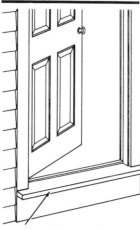
Threshold

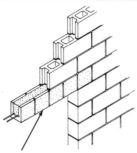
Through Lintel

of cuttings but represents a small portion of a mill's total shop production.

thixotropy The property of a material that enables it to stiffen quickly while standing, but acquire lower viscosity with mechanical agitation, such as certain gels. Material having this property is called thixotropic. See **rheology**.

thoroughly air dried (TAD) Lumber that is air dried sufficiently to meet the grading rule requirements for dry lumber.

thread A ridge, of uniform cross section, following a helix on the external or internal surface of a cylinder.

threaded anchorage An anchorage device which is provided with threads to facilitate attaching the jacking device and to effect the anchorage.

three-coat work The application of three coats of plaster, scratch coat, brown coat, and finish coat.

three-hinged arch A glue-laminated wood arch designed in two parts which are pin connected, or hinged, to each other and to their supports. Such arches are designed for convenience of transportation and erection.

three-point lock An assembly that latches the active leaf or a pair of doors at three points.

three-quarter bat See **three-quarter brick**.

three-quarter brick A brick which has a length equal to approximately three-quarters of that of a normal brick.

three-quarter closer See **king-closer**.

three-quarter header A header the length of which is approximately three-quarters of the thickness of the wall.

three-quarter turn Descriptive of a stair which turns through 270 degrees in its progress from top to bottom.

three-quarter view A view of an object which is midway between a front view and a side view.

three-way strap A metal strap used to tie three members of a wood truss together at a joint.

three-way switch An electric switch used to control lights from two different points, as from two different ends of a hallway.

three-wire system A system of electric power supply consisting of three conductors, one of which, the neutral wire, is maintained at a potential midway between the other two.

threshold (1) A shaped strip on the floor between the jambs of a door, used to separate different types of flooring or to provide weather protection at an exterior door. (2) The level of lighting or volume of illumination which permits an object to be seen a specified percentage of the time with specified accuracy.

throat (1) See **chimney throat**. (2) A groove cut in the underside of an exterior projecting piece, such as a sill or coping, to prevent water running back across the underside to a wall.

through bolt A bolt which passes completely through the members it connects.

through bond The transverse bond formed by masonry units extending through a wall.

through lintel A lintel having thickness equal to that of the wall in which it is placed.

through shake A shake extending through the thickness of the timber.

through stone A bond stone which extends through the full thickness of a wall.

through tenon A tenon extending completely through the part in which the mortice is cut.

through-wall flashing A flashing which extends completely through a wall, as at a parapet.

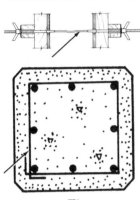

Tie

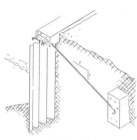

Tieback

throw (1) The distance a latch bolt extends. (2) The horizontal or vertical distance an airstream travels after leaving an outlet until its velocity is reduced to a specific value. (3) The effective distance between a fixture and the area being illuminated. (4) The scattering of fragments from a blast. (5) The longest straight distance traveled in a complete stroke of a rotary part.

throwout bearing A bearing which slides on the clutch jackshaft, that carries the engage-and-disengage mechanism.

thrust (1) The amount of force or push exerted by or on a structure, sometimes the horizontal component of that force. (2) In an arch, the resultant force normal to any cross section of the arch.

thrust bearing A support for a shaft which is designed to resist its end thrust.

thumb knob See **turn knob**.

thumb latch See **latch**.

thumb nut See **wing nut**.

thumb piece A small pivoted part above a door handle, pressure on this part by a thumb operates the latch.

thumbscrew A screw which has a head that is either curled or flattened so it can be turned with a thumb and fingers.

Tideland Spruce Sitka Spruce.

Tidewater Red Cypress Bald Cypress.

tie (1) Loop of reinforcing bars encircling the longitudinal steel in columns. (2) A tensile unit adapted to holding concrete forms secure against the lateral pressure of unhardened concrete, with or without provision for spacing the forms a definite distance apart, and with or without provision for removal of metal to a specified distance from the finished concrete surface. See **cross tie**.

tieback A rod fastened to a deadman, a rigid foundation, or a rock or soil anchor to prevent lateral movement of formwork, sheet pile walls, retaining walls, and bulkheads.

tie bar (1) Bar at right angles to and tied to minimum reinforcement to keep it in place. (2) Bar extending across a construction joint.

tie beam (1) A concrete beam that connects individual pile caps or spread footings. (2) A horizontal timber that connects the lower end of two opposite rafters to prevent spreading.

tied column A column laterally reinforced with ties.

tie iron See **wall tie**.

tie rod See **form tie** and **tieback**.

tier See **lift**.

tie wall A wall built perpendicular to a spandrel wall for lateral stability.

tie wire (1) A wire used to hold forms together so they will not spread when filled with concrete. (2) A single-strand wire used to tie reinforcing in place or metal lath to a column.

tight knot A knot in a piece of lumber which is sound and poses no detriment to the proper use of the wood member.

tight sheathing (1) Tongue-and-groove or matched boards nailed to rafters or studs which may run at an angle to provide stiffness to the roof or wall. (2) Excavation sheathing with the vertical planks interlocked for use in saturated soils.

tight sheething See **closed sheeting**.

tile A thin rectangular unit used as a finish for walls, floors or roofs, such as ceramic tile, structural clay tile, asphalt tile, cork tile, resilient tile, and roofing tile.

tile-and-a-half tile A roof tile which is the same length as the

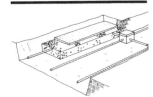

Tile Field

Tilting Mixer

Tilt-up

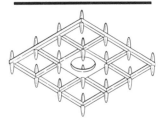

Timber Connector

other tiles on a roof, but 1 1/2 times as wide.

tileboard (1) A wall board with a factory applied facing which is hard, glossy, and decorated to simulate tile. (2) A square or rectangular board of compressed wood or vegetable fibers, used for ceiling or wall facings.

tile creasing A water shedding barrier at the top of a brick wall consisting of two courses of tile which project beyond both faces of the wall.

tile field A system of distribution tile.

till, glacial till, boulder clay An unstratified glacial deposit consisting of compacted pockets of clay, silt, sand, gravel, and boulders, usually having good bearing capacity.

tilting concrete mixer See **tilting mixer**.

tilting drum mixer See **tilting mixer**.

tilting fillet, cant strip, doubling piece, tilting piece A thin wedge placed under the eave course of shingles or tiles to shed water more effectively.

tilting level A surveyor's level with a bubble mounted on the telescope and a provision for slight tilting of the telescope and level. The upright axis of the unit does not need to be vertical but the level and telescope must be precisely aligned.

tilting mixer A small mixer for concrete or mortar that is emptied by tilting the mixer about a horizontal pivot.

tilt-up, tilt-up construction A method of concrete construction in which members are cast horizontally at a location adjacent to their eventual position and tilted into place after removal of molds.

timber (1) Uncut trees or logs suitable for cutting into lumber. (2) Wood sawn into balks and

planks, suitable for use in carpentry or construction. (3) Square-sawn lumber having a minimum nominal dimension of 5 inches in USA and approximately equal cross dimension greater than 4 x 4 1/2 inches or 10 x 11 cm in Britain. (4) Any heavy wood beam used for shoring or bracing.

timber connector One of a variety of metal connectors used in conjunction with bolts to form connections of timbers. Usually the bolt holds the timbers together while the connector prevents slippage.

timber-framed building A building which has timbers for structural elements except foundations.

time Term defined in reference to a construction contract as time limits or periods stated in the contract. A provision in a construction contract that "time is of the essence of the contract" signifies that the parties consider punctual performance within the time limits or periods in the contract to be a vital part of the performance. Failure to perform on time is a breach for which the injured party is entitled to damages in the amount of loss sustained.

time-delay fuse A fuse in an electric circuit that takes more than 12 seconds to open at 200% load.

time-dependent deformation Combined effects of autogenous volume change, contraction, creep, expansion, shrinkage, and swelling occurring during an appreciable period of time, not synonymous with inelastic behavior or volume change.

timekeeper A representative of the contractor who keeps records of hours worked by employees of the contractor and allocates the hours to various parts of the work, may

Tin-clad Fire Door

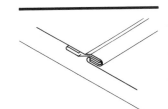

Tin Roofing

also keep records of work completed by subcontractors.

timely completion Completion of the work or designated portion thereof on or before the date required.

time of completion Date established in the contract, by calendar date or by number of days, for substantial completion of the work. See **date of substantial completion**.

time of haul In production of ready-mixed concrete, the period from first contact between mixing water and cement until completion of discharge of the freshly mixed concrete.

time of set See **initial setting time** and **final setting time**.

time of setting See **initial setting time** and **final setting time**.

time system A system of clocks and control devices, with or without a master timepiece, which will display current time at various locations and may include devices to program other systems, such as bells.

tin (1) A lustrous white, malleable metal with a low melting point, highly resistant to corrosion, used to make alloys and solder, and to coat sheet metal. (2) To coat with a thin layer of tin or other protective metal.

tinbender A sheet metal worker.

tin-canning See **oil-canning**.

tin cap A small, flat metal washer used under roofing nails.

tin-clad fire door A door of two or three plywood plank core covered with metal sheets and constructed in accordance with specifications of labeling authorities.

tingle A flexible metal clip designed to hold a sheet of metal or glass.

tin-knocker A sheet metal worker.

tinning Coating metal with a tin alloy for corrosion protection or as a presoldering procedure.

tinplate Sheet steel or iron which has been coated with tin as protection against corrosion.

tin roofing A roof covering of tinplate or terneplate.

tin snips Strong shears with a blunt nose, used to cut sheet metal.

tint A light color made by diluting a color with white.

T-iron See **tee-iron**.

title Legal documents which indicate right of ownership of real property.

title insurance Insurance, offered by a company, that a title to property is clear or that it can be cleared by resolving certain defects.

title search A search into the historical ownership record of a property to establish its true ownership and check the existence of liens or easements which might affect the sale of the property.

T-joint See **tee joint**

tobermorite A mineral found in northern Ireland and elsewhere, having the approximate formula Ca (Si O H). Ca:4H O identified approximately with the artificial product tobermorite, G, of Brunauer, a hydrated calcium silicate having CaOSiO ratio in the range 1.39 to 1.75 and forming minute layered crystals that constitute the principal cementing medium in portland cement concrete, a mineral with 5 mols of lime to 6 mols of silica, usually occurring in plate-like crystals which is easily synthesized at steam pressures of about 100 psi and higher. Tobermorite is the the binder in several properly autoclaved products.

tobermorite gel The binder of concrete cured moist or in atmospheric-pressure steam, a lime-rich gel-like solid containing

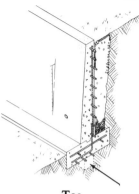

Toe

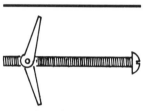

Toggle Bolt

Toilet Enclosure

Tongue-And-Groove Joint

1.5 to 2.0 mols of lime per mol of silica.

toe (1) Any projection from the base of a construction or object to give it increased bearing and stability. (2) That part of the base of a retaining wall that projects beyond the face away from the retained material. (3) The lower portion of the lock stile. (4) The junction between the base metal and the face of a filled weld. (5) To drive a nail at an oblique angle. (6) That portion of sheeting below the excavated material. (7) The part of a blasting hole furthest from the face.

toeboard As defined by OSHA, a vertical barrier at floor level erected along exposed edges of a floor opening, wall opening, platform, runway or ramp to prevent falls of materials.

toed Descriptive of a board or strut that has its ends secured by nailing at an oblique angle.

toehold A batten or board temporarily nailed to a sloping roof as a footing for workmen.

toe joint A joint between a horizontal timber and another angle from the horizontal, as between a rafter and a plate.

toenail To drive a nail at an angle.

toenailing Fastening a piece of lumber by driving nails obliquely to the surface, alternate nails may be opposing to increase holding power.

toeplate (1) A metal bar fastened to an outer edge of grating or the rear of a tread, and projecting above the surface to form a lip. (2) See **kickplate**.

toe wall A low wall built at the bottom of an embankment for greater stability.

toggle bolt A bolt and nut assembly used to fasten objects to a hollow wall or a wall accessible from only one side. The nut has pivoted wings that close against a spring when the nut end of the

assembly is pushed through a hole and open on the other side.

toggle switch A lever actuated snap switch.

toilet (1) The room housing one or more water closets. (2) See **water closet**.

toilet enclosure A compartment placed around a water closet for privacy.

toilet partition One of the panels forming a toilet enclosure.

tolerance (1) The permitted variation from a given dimension or quantity. (2) The range of variation permitted in maintaining a specified dimension. (3) A permitted variation from location or alignment.

tom See **shore**.

ton A measure of weight equal to 2,000 pounds or 907.2 kg. Also see **metric ton**.

Tonawanda Pine *Pinus strobus*. Eastern White Pine.

tongue One edge of a piece of lumber that has been rabbeted from opposite faces, leaving a projection intended to fit into a groove cut into another board.

tongue and groove (1) Lumber machined to have a groove on one side and a protruding tongue on the other, so that pieces will fit snugly together, with the tongue of one fitting into the groove of the other. (2) A type of lumber or precast concrete pile having mated projecting and grooved edges to provide a tight fit, abbreviated "T & G".

tongue-and-groove joint, T and G joint A joint made by fitting a projecting rib on one piece into a groove in another. If the pieces are metal, the joint may also be welded. For plastic or wood pieces, the joint may be glued.

tongue-and-groove material See **dressed-and-matched-boards**.

tongue-and-lip joint A type of tongue-and-groove joint except

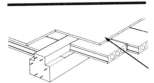

Topping

the tongue is wedge shaped and the groove is tapered to receive it.

tongued miter A miter joint incorporating a tongue.

tongue joint Similar to a tongue-and-groove joint except the tongue is wedge shaped and the groove is tapered to receive it.

ton of refrigeration A measure of refrigerating effect equal to 12,000 Btu per hour.

tooled ashlar Ashlar masonry with a tooled finish.

tooled finish, tooled surface A stone finish with the surface having 2 to 12 concave grooves per inch or 5 to 30 grooves per centimeter.

tooled joint A masonry joint in which the mortar has been shaped or worked before it sets.

tooling (1) Shaping and compacting a mortar joint. (2) The work involved in making a tooled finish. (3) See **batted work**.

tool pad A tool, consisting of a handle and a chuck, for holding small tool bits, such as awls or screwdriver blades. The handle may be hollow for storage of a collection of bits.

tooth A fine texture in a paint film provided by pigments or by abrasives used in sanding, provides a base for adhesion of a second coat.

toothed plate, bulldog plate A punched metal plate in which the punched metal protrudes from one side forming teeth. Toothed plates are used for timber connections.

toothed ring A toothed ring that serves as a timber connector, generally used in the manufacture of large member wood trusses. See **toothed plate**.

toothing Cutting or chipping out courses in old work as a bond for new work.

toothing plane A plane with a serrated blade mounted nearly vertical, used to roughen a surface prior to application of a veneer.

top-and-bottom cap One of the metal channels attached, on the jobsite, to the top or bottom of a hollow metal door that is not so finished at the factory.

top beam A collar beam.

top car clearance The clearance between the top of an elevator car, or crosshead if provided, and the lowest overhead obstruction when the car is level with the top terminal landing.

topcoat The final coat in a paint system.

top cut The vertical cut at the top of a rafter.

top dressing A layer of manure, humus, or loam placed thin to improve soil conditions in planted areas.

top form Form required on the upper or outer surface of a sloping slab or thin shell.

topographic survey The configuration of a surface, including its relief and the locations of its natural and man-made features, usually recorded on a drawing showing surface variations by means of contour lines indicating height above or below a fixed datum.

top out To install the highest structural member or complete the highest course in a construction.

topping (1) A layer of concrete or mortar placed to form a floor surface on a concrete base. (2) A structural, cast-in-place surface for precast floor and roof systems. (3) The mixture of marble chips and matrix which, when properly processed, produces a terrazzo surface.

topping joint A joint, in a topping layer, which is directly over a joint in the base material.

Top Plate

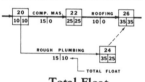

Total Float

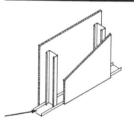

Track

top plate A member on top of a stud wall on which joists rest to support an additional floor or form a ceiling.

topsoil (1) The surface layer of soil, usually contains organic matter. (2) See **loam**.

torch brazing A brazing process in which the heat is supplied by a torch.

torching The application of a lime mortar under the up-slope edges of roof tiles or slates. In full torching the exposed under side of slates between battens is mortared.

torch soldering Soldering with a gas flame supplying the required heat.

torn grain Part of the wood torn out in surfacing. Torn grain is classified for grading purposes as slight, medium, heavy, or deep.

torque (1) Turning or twisting energy, measured as the product of a force and a lever arm. (2) That which tends to produce rotation.

torque viscometer An apparatus used for measuring the consistency of slurries in which the energy required to rotate a device suspended in a rotating cup is proportional to viscosity.

torsel A piece of timber, steel, or stone placed under one end of a beam or joist to distribute its load.

torsion The twisting of a structural member by two equal and opposite torques.

torsional strength The resistance of a material to twisting about an axis.

torus roll A joint, in sheet metal or lead roofing, at the intersection of two surfaces with different slopes. The joint is formed to allow for differential movement.

total float In CPM terminology, the difference between the time available to accomplish an activity and the estimated time required.

total rise of a roof The vertical distance between the plate and the ridge of a roof.

total run The horizontal span of a rafter.

touch sanding A light surface sanding.

Tough Ash *Fraxinus excelsior.* European White Ash, widely used in bent-wood products.

toughness The property of matter which resists fracture by impact or shock.

tower A composite structure of frames, braces, and accessories.

tower crane A crane with a fixed vertical mast which is topped by a rotating boom, equipped with a winch for hoisting and lowering loads. The winch can be moved along the boom so that any location within the diameter of the boom can be reached.

T-plate A flat metal plate in the shape of a "T", used to join two timbers, one of which butts against the other, at a right angle, or to strengthen such a joint.

trabeated system A system of building construction using beams or lintels supported by columns.

trabeation A type of construction utilizing horizontal members supported by columns. Trabeation is a common method of construction in Japan. It is also called "entablature"

tracing cloth A smooth linen fabric impregnated and coated with size to make it transparent, used for tracing.

track (1) A light gauge U-shaped metal member attached to a floor and used to anchor studs for a partition. (2) A U-shaped member attached to a floor, ceiling, or door or window header and used as a guide for a sliding or folding partition, door, or curtain. (3) A pair of special structural

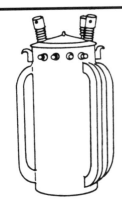

Transformer

shapes with fastenings or ties for a craneway, moveable wall, or railroad.

track roller In a crawler machine, the small wheels which are under the track frame and which rest on the track.

tractor A vehicle on tracks or wheels used for towing or operating equipment.

trade (craft) (1) Occupation requiring manual skill.
(2) Members of a trade organized into a collective body.

trade discount The difference between the seller's list price and the purchaser's actual cost, excluding discounts for prompt payment.

traffic cone A pliable and highly visible cone used as a temporary marker to direct traffic away from a work area.

traffic paint Paint formulated to withstand vehicular traffic and to be highly visible at night, used to mark traffic lanes and pedestrian crossings.

trailer on flat car (TOFC) A truck trailer carried on a flat car, piggyback.

train A string of connected or unconnected vehicles or mobile equipment, such as a paving train, which consists of mobile machines to lay the various courses of a pavement.

trajectory of prestressing force The path along which the prestress is effective in a structure or member. It is coincident with the center of gravity of the tendons for simple flexural members and statically indeterminate members which are prestressed with concordant tendons, but it is not coincident with the center of gravity of the tendons of a statically indeterminate structure which is prestressed with nonconcordant tendons.

transducer A substance or device that converts input energy into output energy of a different form, such as a photoelectric cell.

transfer The act of transferring the stress in prestressing tendons from the jacks or pretensioning bed to the concrete member.

transfer bond In pretensioning, the bond stress resulting from the transfer of stress resulting from the tendon to the concrete.

transfer case A transmission or gear set in an all-wheel drive vehicle that provides drive to the front shaft.

transfer column A column, in a multistory framed building that is not continuous to the building foundation. At some floors the column is supported by a girder or girders and its load transferred to adjacent columns.

transfer girder A girder which supports a transfer column.

transfer grille A grill or pair of grills that allow air to move from one space to another, installed in locations such as a wall or door.

transfer length See **transmission length**.

transfer molding An injection molding using a thermosetting material.

transfer strength In prestressed concrete, the concrete strength required before stress is transferred from the stressing mechanism to the concrete.

transformed section A hypothetical section of one material arranged so as to have the same elastic properties as a section of two materials.

transformer An electric device with two or more coupled windings, with or without a magnetic core, for introducing mutual coupling between circuits, generally used to convert a power supply at one voltage to another voltage.

Transit Mix

Transom

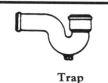

Trap

transit A surveyor's instrument used to measure or lay out horizontal or vertical angles, or measure distance or difference in elevation.

transit-and-stadia survey A type of survey in which angles are measured with a transit and distances are measured with a transit and stadia rod.

transite A product sold in the form of flat and corrugated sheets, and pipe, made of asbestos reinforced cement, trade named "Transite".

transit line Any line of a survey traverse projected by use of a transit or similar instrument.

transit mix Concrete which is wholly or mainly mixed in truck mixer, usually while in transit to the job site.

transit-mixed concrete Concrete, the mixing of which is wholly or principally accomplished in a truck mixer.

transit-mix truck See **truck mixer**.

translucent Descriptive of a material that transmits light but diffuses it sufficiently that an object can not be seen clearly through the material.

transmission A gear set or similar device that permits changes in speed/power ratio and/or direction of rotation.

transmission length The distance at the end of a pretensioned tendon necessary for the bond stress to develop the maximum tendon stress, sometimes called transfer length.

transom A glazed or solid panel over a door or window, usually hinged and used for ventilation. The transom and bar may be removable for passage of large objects.

transom bar (1) A horizontal member separating a transom from the door or window. (2) Any intermediate horizontal member of the frame of a door or window.

transom bracket A bracket that supports an all glass transom over an all glass door.

transom catch A lock, used on a transom, with the latch bolt operated by a ring shaped part. It may be opened using a pole with a hook on its end.

transom lift A linkage system attached to a doorframe and used to open a transom.

transom light A glazed light above the transom bar of a door.

transverse cracks Cracks that develop at right angles to the long direction of the member.

transverse joint A joint parallel to the intermediate dimension of a structure.

transverse load A load applied at right angles to the longitudinal axis of a structural member, such as a wind load.

transverse prestress Prestress that is applied at right angles to the principal axis of a member.

transverse reinforcement Reinforcement at right angles to the principal axis of a member.

transverse rib A rib in vaulting that spans the nave, cross aisle or an aisle, at right angles to the longitudinal axis of the area spanned.

transverse shear A shearing action or force perpendicular to the main axis of a member.

transverse strength See **flexural strength** and **modulus of rupture**.

trap (1) A plumbing fixture so constructed that, when installed in a system, a water seal will form and prevent backflow of air or gas

Travel, Rise

Tread Width

but permit free flow of liquids. (2) A removable section of stage floor. (3) See **traprock**.

trapdoor A door set in a floor, ceiling or roof.

traprock Any of various fine-grained, dense, dark-colored igneous rocks, typically basalt or diabase, also called "trap".

trap seal The vertical distance between the crown weir and the dip of a trap fixture, the head, expressed linearly, resisting back pressure.

trash chute (1) A smooth clear shaft in a multistory building, used to convey trash from upper floors to a collection room. (2) A temporary chute used for trash removal during the construction of a multistory building.

trash rack A grid of metal bars placed in front of an inlet for water to collect larger solids transported by the water.

trass A natural pozzolan of volcanic origin found in Germany.

T-rated switch A switch rated in accordance with the National Electric Code for use in circuits containing tungsten-filament lamps.

trave (1) A beam or timber crossing a building. (2) A panel in a ceiling delineated by beams or timbers.

travel, rise The vertical distance between the bottom landing of an elevator or escalator and the top landing.

traveler An inverted-U shaped structure usually mounted on tracks which permit it to move from one location to another to facilitate the construction of an arch, bridge, or building.

traveling cable A cable, made of electric conductors, which connects an elevator or dumbwaiter car with a fixed electric outlet in the hoistway.

traveling crane A tower crane

mounted on tires, crawlers, or rails.

traveling form See **slip form**.

travel time Wages paid to workmen under certain union contracts and under certain job conditions for time spent traveling between their home and the work site.

traverse (1) To plane wood across the grain. (2) A structural crosspiece, such as a transom bar. (3) A gallery or left crossing a building. (4) A barrier to allow passage by an official or dignitary but discourage unauthorized passage. (5) See **survey traverse**.

traverse closure The calculated line that closes a measured closed traverse, the resultant error of measurements.

travertine A variety of limestone deposited by running water, usually stratified, used for interior walls and for floors.

tray ceiling A horizontal ceiling constructed part way up the slope of a gabled roof.

tread The horizontal part of a stair. Historically, treads have been made from 5/4x12" vertical grain lumber called stepping. In recent years, however, most stepping has been made from particleboard that is given a bullnosed edge.

tread length The length of a tread measured perpendicular to the travel line of the stair.

tread plate A fabricated metal tread with a slip resistant surface.

tread return The projection of a tread beyond the stringer of an open stair.

tread run The horizontal distance between nosings of stair treads.

tread width The horizontal distance from the nosing of a stair tread to the riser above the tread or the tread run plus nosing.

treated Wood products infused or coated with any of a variety of stains or chemicals designed to

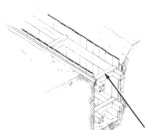

Trench Brace

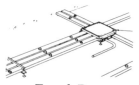

Trench Duct

Trenching Machine

retard fire, decay, insect damage, or deterioration due to weather.

treated wood (1) Wood which has been treated with a wood preservative. (2) See **fire retardant wood**.

tree belt A strip of soil between a sidewalk and a street, planted with grass, and sometimes, trees.

tree-dozer An attachment, for a tractor or bulldozer, consisting of metal bars and a cutting blade, used to clear bushes and small trees.

tree grate A metal grating set around a tree and flush with a pavement.

treenail A hardwood pin formerly used as a wood fastener.

trellis A latticework of wood or metal, usually used to support vines.

tremie A pipe or tube through which concrete is deposited under water, having at its upper end a hopper for filling and a bail for moving the assemblage.

tremie concrete Subaqueous concrete placed by means of a tremie.

tremie seal The depth to which the discharge end of the tremie pipe is kept embedded in the fresh concrete that is being placed. The tremie seal is a layer of tremie concrete placed in a cofferdam for the purpose of preventing the intrusion of water when the cofferdam is dewatered.

trenail See **tree nail**

trench box, trench shield Box shaped sheathing made of wood or steel, permanently braced across a trench for excavation and pipe laying. The trench box unit is pulled along the trench as excavation and pipe laying proceeds.

trench brace A device, usually adjustable, used as crossbracing to support sheeting in a trench.

trench duct A trough, with

removable covers, in which are run electric power and control cables. It can be a metal unit that is set in concrete or formed in a concrete slab. The top of the covers are level with the floor.

trencher See **trench excavator**.

trench excavator A self-propelled machine with a side-mounted shovel or chain of buckets, used to excavate trenches.

trenching machine See **trench excavator**.

trench jack A hydraulic or screw jack used as a cross brace in a trench bracing system.

trench shield See **trench box**.

trestle (1) A framework used to support a bridge. (2) See **sawhorse**.

trial batch A batch of concrete prepared to establish or check proportions of the constituents.

trial pit A small pit dug to investigate the soil, sometimes dug to bed rock or other dense material.

triangular scale A drafting instrument which is triangular in cross section and has a different scale on each edge. Some edges have two opposing scales, one two times the other, such as 1/8 and 1/4 inches equals 1 foot.

triangular truss A light wood, roof truss used for short spans.

triangulation A method of surveying over long distances by establishing a network of triangles. Most sides in the network are computed from a known side which may be calculated, and two measured angles. Lengths are measured periodically as a check.

triaxial compression test A test whereby a specimen is subjected to a confining hydrostatic pressure and then loaded axially to failure.

triaxial test A test in which a specimen is subjected simultaneously to lateral and axial loads.

Trim

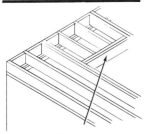

Trimmer

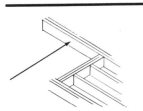

Trimming Joist

tricalcium aluminate A compound having the composition 3CaO.A O , 12 3 abbreviated C A.

tricalcium silicate A compound having the composition 3CaO.SiO ,2 abbreviated C3S, an impure form of which (alite) is a main constituent of portland cement. See **alite**.

trilateration A method of surveying similar to triangulation except that distances are measured by electronic instruments and the two angles are calculated. Quadrilaterals are used in addition to triangles.

trim Millwork, primarily moldings and/or trim to finish off and cover joints around, window and door openings.

trim band A metal strip used as a closing band on sides or ends of grating panels.

trim hardware Decorative finish hardware which is functional or used to operate functional hardware.

trimmed opening See **cased opening**.

trimmer (1) A short beam that supports one or more joists or beams at an opening in the floor, a header. (2) A beam or joist inserted in a floor on the long side of a stair opening and supporting a header. (3) Shaped ceramic tile used as bases, caps, corners, moldings, and angles.

trimming joist A joist, parallel to the common joists but of larger, cross section, possibly two pieces nailed together, that supports a trimmer.

trimming rafter A rafter supporting the end of a header. See **trimming joist**.

trimstone, trim Decorative masonry members on a structure build or faced largely with other masonry, includes sills, jambs, lintels, coping, cornices, and quoins.

trip A device to release a mechanism, such as a pawl. A trip is a release catch.

trip coil A coil used in the electromagnet that activates a trip.

tripod A three-legged adjustable stand for an instrument.

trivet A low support for a surveying instrument used where a tripod can not be used.

troffer A long, recessed lighting unit, usually installed at the ceiling line.

trolley beam An exposed steel beam on the underside of a structure, used to support a trolley crane.

trough A channel used to contain electric power or control cables.

trough gutter See **box gutter**.

trough mixer See **open-top mixer**.

trough roof See **M-roof**.

trowel (1) A flat, broad-blade steel hand tool used in the final stages of finishing operations to impart a relatively smooth surface to concrete floors and other unformed concrete surfaces. (2) Also a flat triangular-blade tool used for applying mortar to masonry.

trowel finish The smooth finish surface produced by troweling.

troweling Smoothing and compacting the unformed surface of fresh concrete by strokes of a trowel.

troweling machine A motor driven device that operates orbiting steel trowels on radial arms from a vertical shaft.

truck crane A crane mounted on a wheeled vehicle.

truck-mixed concrete See **transit-mixed concrete**.

truck mixer A concrete mixer suitable for mounting on a truck chassis and capable of mixing

Truss

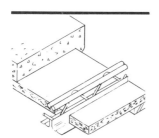

Trussed Joist

Trussed Purlin

**Tubular-Welded-Frame
Scaffold**

concrete in transit. See
**inclined axis mixer,
open-top mixer,** and **agitator.**

true bearing The clockwise angle
between a line and a meridian
which is referenced to the
geographic North Pole.

True Cedar *Cedrus libanotica.*
Lebanon Cedar.

true firs A collective term for a
group of firs of the species *Abies,*
including White Fir and Grand
Fir.

true wood Heartwood.

trunk (1) The main wood shaft of
a tree. (2) The shaft portion of a
column. (3) Descriptive of the
main body of a system, as a trunk
sewer.

trunk sewer A main sewer which
receives flow from many
tributaries, covering a large area.

trunnion A pivot consisting of
two cylinders or pins projecting
from the body of the pivoted
object.

truss A structural component
composed of a combination of
members, usually in a triangular
arrangement, to form a rigid
framework; often used to support
a roof.

truss beam See **trussed beam.**

trussed beam A beam, usually
timber, that has been reinforced
by a center post beneath the beam
and two rods running from the
bottom of the post to the ends of
the beam.

trussed joist A joist in the form of
a truss, an open web joist. See **bar
joist.**

trussed partition A framed
partition which is free standing
and does not depend on
intersecting panels for support.

trussed purlin A lightweight
trussed beam used as a purlin.

trussed-rafter roof A roof system
in which the cross framing

members are some form of light
wood truss.

trussed ridge roof A pitched roof
in which the upper support for
the rafters is a truss.

trussed-wall opening Any
opening in a framed structure with
a truss system used to span the
opening.

truss rod (1) A metal rod used as a
tension member in a truss. (2) A
metal rod used as a diagonal tie.

try square A square whose legs are
at 90 degrees usually graduated.

T-shore A shore with a T-head.

T-square, tee square A guide, in
the shape of a "T", used in
engineering and architectural
drawing. The short arm slides
along the edge of a drawing board
keeping the long arm in a parallel
state.

tube-and-coupler scaffold A
scaffold system using tubes for
posts, bearers, braces, and ties and
special couplers to connect the
parts.

tube-and-coupler shoring A
load-carrying assembly of tubing
or pipe which serves as posts,
braces, and ties, a base supporting
the posts, and special couplers
which connect the uprights and
join the various members.

tubing Any material in the form of
a tube.

tub mixer See **open-top mixer.**

tubular saw See **crown saw.**

tubular scaffolding Scaffolding
manufactured from galvanized
steel or aluminum tube and
connected by clamps.

tubular-welded-frame scaffold A
scaffold system using prefabricated
welded sections that serve as posts
and horizontal bearers. The
prefabricated sections are braced
laterally with tubes and bars.

tuck A recess in a horizontal
masonry joint formed by raking

Turnbuckle

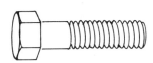

Turned Bolt

out mortar in preparation for tuck pointing.

tuck and pat pointing See **tuck pointing**.

tuck-in That part of a counterflashing, skirting, or roofing felt that is inserted in a reglet.

tuck pointing A method of refinishing old mortar joints. The loose mortar is dug out and the tuck is filled with fine mortar which is left projecting slightly or with a fillet of putty or lime.

tulipwood (1) A soft, close-textured wood, yellowish in color, used for veneers and millwork. (2) A very hard wood, rose-colored, used in inlay work.

tumbler The mechanism in a lock holding the bolt until operated by a key.

tumbler switch A lever operated electric snap switch.

tumbling course A sloping course of brickwork which intersects a horizontal course.

tungsten-halogen lamp An incandescent lamp which consists of a tungsten filament, a gas containing halogens and an envelope of a high-temperature resistant material such as quartz. The lamp is small compared with lamps of similar wattage.

tungsten steel A very hard, heat resistant carbon steel containing tungsten.

Tupelo *Nyssa aquatica*, Water Tupelo, *N. sylvatica*, Black Tupelo. A light, tough hardwood of the southern and eastern United States, used for some paneling and turnings.

turbidimeter A device for measuring the particle-size distribution of a finely divided material by taking successive measurements of the turbidity of a suspension in a fluid.

turbidimeter fineness The

fineness of a material such as portland cement, usually expressed as total surface area in square centimeters per gram, as determined with a turbidimeter. See **Wagner fineness**.

turbine Any of various machines which convert the kinetic energy of a moving fluid to mechanical energy. The turbine is often used for driving an electric generator.

turbine mixer See **open-top mixer**.

turf The uppermost layer of soil containing roots of grass.

turning angles The process of measuring angles between lines with a transit or similar instrument.

turnbuckle A device for adjusting the length of a rod or cable, consisting of a right screw and a left screw coupled by a link.

turned bolt A machine bolt, usually with a hexagonal head, the shank of which is finished to a close tolerance.

turned work Pieces of stone or wood work having a circular cross section, such as posts and balusters. Turned work is usually cut on a lathe.

turning Shaping objects by use of cutting tools while the piece to be shaped is rotated on a lathe.

turning piece A wood template used by a mason to form a small arch which does not require centering.

turnkey contract A contract similar to design and construct except the contractor is responsible for all financing and owns the work until the project is complete and turned over to the owner.

turnkey job A project constructed under a turnkey contract.

turn knob A small doorknob on the inside of a door, used to control the bolt.

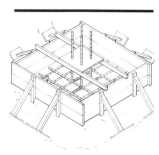

Two-way Footing

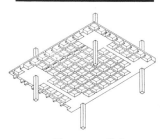

Two-way Joist

turn piece A small knob used to control a dead bolt from the inside of a door, usually crescent or oval shaped for gripping with thumb or fingers.

turnstile A barrier rotating on a vertical axis, usually allowing movement in one direction only and admitting one person at a time. A turnstile is sometimes coin or token operated.

turn tread A wedge shaped tread used where a stair changes direction.

turnup That edge of roofing material turned up along a vertical surface.

turpentine A thin, volatile oil obtained by steam distillation from the wood or exuded resin of certain pine trees. Turpentine was once widely used as a paint thinner and solvent, but now replaced by solvents derived from petroleum.

twin brick A double-sized brick.

twin cable A cable consisting of two parallel insulated conductors fastened side-by-side through the insulation or by common wrapping.

twin-filament lamp An incandescent lamp with two filaments that are wired independently, used as double-function lamps as in automobile stop lights or as three level wattage lamps.

twin pug A mill with two blade-supporting shafts turning in opposite directions, a pugmill.

twin tenons See **double tenons**.

twist drill A drill with one or more helical cutting grooves, used to drill holes in metal, wood, and plastic.

two-by-four A commonly used piece of timber with nominal dimensions 2 in. thick by 4 in. wide or 5 cm thick by 4 cm wide.

two-coat work The application of two coats of plaster, a base coat followed by a finish coat.

two-core block A concrete masonry unit with two hollow cells.

two-four-one (2-4-1) Structural wood panels, at least 1-1/8" thick, designed for single-floor applications over joists spaced 48 inches apart and also used as roof sheathing in heavy timber construction. The term is synonymous with APA Rated Sturd-I-Floor and is a registered trade name of the American Plywood Association.

two-handed saw A large saw with two handles and intended to be worked by two men, used for felling trees or cross cutting logs, largely replaced by chain saws.

two-light window (1) A window which is two panes high or wide. (2) See **gemel window**.

two-man rip rap Stones suitable for rip rap but of a size that can only be handled by two men.

two-part line A single strand, rope, or cable which is doubled back around a sheave to double the capacity.

two-point suspension scaffold See **swinging scaffold**.

two-stage curing A process whereby concrete products are cured in low-pressure steam, stacked, and then autoclaved.

two-way footing See **two-way reinforced footing**.

two-way joist construction Floor or roof construction in which the floor or roof is supported on two mutually perpendicular systems of parallel joists.

two-way reinforced footing A footing having reinforcement in two directions generally perpendicular to each other.

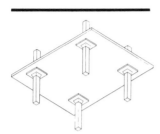

Two-way Slab

two-way reinforcement, two-way system A system of reinforcement. Bars, rods, or wires are placed at right angles to each other in a slab and are intended to resist stresses due to bending of the slab in two directions.

two-way slab A reinforced concrete slab in which the main reinforcing runs in two directions, parallel to the length and width of the panel.

two-way system See **two-way reinforcement**.

U ABBREVIATIONS

u unit

U uranium

UBC Uniform Building Code

UDC universal decimal classification

U/E unedged

uhf ultrahigh frequency

UL Underwriters' Laboratories, Inc.

ult ultimate

unins. uninsurable

uns unsymmetrical

up upper

ur, UR urinal

USASI American National Standards Institute

USG United States gauge

UV ultraviolet

U DEFINITIONS

U-bolt

U-bolt A bolt formed in the shape of the letter U, with threads on the ends to accommodate nuts.

UL Label A seal of certification attached by Underwriters' Laboratories, Inc. to building materials, electrical wiring and components, storage vessels, and other devices, attesting that the item has been rated according to performance tests on such products, is from a production lot that made use of materials and processes identical to those of comparable items that have passed fire, electrical hazard, and other safety tests, and is subject to the UL reexamination service.

ultimate bearing capacity The average load per unit of area required to cause the rupture and subsequent failure of a supporting mass.

ultimate design resisting moment The theoretical, applied bending moment which will cause failure in a reinforced concrete member through yield in the tensile reinforcing steel or crushing of concrete.

ultimate load (1) The maximum load a structure can bear before its failure due to buckling of column members or failure of some component. (2) The load at which a unit or structure fails.

ultimate set The final degree of firmness obtained by a plastic compound after curing.

ultimate shear stress The stress at a section loaded to its maximum in shear. See also **shear strength**.

ultimate strength The maximum resistance to load that a member or structure is capable of developing before failure occurs. With reference to cross sections of members, the largest moment, axial force, torsion, or shear material can sustain without failure.

ultimate strength design See **strength design method**.

ultrasonic soldering A soldering process, usually performed without a flux, wherein unwanted surface films are removed from the base metal by transmitting high frequency sounds through molten solder, resulting in a more effective welding of the base metal with the solder.

ultrasonic welding A process in solid-state welding wherein the metal parts are held together under pressure and joined by applying high frequency sound waves to their surface.

umbrella liability insurance Insurance providing excess liability coverage over existing liability policies such as employer's liability, general liability, or automobile liability, and providing direct coverage for many losses uninsured under the existing policies after a specified deductible is exceeded. See also **liability insurance.**

unbalanced bid A contractor's bid based on increased unit costs for tasks to be performed early and decreased unit costs for later tasks. The unbalanced bid is used in an attempt to get money early to finance later parts of a job.

unbonded member A post-tensioned, prestressed concrete element in which tensioning force is applied against end anchorages only, tendons being free to move within the elements.

unbonded posttensioning Post-tensioning in which the tendons are not grouted after stressing.

unbonded tendon A tendon not bonded to the concrete section.

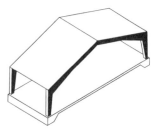

Unbraced Frame

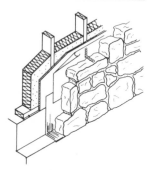

Uncoursed

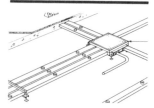

Underfloor Raceway

Underground Piping

unbraced frame A structural frame whose resistance to the lateral load carried is achieved by the ability of its members and connections to withstand bending and sheer stresses without additional diagonal bracing, K-bracing, or other extra supporting devices.

unbraced length The greatest length between points on a compression member not restrained against lateral movement by beams, slabs, or bracing.

unbraced length of column Distance between adequate lateral supports.

unbuttoning Unfastening rivet steel connections by breaking off the heads of rivets.

uncoursed Descriptive of irregularly placed masonry, which is not laid in courses with continuous horizontal joints but in a seemingly random pattern.

underbed The base mortar, usually horizontal, into which strips are embedded and on which terrazzo topping is applied.

undercloak (1) The part of a lower sheet in sheet metal roofing that serves as a seam. (2) A course of tiles or slate used in roofing to provide an under layer for the first course installed at the eaves.

undercoat (1) A coat of paint that improves the seal of wood or of a previous coat of paint, and provides a superior adhesive base for the topcoat. (2) A paint used as a base for enamel. (3) A colored primer paint.

undercourse Low grade, usually #4, shingles used as the initial layer of material at the eaves of a roof. See **undercloak**.

undercured Descriptive of concrete, paint, sealant, or other substances applied in wet or elastic form which have not had time to harden properly because of unsuitable environmental conditions.

undercut To cut away a lower portion of architectural stonework, creating a projection above it that functions as a drip.

undercut door A door with greater than normal clearance at the floor to give more ventilation to an area.

underdrain A drain installed in porous fill under a slab to drain off ground water.

underfloor raceway A raceway suitable for use in a concrete floor and in carrying electric conductors.

underground Descriptive of items or installations that are below grade or ground level.

underground piping Piping which has been or will be laid beneath the surface of the ground.

underlay (1) A material, such as asphaltic felt which isolates a roof covering from the deck. (2) See **underlayment**. (3) See **carpet underlayment**.

underlayment Structural wood panels designed to be used under the finished flooring to provide a smooth surface for the finish material.

underlining felt The material, usually a Number 15 felt, applied to a wood roof deck before shingles are laid.

underpass A roadway which crosses under another roadway.

underpinning To provide new substructure support beneath a column or a wall, without removing the superstructure, in order to increase the load capacity or return it to its former design limits.

undersanded With respect to concrete, containing an insufficient proportion of fine aggregate to produce optimum properties in the fresh mixture, especially workability and finishing characteristics.

Uniform Construction Index

underslung car frame The car frame of an elevator with fastening sheaves for hoisting cables attached at or below its platform.

Underwriters' Laboratories, Inc. A private nonprofit organization that tests, inspects, classifies, and rates devices and components to ensure that manufacturers comply with various UL standards.

Underwriters' loop See **Hartford loop**.

undisturbed sample A sample taken from a soil in such a manner that the soil structure is deformed as little as possible.

undressed Descriptive of lumber products that have not been surfaced.

uneven grain Wood grain showing a distinct difference in appearance between springwood and summerwood. Examples are ring-porous hardwoods, such as oak, and softwoods, such as yellow pine, that have soft springwood and hard, dense summerwood.

unglazed tile A hard ceramic tile of homogeneous composition throughout, deriving its color or texture from the materials used and the method of manufacture. The unglazed tile is used for floors or walls.

Uniform Construction Index A published system for coordination of specification sections. This filing of technical data and product literature, and construction cost accounting is organized into sixteen divisions.

uniform grading A particle-size distribution of aggregate in which pan fractions are approximately uniform with no one size or group of sizes dominating.

uniformity coefficient A coefficient related to the size distribution of a granular material, obtained by dividing the size of the sieve of which 60% of the weight passes by the size of the sieve of which 10% of the weight passes.

uniform load A load distributed uniformly over a structure or a portion of a structure.

uniform system A coordinated system developed by the AIA whereby construction specifications, product literature, technical data and cost accounting are organized into several divisions based on an interrelationship of place, trade, function, or material.

union (1) A confederation of individuals who share the same trade or similar trades and who have joined together for a common purpose. (2) A pipe fitting used to join two pipes without turning either pipe, consisting of collar piece which is slipped on one pipe, a shoulder piece which is threaded or soldered on that pipe and against which the collar piece bears, and a thread piece which is fixed to the other pipe. An outside thread on the thread piece and an internal thread on the collar piece are used to make the joint. A gasket is sometimes incorporated as a fluid seal.

union bend See **union elbow**.

union clip A fitting used to connect two rainwater gutters into one functioning unit.

union elbow A pipe elbow outfitted with a union coupling at one end that makes it possible for the coupling end to connect with the end of a pipe without turning or disturbing the pipe.

union fitting See **union elbow** or **union tee**.

union joint A pipe joint made using a union.

union tee A pipe tee with a union-type joint on one end.

union vent See **dual vent**.

unit air conditioner See **package air conditioner**.

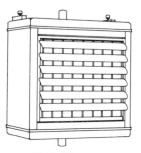

Unit Heater

Unit-type Vent

unitary air conditioner A fabricated assembly of equipment to move, clean, cool, dehumidify, and sometimes heat the air, consisting of a fan, cooling coil, compressor, and condenser.

unit construction See **modular construction**.

unit cooler See **package air conditioner**.

united inches The sum of the length and width of a piece of rectangular glass, each in inches.

unit heater A factory assembled heating unit consisting of a housing, a heating element, a fan and motor, and a directional outlet.

unitized Wood product securely gathered into large standard packages or units, usually fastened with steel straps and often covered by tough paper or plastic.

unit lock A preassembled lock.

unit masonry See **masonry unit**.

unit price Amount stated in the bid as a price per unit of measurement for materials or services as described in the bidding documents or in the proposed contract documents.

unit price contract A construction contract in which payment is based on the work done and an agreed on unit price. The unit price contract is usually only used where quantities can be accurately measured.

unit-type vent One of several relatively small openings on the roof of a structure, equipped with a metal frame and housing as well as manual or automatic hinged dampers which are opened in case of fire.

unit vent See **dual vent**.

unit ventilator A unit with operable air inlets, and often with heating and/or cooling coils, that conveys outdoor air into an interior room.

unit water content (1) The quantity of water per unit volume of freshly mixed concrete, often expressed as pounds or gallons per cubic yard. (2) The quantity of water on which the water-cement ratio is based, not including water absorbed by the aggregate.

unit weight See **bulk density** and **specific gravity**.

universal Descriptive of a door lock, door closer or similar piece of hardware which can be used on either a left-hand or right-hand swing door.

universal motor A motor that can operate on either alternating or direct current and is usually less than 1 horsepower.

unloader A control for an electric-motor-driven compressor. The unloader controls the pressure head of the compressor and allows the motor to be started at low torque by disconnecting one or more cylinders during the initial period of operation.

unmerchantable Logs or products that are faulty and not saleable.

unprotected corner Corner of a slab with no adequate provision for load transfer, so that the corner must carry over 80 percent of the load. See **protected corner**.

unreinforced concrete See **plain concrete**.

unrestrained member A structural member which is allowed to rotate freely about its supports.

unseasoned Lumber that has not been dried to a specified moisture content before surfacing. The American Softwood Lumber Standard defines unseasoned lumber as that having a moisture content above 19%.

unsound Not firmly made, placed, or fixed so as to be subject to deterioration or disintegration during service exposure.

Upright

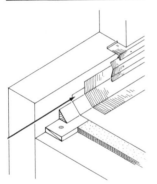

Upstand, Upturn

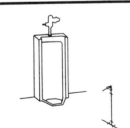

Urinal

unsound plaster Hydrated lime, plaster or mortar which contains particles that are unhydrated and may expand later causing popping or pitting.

unstable soil Defined by OSHA as earth material, other than running, that because of its nature or the influence of related conditions, cannot be depended upon to remain in place without extra support, such as would be furnished by a system of shoring.

unstiffened member A structural member or portion thereof that must withstand compressive force, but is not reinforced in the direction perpendicular to that in which it bends most readily.

upcharge An additional charge for performing some extra service or providing special processing.

upright A vertical length of stone or timber. Also any vertical structural member, such as a stanchion.

upset (1) To make an object or part of an object shorter and thicker by hammering on its end. (2) A flaw in timber caused by a heavy blow or impact that splits fibers across the grain. (3) In welding, an increase in volume at the point of the weld caused by applied pressure.

upset price See **guaranteed maximum cost**.

upset welding A process of resistance-welding, making use of both pressure and of the heat generated by the flow of current as it passes through the resistance provided at the contact point of the surfaces being welded.

upstairs The floors of a house located above the ground or main floor.

upstand, upturn That portion of a flashing or roof covering that is run up a wall without being tucked in and which is usually covered with stepped flashing.

upstanding beam A beam projecting above a concrete floor rather than being beneath it.

urban area An area within the city limits or closely linked to the city by use of common services or utilities.

urban planning See **city planning**.

urban renewal The improvement of deteriorated and underused portions of a city. Urban renewal usually implies improvement through city, state, and federal programs, including demolition of slums and sales of properties to others, rehabilitation of relatively sound structures, and control measures to prevent further spread of blight.

urea resin adhesive A powder which is mixed with water before use and has high early strength and good heat resistance. Urea resin adhesive is not recommended for exposure to moisture or use in poorly fitted joints.

urea resin glue Urea-formaldehyde resin, an adhesive used in the manufacture of hardwood plywood and interior particleboard panel products. Urea is a soluble, crystalline material found in the urine of mammals but it is also produced synthetically for the manufacture of plastics and adhesives.

urinal A plumbing fixture designed for the collection of urine and equipped with a water supply for flushing.

usable life See **pot life**.

U-stirrup A rod shaped like a U, used in reinforced concrete construction.

usury laws Laws that limit the amount of interest that can be charged on a loan. These limits vary from state to state. In some states, the limits are applied only to certain types of loans. Other states have no usury laws.

U-trap

U-tie A wall tie made of heavy wire bent into a U-shape.

utility (1) A grade of softwood lumber used when a combination of strength and economy is desired. It is suitable for many uses in construction but lacks the strength of Standard, the next highest grade in light framing, and is not allowed in some applications. (2) A grade of Idaho White Pine boards, equivalent to #4 Common in other species. (3) A grade of fir veneer that allows white speck and more defects than are allowed in D grade. Utility grade veneer is not permitted in panels manufactured under Product Standard PS-1-83.

utility and better (Util&Btr) A mixture of light framing lumber grades with the lowest being utility. The "and better" signifies that some percentage of the mixture is of a higher grade than utility, but not necessarily of the highest grade. In joist and plank grades, the corresponding term is #3&Btr.

utility knife See **board knife**.

utility pole An outdoor pole installed by a utility company for the support of telephone, electric and other cables.

utility sheet Metal sheeting that is mill-finished and cut into numerous widths and lengths for general use within the building construction industry.

utility survey A survey showing existing site utilities.

utility tractor A tractor of low to moderate horsepower used in construction to tow auxiliary equipment and other site preparation work.

utility vent A pipe that helps provide an air supply within a drainage system or fixture to prevent siphonage. The vent rises above the highest water level of the fixture, and turns downward before connecting to the main vent.

utility window A hot-rolled steel window, generally inexpensive, equipped with a hopper light and a fixed light and used principally in garages, shops, and basements.

utilization equipment Equipment powered by electric energy and used in heating, lighting, and numerous mechanical operations.

U-trap A running trap, built in the shape of a U, that forms a seal against the passage of gasses in a pipe while allowing liquid to flow freely.

U-tube A U-shaped glass tube, also called a manometer, which is filled with water or mercury and used to measure pressure by liquid displacement.

U-value The time rate of heat flow per unit area between fluids on the warm side and cold side of a barrier, calculated in accordance with the difference of unit temperature between the two test fluids. The U-value is the same as *thermal transmittance*.

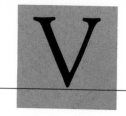

ABBREVIATIONS

The abbreviations listed below are those most commonly used in the construction industry. Alternative forms (usually nonstandard) are shown in parentheses.

V volt, valve, vacuum, v-groove

val value, valuation

van vanity

VAP vapor

var variation, varnished

VAR visual-aural range, volt-ampere reactive

VAT vinyl-asbestos tile

VD vapor density

vel velocity

ven veneer

vent. ventilator

VENT. ventilate

vert, VERT vertical

VF video frequency

VG vertical grain

vhf very high frequency

vic vicinity

VIF verify in field

vil village

vis visibility, visual

VIT vitreous

vit. ch. vitreous china

v.j. V-joint

vlf very low frequency

VLR very long range

vol, VOL volume

vou voussoirs

VP vent pipe

VS versus, vent stack, vapor seal

VT vacuum tube, variable time

VU volume unit

V1S vee one side

V DEFINITIONS

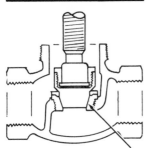

Valve Seat

Vaneaxial Fan

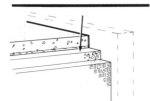

Vapor Barrier

vacuum concrete Concrete from which water and entrapped air are extracted by a vacuum process before hardening occurs.

vacuum pump A pump that removes air or steam from a space or system by producing a partial vacuum in the area treated.

vacuum saturation A process for increasing the amount of filling of the pores in a porous material, such as lightweight aggregate, with a fluid, such as water, by subjecting the porous material to reduced pressure in the presence of the fluid.

valance (1) A board at the top of a window, used to conceal the hanging mechanism for draperies. (2) Draperies for a window.

valance lighting Lighting from sources that are concealed and shielded by a board or panel at the wall-ceiling intersection. This lighting may be directed either upward or downward.

valley The place where two planes of a roof meet at a downward, or V, angle.

valley board The board nailed to the valley rafter of a roof to accommodate the metal gutter.

valley flashing Sheet metal with which the valley of a roof is lined.

valley gutter The exposed open gutter in the valley of a roof, constructed with sloping sides.

valley jack A rafter in a roof system that is cut shorter than a common rafter and connects the ridge and a valley rafter.

valley rafter In a roof frame, the rafter that follows the line of the valley and connects the ridge to the wall plate along the line where the two inclined, perpendicular sides of the roof meet.

valley roof A pitched roof having one or more valleys.

valley shingle A specially-cut shingle, designated for placement next to a valley, with grains that run parallel to the valley.

value engineering A process of reviewing plans and specifications for the purpose of reducing the final cost without changing the intended utility or overall appearance.

valve bag Paper bag for cement or other material, either glued or sewn, made of four or five plies of kraft paper and completely closed except for a self-sealing paper valve through which the contents are introduced.

valve motor An electric or pneumatic control that operates a valve within an air-conditioning system, regulating it at a location remote from the unit.

valve seat The stationary portion of the valve which, when in contact with the movable portion, stops the flow.

vandalism and malicious mischief insurance Insurance against loss or damage to the insured's property caused by willful and malicious damage or destruction.

vaneaxial fan A fan which consists of a disk type wheel within a cylinder and a set of air guide vanes placed on one side of the wheel. The fan is either belt driven or often times connected directly to a motor.

vanishing point A point on a perspective drawing towards which a series of normally parallel lines converge.

vapor barrier Material used to prevent the passage of vapor or moisture into a structure or another material thus preventing condensation within them.

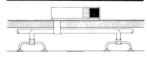

**Variable-volume
Air System**

vapor heating system A system of steam heating functioning at or near atmospheric pressure, wherein the condensed liquid is returned to the boiler by gravity.

vapor lock The formation of vapor in a pipe carrying liquids preventing normal flow.

vapor migration Penetration of vapor through walls or roofs, caused by the vapor pressure differential between the structure and the outside.

vapor pressure A component of atmospheric pressure caused by the presence of vapor. Vapor pressure is expressed in inches, centimeters, or millimeters of height of a column of mercury or water.

vapor seal See **vapor barrier**.

variable-volume air system An air-conditioning system that automatically regulates the quantity of air supplied to each controlled area according to the needs of the different zones, with preset minimum and maximum values based on the load in each area.

variance A written authorization from the responsible agency permitting construction in a manner which is not allowed by a code or ordinance.

variegated Descriptive of a material or surface having streaks, marks, or patches of a different color or colors.

varnish A clear, colorless substance used chiefly on wood to provide a hard, glossy, protective film. Varnish is manufactured from resinous products dissolved in oil, alcohol, or a number of volatile liquids.

varnish stain A colored transparent varnish with a lesser power of penetration than a true stain.

varved clay Sedimentary soil with alternating layers of clay and silt or fine sand that display contrasting colors as they dry, formed during the differing sedimentation conditions in various seasons of the year.

vats Large containers used for steaming logs or submerging them in hot water prior to peeling, slicing them into veneer, or cutting them into wafers. The steaming or heating of logs makes them easier to process. Vats were first widely used in hardwood veneer manufacturing, but are now frequently used in the manufacture of softwood plywood panels, waferboard, and oriented strand board.

vault (1) A storage enclosure built above or below ground, large enough to accommodate human entry. The vault is used to install, operate, and maintain electrical cables and equipment. (2) A masonry structure with an arched ceiling. (3) A room used for storage of valuable records and/or computer tapes, that is of fire resistive construction, has safe electric components, and has a controlled atmosphere.

V-beam sheeting Corrugated sheeting formed with flat, V-angled surfaces instead of rippled or curved surface.

V-brick Brick that is vertically perforated.

V-cut (1) Descriptive of a style of lettering carved into stone in acute, triangular cuts. (2) A V-shaped saw cut or incision in wood.

vehicle A substance, as oil, in which pigments are mixed for application.

vein A thin layer or deposit of one material in another, usually in an approximate plane.

velocity head A measurement of the velocity of fluid through a pipe or watercourse. The velocity head is equal to the height that the fluid must fall to achieve that same velocity.

Veneer

vendor A person or entity who furnishes materials or equipment not fabricated to a special design for the work. See also **supplier**.

veneer (1) A masonry facing which is attached to the backup but not so bonded as to act with it under load. (2) Wood peeled, sawn, or sliced into sheets of a given constant thickness and combined with glue to produce plywood. Veneers laid up with the grain direction of adjoining sheets at right angles produce plywood of great stiffness and strength, while those laid up with grains running parallel produce flexible plywood most often used in furniture and cabinetry. See **laminated wood**.

veneer adhesives Several basic substances are used in the gluing of veneers to produce plywood. These include blood, soybean, and phenolic resins. Other adhesives made from urea, resorcinol, polyvinyl, and melamine are sometimes used in edge gluing, patching, and scarfing. Among the principal adhesives are (1) soybean glue, a protein-type adhesive made from soybean meal and usually blended with blood and used in certain panels for interior use, (2) blood glue, made from animal blood from slaughter houses, dried and supplied in powder form, and intended for interior uses, and
(3) phenolic resin, produced from synthetic phenol and formaldehyde. Phenolic resin is cured only under heat and undergoes chemical changes, which makes it impervious to attack by micro-organisms. It is used in undiluted form for the production of exterior plywood. However, it may be extended by the addition of other substances in the production of interior plywood.

veneer base Gypsum lath sheeting, ordinarily cut in widths of 4 feet and in varying lengths and

thicknesses, with a core of gypsum and a special paper facing that is receptive to veneer plaster.

veneered construction Construction of wood, reinforced concrete, or steel faced with a thin layer of another material, such as structural glass or marble.

veneered door A hollow or solid core door with veneer faces.

veneered plywood Plywood which is faced with a decorative veneer; usually wood or plastic.

veneer holddown The part of a veneer clipper that holds down the veneer to assure proper alignment with the clipper blade.

veneer plaster A mill-mixed gypsum plaster made up of one or two components and desired for its bond, strength, and ease of installation.

veneer tie A wall tie designed to hold a veneer to a wall.

veneer wall A wall with a facing which is attached to but not bonded to the wall.

veneer wall tie See **veneer tie**.

Venetian A type of terrazzo topping in which large chips of stone are incorporated.

Venetian blind (1) A blind made of thin slats so mounted as to overlap when closed and provide spaces for admitting light and/or air when open. The mounting usually consists of strips of webbing. The unit is operated by cords. (2) Adjustable exterior slatted shutters.

vent (1) A pipe built into a drainage system to provide air circulation, thus preventing siphonage and back pressure from affecting the function of the trap seals. (2) A stack through which smoke, ashes, vapors, and other airborne impurities are discharged from an enclosed space to the outside atmosphere.

vent connector A metal pipe connecting the exhaust of a burner to a chimney.

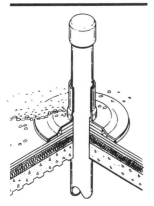

Veneered Door

Vent

Ventilator

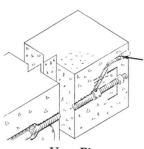

Vent Pipe

Vent Sash

vented form A concrete form so constructed as to retain the solid constituents of concrete and permit the escape of water and air.

vent flue See **vent**.

ventilating bead See **draft bead**.

ventilating brick A brick with holes in it for the passage of air.

ventilating eyebrow See **eyebrow**.

ventilating jack A sheet metal hood over the inlet to a vent pipe to direct flow into the pipe.

ventilation A natural or mechanical process by which air is introduced to or removed from a space, with or without heating, cooling, or purification treatment.

ventilator (1) A device or opening in a room or building through which fresh air enters the enclosure and stale air is expelled. (2) A pivoted sash or framework, outfitted with hinged panes of glass that may be opened without the need of opening the sash.

ventilator frame An assembly designed to accommodate a pivoted sash, with two rails and two stiles into which the operable panes are set.

vent pipe A small-diameter pipe used in concrete construction to permit the escape of air in a structure being concreted or grouted.

vent sash A small operable light in a window which may be swung open to allow some ventilation without opening the entire sash, usually hinged on its upper edge.

vent stack, main vent A vertical vent pipe whose functions are to provide air circulation to or from any part of a building drainage system and to protect its trap seals from siphonage.

vent system A chimney or vent, combined with a vent connector to form a clear passageway for expulsion of vent gases from gas-burning equipment to the outside air.

Venturi (1) A short tube with a constriction used to measure fluid velocity by the differential pressures as the fluid flows. (2) A constricted throat in an air passage of a carburetor used to mix fuel and air.

veranda A covered porch or balcony along the outside of a building, intended for leisure.

verge (1) The edge that projects over the gable of a roof. (2) The shaft of a column. The verge is also a small shaft employed for ornamental effect.

vergeboard See **bargeboard**.

verge fillet A strip of wood that is fastened to the roof battens of a gable to cover the upper edges of the gable wall.

vermiculated work A form of incised masonry surface with ornamental track-like grooves, frets, or knots, employed architecturally on wall surfaces and pavements.

vermiculite A group name for certain clayey minerals, hydrous silicates or aluminum, magnesium, and iron that have been expanded by heat. Vermiculite is used for lightweight aggregate in concrete and as a loose fill for thermal insulating applications.

vermiculite concrete Concrete in which the aggregate consists of exfoliated vermiculite.

vermiculite plaster A fire-retardant plaster covering for steel beams, concrete slabs, and other heavy construction materials, constituted with an aggregate of very fine exfoliated vermiculite.

vernier An auxiliary scale which slides along a larger scale and allows more accurate interpretation of subdivisions.

vertical angle An angle in a vertical plane.

vertical bar An upright muntin.

vertical bond Same as stack bond, with one brick or block placed directly on top of the next.

Vertical Siding

Vertical Sliding Window

vertical circle A graduated circle mounted on an instrument so that the plane of the circle will be in a true vertical plane when the instrument is leveled.

vertical curve A parabolic curve in the vertical plane used to connect two highway grades in order to form a smooth transition.

vertical exit A stairway, ramp, fire escape, escalator, or any other route of movement that serves as an exit from the floors of a building above or below street level.

vertical-fiber brick A paving brick cut with wire during the manufacturing process, and laid on pavement with the wire-cut side exposed to view.

vertical firing Mechanical arrangement of gas, oil, or coal burners in a furnace so that fuel is vertically discharged up from burners below or down from burners on top.

vertically pivoted window A window having a sash which pivots about a vertical axis near its center so the outside of the glass can be conveniently cleaned.

vertical photograph A photograph taken from an airplane looking straight down. The vertical photograph is used in aerial surveying.

vertical pipe Any pipe or fitting which makes an angle of 45 degrees or less with the vertical.

vertical sand drain Vertical columns of sand formed by replacing a pipe pile with the sand and used to speed drainage and consolidation of thick deposits of clay.

vertical sash See **vertical sliding window**.

vertical section An illustration of how an object would appear if a vertical plane were cut through it. An example of a vertical section is a drawing of sedimentary soil layers or of the wood layers on a sheet of plywood.

vertical siding A type of exterior wall cladding consisting of wide matched boards.

vertical sliding window A window with one or more sashes that move only in a vertical direction and make use of friction or a ratchet device to remain in an open position.

vertical slip form A form which is jacked vertically during construction of a concrete structure. Movement may be continuous with placing or intermittent with horizontal joints.

vertical spring-pivot hinge A spring hinge that is mortised into the heel of a door, which in turn is fastened with pivots to its door head and the floor.

vertical tiling Tile that is hung on the surface of a wall in a vertical arrangement and protects the wall against moisture.

vertical tray conveyors A vertical conveying system designed to carry trays or boxes.

vertical-vision-light door See narrow-light door.

very-high-output fluorescent lamp A rapid-start fluorescent lamp designed to operate on a very high current, providing a light flux per unit length of lamp higher than that obtained from a high-output fluorescent lamp.

vestibule An anteroom or foyer which leads to a larger space.

V-groove Any of several longitudinal cuts made on the faces of pieces of lumber or plywood. The face veneer of plywood paneling is V-grooved to relieve the flat appearance of the surface. The grooving usually creates a pattern resembling random width boards placed side by side. Usually, V-grooves in paneling are stained darker than the surface. In lumber, edges are sometimes chamfered to create a

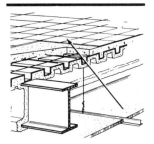

Vinyl Composition Tile

V where pieces are placed edge to edge. A V may also be machined the length of the piece to provide decoration. A V pattern also may be used to form tongue and groove connections on either lumber or plywood.

V-gutter See **valley gutter**.

vibrated concrete Concrete compacted by vibration during and after placing.

vibrating pile driver See **sonic pile driver**.

vibrating roller A soil compacting machine consisting of a roller with a motor-driven vibrating mechanism.

vibrating screed A machine designed to act as a vibrator while leveling freshly placed concrete.

vibration Energetic agitation of freshly mixed concrete during placement by mechanical devices, either pneumatic or electric, that create vibratory impulses of moderately high frequency that assist in evenly distributing the concrete in the formwork.

   external Employs vibrating devices attached at strategic positions on the forms and is particularly applicable to manufacture of precast items and for vibration of tunnel-lining forms. In manufacture of concrete products, external vibration or impact may be applied to a casting table.

   internal Employs one or more vibrating elements to be inserted into the concrete at selected locations, and is more generally applicable to in-place construction.

   surface Employs a portable horizontal platform on which a vibrating element is mounted.

vibration isolator A flexible support for any form of vibrating or reverberating machinery, piping, or ductwork, serving to reduce the vibrations that are carried to the remainder of the building structure.

vibration limit That time at which fresh concrete has hardened sufficiently to prevent its becoming mobile when subjected to vibration.

vibration meter Any instrument which measures the displacement, velocity, or acceleration of a vibrating body.

vibrator An oscillating machine used to agitate fresh concrete so as to eliminate gross voids, including entrapped air, but not entrained air, and to produce intimate contact with form surfaces and between embedded materials.

Vicat apparatus A penetration device used in the testing of hydraulic cements and similar materials.

Vicat needle A weighted needle for determining setting time of hydraulic cements.

vice See **vis**.

vine Any plant having a flexible stem which is supported by climbing, twining or creeping along a surface.

vinyl A thermoplastic compound made from polymerized vinyl chloride, vinylide chloride, or vinyl acetate. Vinyl is typically tough, flexible, and shiny.

vinyl-asbestos tile A resilient, semiflexible floor tile, composed of ground limestone, plasticizers, pigments, polyvinyl chloride binder, and asbestos fiber reinforcing. This product has been replaced by vinyl composition tile.

vinyl composition tile A floor tile similar to vinyl-asbestos floor tile except the asbestos has been replaced by glass fiber reinforcing.

vinyl overlay An overlay applied to panel products and moldings and usually printed with a color and grain, needing no further finishing.

vinyl tile A floor tile similar to vinyl-asbestos floor tile but does

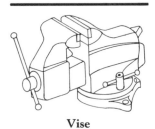

Vise

Vitrified-clay Pipe

not contain mineral or other fibers.

Virginia Pine *Pinus virginiana.* This is another minor member of the Southern Yellow Pine group. It is grouped for grading purposes with Pond Pine, under the stamp "Mixed Pine Species".

vis, vice, vise (1) A spiral staircase, generally of stone, with steps winding around a central shaft. (2) A screw stair.

viscometer Instrument for determining viscosity of slurries, mortars, or concretes.

viscosity The degree to which a fluid resists flow under an applied force.

viscous filter An air-cleaning filter that employs a surface covered with viscous oil or fluid, to which dirt particles and other airborne impurities cling as the air passes through.

vise (1) A gripping tool with adjustable jaws controlled by a lever or screw, used to clamp an object firmly in place while work is being done on it. (2) A *vis*.

visual concrete See **exposed aggregate**.

visual inspection An inspection made without the use of instruments.

vitreous Descriptive of the degree of impermeability in a material at which it is characterized by low water absorption. Vitreous is less than 3.0 percent in floor and wall tile and low voltage electrical porcelain, and less than 0.3 percent absorption in other materials.

vitreous sand See **smalt**.

vitrification The fusion of grains in clay products under high kiln temperatures, resulting in closure of pores and an impervious material mass.

vitrified brick A form of glazed brick that is impervious to moisture and highly resistant to chemical corrosion.

vitrified-clay pipe Glazed earthenware pipe favored for use in sewage and drainage systems because it is impervious to water and resistant to chemical corrosion.

V-joint See **V-shaped joint**.

void-cement ratio Volumetric ratio of air plus net mixing water to cement in a concrete or mortar mixture.

void ratio In a mass of granular material, the ratio of the volume of voids to the volume of solid particles.

voids The air spaces between particles in granular material such as sand or a paste such as mortar.

void-solid ratio The ratio of the sum of the areas of window and door openings to the gross area of an exterior wall of a building.

volatile material (1) Material that is subject to release as a gas or vapor. (2) Liquids that evaporate readily.

volt The unit of voltage or potential difference equal to the voltage between two points of a conducting wire carrying a constant current of one ampere, when the power dissipated between the points is one watt.

voltage The greatest root-mean-square potential difference between any two points in an electric circuit.

voltage drop The difference in voltage between any two points in an electric circuit.

voltage regulator A control device within an electrical system that automatically keeps the voltage supply constant despite variable line voltage at the point of input.

voltage-to-ground (1) The voltage between a conductor in a grounded electric circuit and the point of the circuit that is grounded. (2) The maximum voltage between a conductor in an ungrounded electric circuit and another conductor.

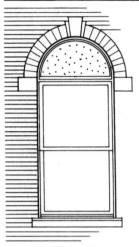

Voussoir

volt-ampere The product obtained by multiplying one volt times one ampere, equivalent to one watt in direct-current circuits and to one unit of apparent power in alternating current circuits.

voltmeter An instrument used to measure the voltage drop between any two points in an electric circuit.

volume batching The measuring of the constituent materials for mortar or concrete by volume.

volume change An increase or decrease in volume. See **deformation**.

volume method (of estimating cost) Method of estimating probable construction cost by multiplying the volume of the structure by an estimated current cost per unit of volume. See **architectural volume of buildings, architectural area or buildings**, and **area method**.

voussoir A wedge-shaped masonry unit in an arch or vault, the converging sides of which are cut along radii of the vault or arch.

voussoir brick See **arch brick**.

V-shaped joint, V-joint, V-tooled joint (1) A V-shaped horizontal joint in mortar, formed with a steel jointing tool, that serves to resist rainwater penetration. (2) Two adjacent wood boards, with beveled or chambered edges, that form a joint in the same plane.

V-tool A gauge with a V-shaped cutting edge. See **parting tool**.

V-tooled joint See **V-shaped joint**.

vulcanization An irreversible chemical process in which a rubber compound becomes less plastic, more resistant to swelling by organic liquids, and more elastic (or its elastic properties are extended over a greater temperature range).

ABBREVIATIONS

The abbreviations listed below are those most commonly used in the construction industry. Alternative forms (usually nonstandard) are shown in parentheses.

w water, watt, weight, wicket, wide, width, work, with

W watt, west, western, width

W/ with

WA with average

WAF wiring around frame

WB welded base, water ballast, waybill

WBT wet-bulb temperature

WC, W.C. water closet

wd wood, window

Wdr wider

WF wide flange

wfl waffle

wg wing, wire gauge

WG wire gauge

wh watt-hour

WH water heater

WHP water horsepower

whr watt-hour

whse, WHSE warehouse

WI wrought iron

WK week, work

wm wattmeter

WM wire mesh

W/M weight or measurement

w/o water-in-oil, without

W/O without

WP waterproof, weatherproof, white phosphorus

wpc watts per candle

w proof waterproofing

wrt wrought

WS weather strip

wsct/wains wainscoting

wt., Wt. weight

WT watertable, watertight

ww white wash

WWM welded wire mesh

DEFINITIONS

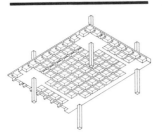

Waffle Slab

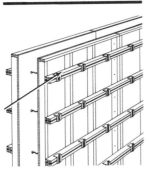

Wale

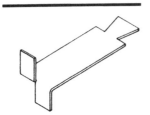

Wall Anchor

Wall Bracket

waferboard A panel product made of discrete wafers of wood bound together by resin, heat, and pressure. Waferboard can be made of timber species, such as Aspen, that are not suitable for lumber or plywood manufacture.

waffle See **dome**.

waffle floor See **waffle slab**.

waffle slab A reinforced concrete slab with equally spaced ribs parallel to the sides, having a waffle appearance from below.

Wagner fineness The fineness of portland cement, expressed as total surface area in square centimeters per gram, determined by the Wagner turbidimeter apparatus and procedure.

wagon drill A movable rig for positioning and holding a pneumatic drill, consisting of a mast with a carrier for the drill and a supporting wheeled carriage.

wainscot (1) A lower interior wall surface, usually extending three to four feet up from the floor, that contrasts with the wall surface above it. (2) The lower wall surface of an interior wall composed of two wall surfaces.

wainscot cap The finish molding at the top of wainscot.

wainscot oak Selected oak planks used in wainscoting.

waiver of lien An instrument by which a person or organization who has or may have a right of mechanic's lien against the property of another relinquishes such right. See also **mechanic's lien** and **release of lien**.

wale, waler, whaler (1) Timber placed horizontally across a structure to strengthen it. (2) Horizontal bracing used to stiffen concrete form construction and to hold studs in place.

walk A paved or plank path for foot traffic.

walk-in box A refrigerator or freezer large enough for one or more persons to enter to deposit or retrieve food.

wall A vertical element used primarily to enclose or separate spaces.

wall anchor A steel strap used to attach a back-up surface to the masonry fascia.

wall arcade A blind arcade used as an ornamental dressing to a wall.

wall base See **base**.

wall beam (1) A special form of reinforced concrete framing, used for some apartment buildings, in which walls between apartments are used as deep, transverse beams supporting one edge of a floor slab below and edge above. The deep beams alternate between floors and between columns. (2) A header bolted to a wall and used to support joists or beams.

wall-bearing partition See **load-bearing partition**.

wallboard A manufactured sheet material used to cover large areas. Wallboards are made from many items, including wood fibers, asbestos, and gypsum. In North America, the most common is "sheet rock", a gypsum-based panel bound by sheets of heavy paper. It is used to seal interior walls and ceilings in place of wet plaster.

wall box, beam box, wall frame (1) A bracket fixed to a wall to support a structural member. (2) A metal box set in a wall to house an electric switch or receptacle.

wall bracket A bracket fastened to a wall and used to support a structural member, pipe, an

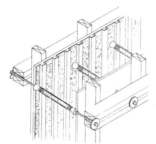

Wall Form

Wall-hung Water Closet

electric insulator, an electric fixture, or a section of scaffolding.

wall chase See **chase**.

wall clamp A brace or tie used to connect two parallel walls to each other.

wall column A column fully or partly embedded in a wall.

wall coping See **coping**.

wall covering Any material or assembly used as a wall finish and not an integral part of the wall.

wall crane A crane with a horizontal arm supported from a wall or columns of a building. The arm may support a trolley.

wall form A retainer or mold erected to give the necessary shape, support, and finish to a concrete wall.

wall furring Strips of wood or shaped sheet metal attached to a rough wall to provide a plane on which lath and plaster, paneling, or wainscoting may be installed.

wall grille A perforated plate, casting, punched sheet, or frame used to conceal an opening, radiator, or the like, while allowing a passage for air.

wall guard A protective, resilient strip attached to a wall to protect the surface from carts, transporters, or other movable conveyors.

wall handrail A rail attached to a wall at a flight of stairs and sloped parallel with the stairs for use as a handrail.

wall hanger A stirrup or bracket fixed to a wall and used to support one end of a horizontal member.

wall height The vertical distance from the top of a wall to its support, such as a foundation or support beam.

wall hook (1) A special large nail or hook used as a beam anchor or for supporting a beam plate. (2) A hook or bracket fixed in a masonry wall to hold downspouts,

lightening rods, etc.

wall-hung water closet A water closet mounted on a wall so the area beneath is clear for cleaning.

wall opening (OSHA) An opening at least 30 inches high and 18 inches wide, in any wall or partition, through which persons may fall, such as a chute opening.

wall outlet An electric outlet mounted in a wall with a decorative cover.

wallpaper Paper, paper-like material, or plastic film used as a decorative facing for walls or ceilings.

wall plate The top plate in construction, placed on top of studs and bearing the joists of the next floor above.

wall plug (1) See **plug**. (2) See **wall outlet**.

wall post A post which is next to, and may be fastened to, a wall.

wall rib A longitudinal rib against an exterior wall of a vaulting compartment.

wall shaft A small column, supported on a corbel or bracket, which appears to support a rib of a vault.

wall sign (1) A sign mounted on, or fastened to, a wall. (2) A sign attached to the exterior wall of a building and projecting not more than a code-defined distance. Most codes specify 15 inches.

wall socket See **wall outlet**.

wall spacer A metal tie used to hold forms in position until concrete has set.

wall tie A metal strip or wire used to tie wythes together or tie a masonry veneer to a wood frame.

wall tile A glazed tile used in a wall facing.

wall-wash luminaire A luminaire located adjacent to or on a wall with most of its light directed on the wall.

Walnut *Juglans regia*, English Walnut; *J. nigra*, Black Walnut; *J.*

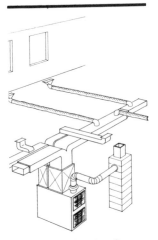

Warm-air Heating System

Washer

cinerea, Butternut. The last two, native to North America, are used for veneer and furniture stock, in addition to their edible nuts.

Walter Pine Spruce Pine.

wane A defective edge of a board due to remaining bark or a beveled end. A wane is usually caused by sawing too near the surface of the log.

ward (1) A baffle in a lock to prevent use of an unauthorized key. (2) A division of a hospital or jail.

warehouse set The partial hydration of cement stored for a time and exposed to atmospheric moisture, or mechanical compaction occurring during storage.

warm-air furnace A furnace that generates warm air for a heating system.

warm-air heating system A heating system in which warm air is distributed through a single register or series of ducts. Circulation may be by convection (gravity system) or by a fan in the duct work (forced system).

warning pipe An overflow pipe with an outlet located so that discharge can be readily observed.

warp Distortion in shape of a plane surface, such as that in lumber as a result of a change in moisture content.

warping A deviation of a slab or wall surface from its original shape, usually caused by temperature and/or moisture differentials within the slab or wall. See also **curling**.

warping joint A longitudinal or transverse joint, with bonded steel or tie bars passing through it, with the sole function of permitting warping of pavement slabs when moisture and temperature differentials occur in the pavement.

warranty Legally enforceable assurance of quality or performance of a product or work, or of the duration of satisfactory performance. Warranty, guarantee, and guaranty are substantially identical in meaning; nevertheless, confusion frequently arises from supposed distinctions attributed to guarantee (or guaranty) being exclusively indicative of duration of satisfactory performance or of a legally enforceable assurance furnished by a manufacturer or other third party. The Uniform Commercial Code provisions on sales (effective in all states except Louisiana) use warranty but recognize the continuation of the use of guarantee and guaranty.

warranty deed A deed conveying real property, in which the grantor makes binding representations concerning the quality of his title and its freedom from encumbrances.

Warren Truss, Warren girder A truss consisting of horizontal top and bottom chords, separated by sloping members, and without vertical pieces.

wash (1) The top slope of a building element, or portion thereof, intended to shed water. (2) A manner of applying water colors to a rendering.

washable Descriptive of a material that may be washed repeatedly without a noticeable deterioration in appearance or function.

wash coat A thin, almost transparent coat of paint applied to a surface as a sealer or stain.

washer (1) A flat ring of rubber, plastic, or fibrous material, used as a seal in a faucet or valve or to minimize leakage, as in a threaded connection. (2) A flat ring of steel, which may be split, toothed, or embossed, used in threaded connections to distribute loads, span large openings, relieve friction, or prevent loosening.

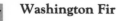

Water Closet, W.C.

Water Cooling Tower

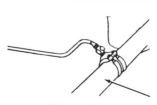

Water Main

(3) Any machine used to wash objects such as clothes or dishes.

Washington Fir Douglas Fir.

wash primer Any thin paint that promotes adhesion of the subsequent coat.

wash water, flush water Water carried on a truck mixer in a special tank for flushing the interior of the mixer after discharge of the concrete.

waste-disposal unit An electric-operated device for grinding waste food and disposing of it through the plumbing drainage pipes.

waste pipe A pipe to convey discharge from plumbing fixtures.

waste plug A tapered, rubber-like object used to close the drain of a wash basin or tub.

wasting Splitting of excess stone with a hand tool so that the resulting surfaces of block are nearly plane.

water analysis Chemical and bacteriological analysis of water.

water bar, weather bar A wood or metal strip fixed to the exterior sill of a window or door to resist the penetration of water.

waterborne preservative Preservative salts dissolved in water and transferred to the wood during the treating process.

water-cement ratio The ratio of the amount of water, exclusive only of that absorbed by the aggregates, to the amount of cement in a concrete or mortar mixture. The ratio is preferably stated as a decimal by weight.

water closet, W.C. (1) A plumbing fixture used to receive human wastes and flush them to a waste pipe. Also called a "flushable toilet". (2) A room that contains a water closet.

water content See **moisture content**.

water cooling tower See **cooling tower**.

water crack A fine crack in a coat of plaster, caused by applying a coat over a previous coat that had not dried sufficiently or by using plaster with a water content that is too high.

water gain See **bleeding**.

water gauge A manometer filled with water.

water-gel explosive A wide variety of explosives, all containing water and ammonium nitrate, used for blasting. The explosives are sensitized with an explosive such as TNT or smokeless powder, or with metals such as aluminum, or with other fuels. They may be premixed at a plant or site-mixed immediately prior to loading in a blast hole.

water hammer (1) A loud thumping noise in a water service line due to the surge of suddenly checked water. (2) A banging in steam lines occurring at the time of steam flow. The steam, traveling at high velocity picks up drops of condensed water and slams them against the piping at a bend.

water joint See **saddle joint**.

water-level control A control, on a boiler, used to maintain the water at a safe level.

water lime A hydraulic lime or cement that will set under water.

waterline The highest water level, in a cistern or flush tank, to which the shut-off should be adjusted.

water main A main supply pipe in a water system supplying water for public or community use.

water outlet (1) Any opening or end of pipe used to discharge water to a plumbing fixture, boiler, or other device. (2) An opening through which water is discharged to the atmosphere.

waterproof Any material, treatment, or construction which resists flow or penetration of water.

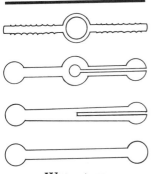

Waterstops

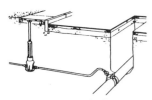

Water-supply System

waterproof adhesive An adhesive which, when properly cured, is not affected by water.

"waterproofed" cement Cement interground with a water-repellent material such as calcium stearate.

waterproof paper A paper made water-impervious by adding a resin with the manufacturing process.

water-reducing agent A material that either increases workability of freshly mixed mortar or concrete without increasing water content or maintains workability with a reduced amount of water, the effect being due to factors other than air entrainment.

"water-repellent" cement A hydraulic cement having a water-repellent agent added during the process of manufacture, with the intention of resisting the absorption of water by the concrete or mortar.

water retentivity The property of mortar that prevents the rapid loss of mix water by absorption.

water ring A perforated manifold in the nozzle of dry-mix shotcrete equipment through which water is added to the materials.

water seal The water in a trap acting as a seal against the passage of gases.

water seasoning The seasoning of lumber by soaking it in water for a period, usually two weeks, and air drying it.

water-service pipe The part of a water-service main which is owned by the water department.

watershed The area of land that drains naturally into a stream or complex of streams. The management of watersheds in the national forests is a principal function of the Forest Service.

water softener A device to remove calcium and magnesium salts from

a water supply, usually by ion exchange.

waterstop A thin sheet of metal, rubber, plastic, or other material inserted across a joint to obstruct the seeping of water through the joint.

water-supply system (1) The system that supplies water throughout a building, including the service pipe(s), distribution and connecting pipes, fittings, and control valves. (2) The system that supplies water to units in a community, including reservoirs, tunnels, and pipelines.

water table (1) The top surface of groundwater. (2) A slight projection of exterior wall construction outside and at the foundation line.

water tap An outlet valve for water. Also called a "faucet".

watt A unit of power equal to the power dissipated in an electric circuit in which a potential difference of 1 volt causes a current flow of 1 ampere, or the power required to do work at the rate of 1 joule per second.

watt-hour A unit of work equivalent to the power of 1 watt operating for 1 hour, which is equal to 3600 joules.

wattle A framework consisting of interwoven rods, poles, branches, etc.

wax A material obtained from vegetable, mineral, and animal matter which is soluble in organic solvents and solid at room temperature. Wax is applied in liquid or paste form on wood and metal surfaces to provide gloss and to protect the surface.

waxing The filling of cavities, in a finished piece of marble, with materials to match the stone in color and texture.

weakened-plane joint See **groove joint**.

wearing course A topping or surface treatment to increase the

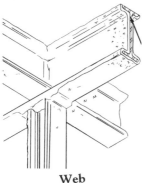

Web

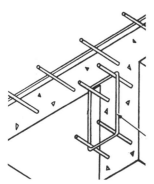

Web Reinforcement

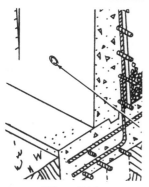

Weep Hole

resistance of a concrete pavement or slab to abrasion.

weather (1) The length of shingle or tile that is exposed, as measured along the slope of a roof. (2) To deteriorate or discolor when exposed to the weather. (3) To slope a surface for the purpose of shedding rainwater.

weatherboard (1) Boards or siding lapped to shed water. (2) A projecting member attached to the bottom rail of an external door to divert water from the sill or threshold.

weathered (1) Descriptive of a surface exposed long enough that distinct weathering has taken place. (2) Having an upper surface that is sloped to shed water.

weathered joint See **weather-struck joint.**

weathering (1) Changes in color, texture, strength, chemical composition, or other properties of a natural or artificial material due to the action of the weather. (2) The mechanical or chemical disintegration and discoloration of the surface of wood, caused by exposure to light, the action of dust and sand carried by the wind, and the alternate shrinking and swelling of the surface fibers due to continual variation in moisture content brought on by changes in the weather.

weather resistance The ability of a material or coating to resist **weathering, def. 1.**

weatherseal channel A channel installed, flanges downward, on the top of an exterior door.

weather strip A strip of wood, metal, felt, plastic, or other material applied at an exterior door or window to seal or cover the joint made by the door or window with the sill, casings, or threshold.

weather-struck joint, weathered joint A horizontal masonry joint

sloped from a point inside the face of the wall to the face of the wall so the joint will shed water.

weaving (1) The alternate lapping of shingles on opposite surfaces, when two adjacent roofs intersect. (2) The process of making a rug by interlacing surface and backing yarns.

web (1) That part of a beam or truss between the flanges or cords, used mainly in resisting shear stresses. (2) The walls connecting the face shells of a hollow concrete masonry unit.

web bar See **web reinforcement.**

web clamp A device consisting of a fabric tape with a fastener that is secured with a wrench or screwdriver, used to hold carpentry work while gluing.

web crippling Local failure of the web of a beam or girder caused by a concentrated load.

web plate A steel plate forming the web of a built-up girder struss, or beam.

web reinforcement Reinforcement placed in a concrete member to resist shear and diagonal tension.

web splice A splice joining two web plates.

web stiffener A vertical steel shape attached to the web of a plate girder or rolled beam and used to prevent web buckling.

wedge A piece of wood or metal tapering to a thin edge, used to adjust elevation or tighten formwork.

wedge anchor A wedging device used in the anchorage of a tendon in posttensioned, prestressed concrete.

weep hole (1) A small hole in a wall or window member to allow accumulated water to drain. The water may be from condensation and/or surface penetration. (2) A small hole in a retaining wall located near the lower ground surface. The hole drains the soil

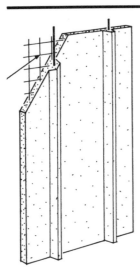

Welded-wire Fabric

behind the wall and prevents build-up of water pressure on the wall.

Weeping Pine *Pinus excelsa.* Himalayan Pine, so called because of its long, drooping needles.

weight batching Measuring the constituent materials for mortar or concrete by weight.

weight box The channel in a window frame housing the sash weights.

weighted average An average in which certain entries, usually at the extremes of the range of values, are manipulated so a number is obtained which is believed to be more representative of the true mean than that obtained by straight averaging.

weir A structure across a ditch or stream, used for measuring or diverting the flow of water.

weir head The depth of water above the top of a flat weir or above the bottom of a notch in a notched weir.

weld (1) To build up or fasten together, as with cements or solvents. (2) To fasten two pieces of metal together by heating them until there is a fusing of material, either with or without a filler metal.

weld decay Local corrosion at or adjacent to a weld.

welded butt splice A reinforcing bar splice made by welding the butted ends.

welded cover plate A cover plate welded to the flange of a beam or girder, usually to increase strength in the middle length of a span.

welded joint A joint connecting two metal parts by welding.

welded reinforcement Concrete reinforcing steel joined by welding and most often used in columns to extend vertical bars.

welded truss A truss of metal members in which the joints are made by welding.

welded-wire fabric, welded-wire mesh A series of longitudinal and transverse wires arranged substantially at right angles to each other and welded together at all points of intersection.

welded-wire fabric reinforcement Welded-wire fabric in either sheets or rolls, used to reinforce concrete.

welded wire lath See **wire lath**.

welding cables The pair of cables supplying electric energy for use in welding. One lead connects a welding machine with an electrode; the other lead connects the machine with the work.

welding nozzle A stub-pipe, shop welded to a vessel, to facilitate welding a connecting pipe in the field.

welding rod Filler metal in the form of wire or rod used in gas welding and brazing and in some arc welding processes.

welding screw A screw with lugs or weld projections on the top or underside of the head to facilitate attaching the screw to a metal part by resistance welding.

weldment Any assembly made by welding parts together.

weld metal That part of a weld that was melted while welding.

weld nut A nut designed to be attached to a part by resistance welding.

well, wellhole (1) Any enclosed space of considerable height, such as an air shaft or the space around which a stair winds. (2) A collection device for ground water. (3) A wall around a tree trunk to hold back soil. (4) A slot in a machine or device into which a part fits.

well drill A churn-type drill mounted on a truck and used to drill for water. A well drill usually has a limited operating depth.

well-graded aggregate Aggregate with a particle-size distribution

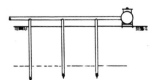

Well-point System

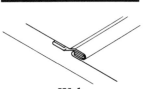

Welt

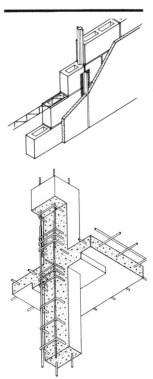

Wet Construction

producing maximum density, or minimum void space.

well point A perforated pipe sunk into granular soil to permit the pumping of ground water.

well-point system A series of well points connected to a header and used to drain an area or to control groundwater seepage into an excavation.

welt (1) A seam in sheet metal, formed by folding over the edges of two sheets, interlocking the folded portions, and flattening the formed seam. (2) A strip of wood fastened over a seam or joint or a shaped piece fastened over an angle for reinforcing.

welting strip A strip of sheet metal at the intersection of the roof and a vertical surface. One edge of the strip is fastened to the roof and the other edge is bent to lock with the lower edge of a vertical sheet.

West Coast Hemlock See **Hemlock**.

Western Balsam See **Grand Fir**.

Western Jackpine See **Lodgepole Pine**.

Western Larch See **Larch**.

Western Pine Any of several pines growing in the western United States or Canada, including Ponderosa, Sugar, and Western (Idaho) White Pine.

Western Red Cedar *Thuja plicata.* This species is found principally along the western edges of British Columbia, Washington, and Oregon. The wood is soft, straight-grained, and extremely resistant to decay and insect damage. It is used extensively in roof coverings, exterior sidings, fences, decks, and other outdoor applications.

Western S-P-F Lumber of the Spruce-Pine-Fir group produced in British Columbia or Alberta. See **Spruce-Pine-Fir**.

Western Spruce Sitka Spruce.

Western Tamarack Larch.

Western White Pine *Pinus monticola.* A species of pine found in a wide range throughout the western U.S. and British Columbia. It is easily worked and is favored for shelving, cabinets, and a variety of specialized uses. Commonly called "Idaho White Pine".

Western Yellow Pine Ponderosa Pine (rarely used).

westside (1) An unofficial division of the Southern-Yellow-Pine-producing region, consisting of that part of the region located west of the Mississippi-Alabama state line, and including southwestern Alabama. Products from this region are most often marketed in the central southern states and the midwest. (2) The area west of the Cascade Mountains in Oregon and Washington, contrasted with eastside, the pine country.

West Virginia Spruce Red Spruce.

wet bulb A thermometer utilizing evaporation of moisture from a water-saturated cloth on its bulb to measure temperature.

wet-bulb depression The difference between the dry-bulb and the wet-bulb temperatures.

wet-bulb temperature (1) The reading on a thermometer whose bulb is enclosed in a layer of wet fabric. (2) The ambient temperature of an object cooled by evaporation.

wet-bulb thermometer A thermometer in which the bulb is enclosed in a layer of wet fabric.

wet construction Any construction using materials that are placed or applied in other than a dry condition, as in concreting or plastering.

wet glazing Sealing glass in a frame using a glazing compound or

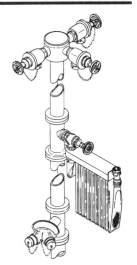

Wet Standpipe System

sealant applied with a blade or gun.

wet mix Concrete containing too much water, immediately evidenced by a runny consistency.

wet-mix shotcrete Shotcrete in which all ingredients, including mixing water, are mixed before introduction into the delivery hose. The shotcrete may be pneumatically conveyed or moved by displacement. See also **pneumatic feed** and **positive displacement**.

wet-on-wet painting A procedure of spraying a second coat of paint over a first coat that is not completely dry.

wet process In the manufacture of cement, the process in which the raw materials are ground, blended, mixed, and pumped while mixed with water. The wet process is chosen where raw materials are extremely wet and sticky, making drying before crushing and grinding difficult. See also **dry process**.

wet process hardboard Hardboard manufactured by a process in which a wood slurry is combined with a resin binder and is then dried, first on a screen and later under pressure. The pressure imparts high density to the board and also sets the binder.

wet rot Decay of wood, in the presence of moisture and warmth, as a result of attack by fungi.

wet screening Screening to remove from fresh concrete all aggregate particles larger than a certain size.

wet sieving Use of water during sieving of a material on a No. 200 or No. 325 sieve.

wet standpipe system A standpipe system filled with water at design pressure, ready for immediate use.

wet strength The strength of an adhesive joint as measured immediately after it has been removed from an adhesive in which it was immersed.

wettest stable consistency The condition of maximum water content at which cement grout or mortar will adhere to a vertical surface without sloughing.

wetting Coating a base metal with filler metal prior to soldering or brazing.

wetting agent A substance capable of lowering the surface tension of liquids, facilitating the wetting of solid surfaces and permitting the penetration of liquids into the capillaries.

wet-use adhesive Adhesives, used in glue-laminated timber, which will perform satisfactorily under a wide variety of exposures, including weather, dry atmosphere, and marine use.

wet wall A plaster or stucco wall that sets up as the material dries.

whaler See **wale**.

wheelbarrow A hand cart with one wheel in front and two legs and two handles in back, used to move materials short distances.

wheel ditcher A trench digger consisting of a vehicle equipped with a large wheel on which buckets are mounted.

wheel load The portion of the gross weight of a loaded vehicle transferred to a supporting structure under a given wheel of the vehicle.

whetstone A piece of natural or manufactured abrasive stone used to sharpen cutting tools.

White Balsam White Fir.

White Cedar Port Orford Cedar.

white coat A lime-putty plaster coat with a troweled finish. Also called the "finish coat".

White Fir *Abies concolor*. The most important of the true firs, this species is found in a wide range in the western U.S. Northern California accounts for the majority of White Fir (or Inland

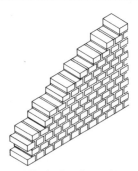

Whole-brick Wall

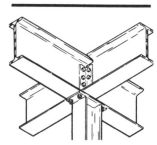

Wide-flange beam

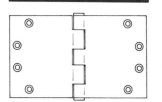

Wide-throw Hinge

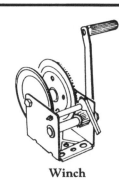

Winch

Hem-Fir) lumber produced in the U.S. The wood is straight-grained, fine-textured, and relatively light. It is used in general construction and for such specialized uses as moldings and doors.

White Hemlock Eastern Hemlock.

white lauan See **Philippine mahogany**.

white lead Basic lead carbonate, white in color, and used as a pigment in exterior paints, ceramics, and putty.

white mahogany See **avodire**.

whitening A white appearance, in the grain of finished wood, usually due to spotty adhesion of the coating or improper finishing methods.

White Oak Any of several species of American Oaks, principally *Quercus alba*, whose wood is more white than red.

White Pine *Pinus strobus*, Northern White Pine; *P. monticola*, Idaho White Pine; *P. lambertiana*, Sugar Pine.

white portland cement A white portland cement made from materials with low iron content, used to produce concrete or mortar that is white in color.

white rot A type of decay in wood caused by a fungus that leaves a white deposit.

White Spruce *Picea glauca*. A species of spruce found predominantly in British Columbia and Alberta. It is marketed as part of the Spruce-Pine-Fir group and is an important source of general construction lumber. It is also used extensively in pulp production.

White Walnut See **butternut**.

whitewash A white, paint-like coating by adding water to quicklime or slaked lime, whiting, and glue.

Whitewood See **tulipwood, def. 1**.

whiting Calcium carbonate pigment, used as an extender in paint, putty, and whitewash.

Whitney strain diagram The assumed strain distribution of a reinforced concrete member in the yield state which is used in Ultimate Strength Design.

whole-brick wall A brick wall having a thickness equal to the length of one brick.

wicket A small door or gate, especially one which is mounted in a larger one.

wide-flange beam A hot-rolled steel beam resembling an "H" on its side, and having parallel flanges.

wide-throw hinge A rectangular hinge having extra wide leaves.

widow's walk A narrow walkway on a roof of a house.

wiggle nail See **corrugated fastener**.

wigwam A sheet metal structure in the shape of a conical pyramid, used as a large incinerator.

wigwam burner A metal structure, shaped roughly like an Indian wigwam or tepee, in which mill wastes are burned. Although still used in some areas, such burners are being phased out, partly because mill wastes are being utilized to a greater degree, and partly because the burners can be a major source of air pollution.

Williot diagram A graphical method used to determine deflections of trusses.

winch A stationary hoisting machine having a rotating drum around which a cable, rope, or chain is wrapped.

windage loss Water removed by circulating air, as in a cooling tower.

wind-brace A brace provided in a frame to support the frame against the wind loads.

windbreak A dense growth of trees, fence, wall, etc., provided as

Window

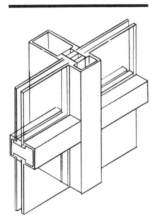

Window Frame

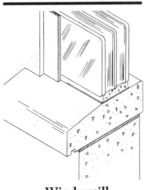

Windowsill

protection against the wind.

winder A tread, used where a stair turns a horizontal angle or in a spiral stair, in the shape of a wedge or truncated wedge.

winding stair A stair constructed mostly of winders.

windlass A device for lifting heavy objects, usually consisting of a horizontal drum turned by a lever or crank. A cable runs from the object over a pulley to the drum where lengths of cable are wound or unwound to move the object.

wind load The horizontal load used in the design of a structure to account for the effects of wind.

window (1) An opening, usually glazed, in an external wall to admit light and, in buildings without central air conditioning, air. (2) An assembly consisting of a window frame, glazing, and necessary appurtenances. (3) A small opening in a wall, partition, or enclosure for transactions, such as a ticket window or information window.

window bead See **inside stop, draft bead**.

window board See **window stool**.

window box See **weight box**.

window catch A fastening device fixed to a window sash to secure it in the closed position.

window cleaner's anchor An anchor, fastened near a window on an outside wall of a multistory building, to which a person washing the window from the outside can attach a safety belt.

window cleaner's platform A platform suspended from a trolley on the roof of a multistory building, used for outside window washing and maintenance. The entire assembly is custom-made and may be either manually or mechanically operated.

window frame The fixed part of a window assembly attached to the wall and receiving the sash or casement and necessary hardware.

window glass, sheet glass A soda-lime-silica glass made in continuous sheets of varying thickness and cut to size as required.

window glazing bar See **muntin**.

window hardware All the devices, fittings, or assemblies necessary to operate a window as intended. Window hardware may include catches, cords, fasteners, hinges, handles, locks, pivots, pulls, pulleys, and sash weights.

window jack scaffold (OSHA) A scaffold, the platform of which is supported by a bracket or jack projecting through a window opening.

window lead A slender lead rod with grooves used to hold glass in a window, usually in a decorative window.

window lift, sash lift A handle fastened to the lower sash of a window for use in moving the sash.

window lock See **sash lock**.

windowpane A segment of glass in a window.

window post One of the wood uprights between which a window frame is set, usually consisting of two studs nailed together.

window pull See **sash pull**.

window sash See **sash**.

window schedule A tabulation, usually on a drawing, listing all windows on a project and indicating sizes, number of lights, type of sash and frame, and hardware required.

window screen (1) An ornamental grille or lattice set in a window opening. (1) See **insect screen**.

window seat A built-in seat at a window.

windowsill See **sill**.

window spring bolt A spring bolt, used on a sash that is not counter-

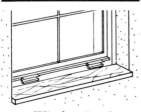

Window Stool

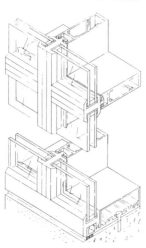

Window Wall

Wing

Wing Wall

balanced to hold the sash in a desired position.

window stool A horizontal board or plate at the windowsill on the inside of the window, fitted against the bottom rail of the lower sash to form a base for the casing.

window stop See **sash stop**.

window trim The finished casing around a window.

window wall An exterior envelope using a frame containing windows that may be fixed or operable. The glazing may be clear, tinted, and/or opaque.

window weight See **sash weight**.

wind pressure The pressure produced when wind blows against a surface.

windrow (1) A ridge of loose soil, such as that produced by the spill off of a grader blade. (2) A row of leaves or snow heaped up by the wind.

wind sock A bright-colored cloth tube, used to indicate wind direction.

wind stop (1) A weather strip used around a door or window. (2) A strip of wood or metal which covers the joint where a sash or casement meets a stile. (3) Any wood or metal strip used to cover a crack in an exterior wall of a building.

windtight Construction terminology to indicate that all openings and cracks in exterior walls have been sealed.

wind tunnel A structure through which a controlled stream of wind is directed at a model in order to study the probable effects of wind on a structure.

wind uplift The upward component of the force produced as wind blows around or across a structure or an object.

wine cellar, wine vault A storage room for wine, usually located underground.

wing (1) A section or addition extending out from the main part of a building. (2) The offstage space at a side of a stage. (3) One of the four leaves of a revolving door.

wing compass A compass with an arc-shaped piece permanently fastened to one leg and clamped at the other by a screw. The arc is usually graduated.

wing dividers Dividers made similar to a wing compass.

wing nut A nut provided with wing-like projections to facilitate turning by fingers and thumb.

wing pile A bearing pile, usually of concrete, widened in the upper portion to form part of a sheet pile wall.

wing screw A screw provided with wing-like projections to facilitate turning by fingers and thumb. A wing screw is often used to secure windows against unauthorized entry.

wing wall (1) A short section of wall at an angle to a bridge abutment, used as a retaining wall and to stabilize the abutment. (2) A short section of wall used to guide a stream into an opening.

wiped joint A wiped lead joint connecting two lead pipes.

wire A usually pliable metal strand or filament.

wire brad See **brad**.

wire cloth A stiff fabric of woven wire, usually having a larger mesh than screen material, and used such as reinforcing in plaster, in sieves, and as a leaf catcher in gutters.

wire, cold-drawn Wire made from the rods hot-rolled from billets and then cold-drawn through dies. See also **reinforcement, cold-drawn wire**.

wire comb A tool with long wire teeth, used to scratch a coat of wet plaster to increase the bond of a subsequent coat.

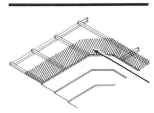

Wire Lath

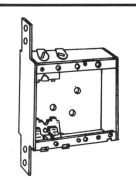

Wiring Box

Wood-cement Concrete

wire-cut brick Brick that has been extruded to shape and then cut to length with wire before firing.

wired glass See **wire glass**.

wire gauge (1) The diameter of a wire as defined by several different systems. Usually, the thicker the wire is the smaller the gauge number is. (2) A device for measuring the thickness of a wire, usually consisting of a metal sheet with standard-sized notches on one or more edges.

wire glass Sheet glass with wire mesh embedded in the glass to prevent shattering.

wire holder An electrical insulator having a screw or bolt for fastening the insulator to a support.

wire lath A welded-wire mesh used as a base for plaster.

wire mesh See **welded-wire fabric**.

wire nail A very thin nail with a small head, usually used as a finishing nail.

wire nut A connector for two or more electric conductors, made in the form of a plastic cap with an internal spring-thread, and turned over the parallel or twisted ends of the conductors.

wire rope A rope made of twisted steel strands laid around a central core.

wire saw A machine for sawing stone using a rapidly moving continuous wire carrying a slurry of sand or other abrasive material.

wire winding Application of high tensile wire, wound under tension by machines, around concrete circular or dome structures to provide tension reinforcing.

wiring box A box used in interior electric wiring at each junction point, outlet, or switch, which serves as protection for electric connections and as a mounting for fixtures or switches.

withe, wythe (1) A partition used to separate two flues in a chimney stack. (2) See **wythe**.

witness corner A marker set on a property line near but not at a corner. Its relation to the corner is recorded.

wobble coefficient A coefficient used in determining the friction loss occurring in posttensioning, which is assumed to account for the secondary curvature of the tendons.

wobble friction In prestressed concrete, the friction caused by the unintended deviation of the prestressing sheath or duct from its specified profile.

wobble saw A drunken saw.

wobble-wheel roller A compactor consisting of a weighted bed mounted on a series of wheels that are mounted loosely and allowed to work the surface of a soil.

wood block floor A finished floor consisting of rectangular blocks of a tough wood such as oak, set in mastic with the end grain exposed, usually over concrete slab. Wood block floor is used where very heavy traffic and heavy loads are expected.

wood brick, fixing brick, nailing block A block of wood, the same size as a brick, inserted into brickwork as a base for attaching nailed or screwed objects.

wood-cement concrete A portland-cement concrete using sawdust and wood chips as aggregate.

wood cement particleboard (WCP) A high-density board manufactured in Europe for use on exteriors, or where fire resistance is needed. Wood particles are combined with portland cement, or other mineral, as a binder.

Wood Chisel

Wood Gutter

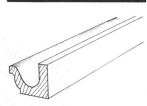

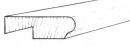

Wood Molding

Wood Screw

wood chipboard See **particleboard**.

wood chisel Any of a number of firmer chisels used in carpentry and woodworking.

wood dough A synthetic putty containing wood fibers, used as a filler.

wood-fibered plaster A gypsum plaster mix containing wood fibers.

wood-fiber insulation Thermal insulation using wood fibers and an asphalt binder.

wood filler A liquid or paste compound used to fill pores of a wood before finishing.

wood finishing The final finishing of a wood surface, including planing, sanding, staining, varnishing, waxing and/or painting.

wood fire-retardant treatment The impregnation of wood or wood products with special solutions to reduce the flame spread of the finished product.

wood flooring Flooring consisting of dressed and matched boards.

wood flour A finely ground dried wood powder, used in plastic wood, as an extender in some glues and in the molding of plastics.

wood-frame construction A type of construction in which floors, roofs, and exterior and other bearing walls are of wood, as opposed to masonry.

wood-grain print A simulated-wood-grain pattern used on plastic veneer, hardboard, and low-grade plywood.

wood gutter A gutter, under the eaves of a roof, made from a solid piece of wood or built up boards.

wood lath Narrow strips of wood used as a base for plaster.

wood molding Wood strips factory-shaped to commercially available patterns.

wood oil (1) Tung oil used in the manufacture of varnish, putty, etc. (2) An oleoresin used for caulking and waterproofing.

wood preservative Any chemical preservative for wood, applied by washing on or pressure-impregnating. Products used include creosote, sodium fluoride, copper sulfate, and tar or pitch.

wood rasp See **rasp**.

wood screw A helically threaded fastener that cuts its way into wood when turned under axial pressure.

wood sill See **sill**.

wood slip A wooden nailing strip fixed in a masonry or concrete wall as a means of attaching wood trim or furring strips.

wood stud anchor, nailing anchor A metal clip attached to the inside of a door frame and used to secure the frame to a wood stud partition.

wood treatment (1) Treatment of wood with a preservative to prevent or retard its decay. (2) See **fire-retardant wood**.

wood veneer See **veneer**.

wood-wool See **excelsior**.

woodwork Work produced by carpenters and woodworkers, especially the finished work.

woodworker's vise Any of a number of tools consisting of two jaws and a mechanism to hold a piece of wood while it is being worked on.

work (1) All labor and materials required to complete a project in accordance with the contract documents. (2) The product of a force times the distance traveled.

Work (capital "W") As used in AIA Documents, the completed construction required by the contract documents, including all labor necessary to produce such construction, and all materials and equipment incorporated or to be incorporated in such construction. The word "work", as contrasted

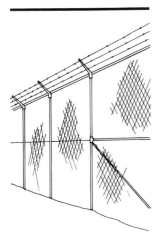

Woven-wire Fabric

Wrench

with capitalized "Work", is used in its ordinary sense.

workability That property of freshly mixed concrete or mortar which determines the ease and homogeneity with which it can be mixed, placed, compacted, and finished.

worker's compensation insurance See **insurance, worker's compensation**.

working The alternate swelling and shrinking of wood caused by changes in its moisture content induced by variations in the humidity of its environment. Also called "movement".

working capital The excess of current assets over current liabilities.

working drawings See **drawings**.

working life The length of time a liquid resin or adhesive remains useful after the ingredients have been mixed.

working load Forces normally imposed on a member in service.

working stage A section of an assembly room or auditorium partially cut off from the audience section by a proscenium wall and equipped with some or all of the following: scenery loft, gridiron, fly gallery, and lighting equipment.

working stress Maximum permissible design stress using working stress design methods.

working stress design A method of proportioning structures or members for prescribed working loads at stresses well below the ultimate, and assuming linear distribution of flexural stresses.

work light (1) A light in a theater used to provide illumination for rehearsing, scene shifting, or other work onstage or backstage. (2) A lamp in a protective cage and having a long, heavy, flexible cord, used to provide temporary illumination in work areas.

workmanship The quality of work performed.

workmen's compensation insurance See **insurance, worker's compensation**.

work order See **notice to proceed**.

work plane The plane at which work is usually done and at which the level of illumination is specified and measured.

woven-wire fabric A prefabricated steel reinforcement composed of cold-drawn steel wires mechanically twisted together to form hexagonally shaped openings.

woven-wire reinforcement See **woven-wire fabric**.

wrapping A method of applying narrow strips of veneer around a curved surface, such as a piece of furniture.

wrecking The process of demolishing a structure.

wrecking ball A heavy steel ball or concrete mass on a heavy chain, swung by a crane to demolish parts of a structure.

wrecking bar See **pinch bar**.

wrecking strip A small piece or panel fitted into a formwork assembly in such a way that it can be easily removed ahead of main panels or forms, making it easier to strip those major form components.

wrench A hand tool consisting of a handle and a jaw at one end used to turn or hold a bolt, nut, pipe, or fitting. The jaw may be shaped for a specific-sized object or may be adjustable.

wrinkling, crinkling, riveling (1) The distortion in a paint film which appears as ripples and may be deliberately induced as a decorative effect or accidentally caused by drying conditions or too thick an application of a paint film. (2) The rippling or crinkling of the surface of an adhesive,

Wye Fitting

usually not affecting performance. (3) The rippling of an area of veneer, caused by lack of contact with the adhesive.

wrought Descriptive of metals or metalwork shaped by hammering with tools.

wrought iron (1) A material consisting of iron with strands of silica throughout. Once the surface iron decomposes, the silica surfacing prevents further oxidation. Wrought iron is no longer commercially available. (2) A number of easily welded or wrought irons with low impurity content used for water pipes, tank plates, or forged work.

wrought nail A nail wrought by hand, often having a head with a decorative pattern.

W-truss A wood root truss with the web members in the shape of a "W".

wye (1) A Y-branch. (2) A Y-fitting.

wye branch See **Y-branch**.

wye fitting See **Y-fitting**.

wye level See **Y-level**.

wythe (leaf) Each continuous vertical section of a wall one masonry unit in thickness.

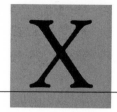

ABBREVIATIONS

The abbreviations listed below are those most commonly used in the construction industry. Alternative forms (usually nonstandard) are shown in parentheses.

X experimental
XBAR crossbar
XH, X HVY extra heavy
XL extra large
xr without rights
X STR extra strong
xw without warrants
XXH double extra heavy

 DEFINITIONS

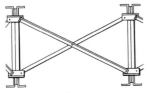

X-brace

X-brace, cross brace Paired set of sway braces.

XCU (insurance terminology) Letters which refer to exclusions from coverage for property damage liability arising out of (1) explosion or blasting (designated by "X"), (2) collapse of or structural damage to any building or structure (designated by "C"), and (3) underground damage caused by and occurring during the use of mechanical equipment (designated by "U").

xonolite 5-calcium-r-silicate monohydrate (C3S5H), a natural mineral that is readily synthesized at 150 to 350 degrees C under saturated steam pressure. Xonolite is a constituent of sand-lime masonry units.

x-ray diffraction (1) The diffraction of x-rays by substances having a regular arrangement of atoms. (2) A phenomenon used to identify substances having such structure.

x-ray fluorescence Characteristic secondary radiation emitted by an element as a result of excitation by x-rays, used to yield chemical analysis of a sample.

xylol An aromatic hydrocarbon distilled from coal tar, used as a solvent for paints and varnishes.

xyst (1) A tree-shaded walk or promenade. (2) A roofed colonnade for exercise in bad weather.

ABBREVIATIONS

The abbreviations listed below are those most commonly used in the construction industry. Alternative forms (usually nonstandard) are shown in parentheses.

y yard
Y yttrium, wye, Y-branch
yd yard
y.p. yellow pine
YP yield paint
YR year
YS yield strength

Y

DEFINITIONS

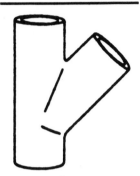

Y-fitting

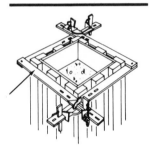

Yoke

Yakima Pine Ponderosa Pine.

Yankee screwdriver See **spiral ratchet screwdriver**.

yard (1) A unit of length in the English System equal to three feet. (2) A term applied to that part of a plot not occupied by the building or driveway.

yardage (1) An amount of excavated material equal to the volume in cubic yards. (2) An area of surface measured in square yards.

yard drain A drain in a pavement or earth surface, used to drain surface water.

yard lumber Lumber graded according to its size, length, and intended use, stockpiled in a lumber yard. Also called "general building construction lumber".

yarn A continuous strand of twisted threads of natural or artificial material, such as wool or nylon, used in making carpets.

Y-branch, wye branch A branch, in a plumbing system, in the shape of a "Y".

yellowing Development of yellow color or cast in white or clear coatings, after aging.

Yellow Pine The hard resinous wood from the long leaf pine tree, used as flooring and in general construction.

Yellow Poplar, Poplar A moderately low density, even-textured hardwood from central and southern United States, used for veneer, plywood, and cabinet work.

yelm A bundle of reeds or straw used as thatching material for a roof.

Y-fitting, wye fitting A pipe fitting in the shape of a "Y". One arm is usually at 45 degrees to the main fitting and may be of reduced size.

yield (1) The volume of freshly mixed concrete produced from a known quantity of ingredients. (2) The total weight of ingredients divided by the unit weight of the freshly mixed concrete. (3) The number of product units, such as block, produced per bag of cement or per batch of concrete.

yield point (1) The point at which a stressed material begins to exhibit plastic properties and deformation increases faster than applied loads. (2) The point beyond which the material will not return to its original length.

yield strength The stress, less than the maximum attainable stress, at which the ratio of stress to strain has dropped well below its value at low stresses, or at which a material exhibits a specified limiting deviation from the usual proportionality of stress to strain.

Y-level A surveyor's level with the telescope and level supported by Y-shaped fittings. The telescope can be removed and reversed in order to average out internal sources of error.

yoke (1) A tie or clamping device around column forms or over the top of wall or footing forms to keep them from spreading from the lateral pressure of fresh concrete. (2) Part of a structural assembly for slipforming which keeps the forms from spreading and transfers form loads to the jacks.

yoke vent (1) A pipe connecting upward from a soil or waste stack to a vent stack for venting purposes. (2) A vertical or 45-degree relief vent of the continuous waste-and-vent type formed by the extension of an upright wye-branch inlet of the horizontal branch to the stack. A yoke vent is called a "dual yoke vent" when two horizontal branches are thus vented by the same relief vent.

Young's modulus See **modulus of elasticity**.

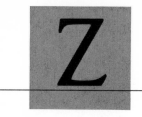

ABBREVIATIONS

The abbreviations listed below are those most commonly used in the construction industry. Alternative forms (usually nonstandard) are shown in parentheses.

z zero, zone
Z modulus of section
ZI zone of interior
Zn azimuth, zinc

Z DEFINITIONS

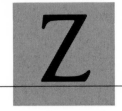

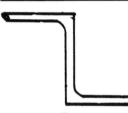

Z-Bar

Zee

Z-bar A Z-shaped member that is used as a main runner in some types of acoustical ceiling.

zebrawood *Connarus guianensis.* A tropical hardwood with strikingly marked grain, used for decorative purposes in cabinetry and paneling.

zee A tight-gauge member with a Z-like cross section. The flanges of the Z are approximately at right angles to the web.

zeolite A group of hydrous aluminum silicate minerals or similar synthetic compounds used in water-softening equipment.

zero-slump concrete Concrete of stiff or extremely dry consistency showing no measurable slump after removal of the slump cone. See also **slump** and **no-slump concrete.**

zeta (1) A small closed room. (2) Originally a room over the porch of a Christian church, used as living quarters for a porter or sexton and for storage of documents.

z flashing Z-shaped metal flashing applied between panels of plywood siding to shed water.

zigzag bond Similar to herring-bone bond.

zigzag rule A folding rule made of pieces connected by a pivot.

zinc A metallic element used for galvanizing steel sheet and steel or iron castings, as an alloy in various metals, as an oxide for white paint pigment, and as a sacrificial element in a cathodic protection system.

zinc chromate A bright yellow pigment, used in paints, which has good rust-inhibiting properties.

zinc coating See **galvanize.**

zinc dust Zinc ground to a fine powder and used as a pigment in priming paints, especially those for use on galvanized surfaces.

zinc oxide, zinc white A pigment used in paints to provide durability, color retention, and hardness, and to improve sag resistance.

zinc white See **zinc oxide.**

zinc yellow Bright yellow pigments of zinc chromates used in primers and paints as a rust inhibitor.

zone A space or group of spaces in a building with common control of heating and cooling.

zoning (BOCA) The reservation of certain specified areas within a community or city for buildings and structures, or of use of land, for certain purposes with other limitations, such as height, lot coverage, and other stipulated requirements.

zoning permit A permit issued by appropriate governmental officials authorizing land to be used for a specific purpose.

zoological garden A park used to exhibit wild animals.

577

NOTES

NOTES

NOTES